# Structural Dynamics by Finite Elements

**William Weaver, Jr.**
*Stanford University*

**Paul R. Johnston**
*Failure Analysis Associates*

PRENTICE-HALL, INC., Englewood Cliffs, New Jersey 07632

*Library of Congress Cataloging-in-Publication Data*

WEAVER, JR., WILLIAM, (date)
    Structural dynamics by finite elements.

    (Prentice-Hall civil engineering and engineering mechanics series)
    Includes bibliographies and index.
    1. Structural dynamics.  2. Finite element method.
I. JOHNSTON, PAUL R., (date)  II. Title.
III. Series.
TA654.W42 1987    624.1′71    86-12257
ISBN 0-13-853508-6


Editorial/production supervision
and interior design: **Kathryn Pavelec**
Cover design: **Edsal Enterprises**
Manufacturing buyer: **Rhett Conklin**


**PRENTICE-HALL CIVIL ENGINEERING AND ENGINEERING MECHANICS SERIES**
*W. J. Hall, editor*


© 1987 by **Prentice-Hall, Inc.**
A Division of Simon & Schuster
Englewood Cliffs, New Jersey 07632


*All rights reserved. No part of this book may be
reproduced, in any form or by any means,
without permission in writing from the publisher.*


Printed in the United States of America

10  9  8  7  6  5  4  3  2  1


ISBN  0-13-853508-6  025


PRENTICE-HALL INTERNATIONAL (UK) LIMITED, *London*
PRENTICE-HALL OF AUSTRALIA PTY. LIMITED, *Sydney*
PRENTICE-HALL CANADA INC., *Toronto*
PRENTICE-HALL HISPANOAMERICANA, S.A., *Mexico*
PRENTICE-HALL OF INDIA PRIVATE LIMITED, *New Delhi*
PRENTICE-HALL OF JAPAN, INC., *Tokyo*
PRENTICE-HALL OF SOUTHEAST ASIA PTE. LTD., *Singapore*
EDITORA PRENTICE-HALL DO BRASIL, LTDA., *Rio de Janeiro*
WHITEHALL BOOKS LIMITED, *Wellington, New Zealand*

# Contents

Preface    ix

**1**  Introduction to Structural Dynamics    1

    1.1 Structural Dynamics Concepts    1
    1.2 Dynamic Influences    4
    1.3 Discretization by Finite Elements    6
    1.4 Computer Programs    9
        References    9

**2**  Systems with One Degree of Freedom    10

    2.1 Introduction    10
    2.2 Free, Undamped Vibrations    12
    2.3 Harmonic Forcing Functions    19
    2.4 Effects of Damping    25
    2.5 Periodic Forcing Functions    35
    2.6 Arbitrary Forcing Functions    38
    2.7 Step-by-Step Response Calculations    45
    2.8 Response Spectra    51
        References    59
        Problems    59

## 3  Finite Elements and Vibrational Analysis    73

3.1  Introduction    73
3.2  Stresses and Strains    75
3.3  Equations of Motion for Finite Elements    78
3.4  One-Dimensional Elements    82
3.5  Transformation and Assemblage of Elements    95
3.6  Vibrational Analysis    105
3.7  Symmetric and Antisymmetric Modes    112
3.8  Program VIB for Vibrational Analysis    118
    References    124
    Problems    125

## 4  Normal-Mode Method of Dynamic Analysis    138

4.1  Introduction    138
4.2  Principal and Normal Coordinates    139
4.3  Normal-Mode Response to Initial Conditions    147
4.4  Normal-Mode Response to Applied Actions    152
4.5  Normal-Mode Response to Support Motions    157
4.6  Damping in MDOF Systems    164
4.7  Damped Response to Periodic Forcing Functions    168
4.8  Damped Response to Arbitrary Forcing Functions    172
4.9  Step-by-Step Response Calculations    175
4.10 Program NOMO for Normal-Mode Response    177
    References    192
    Problems    192

## 5  Direct Numerical Integration Methods    195

5.1  Introduction    195
5.2  Extrapolation with Explicit Formulas    197
5.3  Iteration with Implicit Formulas    203
5.4  Direct Linear Extrapolation    211
5.5  Newmark's Generalized Acceleration Method    217
5.6  Numerical Stability and Accuracy    223
5.7  Program DYNA for Dynamic Response    225
    References    236
    Problems    237

## 6 Framed Structures 241

6.1 Introduction 241
6.2 Plane Frames 244
6.3 Grids 249
6.4 Space Trusses 253
6.5 Space Frames 259
6.6 Programs for Framed Structures 264
6.7 Guyan Reduction 282
6.8 Constraints Against Axial Strains 290
6.9 Programs DYPFAC and DYSFAC 299
References 303
Problems 303

## 7 Two- and Three-Dimensional Continua 310

7.1 Introduction 310
7.2 Stresses and Strains in Continua 310
7.3 Natural Coordinates 318
7.4 Numerical Integration 326
7.5 Isoparametric Quadrilaterals for Plane Stress and Plane Strain 333
7.6 Program DYNAPS for Plane Stress and Plane Strain 340
7.7 Isoparametric Hexahedra for General Solids 345
7.8 Program DYNASO for General Solids 351
7.9 Isoparametric Elements for Axisymmetric Solids 357
7.10 Program DYAXSO for Axisymmetric Solids 365
References 369

## 8 Plates and Shells 370

8.1 Introduction 370
8.2 Element for Plates in Bending 371
8.3 Program DYNAPB for Plates in Bending 379
8.4 Element for General Shells 382
8.5 Program DYNASH for General Shells 390
8.6 Element for Axisymmetric Shells 394
8.7 Program DYAXSH for Axisymmetric Shells 406
References 410

## 9 Rigid Bodies within Flexible Structures   411

9.1 Introduction   411
9.2 Rigid Bodies in Framed Structures   413
9.3 Program DYRBPF for Rigid Bodies in Plane Frames   424
9.4 Rigid Laminae in Multistory Buildings   425
9.5 Rigid Bodies in Finite-Element Networks   434
9.6 Program DYRBPB for Rigid Bodies in Plate-Bending Continua   438
References   443

## 10 Substructure Methods   444

10.1 Introduction   444
10.2 Guyan Reduction Methods   445
10.3 Modified Tridiagonal Method for Multistory Buildings   457
10.4 Programs DYMSPF and DYMSTB   463
10.5 Component-Mode Method   468
10.6 Component-Mode Method for Trusses   471
10.7 Programs COMOPT and COMOST   477
References   482

## Notation   483

## General References   491

## Appendix A  Systems of Units and Material Properties   495

A.1 Systems of Units   495
A.2 Material Properties   497

## Appendix B  Eigenvalues and Eigenvectors   498

B.1 Inverse Iteration   498
B.2 Transformation Methods   505
References   517

Appendix C     Flowchart for Program
                DYNAPT      519

Answers to Problems      564

Index     579

# Preface

*Structural Dynamics by Finite Elements* represents a culmination of the two topics identified in its title. Structural dynamics continues to grow as an essential subject for structural engineers, and the best method for handling structural dynamics problems is with finite elements. The solids and structures discussed herein are subjected to time-varying influences that cause accelerations and velocities as well as displacements, strains, and stresses. To analyze such a problem, we discretize the structure (or solid) by dividing it into a network of elements having finite sizes. Then equations of motion are written for the nodes (or joints) of the discretized continuum, which include inertial and damping actions. This finite number of differential equations may be solved on a digital computer to find approximate time-varying nodal displacements and stresses within the finite elements.

This publication is intended to be used as a textbook for a graduate-level course on structural dynamics in civil, mechanical, or aeronautical and astronautical engineering. We have tried to present the material in a clear, forthright manner for either a university student or a structural analyst in industry. As background for studying this book, the user should have had the mathematics and solid mechanics usually offered in schools of engineering at the undergraduate level. Included among the former courses are differential equations, matrix algebra, and computer programming. Other desirable courses are vibration theory, matrix analysis of framed structures, and finite elements. Although it is not necessary to know the theories of elasticity, plates, and shells, previous exposure to those topics will give the reader greater perspective on the present work.

Chapter 1 introduces structural dynamics by comparing time-varying responses to dynamic loads against results for static loads. Next, we describe commonly encountered dynamic influences, which are initial conditions, applied actions, and support motions. Then discretization by finite elements is discussed for framed structures, two- and three-dimensional continua, plates, and shells. Such analytical models are processed by digital computer programs, as mentioned in the last section of this introductory chapter.

In preparation for later work, we describe and analyze systems with only one degree of freedom in Chapter 2. This material would be a useful review for a person who has had a previous course on vibration theory for one-degree systems. It also serves as an introduction to the subject for anyone without such a background. Moreover, this book on structural dynamics becomes more self-contained by including material on one-degree systems.

We present derivations for finite elements and vibrational analysis in Chapter 3, where all of the discretized structures have multiple degrees of freedom. Energy-consistent stiffness, mass, and load matrices are developed for one-dimensional elements for later use in framed structures. From the homogeneous form of the nodal equations of motion for an assembled structure, we can solve the eigenvalue problem for vibrational frequencies and mode shapes. If planes of symmetry exist, appropriate restraints at nodes on those planes allow us to find symmetric and antisymmetric modes, using only part of a structure. At the end of Chapter 3, we describe Program VIB for vibrational analysis of finite-element networks, with specialization to Program VIBPT for plane trusses.

Chapter 4 contains the normal-mode method for dynamic analysis of discretized structures. By this approach, we transform the equations of motion to normal coordinates, where each flexible-body or rigid-body mode has unit mass and responds as if it were a system with only one degree of freedom. We discuss normal-mode responses to initial conditions, applied actions, and support motions for structures with or without damping. Step-by-step response calculations are explained and coded in Program NOMO for normal-mode analysis. This program is specialized to become NOMOPT for plane trusses, which includes VIBPT from Chapter 3.

In Chapter 5 we cover direct numerical integration methods for calculating dynamic responses. These approaches may all be characterized as finite-difference approximations with respect to time. We discuss extrapolation with explicit formulas, iteration with implicit formulas, and direct linear extrapolation, with solution for incremental displacements. Newmark's generalized acceleration method is extended by Hilber's approach and applied in Program DYNA for dynamic responses of structures. Then this program is specialized to become DYNAPT for plane trusses, which includes VIBPT from Chapter 3 and NOMOPT from Chapter 4. Thus, in several stages we construct a program that will handle not only vibrational analysis but also two types of dynamic response calculations.

Concepts developed in previous chapters for plane trusses are extended to all other types of framed structures in Chapter 6. The programs described are DYNACB for continuous beams, DYNAPF for plane frames, DYNAGR for grids, DYNAST for spaces trusses, and DYNASF for space frames. We also discuss methods for reducing the number of degrees of freedom for beams, grids, plane frames, and space frames. Guyan reduction may be used to eliminate joint rotations from the equations of motion for these four types of framed structures. Also, constraints against axial strains can be imposed in plane and space frames; and Programs DYPFAC and DYSFAC have been coded for this technique.

Chapter 7 describes finite elements to be used for dynamic analyses of two- and three-dimensional continua. We employ isoparametric quadrilaterals for solving two-dimensional problems in plane stress or plane strain and present Program DYNAPS for that purpose. Isoparametric hexahedra are applied in calculations for general solids, and we have coded Program DYNASO for obtaining their dynamic responses. Axisymmetric solids require the use of ring elements having cross sections that again are chosen to be quadrilaterals. Program DYAXSO for dynamic analyses of axisymmetric solids completes the set of programs documented in this chapter.

In Chapter 8 the finite elements for analyzing plates in bending and general and axisymmetric shells are based on those for general and axisymmetric solids in Chapter 7. These specializations automatically include the effects of shearing deformations and rotary inertias. The programs described are DYNAPB for plates in bending, DYNASH for general shells, and DYAXSH for axisymmetric shells.

Chapter 9 explains the effects of including rigid bodies in the analytical models for framed structures and other discretized continua. For framed structures the convenient approach is a member-oriented technique with rigid offsets at the ends of each member. On the other hand, for finite elements with more than two nodes, a body-oriented method appears to be mandatory. Sample programs discussed in this chapter are DYRBPF for rigid bodies in plane frames and DYRBPB for rigid bodies connected to plates in bending. We also describe rigid laminae in multistory buildings in preparation for the next chapter.

The topic of Chapter 10 consists of substructure methods for dynamic analysis. We divide the subject into Guyan reduction methods and the component-mode technique. Within the first approach, emphasis is placed on the modified tridiagonal method, which is applied to multistory buildings. Programs DYMSPF and DYMSTB have been coded to handle multistory plane frames and tier buildings with rectangular framing patterns. On the other hand, the theory of the component-mode method is explained and implemented for the analysis of plane and space trusses having only a few members. Programs COMOPT and COMOST represent the codes for such analyses.

At the back of the book we give a list of notation, general references, appendices, and answers to problems. Appendix A describes systems of units (SI and US) and physical properties for various materials. In Appendix B we divide solution of eigenvalue problems into the topics of inverse iteration (for few modes) and transformation methods (for many modes). The latter methods include those of Jacobi, Givens, and Householder. Our final approach consists of Householder transformations of the coefficient matrix to tridiagonal form, followed by iteration with the QR algorithm. Spectral shifting may be used to improve the convergence of either inverse or QR iteration. Last, Appendix C contains the detailed flowchart for Program DYNAPT.

As mentioned in Chapter 1, all of our computer programs are coded in FORTRAN. These codes and data for examples have been assembled on a magnetic tape, a copy of which can be obtained from Paul R. Johnston for a nominal fee. His business address is: Failure Analysis Associates, 2225 East Bayshore Road, Palo Alto, California, 94303.

We wish to thank graduate students and teaching assistants at Stanford who have directly or indirectly contributed ideas for this book. Needless to say, our wives Connie and Terry have shown much patience and consideration while we were engrossed in its development. As before, Suzanne M. Dutcher did an outstanding job of typing the manuscript, and working with her was a great pleasure. Failure Analysis Associates of Palo Alto allowed us free computer usage, for which we are most grateful. Also, a Ford Foundation grant from the Provost at Stanford provided funds to partially offset our expenses.

WILLIAM WEAVER, JR.<br>
PAUL R. JOHNSTON

## Acknowledgement

This book was written in collaboration with **C. Lawrence Loh,** Staff Engineer, Engineering Information Systems, Inc., San Jose, California. Larry composed some of the computer programs, provided computer examples, and checked the Answers to Problems.

# 1

# Introduction to Structural Dynamics

## 1.1 STRUCTURAL DYNAMICS CONCEPTS

If a solid or a structure is loaded very gradually, it is said to be in a state of *static equilibrium*, for which static actions and reactions equilibrate each other. In such a situation, time has no significant influence; and static analysis may proceed without considering this variable. On the other hand, if forces are rapidly applied, the solid or structure is said to experience *dynamic loads*. In this case we can say that a state of *dynamic equilibrium* exists, for which time-varying actions and reactions equilibrate each other at every instant.

To examine some of the differences between static and dynamic analysis, let us consider the prismatic cantilever beam in Fig. 1.1(a). This beam is loaded with a concentrated force $P(t)$ in the $y$ direction at its free end (point 2). If the load is applied slowly, the static displacement at point 2 given by elementary beam theory is

$$(v_2)_{st} = \frac{PL^3}{3EI} \tag{1}$$

in which $EI$ is the flexural rigidity of the cross section. The response curve labeled 1 in Fig. 1.1(c) shows that by gradual application, the load produces an asymptotic value of $(v_2)_{st}$. In addition, the displacement $v(x)$ at any point along the length of the beam is a function of $x$ only.

At the other extreme, suppose that the force in Fig. 1.1(a) were applied instantaneously. In this case the beam will not only displace but will also accelerate at every point along its length. Therefore, the displacement $v(x, t)$

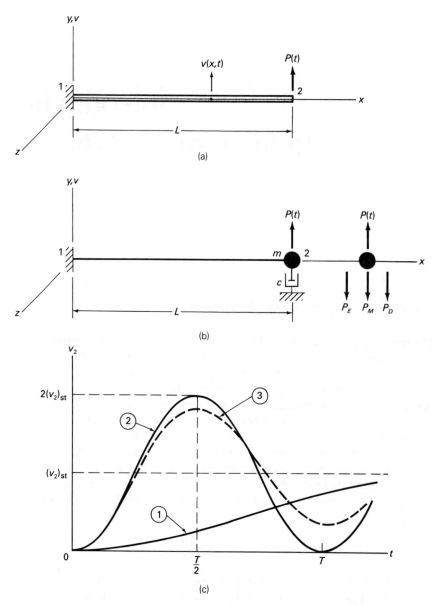

**Figure 1.1** (a) Beam with distributed mass; (b) lumped-mass approximation; (c) static and dynamic responses.

becomes a function of both space ($x$) and time ($t$). Although it is possible to write a partial differential equation of motion for dynamic equilibrium of this simple structure [1],* that approach will not be pursued in this book. Instead, we will always discretize the structure, as discussed later in Sec. 1.3.

A crude discretization of the cantilever beam problem is represented by the analytical model in Fig. 1.1(b). Appearing at point 2 is a concentrated mass $m$, representing some fraction of the distributed mass of the beam. Also shown at point 2 is a hypothetical dashpot damper that generates a dissipative force in proportion to velocity. The three types of forces in the figure opposing the applied load are the *elasticity force*,

$$P_E = kv_2(t) = \frac{3EI}{L^3} v_2(t) \tag{2a}$$

the *inertial force*,

$$P_M = m\ddot{v}_2(t) \tag{2b}$$

and the *dissipative force*,

$$P_D = c\dot{v}(t) \tag{2c}$$

where $c$ is a *damping constant*. From *D'Alembert's principle* we have

$$P(t) - P_E - P_M - P_D = 0 \tag{3a}$$

which expresses dynamic equilibrium of the mass $m$. Substituting Eqs. (2) into Eq. (3a) and rearranging yields

$$m\ddot{v}_2 + c\dot{v}_2 + kv_2 = P(t) \tag{3b}$$

Solution of this ordinary differential *equation of motion* by the theory in Chapter 2 produces the responses in Fig. 1.1(c) labeled 2 and 3. Curve 2 is the response when the damping constant $c$ is zero, and curve 3 represents a case of nonzero damping.

For perspective, let us review what has been shown by this example. First, we replaced a structure having an infinite number of degrees of freedom with an analytical model having only one degree of freedom. That is, in the former case the mass was distributed over an infinite number of points along the length of the beam, whereas in Fig. 1.1(b) the single concentrated mass $m$ exists only at the end. This simplification separates the variables of space and time both physically and mathematically. Therefore, Eq. (3b) is characterized as an ordinary differential equation instead of a partial differential equation.

Second, we described the response at point 2 due to the applied force for both gradual and instantaneous rise times. In Fig. 1.1(c) we can see that the maximum value of curve 2 is twice that of curve 1. This figure dramatically illustrates the difference between responses due to static and dynamic loads of

*Numbers in brackets indicate references at the end of the chapter.

the same magnitude. Curve 2 oscillates about the displaced position $(v_2)_{st}$ with a period $T$ equal to the *natural period of vibration* for the analytical model.

Third, we included the possibility of decaying motion caused by the presence of a hypothetical damping mechanism. Curve 3 in Fig. 1.1(c) shows that the effect of a dissipative force is to suppress the response as time passes. Although this influence is of interest to us, it is not nearly so important as the effect of the inertial force due to the presence of the mass.

This simple example also demonstrates the steps an engineer takes when solving a structural dynamics problem. They are:

1. Define the problem.
2. Compose an analytical model.
3. Calculate the response.

By the word *response* we mean time-varying displacements, stresses, internal actions, and so on, that may be of interest. Such quantities may be expressed as time histories, maximum values, or response spectra, depending on the nature of the problem.

## 1.2 DYNAMIC INFLUENCES

Various natural and man-made influences may cause dynamic responses in structures. The most common types of influences are *initial conditions, applied actions,* and *support motions.* Figures 1.2(a) and (b) show what we mean by initial conditions that result in dynamic responses. In the first case, a crane

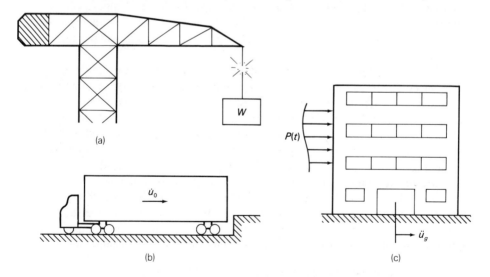

**Figure 1.2** Dynamic influences.

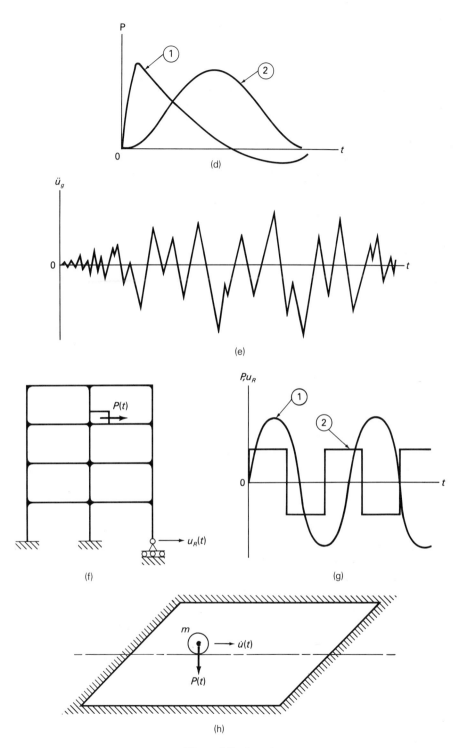

**Figure 1.2** (cont.)

suddenly drops its load and rebounds from an initial condition of static-load displacements. In the second instance, a truck backs into a loading platform at some initial velocity. Consequently, the frame and body of the truck respond dynamically.

In Figs. 1.2(c) and (d) we illustrate two types of applied actions in the forms of lateral loads that act on a building. Curve 1 in Fig. 1.2(d) depicts the force of an air blast that rises very suddenly, decreases more slowly, and also becomes negative. On the other hand, curve 2 represents the force of a wind gust that rises and falls gradually, but still fast enough to induce significant accelerations. Some other types of applied actions appear in Figs. 1.2(f)–(h). The machinery force acting at the third level of the frame in Fig. 1.2(f) follows a simple harmonic function, as shown by curve 1 in Fig. 1.2(g). On the other hand, the moving load on the plate in Fig. 1.2(h) may have any time variation.

Two kinds of support motions are indicated in Figs. 1.2(c) and (e)–(g). The first type consists of rigid-body ground accelerations due to earthquake, against which most modern buildings and other structures are designed. Figure 1.2(e) shows a typical time history of such an aperiodic ground acceleration. A second type of restraint motion is the independent displacement implied at the right-hand support in Fig. 1.2(f). The time history specified for such an induced displacement can be arbitrary. For example, it may vary according to the square-wave pattern labeled 2 in Fig. 1.2(g).

## 1.3 DISCRETIZATION BY FINITE ELEMENTS

To set up an analytical model for dynamic analysis, we will discretize solids and structures using the method of *finite elements*. The applications include framed structures [2], two- and three-dimensional continua, plates, and shells [3]. The finite-element method enables us to convert a dynamics problem with an infinite number of degrees of freedom to one with a finite number in order to simplify the solution process. Ordinary differential equations of motion can be written for selected points (called *nodes*) on the analytical model. The primary objectives of dynamic analysis by finite elements are to calculate approximately the responses at such nodes or at other selected points.

*Framed structures* usually are automatically discretized by virtue of the common definitions for members and joints. That is, a frame member may be considered as a one-dimensional finite element spanning the distance between two joints. On the other hand, a joint (or node) is defined as a point where members join, a point of support, or a free end of a member. However, we need not restrict ourselves to these particular definitions. Figure 1.3 shows a plane frame consisting of three members (indicated by encircled numbers) and four joints. If desired, we could subdivide each member into four parts and view each segment as a new member. Then the more refined analytical model would have 12 members and 13 joints. Such a refinement would provide more points where

Sec. 1.3  Discretization by Finite Elements

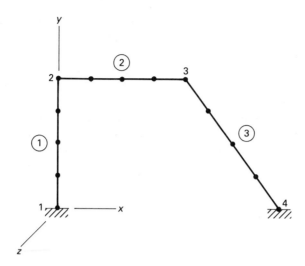

**Figure 1.3** Plane frame with subdivided members.

masses could be located for the purpose of writing more equations of motion. Thus, with these liberalized definitions for members and joints we are able to use the finite-element method more effectively. Although framed structures are discussed throughout the book, Chapter 6 is devoted exclusively to this topic.

Whereas framed structures fit rather neatly into the theory, finite elements have much more powerful applications in two- and three-dimensional continua. For example, the hyperbolic paraboloidal (or hypar) shell structure in Fig. 1.4 may be discretized into quadrilateral elements that are curved in space. We can derive energy-consistent stiffnesses, masses, and equivalent nodal loads for such elements and assemble them into an analytical model having a large but finite number of degrees of freedom. By so doing, we are able to simulate accurately the response of the original structure. The theory of finite elements appears first

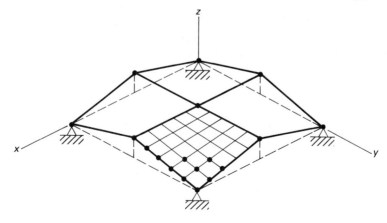

**Figure 1.4** Hypar shell discretized by finite elements.

in Chapter 3 for one-dimensional elements. Later, Chapters 7 and 8 contain applications to two- and three-dimensional continua, plates, and shells.

Occasionally, we may encounter a structure in which one or more parts are very stiff compared to other parts. In such a case, it may be convenient to include *rigid bodies* within the analytical model. For example, the rectangular solid in Fig. 1.5 could be divided into three-dimensional finite elements, while the plate is modeled by two-dimensional elements. However, a more clever modeling procedure would treat the block as a rigid body, which is connected to the flexible plate. The resulting model would have fewer degrees of freedom (due to rigid-body constraints) and better numerical conditioning than the first choice. This topic of rigid bodies within flexible continua is covered in Chapter 9.

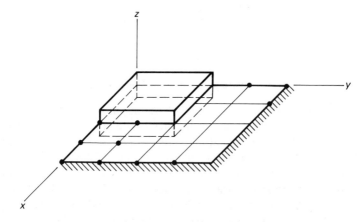

**Figure 1.5** Rigid body supported by discretized plate.

It is also possible to analyze a structure by dividing it into *substructures* that are handled one at a time. Substructuring conserves core storage in a digital computer and also allows several groups of analysts to work on different parts of a structure simultaneously. Figure 1.6 shows half of a symmetric aircraft structure divided into five types of substructures that are joined at common (interface) nodes. Five groups could analyze the substructures individually while a sixth group handles the assembly and solution process for the entire structure.

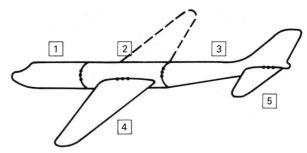

**Figure 1.6** Symmetric aircraft divided into substructures.

The wing group may choose to model the engines as rigid bodies connected to a flexible substructure. Chapter 10 contains descriptions of various substructure methods that have proven to be useful for structural dynamics.

## 1.4 COMPUTER PROGRAMS

Structural dynamics by finite elements is a computer-based method, so we include and explain digital computer programs that perform the calculations. Our programming philosophy consists of starting in Chapter 3 with the relatively simple task of vibrational analysis for plane trusses (Program VIBPT). In Chapter 4 we extend the vibrations program to include dynamic responses by the normal-mode method (Program NOMOPT). Then we further extend the program in Chapter 5 to calculate dynamic responses by a direct numerical integration approach (Program DYNAPT). In subsequent chapters this program is revised many times so that it applies to other types of framed structures and continua that are discretized by finite elements.

All of our coding is written in FORTRAN, and we present FORTRAN-oriented flowcharts for the main programs in VIB, NOMO, and DYNA. We also include a detailed flowchart for Program DYNAPT in Appendix C. Various tables for preparation of data appear throughout the book in order to show how a user must interpret problems to the computer. We show some computer output in the form of line prints, but the amount of information for structural dynamics is so large that we decided to minimize printing. Instead, we put emphasis on computer plots to show time histories of forcing functions, displacements, stresses, and so on. A plot is much more informative than a print, so the more desirable choice of output became obvious. Program notation is included in the list of notation near the end of the book.

## REFERENCES

1. Timoshenko, S. P., Young, D. H., and Weaver, W., Jr., *Vibration Problems in Engineering*, 4th ed., Wiley, New York, 1974.
2. Weaver, W., Jr., and Gere, J. M., *Matrix Analysis of Framed Structures*, 2nd ed., Van Nostrand Reinhold, New York, 1980.
3. Weaver, W., Jr., and Johnston, P. R., *Finite Elements for Structural Analysis*, Prentice-Hall, Englewood Cliffs, N. J., 1984.

# 2

# Systems with One Degree of Freedom

## 2.1 INTRODUCTION

For preliminary or approximate analyses, simple structures may often be idealized as systems with a single degree of freedom (*SDOF systems*). Figure 2.1 shows a few cases where mass $m$ or mass moment of inertia $I_r$ is somehow associated with a single point or axis. In each case the translation of a point or rotation about an axis varies with time. For the beams in Figs. 2.1(a) and (b), the masses may be either attached concentrations or some fractions of the distributed masses of the members. Each of the translational displacements is indicated in the figures by a single-headed arrow.

On the other hand, the rectangular plane frame in Fig. 2.1(c) has a mass $m$ taken to be the tributary mass at the horizontal framing level. Also, in Fig. 2.1(d) we see a massive rigid body connected to framing members in space. In the latter case, the mass moment of inertia $I_r$ of the body is computed with respect to the axis of rotation, for which the displacement is shown by a double-headed arrow. Finally, Fig. 2.1(e) depicts a thin plate in bending with its edges fixed and a single mass $m$ at its center, associated with a translational displacement.

*Flexibilities* for the structures in Fig. 2.1 are obtained as displacements (translations or rotations) due to unit values of the corresponding actions (forces or moments). If the cantilever beam in Fig. 2.1(a) is prismatic, the flexibility $\delta$ of the free end is

$$\delta = \frac{L^3}{3EI}$$

Sec. 1.1  Introduction  11

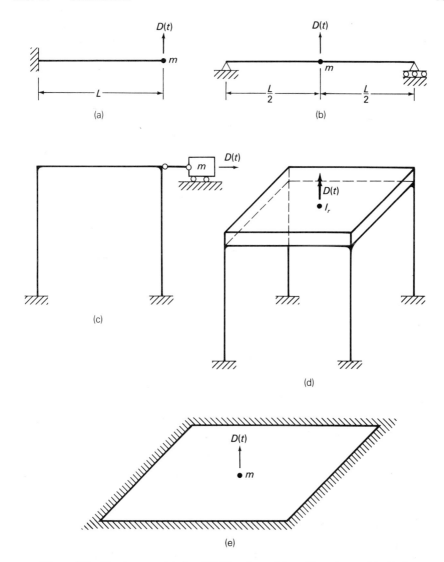

**Figure 2.1** Structures modeled as SDOF systems: (a) cantilever beam; (b) simple beam; (c) plane frame; (d) space frame; (e) fixed plate.

which is the static translation caused by a unit force applied in the positive direction of the displacement $D$. In this expression the symbol $E$ represents Young's modulus of elasticity, $I$ is the moment of inertia* of the cross section of the beam, and $L$ is its length.

*Stiffnesses* are found as static holding actions, corresponding to the displacements shown, caused by inducing unit values of those displacements. At

---
*The moment of inertia of the cross section is a misnomer for the second moment of area of the cross section with respect to the neutral axis.

the free end of the cantilever beam in Fig. 2.1(a), the stiffness $k$ is a static force in the direction of $D$ that is required to produce a unit value of $D$. Thus, we have

$$k = \frac{3EI}{L^3} = \frac{1}{\delta}$$

Note that the stiffness $k$ is the reciprocal of the flexibility $\delta$.

Similarly, if we assume that the simple beam in Fig. 2.1(b) is prismatic, then its flexibility is

$$\delta = \frac{L^3}{48EI}$$

which is the translation at the center due to a unit force corresponding to $D$. Furthermore, the stiffness of this beam has the value

$$k = \frac{48EI}{L^3} = \frac{1}{\delta}$$

which is the holding force at the center required for a unit amount of $D$.

We could also find flexibilities and stiffnesses for the other structures in Fig. 2.1. However, the analyses required would be more complicated than those for the statically determinate beams. Although other motions of the structures are possible, we will restrict our attention in this chapter to systems idealized as having only one degree of freedom. Analytical models having coupled multiple degrees of freedom will be discussed in subsequent chapters.

Topics of interest for SDOF systems consist of free and forced harmonic motions with and without damping, response to arbitrary time-varying loads or support motions, and response spectra for dynamic loads. These subjects are covered in the ensuing sections of this chapter and will be used throughout the book.

## 2.2 FREE, UNDAMPED VIBRATIONS

Figure 2.2(a) shows a *mechanical analogue* for any of the SDOF systems described in the preceding section. The mass $m$ and the spring constant $k$ are determined from the properties of a given elastic structure, and the symbol $u(t)$ denotes the single displacement coordinate. The figure also indicates the acceleration $\ddot{u}(t)$, which is the second derivative of $u(t)$ with respect to time ($\ddot{u} = d^2u/dt^2$). If the displacement is rotational, then $m$, $k$, and $u$ may be replaced by $I_r$, $k_r$, and $\delta\theta$ (a small angle of rotation).

We assume that the system in Fig. 2.2(a) is initially at rest inside or outside a gravitational field [1]. Then let us disturb it in some manner from its position of static equilibrium. Due to the displacement $u$ relative to the static position, a restoring force equal to $ku$ develops in the spring, as shown in Fig. 2.2(b). An

## Sec. 2.2 Free, Undamped Vibrations

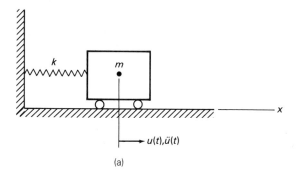

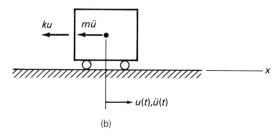

**Figure 2.2** (a) Mechanical analogue for SDOF system; (b) partial free-body diagram.

inertial restoring force $m\ddot{u}$ also arises due to the presence of the mass. From the partial free-body diagram in Fig. 2.2(b), we see that

$$-ku - m\ddot{u} = 0 \quad (1)$$

which is an application of *D'Alembert's principle* for dynamic equilibrium. Rearranging Eq. (1) slightly, we have the following *equation of motion*:

$$m\ddot{u} + ku = 0 \quad (2)$$

This second-order differential equation may be simplified further by introducing the notation

$$\omega^2 = \frac{k}{m} \quad (3)$$

which produces

$$\ddot{u} + \omega^2 u = 0 \quad (4)$$

Such an equation will be satisfied if we take the general solution

$$u = C_1 \cos \omega t + C_2 \sin \omega t \quad (5)$$

where $C_1$ and $C_2$ are arbitrary constants. Thus, the response consists of the sum of two harmonic functions that repeat themselves with time. This oscillatory motion is called free vibration, for which the *angular frequency* $\omega$ (rad/sec) is the constant

$$\omega = \sqrt{\frac{k}{m}} \tag{6}$$

as given by Eq. (3). The *natural frequency f* (cycles/sec) is

$$f = \frac{\omega}{2\pi} = \frac{1}{2\pi}\sqrt{\frac{k}{m}} \tag{7}$$

and the *natural period T* (sec) becomes

$$T = \frac{1}{f} = \frac{2\pi}{\omega} = 2\pi\sqrt{\frac{m}{k}} \tag{8}$$

The last expression is the time for which the vibration repeats itself.

To determine the constants $C_1$ and $C_2$ in Eq. (5), we must consider the *initial conditions* of a given problem. Assume that at the time $t = 0$ the mass $m$ has an *initial displacement* $u_0$ from its position of equilibrium and that its *initial velocity* is $\dot{u}_0$. Substituting $t = 0$ into Eq. (5), we obtain

$$C_1 = u_0$$

If we also take the derivative of Eq. (5) with respect to time and substitute $t = 0$, we find that

$$C_2 = \frac{\dot{u}_0}{\omega}$$

Thus, Eq. (5) becomes

$$u = u_0 \cos \omega t + \frac{\dot{u}_0}{\omega} \sin \omega t \tag{9}$$

Hence, the cosine term depends only on the initial displacement, whereas the sine term depends only upon the initial velocity. Each of these harmonic terms can be represented graphically, as shown in Figs. 2.3(a) and (b), by plotting displacements against time. The total displacement $u$ of the vibrating mass $m$ at any time $t$ is equal to the sum of the ordinates of the two curves, yielding the third harmonic curve shown in Fig. 2.3(c).

An alternative way to represent vibrations is with *rotating vectors*. Imagine two orthogonal axes $u$ and $\dot{u}/\omega$, as shown in Fig. 2.4, which define the *phase plane*. Let a vector **OP** of magnitude $u_0$ rotate with constant angular velocity $\omega$ around the fixed point $O$. If at time $t = 0$ the vector **OP** coincides with the $u$ axis, the angle it makes with the same axis at any later time $t$ is equal to $\omega t$. The projection of this vector on the $u$ axis is equal to $u_0 \cos \omega t$ and represents the first

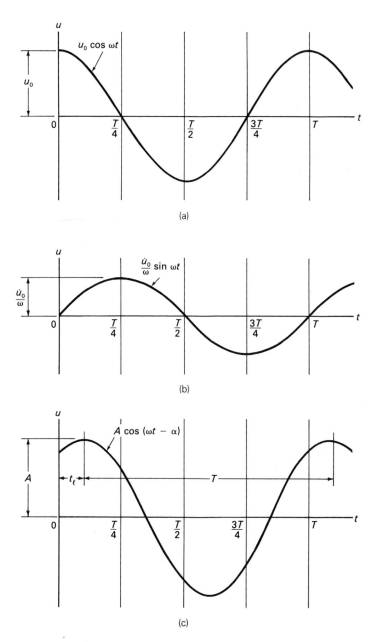

**Figure 2.3** Response functions: (a) cosine; (b) sine; (c) combined.

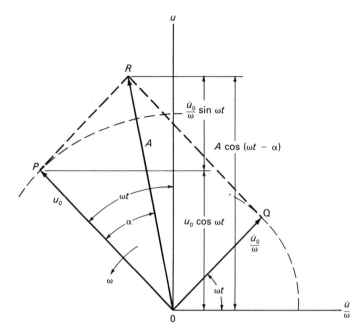

**Figure 2.4** Phase-plane representation.

term in Eq. (9). If we also take another vector **OQ** of magnitude $\dot{u}_0/\omega$ and perpendicular to the vector **OP**, we see that its projection on the $u$ axis gives the second term of Eq. (9). The total displacement $u$ of the vibrating mass is found by adding the projections on the $u$ axis of the two perpendicular vectors **OP** and **OQ**, rotating with angular velocity $\omega$.

The same result will be obtained if we consider the rotating vector **OR** (the sum of vectors **OP** and **OQ**) and take the projection of this vector on the $u$ axis. From Fig. 2.4 the magnitude $A$ of this resultant vector is

$$A = \sqrt{u_0^2 + \left(\frac{\dot{u}_0}{\omega}\right)^2} \tag{10}$$

and the angle that it makes with the $u$ axis is $\omega t - \alpha$, where

$$\alpha = \tan^{-1} \frac{\dot{u}_0}{\omega u_0} \tag{11}$$

Thus, we can express Eq. (9) in the equivalent form

$$u = A \cos(\omega t - \alpha) \tag{12}$$

where the *amplitude of vibration A* is given by Eq. (10), and the *phase angle* $\alpha$ is defined by Eq. (11). Note that the resultant vector **OR** in Fig. 2.4 lags the component vector **OP** by the phase angle $\alpha$. For this reason the total vibration in Fig. 2.3(c) lags the component of the motion in Fig. 2.3(a) by the time

## Sec. 2.2  Free, Undamped Vibrations

$$t_\ell = \frac{\alpha}{\omega} \qquad (13)$$

This *lag time* is indicated at the upper left in Fig. 2.3(c).

### Example 2.1

The rectangular plane frame in Fig. 2.5(a) has a very stiff girder of mass $m$ and rather flexible columns, each having length $L$ and constant bending rigidity $EI$. Neglecting the mass of the columns and their axial strains, find the values of $\omega$, $f$, and $T$ for this structure, treating it as a SDOF system.

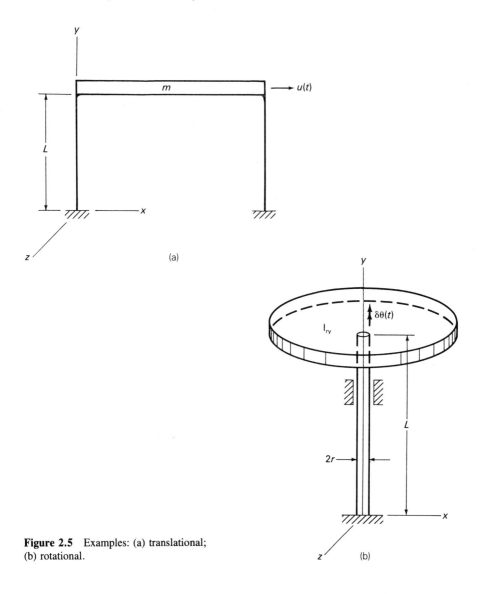

**Figure 2.5** Examples: (a) translational; (b) rotational.

By translating the girder a unit distance ($u = 1$) in the $x$ direction, we find the stiffness constant as the following holding force:

$$k = 2\left(\frac{12EI}{L^3}\right) = \frac{24EI}{L^3} \tag{a}$$

Then the angular frequency $\omega$ from Eq. (6) is

$$\omega = \sqrt{\frac{k}{m}} = \frac{2}{L}\sqrt{\frac{6EI}{mL}} \tag{b}$$

and the natural frequency $f$ from Eq. (7) becomes

$$f = \frac{\omega}{2\pi} = \frac{1}{\pi L}\sqrt{\frac{6EI}{mL}} \tag{c}$$

Also, the natural period $T$ from Eq. (8) is seen to be

$$T = \frac{1}{f} = \pi L \sqrt{\frac{mL}{6EI}} \tag{d}$$

### Example 2.2

Suppose that the rigid disk in Fig. 2.5(b) has a mass moment of inertia $I_{ry}$ with respect to the $y$ axis. The disk is supported at its center by a massless flexible rod of length $L$ having a circular cross section with radius equal to $r$. If the disk can rotate about the $y$ axis without translation, find $\omega$, $f$, and $T$ for this problem as a SDOF system.

Rotation of the disk a unit amount ($\delta\theta = 1$) about the $y$ axis yields the stiffness constant as the holding moment

$$k_r = \frac{GJ}{L} \tag{e}$$

in which $G$ is the shearing modulus and

$$J = \frac{\pi r^4}{2} \tag{f}$$

is the polar moment of inertia of the cross section of the rod. Replacing $m$ and $k$ with $I_{ry}$ and $k_r$ in Eqs. (6), (7), and (8), we find that

$$\omega = \sqrt{\frac{k_r}{I_{ry}}} = r^2 \sqrt{\frac{\pi G}{2I_{ry}L}} \tag{g}$$

and

$$f = \frac{\omega}{2\pi} = \frac{r^2}{2\pi}\sqrt{\frac{\pi G}{2I_{ry}L}} \tag{h}$$

and

$$T = \frac{1}{f} = \frac{2\pi}{r^2}\sqrt{\frac{2I_{ry}L}{\pi G}} \tag{i}$$

## 2.3 HARMONIC FORCING FUNCTIONS

One of the most important types of forcing functions that we must study is the simple harmonic function $P \sin \Omega t$ (or $P \cos \Omega t$), where $\Omega$ is the angular frequency of the function. The first of these functions appears in Fig. 2.6(a), applied to the undamped SDOF system of the preceding section. From the partial free-body diagram in Fig. 2.6(b), we can write the differential equation of motion for this case, as follows:

$$m\ddot{u} + ku = P \sin \Omega t \tag{1}$$

To simplify this equation, let us divide both sides by $m$ to obtain

$$\ddot{u} + \omega^2 u = p_m \sin \Omega t \tag{2}$$

where

$$p_m = \frac{P}{m} \tag{3}$$

Here the symbol $p_m$ represents a force per unit of mass.

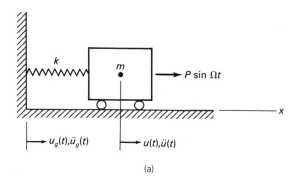

(a)

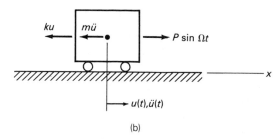

(b)

**Figure 2.6** (a) Harmonic forcing function applied to SDOF system; (b) partial free-body diagram.

The total solution for Eq. (2) consists of the sum of the general solution for the homogeneous equation, plus the particular solution that satisfies the whole equation. The form of such a particular solution is

$$u = C_3 \sin \Omega t \qquad (4)$$

Substituting this expression and its second derivative with respect to time into Eq. (2), we have

$$-C_3 \Omega^2 \sin \Omega t + C_3 \omega^2 \sin \Omega t = p_m \sin \Omega t$$

Therefore,

$$C_3 = \frac{p_m}{\omega^2 - \Omega^2} \qquad (5)$$

Then the total solution becomes

$$u = C_1 \cos \omega t + C_2 \sin \omega t + C_3 \sin \Omega t \qquad (6)$$

The first two terms on the right-hand side of Eq. (6) are called the *free part* of the response (with angular frequency $\omega$), and the third term is referred to as the *forced part* (with angular frequency $\Omega$).

Let us rewrite the forced part of the solution in Eq. (6) as

$$u = \frac{p_m}{\omega^2 - \Omega^2} \sin \Omega t \qquad (7)$$

Then, by using Eq. (3) and $\omega^2 = k/m$, we may cast this expression into the form

$$u = \left[ \frac{1}{1 - (\Omega/\omega)^2} \right] \frac{P}{k} \sin \Omega t \qquad (8)$$

The second factor on the right-hand side of this equation is the displacement of the mass in Fig. 2.6(a) if the forcing function were applied statically. The absolute value of the term in brackets is called the *magnification factor $\beta$*. Thus, we have

$$\beta = \left| \frac{1}{1 - (\Omega/\omega)^2} \right| \qquad (9)$$

which represents the ratio of the dynamic response to the static response. The part of the motion in which this factor appears is called *steady-state forced vibration*.

Figure 2.7 shows values of the magnification factor $\beta$ plotted against the frequency ratio $\Omega/\omega$, which is the ratio of the impressed angular frequency $\Omega$ of the disturbing force to the natural angular frequency $\omega$ of the SDOF system. For small values of the frequency ratio, the magnification factor is approximately unity; and the response is about the same as for the static application of the harmonic load. When this ratio approaches unity, however, the magnification

## Sec. 2.3  Harmonic Forcing Functions

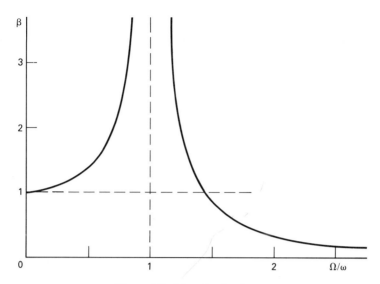

**Figure 2.7** Magnification factor.

factor and the *amplitude of forced vibration* rapidly increase and theoretically become infinite for the case of $\Omega = \omega$. This state is known as the condition of *resonance*. The infinite response at resonance implies that there is no damping to dissipate energy and suppress the response. However, in practical structures there is always some damping, as will be discussed in Sec. 2.4.

When the frequency of the disturbing force increases beyond the frequency of free vibration, the magnification factor again becomes finite. Its absolute value diminishes with the increase of the ratio $\Omega/\omega$ and approaches zero when this ratio becomes very large. Thus, when a harmonically varying force of high frequency acts on a SDOF system, the response is very small; and in such cases the mass may be considered as remaining stationary.

Considering the sign of the expression $1/[1 - (\Omega/\omega)^2]$, we see that for $\Omega < \omega$ the sign is positive. Physically, this means that the displacement of the vibrating mass is in the same positive sense as the disturbing force. On the other hand, when $\Omega > \omega$ the expression is negative, meaning that the displacement of the mass is opposite to that of the force. In the first case the vibration is said to be *in phase* with the excitation, while in the latter case the response is said to be *out of phase*.

It is also possible to cause forced vibrations with *harmonic ground motions* instead of applied forcing functions. Two types of ground motions of interest in structural dynamics are *ground displacement $u_g(t)$* and *ground acceleration $\ddot{u}_g(t)$*, as indicated in Fig. 2.6(a). A harmonic version of the first of these influences may be expressed as

$$u_g = d \sin \Omega t \tag{10}$$

where the symbol $d$ represents the amplitude of ground displacement. For this case we write the following differential equation of motion:

$$m\ddot{u} + k(u - u_g) = 0 \tag{11}$$

in which $u - u_g$ is the displacement of the mass $m$ relative to the ground. Substituting Eq. (10) into Eq. (11) and rearranging the latter gives

$$m\ddot{u} + ku = kd \sin \Omega t \tag{12}$$

Then division of both sides by $m$ yields

$$\ddot{u} + \omega^2 u = p_g \sin \Omega t \tag{13}$$

where

$$p_g = \frac{kd}{m} \tag{14}$$

This value of $p_g$ is an equivalent force per unit of mass, due to ground displacement.

Comparing Eq. (12) to Eq. (1), we see that $kd$ has replaced $P$. Therefore, it is evident that <u>occurrence of the harmonic ground displacement $d \sin \Omega t$ is equivalent to applying a force $kd \sin \Omega t$ directly to the mass $m$</u>. All previous conclusions regarding the harmonic force may now be applied to the case of harmonic ground displacement. That is, if we replace $P$ in Eq. (8) with $kd$ and cancel $k$, the result is

$$u = \left[\frac{1}{1 - (\Omega/\omega)^2}\right] d \sin \Omega t \tag{15}$$

The term $d \sin \Omega t$ in this expression is the motion of the mass when the ground displacement occurs very slowly (or statically). The premultiplier in brackets is the same factor as that discussed before [see Eq. (9)]. Thus, we need only consider the displacement of the mass due to the static displacement of the ground in order to calculate the steady-state forced response of the SDOF system.

<u>In many cases it is more convenient to deal with ground accelerations than ground displacements because a measuring device called an *accelerometer*</u> [2] has been used to obtain information about the ground motion. For example, earthquake ground motions are usually measured and reported in terms of three orthogonal components of ground acceleration, which are north-south, east-west, and vertical. Therefore, we shall reexamine the ground motion problem by specifying a harmonic ground acceleration, as follows:

$$\ddot{u}_g = a \sin \Omega t \tag{16}$$

where the symbol $a$ denotes the amplitude of ground acceleration. In order to use this expression in Eq. (11), we must change that equation to the *relative coordinate*:

### Sec. 2.3  Harmonic Forcing Functions

$$u^* = u - u_g \qquad \ddot{u}^* = \ddot{u} - \ddot{u}_g \tag{17}$$

In these relationships the symbol $u^*$ represents the *relative displacement* of the mass with respect to the ground, and $\ddot{u}^*$ is the *relative acceleration*. Substituting $u - u_g$ and $\ddot{u}$ from Eqs. (17) into Eq. (11) and rearranging, we obtain

$$m\ddot{u}^* + ku^* = -m\ddot{u}_g \tag{18}$$

If we also use Eq. (16) in Eq. (18), the result is

$$m\ddot{u}^* + ku^* = -ma \sin \Omega t \tag{19}$$

Then division of Eq. (19) by $m$ produces

$$\ddot{u}^* + \omega^2 u^* = p_g^* \sin \Omega t \tag{20}$$

where

$$p_g^* = -a \tag{21}$$

This value of $p_g^*$ is an equivalent force per unit of mass, due to ground acceleration.

Equation (19) is of the same mathematical form as Eq. (1), except that $u^*$ has replaced $u$ and $-ma$ has replaced $P$. Thus, the case of harmonic ground acceleration $a \sin \Omega t$ is equivalent to applying a force $-ma \sin \Omega t$ directly to the mass $m$. Using $-ma$ instead of $P$ in Eq. (8), we have

$$u^* = -\left[\frac{1}{1 - (\Omega/\omega)^2}\right] \frac{ma}{k} \sin \Omega t \tag{22}$$

which is the displacement of the mass relative to the ground. This relative motion is useful for finding the force in the structure,* which is represented by the spring in the SDOF system. Note that the minus sign before the brackets in Eq. (22) puts the relative response of $m$ out of phase with the ground acceleration.

**Example 2.3**

Suppose that the rectangular plane frame in Example 2.1 is subjected to a harmonic force $P \sin \Omega t$, as indicated in Fig. 2.8. Calculate the steady-state forced vibration of the mass $m$ if $\Omega = 5\omega/6$.

Recalling that the stiffness constant for this frame is $k = 24EI/L^3$, we substitute the values of $\Omega$ and $k$ into Eq. (8) to obtain

$$u = \frac{1}{1 - (5/6)^2} \frac{PL^3}{24EI} \sin \Omega t$$

$$= \frac{3PL^3}{22EI} \sin \Omega t \tag{a}$$

Note that the value of the magnification factor is $\beta = |36/11|$, in which the positive sign implies that the vibration is in phase with the forcing function.

---

*We will return to this matter of calculating the relative displacement of the mass with respect to the ground in Example 2.12.

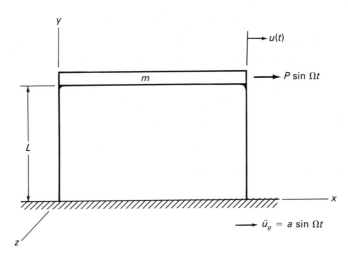

**Figure 2.8** Frame with harmonic forcing functions.

### Example 2.4

If the same frame as that in Example 2.3 has a harmonic ground acceleration $\ddot{u}_g = a \sin \Omega t$ (see Fig. 2.8), find the steady-state forced vibration for $\Omega = 3\omega$.

In this case the relative response, as given by Eq. (22) is:

$$u^* = -\frac{1}{1-(3)^2}\frac{maL^3}{24EI}\sin \Omega t$$

$$= \frac{maL^3}{192EI}\sin \Omega t \tag{b}$$

Here we see that the value of the magnification factor is $\beta = |-1/8|$, so that the vibration is apparently out of phase with the ground acceleration. However, the two minus signs cancel to yield a positive sign in Eq. (b), meaning that in this situation the vibration is actually in phase with the ground acceleration.

In the explanation preceding the examples, only the third term on the right-hand side of Eq. (6) was used for studying forced vibrations. Now let us examine the free vibrations associated with the first two terms, which will be different from those in Sec. 2.2. Substituting Eq. (5) into Eq. (6), we have

$$u = C_1 \cos \omega t + C_2 \sin \omega t + \frac{p_m}{\omega^2 - \Omega^2}\sin \Omega t \tag{23}$$

Differentiating this equation with respect to time yields

$$\dot{u} = -C_1 \omega \sin \omega t + C_2 \omega \cos \omega t + \frac{p_m \Omega}{\omega^2 - \Omega^2}\cos \Omega t \tag{24}$$

From the initial condition $u = u_0$ at time $t = 0$, Eq. (23) gives $C_1 = u_0$, which is the same as before. However, substitution of $\dot{u} = \dot{u}_0$ at time $t = 0$ into Eq.

(24) produces
$$\dot{u}_0 = C_2\omega + \frac{p_m\Omega}{\omega^2 - \Omega^2}$$

Thus,
$$C_2 = \frac{\dot{u}_0}{\omega} - \frac{p_m\Omega/\omega}{\omega^2 - \Omega^2} \tag{25}$$

Then the total solution in Eq. (23) becomes
$$u = u_0 \cos \omega t + \frac{\dot{u}_0}{\omega} \sin \omega t + \frac{p_m}{\omega^2 - \Omega^2}\left(\sin \Omega t - \frac{\Omega}{\omega} \sin \omega t\right) \tag{26}$$

If the initial conditions are $u_0 = \dot{u}_0 = 0$, Eq. (26) reduces to
$$u = \frac{p_m}{\omega^2 - \Omega^2}\left(\sin \Omega t - \frac{\Omega}{\omega} \sin \omega t\right) \tag{27}$$

Equation (27) represents the response of the SDOF system to the forcing function $P \sin \Omega t$, and we see that it consists of two parts. The first part is the steady-state response discussed previously, and the second part is called the *transient response*. This name refers to the fact that the term dies out in the presence of damping, as do all free vibrations. The sum of the two terms is not a harmonic motion even though it is composed of two harmonic functions, because the components have different frequencies, $\Omega$ and $\omega$.

When the forcing function is $P \cos \Omega t$ instead of $P \sin \Omega t$, the term $\cos \Omega t$ replaces $\sin \Omega t$ in Eq. (23). In this case the initial conditions result in the following constants of integration:
$$C_1 = u_0 - \frac{p_m}{\omega^2 - \Omega^2} \qquad C_2 = \frac{\dot{u}_0}{\omega} \tag{28}$$

Substitution of these values into the total solution gives
$$u = u_0 \cos \omega t + \frac{\dot{u}_0}{\omega} \sin \omega t + \frac{p_m}{\omega^2 - \Omega^2}(\cos \Omega t - \cos \omega t) \tag{29}$$

If the initial conditions are taken to be $u_0 = \dot{u}_0 = 0$, this equation becomes
$$u = \frac{p_m}{\omega^2 - \Omega^2}(\cos \Omega t - \cos \omega t) \tag{30}$$

In this instance the transient part of the response has the same amplitude as the steady-state part, regardless of the ratio $\Omega/\omega$.

## 2.4 EFFECTS OF DAMPING

In previous discussions of free and forced vibrations for SDOF systems, we did not consider the effects of dissipative influences, such as friction or air resistance. Consequently, we found that the amplitude of free vibration remains constant with time; but experience shows that the amplitude diminishes with time and that the vibrations are gradually damped out. Similarly, for undamped forced vibrations the theory indicates that the amplitude can grow without limit at resonance. However, we know that because of damping there is always some finite amplitude of steady-state response, even at resonance.

To bring our analytical discussion of vibrations into better agreement with reality, *damping forces* must be included. These forces may arise from several causes, such as friction between dry or lubricated sliding surfaces, air or fluid resistance, electric impedance, internal friction due to imperfect elasticity of materials, and so on. Among all of these sources of energy dissipation, the case where the damping force is proportional to velocity, called *viscous damping*, is the simplest to deal with mathematically. For this reason resisting forces of a complicated nature are usually replaced, for purposes of analysis, by *equivalent viscous damping* [1]. This equivalent damping is found by equating the dissipation of energy per cycle of vibration to that for viscous damping. For example, *structural damping* due to internal friction can be treated by this approach.

We shall now consider the case of a SDOF system that includes viscous damping in the form of a dashpot damper, as shown by Fig. 2.9(a). Assuming that a viscous fluid in the dashpot resists motion in proportion to velocity, we

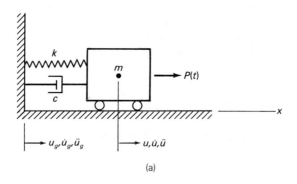

(a)

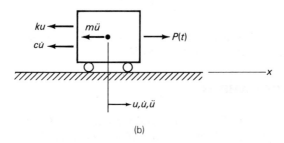

(b)

**Figure 2.9** (a) SDOF system with viscous damping; (b) partial free-body diagram.

## Sec. 2.4  Effects of Damping

write the homogeneous equation for free vibration as

$$m\ddot{u} + c\dot{u} + ku = 0 \tag{1}$$

The coefficient $c$ in this equation denotes the *damping constant*, which has dimensions of force per unit velocity. Note that in the partial free-body diagram [see Fig. 2.9(b)] the damping force acts in the direction opposite to the velocity. Dividing Eq. (1) by $m$ produces

$$\ddot{u} + 2n\dot{u} + \omega^2 u = 0 \tag{2}$$

in which we define the new constant

$$2n = \frac{c}{m} \tag{3}$$

for convenience in the ensuing solution.

To solve Eq. (2), we assume a harmonic function in the following form:

$$u = Ce^{st} \tag{4}$$

in which $C$ and $s$ are constants that satisfy Eq. (2). Substituting Eq. (4) into Eq. (2), we find that

$$s^2 + 2ns + \omega^2 = 0$$

from which

$$s = -n \pm \sqrt{n^2 - \omega^2} \tag{5}$$

If the value of $n$ is less than that of $\omega$, then the quantity

$$\omega_d^2 = \omega^2 - n^2$$

is positive, and we obtain for $s$ two complex roots, as follows:

$$s_1 = -n + i\omega_d \qquad s_2 = -n - i\omega_d$$

where $i = \sqrt{-1}$. Substituting these roots into Eq. (4), we find two solutions of Eq. (2). The sum or difference of these two solutions multiplied by any constant will also be a solution. Thus,

$$u_1 = \frac{C_1}{2}(e^{s_1 t} + e^{s_2 t}) = C_1 e^{-nt} \cos \omega_d t$$

$$u_2 = \frac{C_2}{2i}(e^{s_1 t} - e^{s_2 t}) = C_2 e^{-nt} \sin \omega_d t$$

Adding these formulas, we obtain the general solution of Eq. (2) for *damped free vibrations* as

$$u = e^{-nt}(C_1 \cos \omega_d t + C_2 \sin \omega_d t) \tag{6}$$

The factor $e^{-nt}$ in this equation decreases with time, so the vibrations originally generated will be gradually damped out.

Sine and cosine functions appearing in the parentheses of Eq. (6) are of the same general form that we found before for free vibrations without damping [see Eq. (2.2-5)]. However, now they have the angular frequency

$$\omega_d = \sqrt{\omega^2 - n^2} = \omega\sqrt{1 - \left(\frac{n}{\omega}\right)^2} \tag{7}$$

which is called the *angular frequency of damped free vibrations*. The corresponding *period of damped free vibrations* is

$$T_d = \frac{2\pi}{\omega_d} = \frac{2\pi}{\sqrt{\omega^2 - n^2}} \tag{8}$$

If the damping term $n$ is small, then $\omega_d \approx \omega$ and $T_d \approx T$.

The constants $C_1$ and $C_2$ in Eq. (6) must be determined from known initial conditions. Assume that at time $t = 0$ we have initial displacement $u_0$ and initial velocity $\dot{u}_0$. Substituting these quantities into Eq. (6) and its first derivative with respect to time, we find that

$$C_1 = u_0 \qquad C_2 = \frac{\dot{u}_0 + nu_0}{\omega_d} \tag{9}$$

Putting these constants into Eq. (6) yields

$$u = e^{-nt}\left(u_0 \cos \omega_d t + \frac{\dot{u}_0 + nu_0}{\omega_d} \sin \omega_d t\right) \tag{10}$$

The first harmonic term in this equation depends only on $u_0$, but the second depends on both $\dot{u}_0$ and $u_0$.

Equation (10) can also be written in the equivalent form

$$u = Ae^{-nt} \cos(\omega_d t - \alpha_d) \tag{11}$$

In this expression the maximum value is

$$A = \sqrt{C_1^2 + C_2^2} = \sqrt{u_0^2 + \frac{(\dot{u}_0 + nu_0)^2}{\omega_d^2}} \tag{12}$$

and

$$\alpha_d = \tan^{-1}\frac{C_2}{C_1} = \tan^{-1}\frac{\dot{u}_0 + nu_0}{\omega_d u_0} \tag{13}$$

We may regard Eq. (11) as a pseudoharmonic motion having an exponentially decreasing amplitude $Ae^{-nt}$, an angular frequency $\omega_d$, and a phase angle $\alpha_d$. Figure 2.10 shows this damped free vibration with a lag time of

$$t_\ell = \frac{\alpha_d}{\omega_d} \tag{14}$$

indicated at the upper left.

## Sec. 2.4  Effects of Damping

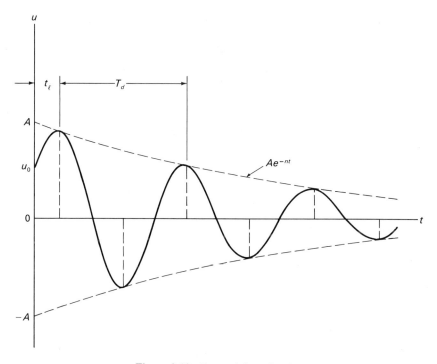

**Figure 2.10** Damped free vibration.

In the foregoing discussion of Eq. (2), we assumed that $n < \omega$. For the opposite case of $n > \omega$, both of the roots in Eq. (5) become real and negative. Substituting them into Eq. (4), we obtain two solutions of Eq. (2), and the general solution becomes

$$u = C_1 e^{s_1 t} + C_2 e^{s_2 t} \tag{15}$$

In this case the solution is not periodic, so it does not represent a vibratory motion. The viscous resistance is so large that when the mass is displaced from its equilibrium position, it merely creeps back to that position. In such a case the system is said to be *overdamped*, and the motion is called aperiodic. The constants $C_1$ and $C_2$ in Eq. (15) may be evaluated by substituting the initial conditions at time $t = 0$ into the equation and its first derivative, yielding

$$C_1 + C_2 = u_0 \qquad s_1 C_1 + s_2 C_2 = \dot{u}_0$$

from which

$$C_1 = \frac{\dot{u}_0 - s_2 u_0}{s_1 - s_2} \qquad C_2 = \frac{s_1 u_0 - \dot{u}_0}{s_1 - s_2} \tag{16}$$

Thus, Eq. (15) becomes

$$u = \frac{\dot{u}_0 - s_2 u_0}{s_1 - s_2} e^{s_1 t} + \frac{s_1 u_0 - \dot{u}_0}{s_1 - s_2} e^{s_2 t} \tag{17}$$

The general appearance of a graph of Eq. (17) depends on the parameters $n$, $u_0$, and $\dot{u}_0$.

Between the underdamped and overdamped cases lies the special case of $n = \omega$, which is the level of damping where the motion first loses its vibratory character. Using Eq. (3) for this condition, we have

$$c_{cr} = 2nm = 2\omega m = 2\sqrt{km} \tag{18}$$

in which the symbol $c_{cr}$ denotes *critical damping*. For the critically damped SDOF system, Eqs. (5) and (7) show that $s_1 = s_2 = -\omega$ and $\omega_d = 0$. Neither Eq. (6) nor Eq. (17) constitutes the solution, which in this particular case of repeated roots takes the form

$$u = e^{-\omega t}(C_1 + C_2 t) \tag{19}$$

Substituting the initial conditions into Eq. (19) and its first derivative, we find that

$$C_1 = u_0 \qquad C_2 = \dot{u}_0 + n u_0 \tag{20}$$

Then the general solution becomes

$$u = e^{-\omega t}[u_0 + (\dot{u}_0 + n u_0)t] \tag{21}$$

which again depends on $n$, $u_0$, and $\dot{u}_0$.

Turning now to *damped forced vibrations*, we take the applied forcing function in Fig. 2.9(a) to be $P \cos \Omega t$. Then the equation of motion for the mass becomes

$$m\ddot{u} + c\dot{u} + ku = P \cos \Omega t \tag{22}$$

Dividing this equation by $m$ produces

$$\ddot{u} + 2n\dot{u} + \omega^2 u = p_m \cos \Omega t \tag{23}$$

for which all of the notation has been defined previously. A particular solution of Eq. (23) can be taken in the form

$$u = Q \cos \Omega t + R \sin \Omega t \tag{24}$$

where $Q$ and $R$ are constants. To determine these constants, we substitute Eq. (24) and its derivatives into Eq. (23) and obtain

$$(-\Omega^2 Q + 2n\Omega R + \omega^2 Q - p_m) \cos \Omega t \\ + (-\Omega^2 R - 2n\Omega Q + \omega^2 R) \sin \Omega t = 0$$

This equation can be solved for all values of $t$ only if the expressions in the parentheses vanish. Thus, for calculating $Q$ and $R$ we have two linear algebraic equations, as follows:

$$-\Omega^2 Q + 2n\Omega R + \omega^2 Q = p_m$$
$$-\Omega^2 R - 2n\Omega Q + \omega^2 R = 0$$

## Sec. 2.4  Effects of Damping

from which the solution is

$$Q = \frac{(\omega^2 - \Omega^2)p_m}{(\omega^2 - \Omega^2)^2 + 4n^2\Omega^2} \qquad R = \frac{2n\Omega p_m}{(\omega^2 - \Omega^2)^2 + 4n^2\Omega^2} \qquad (25)$$

By substituting these constants into Eq. (24), we can obtain the particular solution of Eq. (23).

The total solution of Eq. (23) consists of the sum of the particular solution and the general solution derived previously as Eq. (6). Thus, considering only subcritical damping, we have

$$u = e^{-nt}(C_1 \cos \omega_d t + C_2 \sin \omega_d t) + Q \cos \Omega t + R \sin \Omega t \qquad (26)$$

Because of the factor $e^{-nt}$ in the first part, the free vibrations gradually subside, leaving only the steady-state forced vibrations represented by the last two terms. The harmonic force maintains these latter vibrations indefinitely, so they are of great practical importance.

The steady-state response in Eq. (24) may be written in the equivalent phase-angle form as

$$u = A \cos (\Omega t - \theta) \qquad (27)$$

where

$$A = \sqrt{Q^2 + R^2} = \frac{p_m}{\sqrt{(\omega^2 - \Omega^2)^2 + (2n\Omega)^2}}$$

$$= \frac{p_m/\omega^2}{\sqrt{[1 - (\Omega/\omega)^2]^2 + (2n\Omega/\omega^2)^2}} \qquad (28)$$

and

$$\theta = \tan^{-1}\frac{R}{Q} = \tan^{-1}\frac{2n\Omega}{\omega^2 - \Omega^2}$$

$$= \tan^{-1}\frac{2n\Omega/\omega^2}{1 - (\Omega/\omega)^2} \qquad (29)$$

Thus, we see that steady-state forced vibration with viscous damping is a simple harmonic motion having constant amplitude $A$, phase angle $\theta$, and period $T_f = 2\pi/\Omega$.

Using the values of $\omega^2 = k/m$ and $p_m = P/m$ and introducing the symbol $\gamma$ for the *damping ratio*,

$$\gamma = \frac{n}{\omega} = \frac{c}{c_{cr}} \qquad (30)$$

we may substitute Eq. (28) into Eq. (27) to obtain

$$u = \beta \frac{P}{k} \cos (\Omega t - \theta) \qquad (31)$$

in which the *magnification factor* $\beta$ for damped forced vibrations is

$$\beta = \frac{1}{\sqrt{[1 - (\Omega/\omega)^2]^2 + (2\gamma\Omega/\omega)^2}} \qquad (32)$$

Also, Eq. (29) for the *phase angle* becomes

$$\theta = \tan^{-1} \frac{2\gamma\Omega/\omega}{1 - (\Omega/\omega)^2} \qquad (33)$$

Thus, the amplitude and the phase angle both depend on the damping ratio $\gamma$ as well as the frequency ratio $\Omega/\omega$.

Figure 2.11(a) shows the magnification factor $\beta$ plotted against the ratio $\Omega/\omega$ for various levels of damping. As for the undamped case, the value of $\beta$ is approximately unity for small values of $\Omega/\omega$; and $\beta$ approaches zero for large values of $\Omega/\omega$. However, as the value of $\Omega$ approaches that of $\omega$ (that is, $\Omega/\omega$ approaches unity), the magnification factor grows rapidly. Furthermore, the value of $\beta$ at or near resonance is very sensitive to the amount of damping. Also, note that the maximum values of $\beta$ occur at frequency ratios less than unity. Setting the derivative of $\beta$ with respect to $\Omega/\omega$ equal to zero, we find that the maximum occurs when

$$\frac{\Omega}{\omega} = \sqrt{1 - 2\gamma^2} \qquad (34)$$

For small damping ratios ($\gamma \leq 0.20$) the maximum value of $\beta$ occurs very near to resonance, and taking the value of $\beta$ at resonance as the maximum provides sufficient accuracy for engineering analysis. Then, from Eq. (31) we have

$$A_{max} = \beta_{res}\frac{P}{k} = \frac{1}{2\gamma}\frac{P}{k} = \frac{\omega}{2n}\frac{P}{\omega^2 m} = \frac{P}{c\omega} \qquad (35)$$

Thus, while damping has only a minor effect when the system is remote from resonance, it has a dramatic effect at or near resonance. In structural dynamics the influence of damping is crucial for this case and represents its most important application. In metal structures the damping ratio $\gamma$ is usually in the range 0.01 to 0.05; whereas its range for reinforced concrete structures is about 0.05 to 0.10. The value of $\gamma$ would always be less than 0.20 for practical structures.

While the phase angle is of less consequence, we also show in Fig. 2.11(b) plots of $\theta$ versus the frequency ratio $\Omega/\omega$ for different values of the damping ratio. The physical meaning of the phase angle in damped forced vibrations is that the response of the system lags the forcing function by the angle $\theta$. For the case of zero damping, the forced vibrations are exactly in phase ($\theta = 0$) with the disturbing force for all values of $\Omega/\omega < 1$. On the other hand, they are a half cycle out of phase ($\theta = \pi$) for all values of $\Omega/\omega > 1$. Also, for zero damping the phase angle is indeterminate at resonance, where $\Omega = \omega$. When damping is nonzero, we note a continual change in $\theta$ as the ratio $\Omega/\omega$ varies. But, regardless

## Sec. 2.4    Effects of Damping

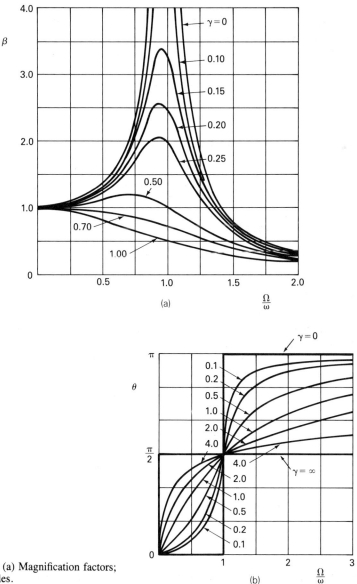

**Figure 2.11** (a) Magnification factors; (b) phase angles.

of the amount of damping, the phase angle is always equal to $\pi/2$ at resonance. That is, at resonance the response lags the force by a quarter cycle.

### Example 2.5

Let the rectangular plane frame in Fig. 2.8 have a harmonic force $P \cos \Omega t$ in place of the force used in Example 2.3. Also, imagine that the ground has an acceleration

$\ddot{u}_g = a \cos \Omega t$ instead of the sine function in Example 2.4. Calculate separately the steady-state forced responses of the mass $m$ due to these influences, assuming that $\gamma = 0.02$ and $\Omega = 0.9\omega$.

For this example the magnification factor in Eq. (32) becomes

$$\beta = \frac{1}{\sqrt{[1 - (0.9)^2]^2 + [(2)(0.02)(0.9)]^2}} = 5.17 \tag{a}$$

Using this value and $k = 24EI/L^3$ in Eq. (31), we calculate the response to the force as

$$u = (5.17)\frac{PL^3}{24EI} \cos(\Omega t - \theta)$$

$$= 0.215 \frac{PL^3}{EI} \cos(\Omega t - \theta) \tag{b}$$

Considering the ground acceleration, we need only replace $P$ in Eq. (31) with $-ma$ to obtain the relative response as

$$u^* = -0.215 \frac{maL}{EI} \cos(\Omega t - \theta) \tag{c}$$

**Example 2.6**

The *transient response* of a SDOF system with subcritical damping may be found by substituting initial conditions into Eq. (26). Determine the free-vibrational response of such a system due to the forcing function $P \cos \Omega t$.

Using $u = u_0$ and $\dot{u} = \dot{u}_0$ (at time $t = 0$) in Eq. (26) and its first derivative with respect to time, we evaluate the constants of integration as

$$C_1 = u_0 - Q \qquad C_2 = \frac{\dot{u}_0 + n(u_0 - Q) - R\Omega}{\omega_d} \tag{d}$$

Substituting these constants into Eq. (26) gives

$$u = e^{-nt}\left(u_0 \cos \omega_d t + \frac{\dot{u}_0 + nu_0}{\omega_d} \sin \omega_d t\right)$$

$$+ Q\left[\cos \Omega t - e^{-nt}\left(\cos \omega_d t + \frac{n}{\omega_d} \sin \omega_d t\right)\right] \tag{e}$$

$$+ R\left(\sin \Omega t - e^{-nt}\frac{\Omega}{\omega_d} \sin \omega_d t\right)$$

If the initial conditions are taken to be $u_0 = \dot{u}_0 = 0$, the transient portion $u_{tr}$ of the remaining response is

$$u_{tr} = -e^{-nt}\left(Q \cos \omega_d t + \frac{Qn + R\Omega}{\omega_d} \sin \omega_d t\right) \tag{f}$$

Expressing this transient in phase-angle form, we have

$$u_{tr} = -e^{-nt}\frac{C}{\omega_d} \cos(\omega_d t - \alpha_d) \tag{g}$$

where

$$C = \sqrt{(Q\omega_d)^2 + (Qn + R\Omega)^2} \tag{h}$$

and

$$\alpha_d = \tan^{-1}\frac{Qn + R\Omega}{Q\omega_d} \tag{i}$$

In the last expression the symbol $\alpha_d$ represents the phase angle for the damped system.

## 2.5 PERIODIC FORCING FUNCTIONS

In previous discussions of forced vibrations, we assumed simple harmonic functions proportional to $\sin \Omega t$ or $\cos \Omega t$. It is also possible to encounter general periodic functions that are more complicated. For example, Fig. 2.12 shows a repetitive triangular function with period $T_f = 2\pi/\Omega$. In this section we learn how to calculate the response of a SDOF system to such a function.

We can represent a periodic dynamic load of any kind by decomposing it into a *trigonometric* (or *Fourier*) *series,* as follows:

$$\begin{aligned}P(t) &= a_0 + a_1 \cos \Omega t + a_2 \cos 2\Omega t + \ldots \\ &\quad + b_1 \sin \Omega t + b_2 \sin 2\Omega t + \ldots \\ &= a_0 + \sum_{i=1}^{\infty} (a_i \cos i\Omega t + b_i \sin i\Omega t) \end{aligned} \tag{1}$$

The period of the applied force is $T_f = 2\pi/\Omega$, and the symbols $a_0$, $a_i$, and $b_i$ represent constants to be determined. To calculate these constants, we may use the procedure described next.

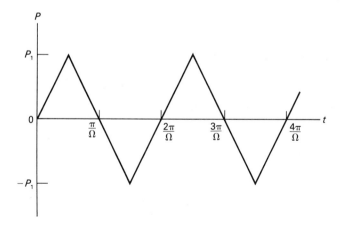

**Figure 2.12** Periodic forcing function.

Assuming that $a_i$ in Eq. (1) is desired, postmultiply both sides of the equation by $\cos i\Omega t\, dt$ and integrate from $t = 0$ to $t = T_f$. Then we see that

$$\int_0^{T_f} a_0 \cos i\Omega t\, dt = 0 \qquad \int_0^{T_f} a_j \cos j\Omega t \cos i\Omega t\, dt = 0$$

$$\int_0^{T_f} b_j \sin j\Omega t \cos i\Omega t\, dt = 0 \qquad \int_0^{T_f} a_i \cos^2 i\Omega t\, dt = \frac{a_i}{2} T_f \tag{2}$$

where $i$ and $j$ are integer numbers 1, 2, 3, . . . , $\infty$. By using these results, we find that

$$a_i = \frac{2}{T_f} \int_0^{T_f} P(t) \cos i\Omega t\, dt \tag{3a}$$

In a similar manner, multiplication of Eq. (1) by $\sin i\Omega t\, dt$ and integrating produces

$$b_i = \frac{2}{T_f} \int_0^{T_f} P(t) \sin i\Omega t\, dt \tag{3b}$$

Finally, multiplying Eq. (1) by $dt$ and integrating yields

$$a_0 = \frac{1}{T_f} \int_0^{T_f} P(t)\, dt \tag{3c}$$

Thus, by using Eqs. (3), we can calculate the *Fourier coefficients* $a_i$, $b_i$, and $a_0$ for any known periodic function.

Assuming that the forcing function has been decomposed into a Fourier series, we can now write the equation of motion for damped forced vibrations as

$$m\ddot{u} + c\dot{u} + ku = a_0 + a_1 \cos \Omega t + a_2 \cos 2\Omega t + \ldots \\ + b_1 \sin \Omega t + b_2 \sin 2\Omega t + \ldots \tag{4}$$

The total solution of this equation consists of the sum of free and forced vibrations associated with initial conditions and the terms on the right-hand side. Because of damping, the free vibrations will disappear with time, leaving only the forced vibrations. Each harmonic term in the Fourier series contributes a steady-state forced vibration of the type described in the preceding section. Therefore, we conclude that forced vibrations with large amplitudes can occur whenever the period of any term in the series is the same as (or close to) the natural period of the SDOF system. This always happens when the period $T_f$ of the forcing function is equal to, or an even multiple of, the natural period $T_d$ of the damped system.

Let us now apply the method of this section to the piecewise-linear forcing function illustrated in Fig. 2.12. To decompose the function into Fourier components, we have the formulas in Eqs. (3). Starting with Eq. (3c), we see that

### Sec. 2.5  Periodic Forcing Functions

the integral

$$\int_0^{T_f} P(t)\, dt$$

is simply the area under the sawtooth diagram for one cycle. Since this area is zero, the constant $a_0$ in Eq. (3c) is also equal to zero. Next, Eq. (3a) involves the multiplication of the forcing function by $\cos i\Omega t\, dt$ and integration from $t = 0$ to $t = T_f = 2\pi/\Omega$. From the antisymmetry of $P(t)$ and the symmetry of $\cos i\Omega t$ with respect to $t = \pi/\Omega$, we conclude again that the integral in Eq. (3a) is zero, so that $a_i = 0$. Finally, considering Eq. (3b), it is apparent that $P(t)$ from $t = 0$ to $t = \pi/\Omega$ is symmetric about $t = \pi/2\Omega$, while from $t = \pi/\Omega$ to $t = 2\pi/\Omega$ it is symmetric about $t = 3\pi/2\Omega$. However, when $i$ is an even integer, corresponding parts of $\sin i\Omega t$ are antisymmetric with respect to $t = \pi/2\Omega$ and $t = 3\pi/2\Omega$. Thus, for all even values of $i$, we conclude that $b_i = 0$.

When $i$ is an odd integer, both $P(t)$ and $\sin i\Omega t$ are antisymmetric with respect to $t = \pi/\Omega$, and Eq. (3b) gives

$$b_i = \frac{\Omega}{\pi} \int_0^{2\pi/\Omega} P(t) \sin i\Omega t\, dt = \frac{4\Omega}{\pi} \int_0^{\pi/2\Omega} P(t) \sin i\Omega t\, dt \qquad (5)$$

In the time interval $0 \le t \le \pi/2\Omega$, the function in Fig. 2.12 has the formula

$$P(t) = \frac{2}{\pi} P_1 \Omega t \qquad (6)$$

where $P_1$ is the maximum value of the forcing function. Substituting this expression into Eq. (5) yields

$$b_i = \frac{8 P_1 \Omega^2}{\pi^2} \int_0^{\pi/2\Omega} t \sin i\Omega t\, dt = \frac{8 P_1}{i^2 \pi^2} \int_0^{i\pi/2} v \sin v\, dv$$

in which the new variable is $v = i\Omega t$. Integrating this expression by parts and substituting the limits produces

$$b_i = \frac{8 P_1}{i^2 \pi^2} \sin \frac{i\pi}{2} = \frac{8 P_1}{i^2 \pi^2} (-1)^{(i-1)/2} \qquad (7)$$

In this formula the odd values of $i$ are 1, 3, 5, 7, and so on. Thus, the Fourier series that approximates the periodic function in Fig. 2.12 is

$$P(t) = \frac{8 P_1}{\pi^2}\left(\sin \Omega t - \frac{1}{3^2} \sin 3\Omega t + \frac{1}{5^2} \sin 5\Omega t - \cdots\right) \qquad (8)$$

and we need only superimpose the sine curves with odd numbers of periods in the interval $t = 0$ to $t = 2\pi/\Omega$.

If we omit damping and use a frequency ratio $\Omega/\omega = 0.9$, the magnification factor for the first term in Eq. (8) is

$$\beta_1 = \frac{1}{1 - (\Omega/\omega)^2} = 5.26$$

whereas that for the second term becomes

$$\beta_3 = \frac{1}{1 - (3\Omega/\omega)^2} = -0.159$$

Furthermore, a multiplier of $1/3^2$ appears in the second term of Eq. (8). Therefore, we conclude that using only the first term for the response

$$u \approx \frac{8}{\pi^2} \frac{P_1}{k} \beta_1 \sin \Omega t \tag{9}$$

causes an error of less than 0.4% in the solution.

Suppose now that the applied force in Fig. 2.12 is replaced by a *periodic ground acceleration* of the same form. If the peak acceleration $a$ replaces the maximum force $P_1$, we need only use $-ma$ instead of $P_1$ in the response calculations. For example, the result in Eq. (9) would be changed to

$$u \approx -\frac{8}{\pi^2} \frac{ma}{k} \beta_1 \sin \Omega t \tag{10}$$

which is only the first term of the series.

## 2.6 ARBITRARY FORCING FUNCTIONS

We now consider dynamic loads that have no periodic character and may vary in any manner with time. Such *arbitrary forcing functions* must be handled in a special way, as described in this section.

Figure 2.13 shows a general forcing function $P(t')$ that is expressed in terms of a new time variable $t'$. The value of $t'$ is less than that of $t$, which is the time when the response is to be calculated. If the function is applied to a damped SDOF system, the differential equation of motion becomes

$$m\ddot{u} + c\dot{u} + ku = P(t') \tag{1}$$

At any instant of time $t'$, we may calculate an *incremental impulse* $P\,dt'$, represented by the hatched strip in Fig. 2.13. That impulse imparts to the mass $m$ an instantaneous increase in velocity (or *incremental velocity*) equal to

$$d\dot{u} = \frac{P\,dt'}{m} = p_m\,dt' \tag{2}$$

This expression is valid regardless of what other forces (such as the spring force) may be acting on the mass, and regardless of its displacement and velocity at the time $t'$. Treating the incremental velocity as if it were an initial velocity (at the time $t'$), we conclude that the *incremental displacement* of the mass at any later time $t$ will be

## Sec. 2.6  Arbitrary Forcing Functions

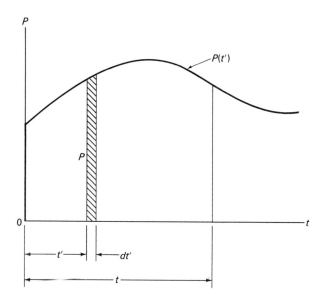

**Figure 2.13**  Arbitrary forcing function.

$$du = e^{-n(t-t')} \frac{P_m \, dt'}{\omega_d} \sin \omega_d (t - t') \tag{3}$$

which is drawn from Eq. (2.4-10). Because each incremental impulse between $t' = 0$ and $t' = t$ has such an effect, we obtain the total displacement due to the applied force as the integral

$$u = \frac{e^{-nt}}{\omega_d} \int_0^t e^{nt'} P_m \sin \omega_d (t - t') \, dt' \tag{4}$$

This mathematical form is known as *Duhamel's integral*.

Equation (4) gives the complete displacement produced by the forcing function $P(t')$, acting during the time interval from 0 to $t$. For periodic forces it includes both steady-state and transient terms and is especially useful in studying the response of a SDOF system to any kind of short-term (or transient) load. If the function $P(t')$ cannot be expressed analytically, we can always evaluate the integral in Eq. (4) by some numerical method. To include initial conditions for the damped system, we need only add their effects to Eq. (4), as follows:

$$u = e^{-nt} \left[ u_0 \cos \omega_d t + \frac{\dot{u}_0 + nu_0}{\omega_d} \sin \omega_d t \right.$$
$$\left. + \frac{1}{\omega_d} \int_0^t e^{nt'} P_m \sin \omega_d (t - t') \, dt' \right] \tag{5}$$

This equation represents the total solution of Eq. (1).

If damping is neglected, we have $n = 0$ and $\omega_d = \omega$, and Eq. (4) reduces to

$$u = \frac{1}{\omega} \int_0^t P_m \sin \omega(t - t') \, dt' \tag{6}$$

Including initial conditions without damping also gives a simpler form of Eq. (5), which becomes

$$u = u_0 \cos \omega t + \frac{\dot{u}_0}{\omega} \sin \omega t + \frac{1}{\omega} \int_0^t P_m \sin \omega(t - t') \, dt' \tag{7}$$

This briefer equation for the total response can be used whenever damping is not significant.

To demonstrate use of the Duhamel integral, assume that a constant force $P_1$ [see Fig. 2.14(a)] is suddenly applied to the mass of the SDOF system in Fig. 2.9(a). This condition of dynamic loading is called a *step function*, where

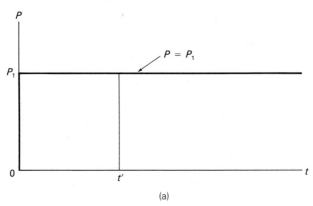

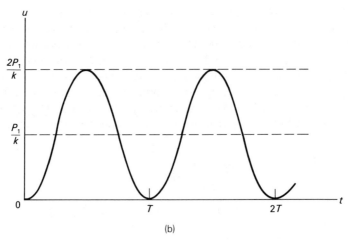

**Figure 2.14** (a) Step function; (b) response.

## Sec. 2.6  Arbitrary Forcing Functions

$p_m = P_1/m$ is constant and acts for an indefinite time. For the case of zero damping, Eq. (6) gives

$$u = \frac{p_m}{\omega} \int_0^t \sin \omega(t - t') \, dt'$$

Integration of this expression yields

$$u = \frac{p_m}{\omega^2}(1 - \cos \omega t) = \frac{P_1}{k}(1 - \cos \omega t) \tag{8}$$

From this result we see that a step force $P_1$ produces free vibrations of amplitude $P_1/k$ superimposed on a static displacement of the same magnitude, as depicted in Fig. 2.14(b). Thus, the maximum displacement due to a suddenly applied force is twice as large as that caused by the same force acting statically.

Next, let us consider a step load that acts only for a period of time $t_1$. This type of forcing function is called a *rectangular impulse* and is illustrated in Fig. 2.15(a). During the time when the force is nonzero, the response of an undamped SDOF system is the same as that given by Eq. (8). On the other hand, the response after time $t_1$ may be found by evaluating the Duhamel integral for two ranges: 0 to $t_1$ and $t_1$ to $t$. Only the integration over the first range will produce nonzero results, because the forcing function is zero in the second range. Altogether, the solution for this case is summarized as follows:

$$u = \frac{P_1}{k}(1 - \cos \omega t) \qquad (0 \le t \le t_1) \tag{8}$$

$$u = \frac{P_1}{k}[\cos \omega(t - t_1) - \cos \omega t] \qquad (t_1 \le t) \tag{9}$$

The same results may be obtained by considering the rectangular impulse in Fig. 2.15(a) to consist of the sum of the two step functions, as indicated in Fig. 2.15(b). The first step function (of magnitude $P_1$) begins at time $t = 0$, while the second step function (of magnitude $-P_1$) begins at time $t = t_1$.

A third method for determining the result given as Eq. (9) involves finding the displacement and velocity of the SDOF system at time $t_1$, using Eq. (8) and its first derivative with respect to time. Thus,

$$u_{t_1} = \frac{P_1}{k}(1 - \cos \omega t_1) \qquad \dot{u}_{t_1} = \frac{\omega P_1}{k} \sin \omega t_1 \tag{10}$$

If these two quantities are treated as initial conditions at time $t_1$, the ensuing free-vibrational response may be calculated from

$$u = u_{t_1} \cos \omega(t - t_1) + \frac{\dot{u}_{t_1}}{\omega} \sin \omega(t - t_1) \tag{11}$$

Substitution of expressions (10) into Eq. (11), followed by trigonometric manipulation, yields the same result as that in Eq. (9).

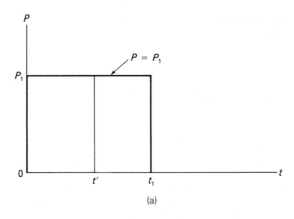

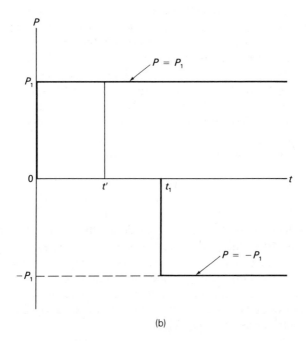

**Figure 2.15** (a) Rectangular impulse; (b) step-load simulation.

Depending on the duration time $t_1$ for the rectangular impulse, the response of a SDOF system may take different forms. However, the maximum response can never be more than the value $2P_1/k$. Instructions for determining loci of response maxima due to various types of forcing functions appear in Sec. 2.8.

If the damped SDOF system in Fig. 2.9(a) is subjected to *arbitrary ground accelerations*, we need only replace the function $p_m(t')$ in Eqs. (4) through (7) with the function $p_m^*(t') = -\ddot{u}_g(t')$. In such cases these equations yield the

## Sec. 2.6  Arbitrary Forcing Functions

displacement $u^*$ of the mass relative to ground. Thus, for a step acceleration $a_1$ of the ground, Eq. (8) becomes

$$u^* = -\frac{a_1}{\omega^2}(1 - \cos \omega t) = -\frac{ma_1}{k}(1 - \cos \omega t) \tag{12}$$

In addition, the second part of the response to a rectangular function for ground acceleration is

$$u^* = -\frac{ma_1}{k}[\cos \omega(t - t_1) - \cos \omega t] \quad (t_1 \leq t) \tag{13}$$

which is a modification of Eq. (9).

### Example 2.7

Find the undamped response of a SDOF system to the linearly increasing force, known as a *ramp function*, given in Fig. 2.16(a). The rate of increase of the force $P(t')$ per unit of time is $P_1/t_1$.

As indicated in the figure, the forcing function for this example is

$$P = \frac{P_1 t'}{t_1} \tag{a}$$

so the force per unit of mass becomes

$$p_m = \frac{P_1 t'}{m t_1} \tag{b}$$

Applying Eq. (6) to this case, we have

$$u = \frac{P_1}{m t_1 \omega} \int_0^t t' \sin \omega(t - t') \, dt'$$

Integrating this expression by parts yields

$$u = \frac{P_1}{k t_1}\left(t - \frac{1}{\omega} \sin \omega t\right) \tag{c}$$

Thus, we see that the response to a ramp function consists of the sum of a linearly increasing static displacement $P_1 t/k t_1$ and a negative sinusoidal free vibration of amplitude $P_1/k t_1 \omega$, as shown in Fig. 2.16(b).

The velocity at any time $t$ is equal to the first derivative of Eq. (c) with respect to time. That is,

$$\dot{u} = \frac{P_1}{k t_1}(1 - \cos \omega t) = \frac{P_1}{k t_1}\left(1 - \cos \frac{2\pi t}{T}\right) \tag{d}$$

From this expression we conclude that the velocity is zero at times $t = 0, T, 2T, 3T$, and so on. Therefore, the slope of the displacement curve in Fig. 2.16(b) is zero at $t = 0$, $T, 2T, 3T$, and so on. Furthermore, the velocity is always positive and has a maximum value of $2P_1/k t_1$ at times $t = T/2, 3T/2, 5T/2$, and so on [see Fig. 2.16(b)].

If we have a ramp ground acceleration instead of an applied force, it may be

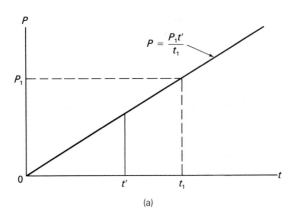

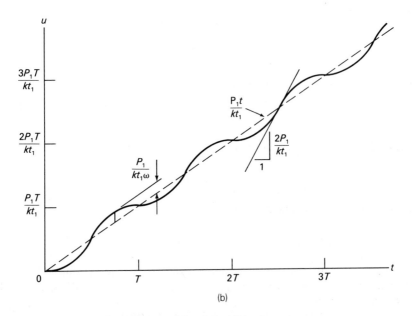

**Figure 2.16** (a) Ramp function; (b) response.

expressed as

$$\ddot{u}_g = \frac{a_1 t'}{t_1} \tag{e}$$

where $a_1$ is the ground acceleration at time $t_1$. Then the response in Eq. (c) is converted to the relative motion

$$u^* = -\frac{ma_1}{kt_1}\left(t - \frac{1}{\omega}\sin \omega t\right) \tag{f}$$

## Sec. 2.7 Step-by-Step Response Calculations

Similarly, the velocity in Eq. (d) is replaced by

$$\dot{u}^* = -\frac{ma_1}{kt_1}(1 - \cos \omega t) \tag{g}$$

The term $-ma_1$ appears in these results instead of $P_1$.

**Example 2.8**

Suppose that the ramp function in Example 2.7 is terminated at time $t_1$, as shown in Fig. 2.17. Determine the response of the undamped SDOF system in the time ranges $0 \le t \le t_1$ and $t_1 \le t$.

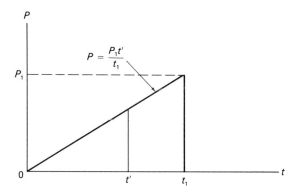

**Figure 2.17** Triangular impulse.

The function appearing in Fig. 2.17 is called a *triangular impulse*. For the first time interval $(0 \le t \le t_1)$, the response is the same as for the ramp function, as given by Eq. (c). In the second time interval, we must integrate over two ranges, as for the rectangular impulse. Doing this and summarizing the results produces

$$u = \frac{P_1}{kt_1}\left(t - \frac{1}{\omega}\sin \omega t\right) \qquad (0 \le t \le t_1) \quad \text{(c)}$$

$$u = \frac{P_1}{kt_1}\left[t_1 \cos \omega(t - t_1) + \frac{1}{\omega}\sin \omega(t - t_1) - \frac{1}{\omega}\sin \omega t\right] \qquad (t_1 \le t) \quad \text{(h)}$$

If the triangular impulse is a ground acceleration with maximum value equal to $a_1$, the responses in Eqs. (c) and (h) have $u^*$ in place of $u$ and $-ma_1$ in place of $P_1$.

## 2.7 STEP-BY-STEP RESPONSE CALCULATIONS

In many practical problems the forcing functions are not analytical expressions but are represented by a series of points on a diagram or a list of numbers in a table. In such cases it may be feasible to replace the data with certain formulas by curve-fitting methods and then to use those formulas in the Duhamel integral. However, a more general method for evaluating the response consists of using some simple interpolation function in a repetitive series of calculations. The latter approach is discussed in this section for *piecewise-linear interpolation functions*.

Figure 2.18 shows a forcing function of general shape that is approximated by a series of straight lines. For a particular interpolating line in the time interval $t_j \leq t \leq t_{j+1}$, the response of a damped SDOF system may be written as the sum of three parts, as follows:

$$u = u_1 + u_2 + u_3 \tag{1}$$

Using the definition $t' = t - t_j$, we have for the first part

$$u_1 = e^{-nt'}\left(u_j \cos \omega_d t' + \frac{\dot{u}_j + nu_j}{\omega_d} \sin \omega_d t'\right) \tag{2a}$$

This equation contains the free-vibrational motion of the system due to the displacement $u_j$ and the velocity $\dot{u}_j$ at time $t = t_j$ (the beginning of the interval). The formula for this portion of the response is drawn from Eq. (2.4-10).

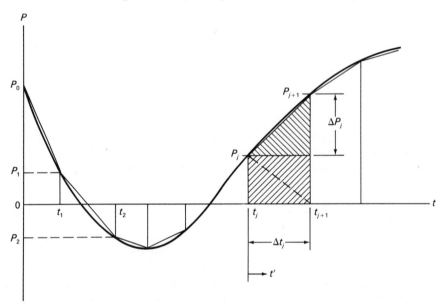

**Figure 2.18** Piecewise-linear interpolation of forcing function.

The other two parts of the response in Eq. (1) are associated with the straight-line forcing function in Fig. 2.18. The one caused by the rectangular impulse of magnitude $P_j$ is

$$u_2 = \frac{P_j}{k}\left[1 - e^{-nt'}\left(\cos \omega_d t' + \frac{n}{\omega_d} \sin \omega_d t'\right)\right] \tag{2b}$$

which comes from the solution of Prob. 2.6-2. On the other hand, the effect of the triangular impulse of magnitude $\Delta P_j = P_{j+1} - P_j$ becomes

## Sec. 2.7 Step-by-Step Response Calculations

$$u_3 = \frac{\Delta P_j}{\Delta t_j k \omega^2}\left[\omega^2 t' - 2n + e^{-nt'}\left(2n \cos \omega_d t' - \frac{\omega_d^2 - n^2}{\omega_d} \sin \omega_d t'\right)\right] \quad (2c)$$

This formula is available as the solution of Prob. 2.6-3.

By differentiating Eqs. (1) and (2) with respect to time, we can also find velocity expressions in three parts. Thus,

$$\dot{u} = \dot{u}_1 + \dot{u}_2 + \dot{u}_3 \quad (3)$$

where

$$\dot{u}_1 = e^{-nt'}\left[-\left(u_j \omega_d + n\frac{\dot{u}_j + nu_j}{\omega_d}\right)\sin \omega_d t' + \dot{u}_j \cos \omega_d t'\right] \quad (4a)$$

and

$$\dot{u}_2 = \frac{P_j \omega^2}{k \omega_d} e^{-nt'} \sin \omega_d t' \quad (4b)$$

Also,

$$\dot{u}_3 = \frac{\Delta P_j}{\Delta t_j k}\left[1 - e^{-nt'}\left(\cos \omega_d t' + \frac{n}{\omega_d}\sin \omega_d t'\right)\right] \quad (4c)$$

At the end of the time interval $\Delta t_j$, the displacement expressions in Eqs. (2) become

$$(u_1)_{j+1} = e^{-n\Delta t_j}\left(u_j \cos \omega_d \Delta t_j + \frac{\dot{u}_j + nu_j}{\omega_d}\sin \omega_d \Delta t_j\right) \quad (5a)$$

$$(u_2)_{j+1} = \frac{P_j}{k}\left[1 - e^{-n\Delta t_j}\left(\cos \omega_d \Delta t_j + \frac{n}{\omega_d}\sin \omega_d \Delta t_j\right)\right] \quad (5b)$$

$$(u_3)_{j+1} = \frac{\Delta P_j}{\Delta t_j k \omega^2}\left[\omega^2 \Delta t_j - 2n + e^{-n\Delta t_j}\left(2n \cos \omega_d \Delta t_j - \frac{\omega_d^2 - n^2}{\omega_d}\sin \omega_d \Delta t_j\right)\right]$$

$$(5c)$$

In addition, the velocity expressions in Eqs. (4) are rewritten as

$$(\dot{u}_1)_{j+1} = e^{-n\Delta t_j}\left[-\left(u_j \omega_d + n\frac{\dot{u}_j + nu_j}{\omega_d}\right)\sin \omega_d \Delta t_j + \dot{u}_j \cos \omega_d \Delta t_j\right] \quad (6a)$$

$$(\dot{u}_2)_{j+1} = \frac{P_j \omega^2}{k \omega_d} e^{-n\Delta t_j} \sin \omega_d \Delta t_j \quad (6b)$$

$$(\dot{u}_3)_{j+1} = \frac{\Delta P_j}{\Delta t_j k}\left[1 - e^{-n\Delta t_j}\left(\cos \omega_d \Delta t_j + \frac{n}{\omega_d}\sin \omega_d \Delta t_j\right)\right] \quad (6c)$$

Equations (5) and (6) constitute *recurrence formulas* that may be used to calculate the damped response at the end of step $j$ and to provide initial conditions at the beginning of step $j + 1$.

If damping is neglected, Eqs. (5) for displacements simplify as follows:

$$(u_1)_{j+1} = u_j \cos \omega \Delta t_j + \frac{\dot{u}_j}{\omega} \sin \omega \Delta t_j \tag{7a}$$

$$(u_2)_{j+1} = \frac{P_j}{k}(1 - \cos \omega \Delta t_j) \tag{7b}$$

$$(u_3)_{j+1} = \frac{\Delta P_j}{\Delta t_j k \omega}(\omega \Delta t_j - \sin \omega \Delta t_j) \tag{7c}$$

and Eqs. (6) for velocities become

$$(\dot{u}_1)_{j+1} = -u_j \omega \sin \omega \Delta t_j + \dot{u}_j \cos \omega \Delta t_j \tag{8a}$$

$$(\dot{u}_2)_{j+1} = \frac{P_j}{k} \omega \sin \omega \Delta t_j \tag{8b}$$

$$(\dot{u}_3)_{j+1} = \frac{\Delta P_j}{\Delta t_j k}(1 - \cos \omega \Delta t_j) \tag{8c}$$

Equations (7) and (8) are simple enough for hand calculations to obtain approximate results.

Of course, we need not take the shaded impulse in Fig. 2.18 as the sum of a rectangle and a triangle. Alternatively, it could be divided into two triangles, as indicated by the dashed diagonal line in the figure. Then it would be possible to express the second and third parts of the response in terms of $P_j$ and $P_{j+1}$. Furthermore, if the time step $\Delta t_j$ is constant, the coefficients of $u_j$, $\dot{u}_j$, $P_j$, and $\Delta P_j$ (or $P_{j+1}$) all become constants for both the displacement and the velocity expressions. Hence, these coefficients need be computed only once and then used repetitively throughout the numerical solution [3].

**Example 2.9**

Figure 2.19(a) shows a forcing function $P = P_1 \sin \Omega t$ that is applied to an undamped SDOF system. The function is discretized by piecewise-linear interpolation into 20 equal time steps of duration $\Delta t = T_f/20$. Using the method of this section, calculate approximately the response of the system. Assume that the initial conditions are $u_0 = \dot{u}_0 = 0$, the values of $P_1$ and $k$ are both unity, and the frequency ratio is $\Omega/\omega = 0.9$.

From Prob. 2.6-1, the exact solution to this problem (with zero damping) is

$$u = \frac{P_1}{k}\left(\sin \Omega t - \frac{\Omega}{\omega} \sin \omega t\right)\beta \tag{a}$$

in which the magnification factor $\beta$ has the value

$$\beta = \left|\frac{1}{1 - (\Omega/\omega)^2}\right| = \frac{1}{1 - (0.9)^2} = 5.263 \tag{b}$$

An approximate solution is found by applying Eqs. (7) and (8) recursively in 20 specified

Sec. 2.7    Step-by-Step Response Calculations    49

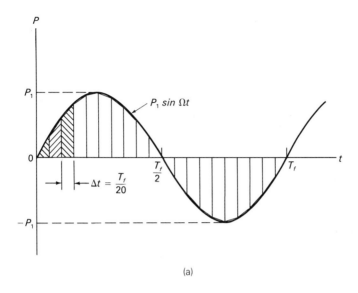

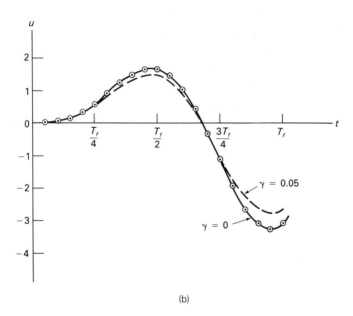

**Figure 2.19** (a) Sinusoidal forcing function; (b) approximate responses.

time steps. Results of such calculations (by hand or computer) are summarized in Table 2.1. Also given in the table are exact displacements obtained from Eq. (a). As expected, the approximate displacements are slightly less than their exact counterparts, because linear interpolation of the sine curve is imperfect. Decreasing the step size would, of course, lead to exact values (except for round-off errors).

**TABLE 2.1** Solution for Example 2.9

| j | $P_j$ | $u_j$ | Exact | j | $P_j$ | $u_j$ | Exact |
|---|---|---|---|---|---|---|---|
| 1 | 0.309 | 0.006 | 0.006 | 11 | −0.309 | 1.407 | 1.418 |
| 2 | 0.588 | 0.048 | 0.049 | 12 | −0.588 | 1.000 | 1.001 |
| 3 | 0.809 | 0.154 | 0.156 | 13 | −0.809 | 0.404 | 0.407 |
| 4 | 0.951 | 0.338 | 0.341 | 14 | −0.951 | −0.338 | −0.341 |
| 5 | 1.000 | 0.593 | 0.598 | 15 | −1.000 | −1.151 | −1.161 |
| 6 | 0.951 | 0.896 | 0.903 | 16 | −0.951 | −1.945 | −1.961 |
| 7 | 0.809 | 1.203 | 1.213 | 17 | −0.809 | −2.616 | −2.634 |
| 8 | 0.588 | 1.461 | 1.473 | 18 | −0.588 | −3.068 | −3.094 |
| 9 | 0.309 | 1.613 | 1.626 | 19 | −0.309 | −3.220 | −3.246 |
| 10 | 0 | 1.607 | 1.620 | 20 | 0 | −3.020 | −3.045 |

The approximate undamped response in Table 2.1 is plotted as the solid curve in Fig. 2.19(b). Also shown in this figure is a dashed curve that results from recursively applying Eqs. (5) and (6) with the damping ratio $\gamma = n/\omega = 0.05$.

**Example 2.10**

Figure 2.20(a) gives a series of plotted points simulating a *blast load* that impinges on an undamped SDOF structure, such as the building frame in Fig. 2.1(c). Note that the blast force rises quickly to the maximum value $P_1$ and then diminishes more slowly (and even becomes negative for awhile). In this case we have 16 equal time steps, each of which has the value $\Delta t = T/30$. Apply the method of this section to find the approximate response of the structure. Let the values of both $P_1$ and $k$ be unity, and assume that the initial conditions are $u_0 = \dot{u}_0 = 0$.

After using Eqs. (7) and (8) recursively to obtain the time history of the response, we then list the calculated displacements in Table 2.2. For this problem there are no exact

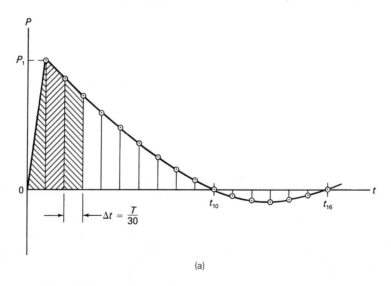

**Figure 2.20** (a) Discretized blast load; (b) approximate responses.

## Sec. 2.8 Response Spectra

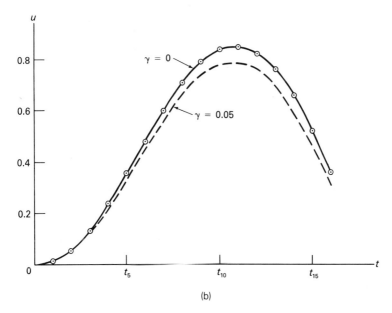

(b)

**Figure 2.20** (cont.)

**TABLE 2.2 Solution for Example 2.10**

| $j$ | $P_j$ | $u_j$ | $j$ | $P_j$ | $u_j$ |
|---|---|---|---|---|---|
| 1 | 1.000 | 0.007 | 9  | 0.070  | 0.780 |
| 2 | 0.850 | 0.050 | 10 | 0      | 0.830 |
| 3 | 0.720 | 0.127 | 11 | −0.050 | 0.844 |
| 4 | 0.590 | 0.230 | 12 | −0.080 | 0.819 |
| 5 | 0.475 | 0.350 | 13 | −0.100 | 0.755 |
| 6 | 0.360 | 0.474 | 14 | −0.080 | 0.654 |
| 7 | 0.250 | 0.594 | 15 | −0.050 | 0.521 |
| 8 | 0.155 | 0.699 | 16 | 0      | 0.363 |

results with which to compare these numbers. Nevertheless, we have great confidence in their validity. Figure 2.20(b) shows a plot of the time history of undamped response (the solid curve), which appears to be quite reasonable. Also appearing in the figure is a dashed curve that gives the effect of using Eqs. (5) and (6) with $\gamma = 0.05$.

## 2.8 RESPONSE SPECTRA

The dynamic loads discussed in this chapter cause vibrational responses in elastic systems, and the maximum values of these responses may be less than, equal to, or greater than the corresponding static responses. For a SDOF system, the natural period (or frequency) is the characteristic that determines its response to a given forcing function. In addition, the shape and duration of the forcing function itself play important roles in the response. Plots (or loci) of maximum response values against selected parameters of the structure or the forcing function are called *response spectra*. Such diagrams are useful in design because they provide the means for predicting the ratio of the maximum dynamic stress in a structure to the corresponding static stress. The time at which the maximum response occurs is also of interest, and plots of this variable usually will accompany the response spectra discussed in this article.

Let us reconsider the rectangular impulse shown in Fig. 2.21(a), which was applied to an undamped SDOF system in Sec. 2.6. If the duration $t_1$ of the impulse exceeds the value $T/2$ (half the natural period), the maximum response is always equal to 2. On the other hand, if the time $t_1$ varies from zero to $T/2$, the maximum response goes from zero to 2. Figure 2.21(b) shows the early stages of a series of such responses for $t_1 = 0.1T, 0.2T, 0.3T, 0.4T$, and $0.5T$. Note that the maximum response always occurs in the first excursion, which is the only part for each curve exhibited in the figure. In all such cases, the maximum displacement occurs after the impulsive action terminates, because the velocity at time $t_1$ is positive.

Thus, to find the maximum value of the response and the time when it occurs, we must examine Eq. (2.6-9), for which $t_1 \leq t$. That equation may be written in dimensionless form as

$$\frac{u}{u_{st}} = \cos \omega(t - t_1) - \cos \omega t \tag{1}$$

where the static displacement due to $P_1$ is

$$u_{st} = \frac{P_1}{k} \tag{2}$$

Differentiating Eq. (1) with respect to time, we obtain

$$\frac{\dot{u}}{u_{st}} = \omega[\sin \omega t - \sin \omega(t - t_1)]$$

By setting the term in brackets equal to zero, we find an expression involving the time $t_m$, at which the maximum response occurs. That is,

$$\sin \omega t_m = \sin \omega(t_m - t_1)$$

The following relationship satisfies this equation:

$$\omega t_m = \frac{\pi}{2} + \frac{\omega t_1}{2} \tag{3}$$

Hence, $t_m$ is a linear function of $t_1$. Substituting the formula for $\omega t_m$ from Eq. (3) into Eq. (1) yields

$$\frac{u}{u_{st}} = 2 \sin \frac{\omega t_1}{2} = \sqrt{2(1 - \cos \omega t_1)} \tag{4}$$

## Sec. 2.8 Response Spectra

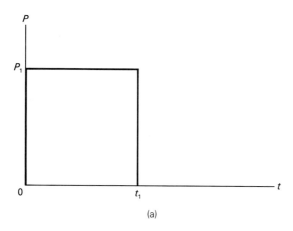

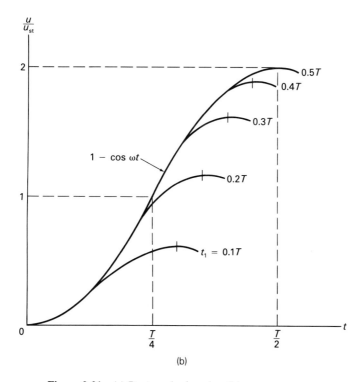

**Figure 2.21** (a) Rectangular impulse; (b) response curves.

Using Eqs. (3) and (4), we can summarize the response spectrum for the rectangular impulse as follows:

$$\frac{u_m}{u_{st}} = 2 \sin \frac{\pi t_1}{T} \qquad \left(0 \le \frac{t_1}{T} \le \frac{1}{2}\right) \qquad (5a)$$

$$\frac{t_m}{T} = \frac{1}{4} + \frac{t_1}{2T} \qquad (5b)$$

and

$$\frac{u_m}{u_{st}} = 2 \qquad \left(\frac{1}{2} \le \frac{t_1}{T}\right) \qquad (5c)$$

$$\frac{t_m}{T} = \frac{1}{2} \qquad (5d)$$

Diagrams of these dimensionless expressions appear in Fig. 2.22, where $u_m/u_{st}$ and $t_m/T$ are plotted against $t_1/T$. From Eq. (5a) we see that if the time $t_1$ is less than $T/6$, the dynamic response is less than that caused by applying the load $P_1$ statically. On the other hand, if the time $t_1$ is between $T/6$ and $T/2$, the value of $u_m/u_{st}$ is between 1 and 2.

At this point we note that plots of magnification factors for forced vibrations constitute response spectra, as defined in this section. Figure 2.11(a) contains a series of curves for $\beta = u_m/u_{st}$ plotted against the frequency ratio $\Omega/\omega$. We recall that these curves represent only the steady-state part of the response and that a different curve is obtained for each level of damping. If the transient parts of the forced vibrations were to be included, the response spectra in Fig. 2.11(a) would be somewhat higher, but this effect is of little significance.

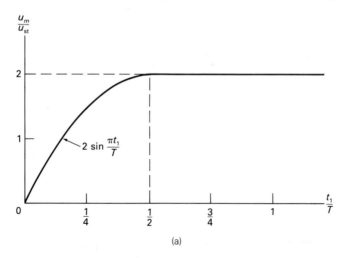

**Figure 2.22** (a) Response spectrum for rectangular impulse; (b) time of maximum response.

## Sec. 2.7  Response Spectra

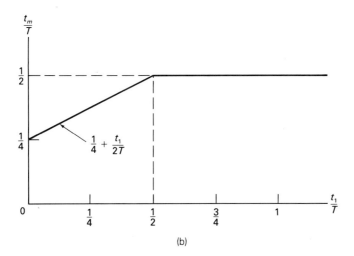

**Figure 2.22** (*cont.*)

Furthermore, while damping is of great importance in the problem of forced vibrations, it is often omitted as a consideration in response spectra due to impulsive excitations. Small values of damping have little effect on such response maxima, which usually occur before much energy is dissipated. However, a group of damped response spectra can always be constructed for any forcing function, with a different curve for each level of damping. This may be accomplished for simple cases by deriving the appropriate analytical functions, but for complicated situations a numerical approach must be used.

### Example 2.11

Figure 2.23(a) shows a *ramp-step function* that increases linearly from zero to $P_1$ in the time $t_1$ and is constant thereafter. From Prob. 2.6-7, the response of an undamped SDOF system to this excitation is

$$u = \frac{P_1}{k}\left(\frac{t}{t_1} - \frac{\sin \omega t}{\omega t_1}\right) \qquad (0 \le t \le t_1) \qquad (a)$$

$$u = \frac{P_1}{k}\left[1 + \frac{\sin \omega(t - t_1) - \sin \omega t}{\omega t_1}\right] \qquad (t_1 \le t) \qquad (b)$$

The objective of this example is to determine the response spectrum and the corresponding time function.

By inspection of Eqs. (a) and (b), we see that the maximum response will always occur after time $t_1$. Therefore, only the latter equation is of interest here, and it may be expressed in the dimensionless form

$$\frac{u}{u_{st}} = \frac{1}{\omega t_1}[\omega t_1 + \sin \omega(t - t_1) - \sin \omega t] \qquad (c)$$

Differentiating Eq. (c) with respect to time, we obtain

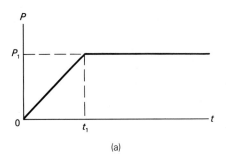

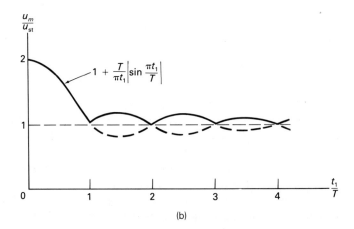

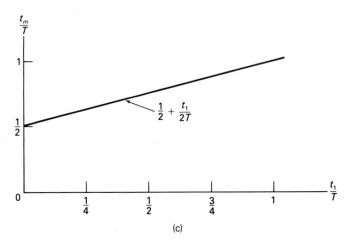

**Figure 2.23** (a) Ramp-step function; (b) response spectrum; (c) time of maximum response.

## Sec. 2.7  Response Spectra

$$\frac{\dot{u}}{u_{st}} = \frac{1}{t_1}[\cos \omega(t - t_1) - \cos \omega t]$$

Setting the term in brackets equal to zero gives an expression containing the time $t_m$. Thus,

$$\cos \omega t_m = \cos \omega(t_m - t_1)$$

This equation is satisfied by

$$\omega t_m = \pi + \frac{\omega t_1}{2} \qquad (d)$$

As for the rectangular impulse, $t_m$ is linearly related to $t_1$. Substitution of $\omega t_m$ from Eq. (d) into Eq. (c) produces

$$\frac{u_m}{u_{st}} = 1 + \frac{2}{\omega t_1}\sin \frac{\omega t_1}{2} = 1 \pm \frac{1}{\omega t_1}\sqrt{2(1 - \cos \omega t_1)} \qquad (e)$$

These expressions represent both maxima and minima, which depend on the value of $\omega t_1$. Summarizing for maxima, we have

$$\frac{u_m}{u_{st}} = 1 + \frac{T}{\pi t_1}\left|\sin \frac{\pi t_1}{T}\right| \qquad (f)$$

$$(t_1 \leq t)$$

$$\frac{t_m}{T} = \frac{1}{2} + \frac{t_1}{2T} \qquad (g)$$

Figure 2.23(b) and (c) contain dimensionless plots of the response spectrum from Eq. (f) and the time of maximum response from Eq. (g). We see that the response spectrum has its highest value of $u_m/u_{st} = 2$ when $t_1 = 0$, for which the input becomes a step function. Whenever $t_1 \geq T$ the value of $u_m$ does not exceed $u_{st}$ very much, and with a large rise time the loading is essentially static.

### Example 2.12

A damped SDOF system with harmonic ground displacement $u_g = d \cos \Omega t$ appears in Fig. 2.24(a). Let us derive an expression for the response spectrum of the steady-state force in the spring due to that influence. For this purpose, we need only find the steady-state relative displacement $u^* = u - u_g$ and multiply it by the spring stiffness $k$.

By taking the second derivative of the ground displacement with respect to time, we obtain the absolute ground acceleration as

$$\ddot{u}_g = -\Omega^2 d \cos \Omega t \qquad (h)$$

Therefore, the equation of motion in the relative coordinate $u^*$ becomes

$$m\ddot{u}^* + c\dot{u}^* + ku^* = -m\ddot{u}_g = m\Omega^2 d \cos \Omega t \qquad (i)$$

Comparing this relationship with Eq. (2.4–22), we see that $u^*$ has replaced $u$ and $m\Omega^2 d$ has replaced $P$. Hence, the steady-state response [see Eq. (2.4–31)] becomes

$$u^* = \beta\frac{m\Omega^2 d}{k}\cos(\Omega t - \theta) \qquad (j)$$

in which the magnification factor $\beta$ and the phase angle $\theta$ are given by Eqs. (2.4–32) and (2.4–33).

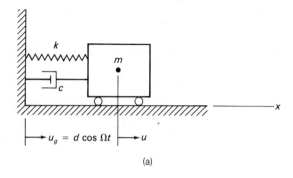

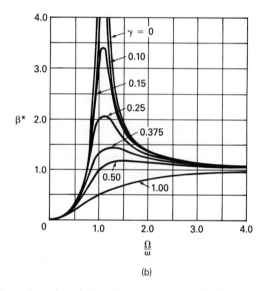

**Figure 2.24** (a) Damped SDOF system with harmonic ground displacement; (b) response spectra for spring force.

To improve the form of Eq. (j), we use $\omega^2 = k/m$ and define the new magnification factor, $\beta^*$, as follows:

$$\beta^* = \beta\left(\frac{\Omega}{\omega}\right)^2 = \frac{(\Omega/\omega)^2}{\sqrt{[1 - (\Omega/\omega)^2]^2 + (2\gamma\Omega/\omega)^2}} \qquad (k)$$

Then rewrite Eq. (j) in the reduced form

$$u^* = \beta^* d \cos(\Omega t - \theta) \qquad (\ell)$$

The steady-state force in the spring may now be taken as

$$F = \beta^* kd \cos(\Omega t - \theta) \qquad (m)$$

Dividing this equation by $F_{st} = kd$ gives

$$\frac{F}{F_{st}} = \beta^* \cos(\Omega t - \theta) \qquad \text{(n)}$$

Of course, the sine function used in place of the cosine function would produce similar results.

Figure 2.24(b) shows dimensionless plots of the response spectrum $\beta^* = F_m/F_{st}$ against $\Omega/\omega$ for various levels of damping. When the frequency ratio $\Omega/\omega$ is zero, the spring force is zero; and when the ratio is large, the spring force approaches the static value, where $\beta^* = 1$. On the other hand, at resonance ($\Omega = \omega$), the magnification factor $\beta^*$ becomes the same as $\beta$.

The forcing functions treated in this section lead to explicit expressions for $u_m/u_{st}$ and $t_m/T$, but it should be mentioned that these are exceptional cases. In general, it is difficult to identify the time range within which the maximum response occurs. In addition, the equation containing $t_m$ is usually transcendental and cannot be solved explicitly. Under these circumstances, values of $u_m/u_{st}$ and $t_m/T$ must be found by exhaustive calculations, using a series of values for the time ratio $t_1/T$. For each ratio the expression for $u/u_{st}$ in terms of $t/T$ may be plotted, and a value of $u_m/u_{st}$ as well as $t_m/T$ may be obtained therefrom.

## REFERENCES

1. Timoshenko, S. P., Young, D. H., and Weaver, W., Jr., *Vibration Problems in Engineering*, 4th ed., Wiley, New York, 1974.
2. Wilson, J. S., "Performance Characteristics and Selection of Accelerometers," *J. Sound Vib.*, Vol. 12, 1978, pp. 24–29.
3. Craig, R. R., *Structural Dynamics*, Wiley, New York, 1981.

## PROBLEMS

**2.2-1.** For the prismatic cantilever beam in Fig. 2.1(a), find approximate expressions for the angular frequency $\omega$, the natural frequency $f$, and the natural period $T$.

**2.2-2.** Determine approximate expressions for the values of $\omega, f$, and $T$ for the simply supported prismatic beam in Fig. 2.1(b).

**2.2-3.** Assume that the overhanging beam shown in Fig. P2.2-3 has constant flexural rigidity $EI$, and find approximate values of $\omega, f$, and $T$.

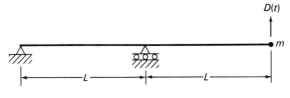

**Figure P2.2-3**

**2.2-4.** Figure P2.2-4 shows a body of mass $m$ hanging from a massless cable having length $L$, cross-sectional area $A$, and effective modulus of elasticity $E_e$. What are the approximate values of $\omega$, $f$, and $T$ for this SDOF system?

Figure P2.2-4

**2.2-5.** A rigid bar $AB$ of mass density $\rho$ has cross-sectional area $A$ and length $L$. At its center it is connected to a flexible, massless rod that has a circular cross section with radius $r$ and length $L$, as indicated in Fig. P2.2-5. Find approximate values of $\omega$, $f$, and $T$ for this problem, assuming that the bar can only rotate about the axis of the rod.

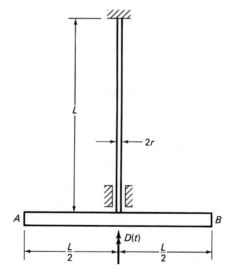

Figure P2.2-5

**2.2-6.** For the fixed beam shown in Fig. P2.2-6, determine approximate expressions for $\omega$, $f$, and $T$. For this purpose let the flexural rigidity $EI$ be constant along the length of the beam.

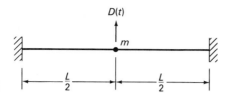
**Figure P2.2-6**

**2.3-1.** Suppose that the SDOF system in Fig. 2.6(a) is subjected to the harmonic ground displacement $u_g = d \cos \Omega t$. What will be the steady-state forced vibration of the mass $m$ due to this influence?

**2.3-2.** If the SDOF system in Fig. 2.6(a) experiences the harmonic ground acceleration $\ddot{u}_g = a \cos \Omega t$, determine the steady-state forced response of the mass $m$ relative to the ground.

**2.3-3.** A harmonic force $P \sin \Omega t$ is applied to the mass $m$ at the center of the fixed beam shown in Fig. P2.3-3. Find the steady-state forced vibration of the mass, assuming that the flexural rigidity $EI$ of the beam is constant along its length and that $\Omega = \omega/2$.

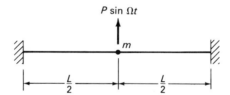
**Figure P2.3-3**

**2.3-4.** Figure P2.3-4 shows a rigid bar $AB$ attached at its center to a flexible rod of length $L$, having a circular cross section with radius $r$. Determine the steady-state rotational response of the bar to the harmonic moment $M \cos \Omega t$, where $\Omega = 2\omega$.

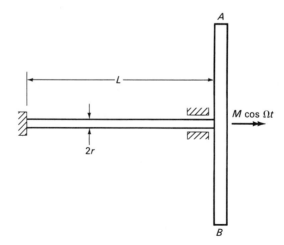
**Figure P2.3-4**

**2.3-5.** Let the prismatic cantilever beam shown in Fig. P2.3-5 be subjected to a ground translation $D_g = d \sin \Omega t$, as indicated in the figure. Calculate the steady-state response of the mass $m$ for $\Omega = 2\omega/3$.

**Figure P2.3-5**

**2.3-6.** Repeat Prob. 2.3-5 with ground acceleration $\ddot{D}_g = a \cos \Omega t$ and $\Omega = 4\omega$.

**2.3-7.** The rigid disk shown in Fig. P2.3-7 is attached at its center to a flexible rod of length $L$ with a circular cross section of radius $r$. Find the steady-state response of the disk caused by a rotational ground displacement $D_g = \delta\theta_g \sin \Omega t$ if $\Omega = 3\omega/4$.

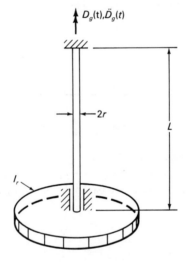

**Figure P2.3-7**

**2.3-8.** Repeat Prob. 2.3-7 with rotational ground acceleration $\ddot{D}_g = \delta\ddot{\theta}_g \cos \Omega t$ and $\Omega = 5\omega$, assuming that the mass moment of inertia of the disk is $I_r$.

**2.3-9.** Let the right-hand support of the simple beam shown in Fig. P2.3-9 oscillate in accordance with the harmonic displacement $d \sin \Omega t$. Determine the steady-state response of the mass $m$, assuming that the beam is prismatic and $\Omega = 7\omega/8$.

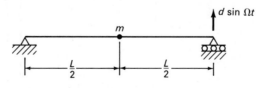

**Figure P2.3-9**

**2.3-10.** If the left-hand support of the overhanging beam shown in Fig. P2.3-10 has the harmonic displacement $d \cos \Omega t$, what is the response of the mass $m$? Assume that the beam has constant flexural rigidity and that $\Omega = 6\omega$.

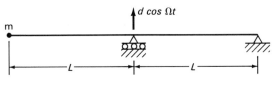

**Figure P2.3-10**

**2.4-1.** Considering a damped SDOF system subjected to the forcing function $P \sin \Omega t$, derive the expression for the steady-state response in phase-angle form.

**2.4-2.** Derive the steady-state response (in phase-angle form) of a damped SDOF system that experiences the ground acceleration $\ddot{u}_g = a \sin \Omega t$.

**2.4-3.** For a SDOF system with subcritical damping, determine the transient response due to the forcing function $P \sin \Omega t$. Give the solution in a form similar to Eq. (2.4-f).

**2.4-4.** Repeat Prob. 2.3-3, assuming that the damping ratio is $\gamma = 0.01$ and that the frequency ratio is $\Omega/\omega = 0.8$.

**2.4-5.** Repeat Prob. 2.3-4, but assume that the damping ratio is $\gamma = 0.03$ and that the frequency ratio is $\Omega/\omega = 1.1$.

**2.4-6.** Repeat Prob. 2.3-6, but let the damping ratio be $\gamma = 0.02$ and take the frequency ratio as $\Omega/\omega = 0.95$.

**2.4-7.** Calculate the magnification factor $\beta$ at resonance for values of the damping ratio $\gamma$ equal to $0.01, 0.02, \ldots, 0.20$.

**2.5-1.** Expand the square-wave function shown in Fig. P2.5-1 into a Fourier (or trigonometric) series.

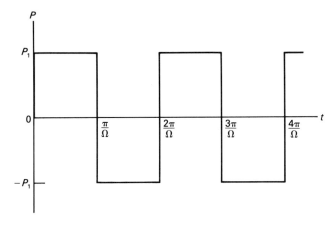

**Figure P2.5-1**

**2.5-2.** In place of the periodic forcing function shown in Fig. P2.5-2, determine a Fourier series.

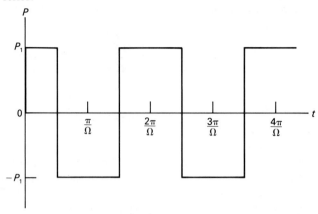

**Figure P2.5-2**

**2.5-3.** Decompose the piecewise-linear periodic function shown in Fig. P2.5-3 into a Fourier series.

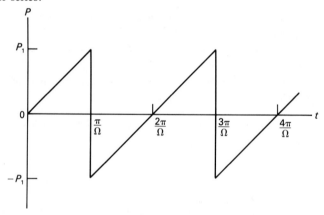

**Figure P2.5-3**

**2.5-4.** For the sawtooth function shown in Fig. P2.5-4, substitute a Fourier series.

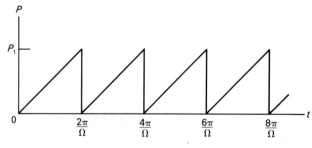

**Figure P2.5-4**

**2.5-5.** Derive the general expression for steady-state forced vibrations of a damped SDOF system due to the forcing function given as the Fourier series in Eq. (2.5-1).

**2.5-6.** Using a frequency ratio of $\Omega/\omega = 0.9$, find the undamped response of a SDOF system to the first term in the series from Prob. 2.5-1.

**2.5-7.** Determine the undamped response of a SDOF system to the first term in the series from Prob. 2.5-2, assuming a frequency ratio of $\Omega/\omega = 0.95$.

**2.5-8.** For a frequency ratio of $\Omega/\omega = 1.05$, calculate the undamped response of a SDOF system to the first term in the series from Prob. 2.5-3.

**2.6-1.** Rederive the expression for undamped forced vibrations of a SDOF system subjected to the harmonic function $P = P_1 \sin \Omega t'$.

**2.6-2.** Find the damped response of a SDOF system to the step function $P = P_1$.

**2.6-3.** Derive the expression for damped response of a SDOF system to the ramp function $P = P_1 t'/t_1$.

**2.6-4.** Determine expressions for the undamped response of a SDOF system to the forcing function shown in Fig. P2.6-4.

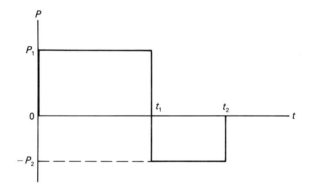

Figure P2.6-4

**2.6-5.** Assuming that the forcing function shown in Fig. P2.6-5 acts on a SDOF system, find expressions for the undamped response.

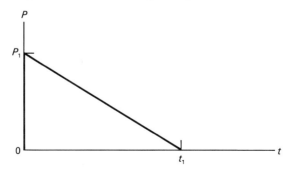

Figure P2.6-5

**2.6-6.** Derive expressions for the undamped response of a SDOF system to the forcing function shown in Fig. P2.6-6.

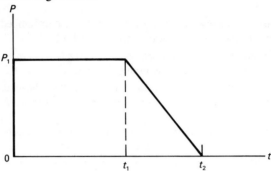

**Figure P2.6-6**

**2.6-7.** Let the ground acceleration shown in Fig. P2.6-7 be imposed on a SDOF system, and find expressions for its undamped response.

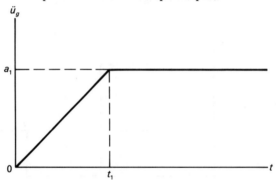

**Figure P2.6-7**

**2.6-8.** For the ground acceleration shown in Fig. P2.6-8, determine expressions for the undamped response of a SDOF system.

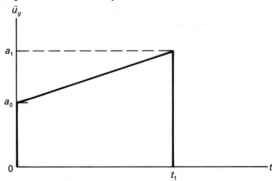

**Figure P2.6-8**

**2.6-9.** Derive expressions for the undamped response of a SDOF system to the parabolic ground acceleration $\ddot{u}_g = a_1(t'/t_1)^2$ given in Fig. P2.6-9.

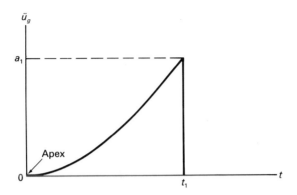

**Figure P2.6-9**

**2.7-1.** For a step function of magnitude $P = P_1$, calculate the undamped response of a SDOF system, using a recursive procedure with 10 equal time steps of duration $\Delta t = T/10$.

**2.7-2.** Assume that a ramp function $P = P_1 t/t_1$ is applied to an undamped SDOF system. With a step-by-step procedure, find the response of the system for 10 equal time steps of duration $\Delta t = T/10 = t_1$.

**2.7-3.** Confirm the approximate results of Example 2.9 in Table 2.1.

**2.7-4.** Confirm the results of Example 2.10 in Table 2.2.

**2.7-5.** Using 20 equal time steps with $\Delta t = T/20$, determine the undamped response of a SDOF system to the forcing function shown in Fig. P2.7-5.

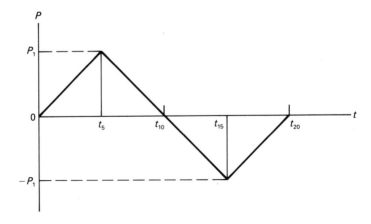

**Figure P2.7-5**

**2.7-6.** Calculate the undamped response of a SDOF system to the forcing function shown in Fig. P2.7-6, using 20 equal time steps with $\Delta t = T/20$.

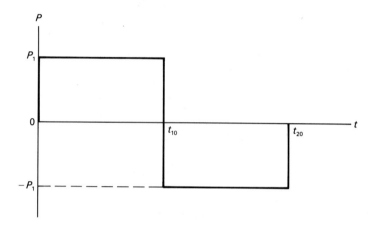

Figure P2.7-6

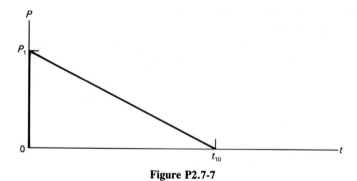

Figure P2.7-7

**2.7-7.** Divide the triangular impulse shown in Fig. P2.7-7 into 10 equal time steps $\Delta t = T/30$, and find the undamped response when it is applied to a SDOF system.

**2.7-8.** Determine the undamped response of a SDOF system to the parabolic forcing function $P = P_1 t^2 / t_{10}^2$ shown in Fig. P2.7-8, using 10 equal time steps of duration $\Delta t = T/30$.

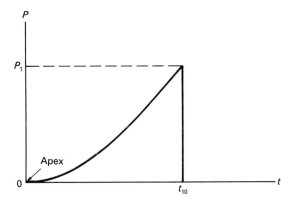

**Figure P2.7-8**

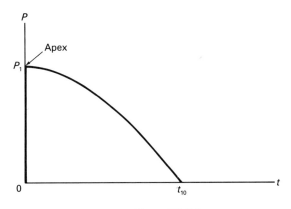

**Figure P2.7-9**

**2.7-9.** The parabolic forcing function shown in Fig. P2.7-9 has the formula $P = P_1(1 - t^2/t_{10}^2)$. With 10 equal time steps of duration $\Delta t = T/25$, obtain the undamped response of a SDOF system.

**2.7-10.** Find the undamped response of a SDOF system to the parabolic forcing function $P = P_1[1 - (t - t_{10})^2/t_{10}^2]$ shown in Fig. P2.7-10, using 10 equal time steps of duration $\Delta t = T/25$.

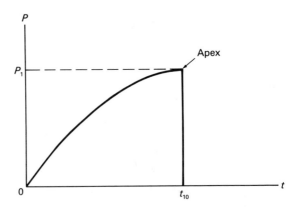

Figure P2.7-10

**2.8-1.** For the forcing function shown in Fig. P2.8-1, plot the response spectrum $u_m/u_{st}$ and the time for maximum response $t_m/t_1$ against the time ratio $t_1/T$. (See Prob. 2.6-4 for response formulas.)

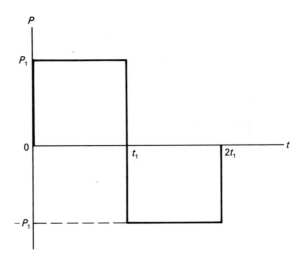

Figure P2.8-1

**2.8-2.** Repeat Prob. 2.8-1 for the forcing function shown in Fig. P2.8-2. (See Prob. 2.6-5 for response formulas.)

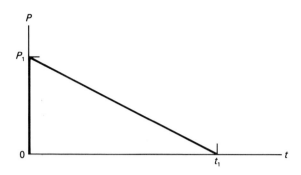

**Figure P2.8-2**

**2.8-3.** Repeat Prob. 2.8-1 for the forcing function shown in Fig. P2.8-3. (See Prob. 2.6-6 for response formulas.)

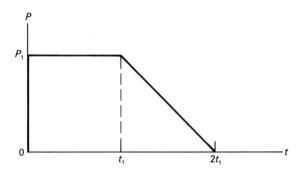

**Figure P2.8-3**

**2.8-4.** Repeat Prob. 2.8-1 for the forcing function shown in Fig. P2.8-4. (See Prob. 2.6-7 for response formulas.)

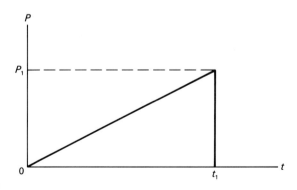

**Figure P2.8-4**

**2.8-5.** Repeat Prob. 2.8-1 for the forcing function shown in Fig. P2.8-5. (See Prob. 2.6-8 for response formulas.)

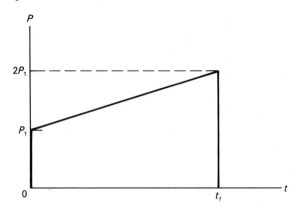

**Figure P2.8-5**

**2.8-6.** Repeat Prob. 2.8-1 for the forcing function shown in Fig. P2.8-6. (See Prob. 2.6-9 for response formulas.)

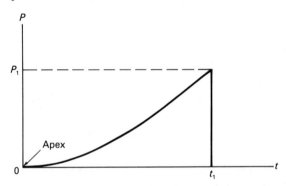

**Figure P2.8-6**

# 3

# Finite Elements and Vibrational Analysis

## 3.1 INTRODUCTION

In this book we use the method of *finite elements* [1–3] to discretize solids and structures for dynamic analysis. The basic concept is to divide a continuum into subregions having simpler geometries than the original problem. Each subregion (or finite element) is of finite size (not infinitesimal) and has a number of key points, called *nodes,* that control the behavior of the element. By making the displacements or stresses at any point in an element dependent on those at the nodes, we need only write a finite number of differential equations of motion for such nodes. This approach enables us to convert a problem with an infinite number of degrees of freedom to one with a finite number, thereby simplifying the solution process. For good accuracy in the solution, the number of nodal degrees of freedom usually must be fairly large; and the details of element formulations are rather complicated. Therefore, it becomes necessary to program this method on a digital computer.

Figure 3.1 shows various examples of solids and structures that are discretized by finite elements, with dots indicating the nodes. In Fig. 3.1(a) we see a continuous beam that is divided into several flexural elements of the type to be described in Sec. 3.4. The space frame with curved members in Fig. 3.1(b) has axial, flexural, and torsional deformations in each of its subdivided members. Figure 3.1(c) depicts a two-dimensional slice of unit thickness, representing the constant state known as plane strain on the cross section of a long, prismatic solid. On the other hand, the discretized general solid in Fig. 3.1(d) has no such restriction. If the thin plate in Fig. 3.1(e) has forces applied in its

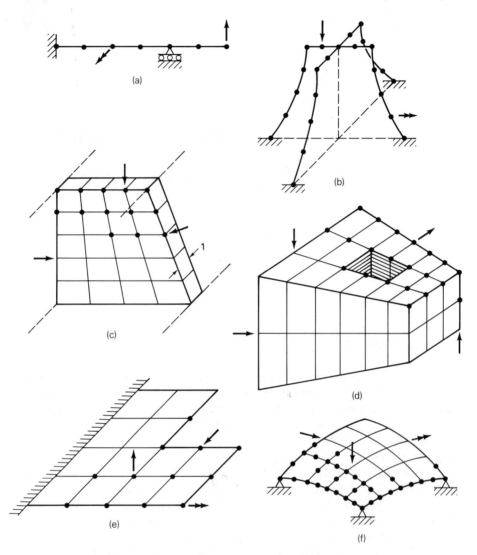

**Figure 3.1** Structures modeled by finite elements: (a) beam; (b) space frame; (c) plane strain; (d) solid; (e) plate; (f) shell.

own plane, it experiences a condition known as plane stress. But if the forces are normal to the plane, it is in a state of flexure, or bending. Finally, a general shell of the type shown in Fig. 3.1(f) can resist any kind of loading. All of the discretized structures in Fig. 3.1 have multiple degrees of freedom and will be referred to as *MDOF systems*.

The finite-element method to be used in this book involves the assumption of *displacement shape functions* within each element. These functions give approximate results when the element is of finite size and exact results at

Sec. 3.2  Stresses and Strains                                                                    75

infinitesimal size. The shape functions make *generic displacements* at any point completely dependent upon *nodal displacements*. Similarly, the local velocities and accelerations are also dependent on the nodal values. With these dependencies in mind, we can devise a procedure for writing *differential equations of motion,* as follows:

1. Divide the continuum into a finite number of subregions (or elements) of simple geometry, such as lines, quadrilaterals, or hexahedra.
2. Select key points on the elements to serve as nodes, where conditions of dynamic equilibrium and compatibility with other elements are to be enforced.
3. Assume displacement shape functions within each element so that displacements, velocities, and accelerations at any point are dependent on nodal values.
4. Satisfy strain-displacement and stress-strain relationships within a typical element for a specific type of problem.
5. Determine equivalent stiffnesses, masses, and nodal loads for each finite element, using a work or energy principle.
6. Develop differential equations of motion for the nodes of the discretized continuum by assembling the finite-element contributions.

From the homogeneous form of the equations of motion, we can perform a *vibrational analysis* for any linearly elastic structure. This type of analysis consists of finding undamped frequencies and corresponding mode shapes for the discretized analytical model. Such information is often useful by itself, and it is essential for the normal-mode method of dynamic analysis described in Chapter 4.

In the present chapter we develop one-dimensional elements that are to be used in subsequent work (especially Chapter 6) for analyzing framed structures. Other discretized continua will be discussed in Chapters 7 and 8, where the applications include two- and three-dimensional solids, plates in bending, and shell structures.

## 3.2 STRESSES AND STRAINS

In this book we assume that the continuum to be analyzed consists of a linearly elastic material with small strains and small displacements. In any case, strains and their corresponding stresses may be expressed with respect to some right-hand orthogonal coordinate system. For example, in a (rectangular) Cartesian set, the coordinates would be $x$, $y$, and $z$. On the other hand, in a cylindrical coordinate system, the symbols $r$, $\theta$, and $z$ would serve as the coordinates.

Figure 3.2 shows an infinitesimal element in Cartesian coordinates, where the edges are of lengths $dx$, $dy$, and $dz$. *Normal* and *shearing stresses* are

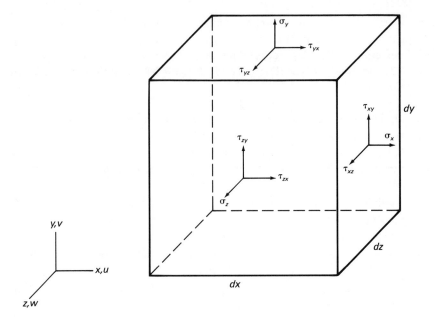

**Figure 3.2** Stresses on an infinitesimal element.

indicated by arrows on the faces of the element. The normal stresses are labeled $\sigma_x$, $\sigma_y$, and $\sigma_z$, whereas the shearing stresses are named $\tau_{xy}$, $\tau_{yz}$, and so on. From equilibrium of the element, the following relationships are known:

$$\tau_{xy} = \tau_{yx} \qquad \tau_{yz} = \tau_{zy} \qquad \tau_{zx} = \tau_{xz} \qquad (1)$$

Thus, only three independent components of shearing stresses need be considered.

Corresponding to the stresses shown in Fig. 3.2 are *normal* and *shearing strains*. Normal strains $\epsilon_x$, $\epsilon_y$, and $\epsilon_z$ are defined as

$$\epsilon_x = \frac{\partial u}{\partial x} \qquad \epsilon_y = \frac{\partial v}{\partial y} \qquad \epsilon_z = \frac{\partial w}{\partial z} \qquad (2)$$

where $u$, $v$, and $w$ are translations in the $x$, $y$, and $z$ directions. Shearing strains, $\gamma_{xy}$, $\gamma_{yz}$, and so on, are given by

$$\gamma_{xy} = \frac{\partial u}{\partial y} + \frac{\partial v}{\partial x} = \gamma_{yx}$$

$$\gamma_{yz} = \frac{\partial v}{\partial z} + \frac{\partial w}{\partial y} = \gamma_{zy} \qquad (3)$$

$$\gamma_{zx} = \frac{\partial w}{\partial x} + \frac{\partial u}{\partial z} = \gamma_{xz}$$

Hence, only three of the shearing strains are independent.

## Sec. 3.2  Stresses and Strains

For convenience, the six independent stresses and the corresponding strains usually will be represented as column matrices (or vectors). Thus,

$$\boldsymbol{\sigma} = \begin{bmatrix} \sigma_1 \\ \sigma_2 \\ \sigma_3 \\ \sigma_4 \\ \sigma_5 \\ \sigma_6 \end{bmatrix} = \begin{bmatrix} \sigma_x \\ \sigma_y \\ \sigma_z \\ \tau_{xy} \\ \tau_{yz} \\ \tau_{zx} \end{bmatrix} \qquad \boldsymbol{\epsilon} = \begin{bmatrix} \epsilon_1 \\ \epsilon_2 \\ \epsilon_3 \\ \epsilon_4 \\ \epsilon_5 \\ \epsilon_6 \end{bmatrix} = \begin{bmatrix} \epsilon_x \\ \epsilon_y \\ \epsilon_z \\ \gamma_{xy} \\ \gamma_{yz} \\ \gamma_{zx} \end{bmatrix} \qquad (4)$$

where the boldfaced symbols denote the vectors shown.

*Strain-stress relationships* for an *isotropic material* are drawn from the theory of elasticity [4], as follows:

$$\epsilon_x = \frac{1}{E}(\sigma_x - \nu\sigma_y - \nu\sigma_z) \qquad \gamma_{xy} = \frac{\tau_{xy}}{G}$$

$$\epsilon_y = \frac{1}{E}(-\nu\sigma_x + \sigma_y - \nu\sigma_z) \qquad \gamma_{yz} = \frac{\tau_{yz}}{G} \qquad (5)$$

$$\epsilon_z = \frac{1}{E}(-\nu\sigma_x - \nu\sigma_y + \sigma_z) \qquad \gamma_{zx} = \frac{\tau_{zx}}{G}$$

where

$$G = \frac{E}{2(1 + \nu)} \qquad (6)$$

In these expressions $E$ is Young's modulus of elasticity, $G$ the shearing modulus, and $\nu$ is Poisson's ratio. With matrix format, the relationships in Eqs. (5) may be written as

$$\boldsymbol{\epsilon} = \mathbf{C}\,\boldsymbol{\sigma} \qquad (7)$$

in which

$$\mathbf{C} = \frac{1}{E}\begin{bmatrix} 1 & -\nu & -\nu & 0 & 0 & 0 \\ -\nu & 1 & -\nu & 0 & 0 & 0 \\ -\nu & -\nu & 1 & 0 & 0 & 0 \\ 0 & 0 & 0 & 2(1+\nu) & 0 & 0 \\ 0 & 0 & 0 & 0 & 2(1+\nu) & 0 \\ 0 & 0 & 0 & 0 & 0 & 2(1+\nu) \end{bmatrix} \qquad (8)$$

Matrix $\mathbf{C}$ is an operator that relates the strain vector $\boldsymbol{\epsilon}$ to the stress vector $\boldsymbol{\sigma}$. By the process of inversion (or simultaneous solution), we can also obtain *stress-strain relationships* from Eq. (7), as follows:

$$\boldsymbol{\sigma} = \mathbf{E}\,\boldsymbol{\epsilon} \qquad (9)$$

where

$\mathbf{E} = \mathbf{C}^{-1}$

$$= \frac{E}{(1+\nu)(1-2\nu)} \begin{bmatrix} 1-\nu & \nu & \nu & 0 & 0 & 0 \\ \nu & 1-\nu & \nu & 0 & 0 & 0 \\ \nu & \nu & 1-\nu & 0 & 0 & 0 \\ 0 & 0 & 0 & \frac{1-2\nu}{2} & 0 & 0 \\ 0 & 0 & 0 & 0 & \frac{1-2\nu}{2} & 0 \\ 0 & 0 & 0 & 0 & 0 & \frac{1-2\nu}{2} \end{bmatrix} \quad (10)$$

Matrix $\mathbf{E}$ is an operator that relates the stress vector $\boldsymbol{\sigma}$ to the strain vector $\boldsymbol{\epsilon}$.

For the elements in this chapter, we will not need the 6 × 6 stress-strain matrix given by Eq. (10). With one-dimensional elements, only one term, such as $E$ or $G$, is required. Later, when we deal with two- and three-dimensional elements, larger matrices, such as that in Eq. (10), will be needed.

## 3.3 EQUATIONS OF MOTION FOR FINITE ELEMENTS

We shall now introduce definitions and notations that pertain to the finite elements to be studied throughout the book. By using the principle of virtual work, we can develop equations of motion for any finite element. Such equations include energy-equivalent stiffnesses, masses, and nodal loads for a typical element. These terms are treated in detail for one-dimensional elements in the next section.

Assume that a three-dimensional finite element with zero damping exists in Cartesian coordinates $x$, $y$, and $z$. Let the time-varying *generic displacements* $\mathbf{u}(t)$ at any point within the element be expressed as the column vector

$$\mathbf{u}(t) = \{u, v, w\} \quad (1)$$

where $u$, $v$, and $w$ are translations in the $x$, $y$, and $z$ directions, respectively.*

If the element is subjected to time-varying *body forces*, such forces may be placed into a vector $\mathbf{b}(t)$, as follows:

$$\mathbf{b}(t) = \{b_x, b_y, b_z\} \quad (2)$$

Here the symbols $b_x$, $b_y$, and $b_z$ represent components of force (per unit of volume, area, or length) acting in the reference directions at a generic point. The

---

*To save space, column vectors may be written in a row enclosed by braces { } and with commas separating the terms.

Sec. 3.3  Equations of Motion for Finite Elements                              79

time variation for each component of body force is assumed to be the same throughout the element. That is, we may have one time function for $b_x$, another for $b_y$, and a third for $b_z$.

Time-varying *nodal displacements* $\mathbf{q}(t)$ will at first be considered as only translations in the $x$, $y$, and $z$ directions. Thus, if $n_{en}$ = number of element nodes,

$$\mathbf{q}(t) = \{\mathbf{q}_i(t)\} \qquad (i = 1, 2, \ldots, n_{en}) \tag{3}$$

where

$$\mathbf{q}_i(t) = \{q_{xi}, q_{yi}, q_{zi}\} = \{u_i, v_i, w_i\} \tag{4}$$

However, other types of displacements, such as small rotations ($\partial v/\partial x$, and so on) and curvatures ($\partial^2 v/\partial x^2$, and so on) will be used later.

Similarly, time-varying *nodal actions* $\mathbf{p}(t)$ will temporarily be taken as only forces in the $x$, $y$, and $z$ directions at the nodes. That is,

$$\mathbf{p}(t) = \{\mathbf{p}_i(t)\} \qquad (i = 1, 2, \ldots, n_{en}) \tag{5}$$

in which

$$\mathbf{p}_i(t) = \{p_{xi}, p_{yi}, p_{zi}\} \tag{6}$$

Time functions for $p_{xi}$, $p_{yi}$, and $p_{zi}$ at each node may be independent and arbitrary. Other types of nodal actions, such as moments, and so on, will be considered later.

For the type of finite-element method to which this book is devoted, certain assumed *displacement shape functions* relate generic displacements to nodal displacements, as follows:

$$\mathbf{u}(t) = \mathbf{f}\,\mathbf{q}(t) \tag{7}$$

In this expression the symbol $\mathbf{f}$ denotes a rectangular matrix containing the functions that make $\mathbf{u}(t)$ completely dependent on $\mathbf{q}(t)$.

*Strain-displacement relationships* are obtained by differentiation of the generic displacements. This process may be expressed by forming a matrix $\mathbf{d}$, called a *linear differential operator*, and applying it with the rules of matrix multiplication. Thus,

$$\boldsymbol{\epsilon}(t) = \mathbf{d}\,\mathbf{u}(t) \tag{8}$$

In this equation the operator $\mathbf{d}$ expresses the time-varying strain vector $\boldsymbol{\epsilon}(t)$ in terms of generic displacements in the vector $\mathbf{u}(t)$ [see Eqs. (3.2-2) and (3.2-3)]. Substitution of Eq. (7) into Eq. (8) yields

$$\boldsymbol{\epsilon}(t) = \mathbf{B}\,\mathbf{q}(t) \tag{9}$$

where

$$\boxed{\mathbf{B} = \mathbf{d}\,\mathbf{f}} \tag{10}$$

Matrix **B** gives strains at any point within the element due to unit values of nodal displacements.

From Eq. (3.2-9) we have the matrix form of *stress-strain relationships*. That is,

$$\boldsymbol{\sigma}(t) = \mathbf{E}\,\boldsymbol{\epsilon}(t) \tag{11}$$

where **E** is a matrix relating time-varying stresses in the vector $\boldsymbol{\sigma}(t)$ to strains in $\boldsymbol{\epsilon}(t)$. Substitution of Eq. (9) into Eq. (11) produces

$$\boldsymbol{\sigma}(t) = \mathbf{E}\,\mathbf{B}\,\mathbf{q}(t) \tag{12}$$

in which the matrix product **E B** gives stresses at a generic point due to unit values of nodal displacements.

***Virtual Work Principle:*** *If a general structure in dynamic equilibrium is subjected to a system of small virtual displacements within a compatible state of deformation, the virtual work of external actions is equal to the virtual strain energy of internal stresses.* When applying this principle to a finite element, we have

$$\delta U_e = \delta W_e \tag{13}$$

where $\delta U_e$ is the virtual strain energy of internal stresses and $\delta W_e$ is the virtual work of external actions on the element. To develop both of these quantities in detail, we assume a vector $\delta\mathbf{q}$ of small virtual displacements. Thus,

$$\delta\mathbf{q} = \{\delta\mathbf{q}_i\} \qquad (i = 1, 2, \ldots, n_{\text{en}}) \tag{14}$$

Then the resulting virtual generic displacements become [see Eq. (7)]

$$\delta\mathbf{u} = \mathbf{f}\,\delta\mathbf{q} \tag{15}$$

Using the strain-displacement relationships in Eq. (9), we obtain

$$\delta\boldsymbol{\epsilon} = \mathbf{B}\,\delta\mathbf{q} \tag{16}$$

Now the *internal virtual strain energy* can be written as

$$\delta U_e = \int_V \delta\boldsymbol{\epsilon}^T \boldsymbol{\sigma}(t)\, dV \tag{17}$$

where integration is over the volume of the element.

For the *external virtual work* we turn to Fig. 3.3, which shows an infinitesimal element with components of *applied body forces* $b_x(t)\, dV$, $b_y(t)\, dV$, and $b_z(t)\, dV$. The figure also indicates *inertial body forces* $\rho\ddot{u}\, dV$, $\rho\ddot{v}\, dV$, and $\rho\ddot{w}\, dV$ due to the accelerations $\ddot{u}$, $\ddot{v}$, and $\ddot{w}$. The symbol $\rho$ in these expressions represents the *mass density* of the material, which is defined as the inertial force per unit acceleration per unit volume. Note that the inertial forces act in directions that are opposite to the positive senses of the accelerations. Thus, we add the external virtual work of nodal and distributed body forces as follows:

$$\delta W_e = \delta\mathbf{q}^T \mathbf{p}(t) + \int_V \delta\mathbf{u}^T \mathbf{b}(t)\, dV - \int_V \delta\mathbf{u}^T \rho\ddot{\mathbf{u}}\, dV \tag{18}$$

Sec. 3.3  Equations of Motion for Finite Elements

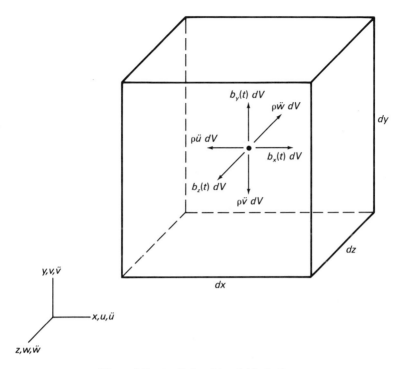

**Figure 3.3** Applied and inertial body forces.

Substitution of Eqs. (17) and (18) into Eq. (13) produces

$$\int_V \delta\boldsymbol{\epsilon}^T \boldsymbol{\sigma}(t)\, dV = \delta\mathbf{q}^T \mathbf{p}(t) + \int_V \delta\mathbf{u}^T \mathbf{b}(t)\, dV - \int_V \delta\mathbf{u}^T \rho \ddot{\mathbf{u}}\, dV \quad (19)$$

Now assume that

$$\ddot{\mathbf{u}} = \mathbf{f}\, \ddot{\mathbf{q}} \quad (20)$$

Then we can substitute Eqs. (12) and (20) into Eq. (19) and use the transposes of Eqs. (15) and (16) to obtain

$$\delta\mathbf{q}^T \int_V \mathbf{B}^T \mathbf{E} \mathbf{B}\, dV\, \mathbf{q} = \delta\mathbf{q}^T \mathbf{p}(t) + \delta\mathbf{q}^T \int_V \mathbf{f}^T \mathbf{b}(t)\, dV - \delta\mathbf{q}^T \int_V \rho \mathbf{f}^T \mathbf{f}\, dV\, \ddot{\mathbf{q}} \quad (21)$$

Cancellation of $\delta\mathbf{q}^T$ and rearrangement of the resulting *equations of motion* gives

$$\mathbf{M}\, \ddot{\mathbf{q}} + \mathbf{K}\, \mathbf{q} = \mathbf{p}(t) + \mathbf{p}_b(t) \quad (22)$$

where

$$\mathbf{K} = \int_V \mathbf{B}^T \mathbf{E} \mathbf{B}\, dV \quad (23)$$

and

$$\mathbf{M} = \int_V \rho \mathbf{f}^T \mathbf{f} \, dV \qquad (24)$$

Also,

$$\mathbf{p}_b(t) = \int_V \mathbf{f}^T \mathbf{b}(t) \, dV \qquad (25)$$

Matrix $\mathbf{K}$ in Eq. (23) is the *element stiffness matrix*, which contains *stiffness coefficients* that are fictitious actions at nodes due to unit values of nodal displacements. Equation (24) gives the form of the *consistent-mass matrix*, in which the terms are energy-equivalent actions at nodes due to unit values of nodal accelerations. Finally, the vector $\mathbf{p}_b(t)$ in Eq. (25) consists of *equivalent nodal loads* due to body forces in the vector $\mathbf{b}(t)$. Other equivalent nodal loads due to initial strains (or stresses) could be derived [1], but analyses for such influences are considered to be statics problems.

## 3.4 ONE-DIMENSIONAL ELEMENTS

In this section we develop properties of one-dimensional elements subjected to axial, torsional, and flexural deformations, starting with the *axial element* in Fig. 3.4(a). The figure indicates a single generic translation $u$ in the direction of $x$. Thus, from Eq. (3.3-1) we have

$$\mathbf{u}(t) = u$$

The corresponding body force is a single component $b_x$ (force per unit length), acting in the $x$ direction. Therefore, Eq. (3.3-2) gives

$$\mathbf{b}(t) = b_x$$

Nodal displacements $q_1$ and $q_2$ consist of translations in the $x$ direction at nodes 1 and 2 [see Fig. 3.4(a)]. Hence, Eq. (3.3-3) becomes

$$\mathbf{q}(t) = \{q_1, q_2\} = \{u_1, u_2\}$$

Corresponding nodal forces at points 1 and 2 are given by Eq. (3.3-5) as

$$\mathbf{p}(t) = \{p_1, p_2\} = \{p_{x1}, p_{x2}\}$$

Figure 3.4(b) and (c) show linear displacement shape functions $f_1$ and $f_2$ that we assume for this element. That is, Eq. (3.3-7) gives

$$u = \mathbf{f}\,\mathbf{q}(t)$$

where

$$\mathbf{f} = \begin{bmatrix} f_1 & f_2 \end{bmatrix} = \begin{bmatrix} 1 - \dfrac{x}{L} & \dfrac{x}{L} \end{bmatrix} \qquad (1)$$

### Sec. 3.4 One-Dimensional Elements

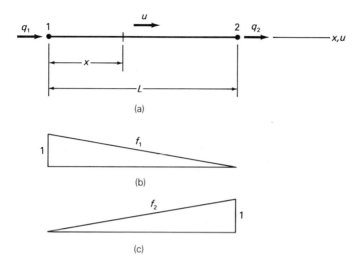

**Figure 3.4** Axial element.

The first function diminishes linearly from 1 to 0, whereas the second increases linearly from 0 to 1.

From Fig. 3.5 we see that the single strain-displacement relationship $du/dx$ for the axial element is constant on the cross section. Thus, Eqs. (3.3-8), (3.3-9), and (3.3-10) yield

$$\boldsymbol{\epsilon}(t) = \epsilon_x = \frac{du}{dx} = \frac{d\mathbf{f}}{dx}\mathbf{q}(t) = \mathbf{B}\,\mathbf{q}(t)$$

Therefore, we have

$$\mathbf{B} = \mathbf{d}\,\mathbf{f} = \frac{d\mathbf{f}}{dx} = \frac{1}{L}\begin{bmatrix} -1 & 1 \end{bmatrix} \tag{2}$$

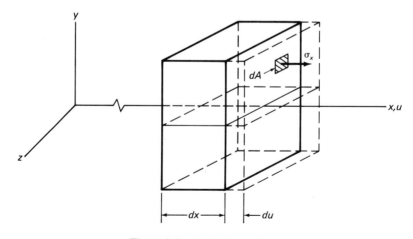

**Figure 3.5** Axial deformations.

which expresses the strain $\epsilon_x$ in terms of the nodal displacements. Similarly, the single stress-strain relationship [see Eqs. (3.3-11) and (3.3-12)] becomes merely

$$\boldsymbol{\sigma}(t) = \sigma_x = \mathbf{E}\,\boldsymbol{\epsilon}(t) = E\epsilon_x = E\mathbf{B}\,\mathbf{q}(t)$$

Hence,

$$\mathbf{E} = E \quad \text{and} \quad E\mathbf{B} = \frac{E}{L}\begin{bmatrix} -1 & 1 \end{bmatrix} \qquad (3)$$

The latter expression gives the stress $\sigma_x$ in terms of the nodal displacements.

The element stiffness matrix $\mathbf{K}$ can now be evaluated from Eq. (3.3-23), as follows:

$$\mathbf{K} = \int_V \mathbf{B}^T \mathbf{E}\, \mathbf{B}\, dV = \frac{E}{L^2}\begin{bmatrix} -1 \\ 1 \end{bmatrix}\begin{bmatrix} -1 & 1 \end{bmatrix} \int_0^L \int_A dA\, dx$$

$$= \frac{EA}{L}\begin{bmatrix} 1 & -1 \\ -1 & 1 \end{bmatrix} \qquad (4)$$

assuming that the cross-sectional area $A$ is constant. Similarly, the consistent-mass matrix $\mathbf{M}$ is found from Eq. (3.3-24) to be

$$\mathbf{M} = \int_V \rho \mathbf{f}^T\, \mathbf{f}\, dV = \frac{\rho}{L^2}\int_0^L \int_A \begin{bmatrix} L - x \\ x \end{bmatrix}\begin{bmatrix} L - x & x \end{bmatrix} dA\, dx$$

$$= \frac{\rho AL}{6}\begin{bmatrix} 2 & 1 \\ 1 & 2 \end{bmatrix} \qquad (5)$$

assuming also that the mass density $\rho$ is constant.

We see that the stiffness matrix $\mathbf{K}$ and the consistent-mass matrix $\mathbf{M}$ are unique for a prismatic axial element of uniform mass density. However, an infinite number of equivalent nodal load vectors $\mathbf{p}_b(t)$ may be derived, depending on the distribution of body forces. For the simplest case, we assume that a uniformly distributed axial load $b_x$ (force per unit length) is suddenly applied to the axial element. Then Eq. (3.3-25) produces

$$\mathbf{p}_b(t) = \int_0^L \mathbf{f}^T b_x\, dx = \frac{b_x}{L}\int_0^L \begin{bmatrix} L - x \\ x \end{bmatrix} dx = b_x \frac{L}{2}\begin{bmatrix} 1 \\ 1 \end{bmatrix} \qquad (6)$$

which shows that the equivalent nodal loads are equal forces at the two ends.

Turning now to the *torsional element* in Fig. 3.6(a), we use a single generic displacement $\theta_x$, which is a small rotation about the $x$ axis (indicated by a double-headed arrow). Thus,

$$\mathbf{u}(t) = \theta_x$$

Corresponding to this displacement is a single body action

$$\mathbf{b}(t) = m_x$$

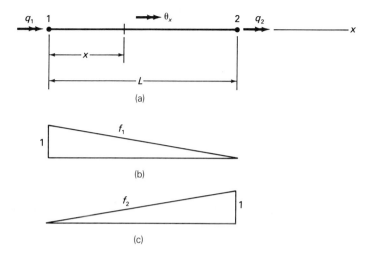

**Figure 3.6** Torsional element.

which is a moment per unit length acting in the positive $x$ sense. Nodal displacements in the figure consist of small axial rotations at nodes 1 and 2. Hence,

$$\mathbf{q}(t) = \{q_1, q_2\} = \{\theta_{x1}, \theta_{x2}\}$$

In addition, the corresponding nodal actions at points 1 and 2 are

$$\mathbf{p}(t) = \{p_1, p_2\} = \{M_{x1}, M_{x2}\}$$

which are moments (or torques) acting in the $x$ direction. As for the axial element, we assume the linear displacement shape functions $f_1$ and $f_2$ shown in Fig. 3.6(b) and (c). Therefore,

$$\theta_x = \mathbf{f}\,\mathbf{q}(t)$$

in which the matrix $\mathbf{f}$ is again given by Eq. (1).

Strain-displacement relationships can be inferred for a torsional element with a circular cross section by examining Fig. 3.7. Assuming that radii remain straight during torsional deformation, we conclude that the shearing strain $\gamma$ varies linearly with the radial distance $r$, as follows:

$$\gamma = r \frac{d\theta_x}{dx} = r\psi \tag{7}$$

where $\psi$ is the *twist*, or rate of change of angular displacement. Thus,

$$\psi = \frac{d\theta_x}{dx} \tag{8}$$

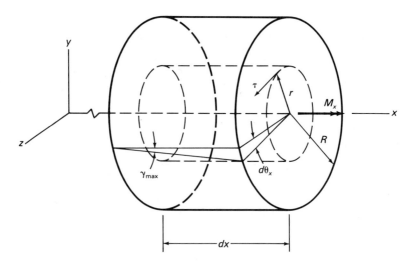

**Figure 3.7** Torsional deformations.

Equation (7) shows that the maximum value of the shearing strain occurs at the surface. That is,

$$\gamma_{max} = R\psi$$

where $R$ is the radius of the cross section (see Fig. 3.7). Also, we see from Eq. (7) that the linear differential operator **d** relating $\gamma$ to $\theta_x$ is

$$\mathbf{d} = r\frac{d}{dx} \tag{9}$$

Thus, the strain-displacement matrix **B** becomes

$$\mathbf{B} = \mathbf{d}\,\mathbf{f} = \frac{r}{L}[-1 \quad 1] \tag{10}$$

which is the same as for the axial element, except for the presence of $r$.

Shearing stress $\tau$ (see Fig. 3.7) is related to shearing strain in a torsional element by

$$\tau = G\gamma \tag{11a}$$

where the symbol $G$ denotes the shearing modulus of the material. Hence,

$$\mathbf{E} = G \quad \text{and} \quad G\mathbf{B} = \frac{Gr}{L}[-1 \quad 1] \tag{11b}$$

These relationships are analogous to Eqs. (3) for the axial element.

We may now find the torsional stiffness matrix **K** by applying Eq. (3.3-23), as follows:

## Sec. 3.4  One-Dimensional Elements

$$\mathbf{K} = \int_V \mathbf{B}^T \mathbf{E} \mathbf{B} \, dV$$

$$= \frac{G}{L^2} \begin{bmatrix} -1 \\ 1 \end{bmatrix} [-1 \quad 1] \int_0^L \int_0^{2\pi} \int_0^R (r^2) r \, dr \, d\theta \, dx$$

$$= \frac{GJ}{L} \begin{bmatrix} 1 & -1 \\ -1 & 1 \end{bmatrix} \tag{12}$$

where $GJ$ is constant. The *polar moment of inertia* $J$ for a circular cross section is defined as

$$J = \int_0^{2\pi} \int_0^R r^3 \, dr \, d\theta = \frac{\pi R^4}{2} \tag{13}$$

If the cross section of a torsional element is not circular, it will warp. Such warping is most severe for elements of open cross sections, such as channel or wide-flanged sections. For most practical cases, the theory of *uniform torsion* described here may be used by substituting the appropriate *torsion constant* [5] for $J$. If a more precise analysis is desired, the theory of *nonuniform torsion* [6] may be applied.

To obtain the consistent-mass matrix $\mathbf{M}$ for a torsional element, we will first integrate over the cross section and then over the length. Due to the small rotation $\theta_x$, the translation of a point on the cross section at distance $r$ from the center is $r\theta_x$. Also, the acceleration of the same point is $r\ddot{\theta}_x$. By integration over the cross section, we find the *inertial moment per unit length* to be $-\rho J \ddot{\theta}_x$, where $J$ is again the polar moment of inertia given in Eq. (13). Use of this inertial moment in conjunction with the corresponding virtual rotation $\delta\theta_x$ leads to

$$\mathbf{M} = \int_0^L \rho J \mathbf{f}^T \mathbf{f} \, dx \tag{14}$$

which is a specialized version of Eq. (3.3-24). Integration of Eq. (14) over the length yields

$$\mathbf{M} = \frac{\rho J}{L^2} \int_0^L \begin{bmatrix} L - x \\ x \end{bmatrix} [L - x \quad x] \, dx$$

$$= \frac{\rho J L}{6} \begin{bmatrix} 2 & 1 \\ 1 & 2 \end{bmatrix} \tag{15}$$

This array is the consistent-mass matrix for a torsional element of constant cross section and uniform mass density.

The simplest case of a body force applied to a torsional element consists of a uniformly distributed axial torque (or moment) $m_x$ per unit length. For this loading Eq. (3.3-25) gives

$$\mathbf{p}_b(t) = \int_0^L \mathbf{f}^T m_x \, dx = \frac{m_x}{L} \int_0^L \begin{bmatrix} L - x \\ x \end{bmatrix} dx = m_x \frac{L}{2} \begin{bmatrix} 1 \\ 1 \end{bmatrix} \tag{16}$$

These equivalent nodal loads are equal moments at the two ends of the element.

Figure 3.8(a) shows a straight *flexural element*, for which the *x-y* plane is a principal plane of bending. Indicated in the figure is a single generic displacement $v$, which is a translation in the $y$ direction. Thus,

$$\mathbf{u}(t) = v$$

The corresponding body force is a single component $b_y$ (force per unit length),

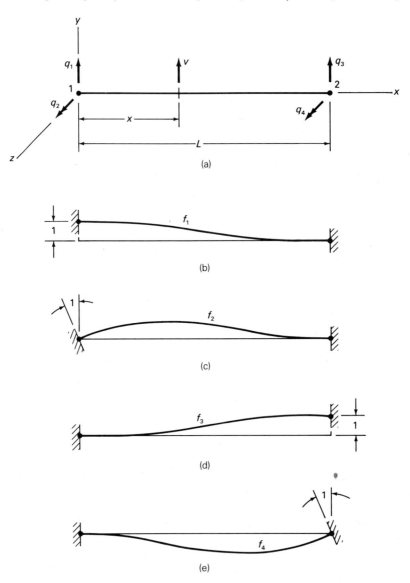

**Figure 3.8** Flexural element.

acting in the $y$ direction. Hence,

$$\mathbf{b}(t) = b_y$$

At node 1 [see Fig. 3.8(a)] the two nodal displacements $q_1$ and $q_2$ are a translation in the $y$ direction and a small rotation in the $z$ sense. The former is indicated by a single-headed arrow, while the latter is shown as a double-headed arrow. Similarly, at node 2 the displacements numbered 3 and 4 are a translation and a small rotation, respectively. Therefore, the vector of nodal displacements becomes

$$\mathbf{q}(t) = \{q_1, q_2, q_3, q_4\} = \{v_1, \theta_{z1}, v_2, \theta_{z2}\}$$

in which

$$\theta_{z1} = \frac{dv_1}{dx} \qquad \theta_{z2} = \frac{dv_2}{dx}$$

These derivatives (or slopes) may be considered to be small rotations even though they are actually rates of changes of translations at the nodes. Corresponding nodal actions at points 1 and 2 are

$$\mathbf{p}(t) = \{p_1, p_2, p_3, p_4\} = \{p_{y1}, M_{z1}, p_{y2}, M_{z2}\}$$

The terms $p_{y1}$ and $p_{y2}$ denote forces in the $y$ direction at nodes 1 and 2, and the symbols $M_{z1}$ and $M_{z2}$ represent moments in the $z$ sense at those points.

For the flexural element we assume cubic displacement shape functions in matrix $\mathbf{f}$, as follows:

$$\mathbf{f} = [f_1 \quad f_2 \quad f_3 \quad f_4]$$
$$= \frac{1}{L^3}[2x^3 - 3x^2L + L^3 \quad x^3L - 2x^2L^2 + xL^3 \quad -2x^3 + 3x^2L \quad x^3L - x^2L^2] \quad (17)$$

These four shape functions appear in Fig. 3.8(b)–(e). They represent the variations of $v$ along the length due to unit values of the four nodal displacements $q_1$ through $q_4$.

Strain-displacement relationships can be developed for the flexural element by assuming that plane sections remain plane during deformation, as illustrated in Fig. 3.9. The translation $u$ in the $x$ direction at any point on the cross section is

$$u = -y\frac{dv}{dx} \quad (18)$$

Using this relationship, we obtain the following expression for flexural strain:

$$\epsilon_x = \frac{du}{dx} = -y\frac{d^2v}{dx^2} = -y\phi \quad (19)$$

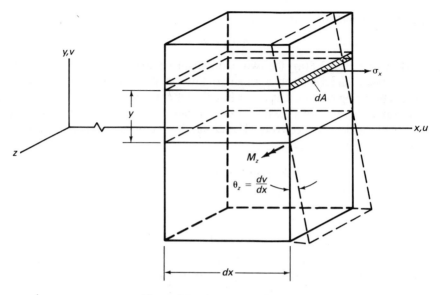

**Figure 3.9** Flexural deformations.

in which $\phi$ represents the *curvature*:

$$\phi = \frac{d^2v}{dx^2} \quad (20)$$

From Eq. (19) we see that the linear differential operator **d** relating $\epsilon_x$ to $v$ is

$$\mathbf{d} = -y\frac{d^2}{dx^2} \quad (21)$$

Then Eq. (3.3-10) gives the strain-displacement matrix **B** as

$$\mathbf{B} = \mathbf{d}\,\mathbf{f} = -\frac{y}{L^3}[12x - 6L \quad 6xL - 4L^2 \quad -12x + 6L \quad 6xL - 2L^2] \quad (22)$$

In addition, flexural stress $\sigma_x$ in Fig. 3.9 is related to flexural strain $\epsilon_x$ simply by

$$\sigma_x = E\epsilon_x \quad (23a)$$

Hence,

$$\mathbf{E} = E \quad \text{and} \quad \mathbf{E}\,\mathbf{B} = E\mathbf{B} \quad (23b)$$

Element stiffnesses may now be obtained from Eq. (3.3-23), as follows:

## Sec. 3.4  One-Dimensional Elements

$$K = \int_V \mathbf{B}^T \mathbf{E} \mathbf{B} \, dV$$

$$= \int_0^L \int_A \frac{Ey^2}{L^6} \begin{bmatrix} 12x - 6L \\ 6xL - 4L^2 \\ -12x + 6L \\ 6xL - 2L^2 \end{bmatrix} [12x - 6L \quad \cdots \quad \cdots \quad 6xL - 2L^2] \, dA \, dx$$

Multiplication and integration (with $EI$ constant) yields

$$\mathbf{K} = \frac{2EI}{L^3} \begin{bmatrix} 6 & 3L & -6 & 3L \\ 3L & 2L^2 & -3L & L^2 \\ -6 & -3L & 6 & -3L \\ 3L & L^2 & -3L & 2L^2 \end{bmatrix} \tag{24}$$

where

$$I = \int_A y^2 \, dA \tag{25}$$

represents the *moment of inertia* (second moment of area) of the cross section with respect to the neutral axis. Additional contributions to matrix **K** due to shearing deformations are given in Ref. 5.

The consistent-mass matrix **M** for a flexural element will be developed in two parts. A typical cross section of this type of member translates in the $y$ direction, as indicated in Fig. 3.8(a). However, the section also rotates about its neutral axis, as shown in Fig. 3.9. The *translational inertia* terms are much more important than the rotational terms, so they will be considered first. Using matrix **f** from Eq. (17) in Eq. (3.3-24), we find that

$$\mathbf{M}_t = \int_V \rho \mathbf{f}^T \mathbf{f} \, dV = \int_0^L \rho A \mathbf{f}^T \mathbf{f} \, dx$$

$$= \frac{\rho AL}{420} \begin{bmatrix} 156 & 22L & 54 & -13L \\ 22L & 4L^2 & 13L & -3L^2 \\ 54 & 13L & 156 & -22L \\ -13L & -3L^2 & -22L & 4L^2 \end{bmatrix} \tag{26}$$

which is the *consistent-mass matrix for translational inertia* in a prismatic beam.

*Rotational inertia* (or *rotary inertia*) terms for a beam can be deduced from Fig. 3.9, where the translation $u$ in the $x$ direction of a point on the cross section is

$$u = -y\theta_z \tag{27}$$

In this expression,

$$\theta_z = v_{,x} = \mathbf{f}_{,x}\mathbf{q}(t) \tag{28}$$

where the symbols $v_{,x}$ and $\mathbf{f}_{,x}$ represent differentiation with respect to $x$. Similarly, the acceleration of the same point in the $x$ direction is

$$\ddot{u} = -y\ddot{\theta}_z \tag{29}$$

where

$$\ddot{\theta}_z = \ddot{v}_{,x} = \mathbf{f}_{,x}\ddot{\mathbf{q}}(t) \tag{30}$$

By integrating the moment of inertial force over the cross section, we find the *inertial moment per unit length* to be $-\rho I \ddot{\theta}_z$, where $I$ is again the moment of inertia given in Eq. (25). Use of this inertial moment in conjunction with the corresponding virtual rotation $\delta\theta_z$ leads to the formula

$$\mathbf{M}_r = \int_0^L \rho I \mathbf{f}_{,x}^T \mathbf{f}_{,x} \, dx \tag{31}$$

which is a modified version of Eq. (3.3-24). Differentiating matrix $\mathbf{f}$ [see Eq. (17)] with respect to $x$, we find that

$$\mathbf{f}_{,x} = \frac{1}{L^3}\begin{bmatrix} 6(x^2 - xL) & 3x^2L - 4xL^2 + L^3 & -6(x^2 - xL) & 3x^2L - 2xL^2 \end{bmatrix} \tag{32}$$

Substitution of this matrix into Eq. (31), followed by integration over the length, produces

$$\mathbf{M}_r = \frac{\rho I}{30L}\begin{bmatrix} 36 & 3L & -36 & 3L \\ 3L & 4L^2 & -3L & -L^2 \\ -36 & -3L & 36 & -3L \\ 3L & -L^2 & -3L & 4L^2 \end{bmatrix} \tag{33}$$

which is the *consistent-mass matrix for rotational inertia* in a prismatic beam. Additional contributions to matrix $\mathbf{M}$ due to shearing deformations have also been developed and are given in Ref. 7.

We now consider the simple loading case of a uniformly distributed body force $b_y$ (per unit length) applied to a flexural element. Equivalent nodal loads at points 1 and 2 [see Fig. 3.8(a)] may be calculated from Eq. (3.3-25) as

$$\mathbf{p}_b(t) = \int_0^L \mathbf{f}^T b_y \, dx = b_y \frac{L}{12}\{6, L, 6, -L\} \tag{34}$$

For this integration the displacement shape functions $f_1$ through $f_4$ were drawn from Eq. (17).

By using *generalized stresses and strains*, we can avoid repetitious integrations over the cross sections of one-dimensional elements. Although this concept is rather trivial for an axial element, it can be more useful for torsional

## Sec. 3.4  One-Dimensional Elements

and flexural elements. Let us reconsider the torsional element in Fig. 3.7 and integrate the moment of the shearing stress $\tau$ about the $x$ axis. Thus, we generate the torque $M_x$, as follows:

$$M_x = \int_0^{2\pi} \int_0^R \tau r^2 \, dr \, d\theta \tag{35}$$

Substitution of the stress-strain and strain-displacement relationships from Eqs. (11a) and (7) produces

$$M_x = G\psi \int_0^{2\pi} \int_0^R r^3 \, dr \, d\theta = GJ\psi \tag{36}$$

If we take $M_x$ as generalized (or integrated) stress and $\psi$ as generalized strain, the generalized stress-strain (or torque-twist) operator $\overline{G}$ becomes

$$\overline{G} = GJ \tag{37}$$

which is the *torsional rigidity* of the cross section. Hence, from Eq. (36) we have

$$M_x = \overline{G}\psi \tag{38}$$

By this method the operator **d** in Eq. (9) does not include the multiplier $r$. Furthermore, the generalized matrix $\overline{\mathbf{B}}$ [Eq. (10) devoid of $r$] is used instead of matrix **B**. That is,

$$\mathbf{B} = r\overline{\mathbf{B}} \tag{39}$$

From this point we can conclude that evaluations of the terms in the stiffness matrix **K** do not require integrations over the cross section. Therefore,

$$\mathbf{K} = \int_0^L \overline{\mathbf{B}}^T \overline{G} \, \overline{\mathbf{B}} \, dx \tag{40}$$

This expression for **K** is equivalent to Eq. (3.3-23) used previously.

Turning now to the flexural element in Fig. 3.9, we integrate the moment of the normal stress $\sigma_x$ about the neutral axis to obtain $M_z$, as follows:

$$M_z = \int_A -\sigma_x y \, dA \tag{41}$$

Then substitute the stress-strain and strain-displacement relationships from Eqs. (23a) and (19) to find

$$M_z = E\phi \int_A y^2 \, dA = EI\phi \tag{42}$$

For this element we can take $M_z$ as generalized (or integrated) stress and $\phi$ as generalized strain. Then the generalized stress-strain (or moment-curvature) operator $\overline{E}$ is

$$\overline{E} = EI \tag{43}$$

which is the *flexural rigidity* of the cross section. Thus, from Eq. (42) we have

$$M_z = \overline{E}\phi \tag{44}$$

With this approach the operator **d** in Eq. (21) is devoid of the multiplier $-y$. In addition, the generalized matrix $\overline{\mathbf{B}}$ [Eq. (22) without the factor $-y$] may be used in place of matrix **B**. That is,

$$\mathbf{B} = -y\overline{\mathbf{B}} \tag{45}$$

Then integration over the cross section for terms in matrix **K** becomes unnecessary. Hence,

$$\mathbf{K} = \int_0^L \overline{\mathbf{B}}^T \overline{E}\, \overline{\mathbf{B}}\, dx \tag{46}$$

which is analogous to Eq. (40).

In Sec. 3.5 we will see how to assemble finite elements, and in later chapters we shall learn methods for calculating the dynamic response $\mathbf{q}(t)$ of the nodes. After those steps, we can find the time-varying stresses $\boldsymbol{\sigma}(t)$ within each element, using the equation

$$\boldsymbol{\sigma}(t) = \mathbf{E}\,\mathbf{B}\,\mathbf{q}(t) \tag{3.3-12}$$

This expression may be converted to a special formula for generalized stress, depending on the application. For example, an axial element has the simple stress-force relationship

$$\sigma_x = \frac{P_x}{A} \tag{47}$$

where $\sigma_x$ is the normal stress on the cross section and $P_x$ is the axial force. If we substitute Eq. (47) and $\mathbf{E} = E$ into Eq. (3.3-12), the result is

$$\frac{P_x}{A} = E\mathbf{B}\,\mathbf{q}(t)$$

Therefore,

$$P_x = \hat{E}\mathbf{B}\,\mathbf{q}(t) \tag{48}$$

where $\hat{E} = EA$ is the *axial rigidity* of the cross section. After the axial force has been determined from Eq. (48), we can find the normal stress using Eq. (47).

A torsional element with a circular cross section bears the following relationship between the shearing stress $\tau$ and the torque $M_x$:

$$\tau = \frac{M_x r}{J} \tag{49}$$

Substituting this term as well as $\mathbf{E} = G$ and Eq. (39) into Eq. (3.3-12) produces

$$\frac{M_x r}{J} = rG\overline{\mathbf{B}}\,\mathbf{q}(t)$$

Thus,
$$M_x = \overline{G}\,\overline{\mathbf{B}}\,\mathbf{q}(t) \qquad (50)$$

Then the shearing stress at any point on the cross section can be found with Eq. (49).

For a flexural element, the relationship between the normal stress $\sigma_x$ and the bending moment $M_z$ is

$$\sigma_x = -\frac{M_z y}{I} \qquad (51)$$

This term as well as $E = E$ and Eq. (45) can be substituted into Eq. (3.3-12), yielding

$$-\frac{M_z y}{I} = -y E \overline{\mathbf{B}}\,\mathbf{q}(t)$$

Hence,

$$M_z = \overline{E}\,\overline{\mathbf{B}}\,\mathbf{q}(t) \qquad (52)$$

Equation (52) is a formula for the bending moment at any point along the length of a flexural element in terms of the nodal displacements $q(t)$. After this calculation, we can find the normal stress at any point on the cross section from Eq. (51).

If we wish to find actions only at the ends of a member, we can simply multiply its stiffness matrix **K** and the vector of nodal displacements $\mathbf{q}(t)$. While this calculation is valid for framed structures, it has no such physical meaning for two- and three-dimensional finite elements, where the resulting actions are fictitious.

## 3.5 TRANSFORMATION AND ASSEMBLAGE OF ELEMENTS

If *local axes* for a finite element are not parallel to *global axes* for the whole structure, *rotation-of-axes transformations* must be used for nodal loads, displacements, accelerations, stiffnesses, and consistent masses. Thus, when the elements are assembled, the resulting equations of motion will pertain to the global directions at each node.

The concept of rotation of axes applies to a force, a moment, a translation, a small rotation, velocities, accelerations, orthogonal coordinates, and so on. Figure 3.10(a) shows a two-dimensional force vector **F** and its components $\mathbf{F}_x$ and $\mathbf{F}_y$ in the $x$ and $y$ directions. The figure also gives the components $\mathbf{F}_{x'}$ and $\mathbf{F}_{y'}$ in the directions of inclined axes $x'$ and $y'$. The scalar values of the components in directions of the primed axes can be computed from those for the unprimed axes, as follows:

$$F_{x'} = (\mathbf{F}_x + \mathbf{F}_y) \cdot \mathbf{i}' = \lambda_{11} F_x + \lambda_{12} F_y \qquad (1a)$$

$$F_{y'} = (\mathbf{F}_x + \mathbf{F}_y) \cdot \mathbf{j}' = \lambda_{21} F_x + \lambda_{22} F_y \qquad (1b)$$

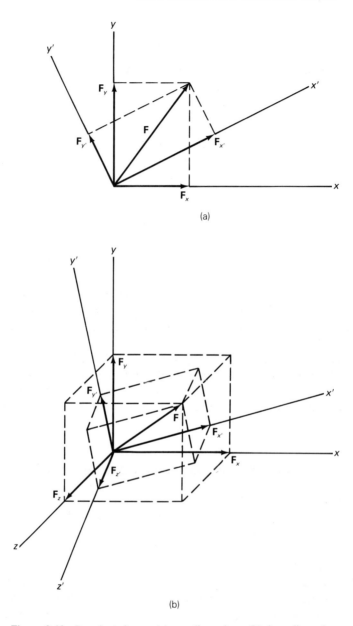

**Figure 3.10** Rotation of axes: (a) two dimensions; (b) three dimensions.

In these expressions the symbols **i**′ and **j**′ denote *unit vectors* in the directions of $x'$ and $y'$. Furthermore, the coefficients $\lambda_{11}$, $\lambda_{12}$, and so on, are *direction cosines* of the primed axes with respect to the unprimed axes. (For example, $\lambda_{12}$ is the direction cosine of axis $x'$ with respect to axis $y$, and so on.) The matrix form of Eqs. (1) is

## Sec. 3.5 Transformation and Assemblage of Elements

$$\begin{bmatrix} F_{x'} \\ F_{y'} \end{bmatrix} = \begin{bmatrix} \lambda_{11} & \lambda_{12} \\ \lambda_{21} & \lambda_{22} \end{bmatrix} \begin{bmatrix} F_x \\ F_y \end{bmatrix} \quad (2)$$

or

$$\boxed{\mathbf{F'} = \mathbf{R}\,\mathbf{F}} \quad (3)$$

The 2 × 2 matrix **R** is called a *rotation matrix*, consisting of direction cosines of the primed axes (with respect to the unprimed axes) listed row-wise. Hence, it is orthogonal, and the reverse transformation becomes

$$\boxed{\mathbf{F} = \mathbf{R}^{-1}\mathbf{F'} = \mathbf{R}^{\mathrm{T}}\mathbf{F'}} \quad (4)$$

Similarly, a three-dimensional force vector **F** appears in Fig. 3.10(b). Also shown in the figure are the force components $F_x$, $F_y$, and $F_z$ in the $x$, $y$, and $z$ directions. Components $F_{x'}$, $F_{y'}$, and $F_{z'}$ in the directions of inclined axes $x'$, $y'$, and $z'$ are given as well. For this case Eqs. (1) can be expanded, as follows:

$$F_{x'} = (\mathbf{F}_x + \mathbf{F}_y + \mathbf{F}_z) \cdot \mathbf{i'} = \lambda_{11} F_x + \lambda_{12} F_y + \lambda_{13} F_z \quad (5a)$$

$$F_{y'} = (\mathbf{F}_x + \mathbf{F}_y + \mathbf{F}_z) \cdot \mathbf{j'} = \lambda_{21} F_x + \lambda_{22} F_y + \lambda_{23} F_z \quad (5b)$$

$$F_{z'} = (\mathbf{F}_x + \mathbf{F}_y + \mathbf{F}_z) \cdot \mathbf{k'} = \lambda_{31} F_x + \lambda_{32} F_y + \lambda_{33} F_z \quad (5c)$$

where $\mathbf{k'}$ is a unit vector in the $z'$ direction. In matrix form, Eqs. (5) become

$$\begin{bmatrix} F_{x'} \\ F_{y'} \\ F_{z'} \end{bmatrix} = \begin{bmatrix} \lambda_{11} & \lambda_{12} & \lambda_{13} \\ \lambda_{21} & \lambda_{22} & \lambda_{23} \\ \lambda_{31} & \lambda_{32} & \lambda_{33} \end{bmatrix} \begin{bmatrix} F_x \\ F_y \\ F_z \end{bmatrix} \quad (6)$$

for which the additional direction cosines are associated with axes $z$ and $z'$. Equation (6) may be stated more concisely as in Eq. (3), but the rotation matrix **R** is of size 3 × 3. In addition, the reverse transformation given by Eq. (4) pertains to the three-dimensional case as well.

Simultaneous transformation of a force vector **F** and a moment vector **M** may be accomplished by

$$\mathbf{A'} = \hat{\mathbf{R}}\,\mathbf{A} = \begin{bmatrix} \mathbf{R} & \mathbf{0} \\ \mathbf{0} & \mathbf{R} \end{bmatrix} \begin{bmatrix} \mathbf{F} \\ \mathbf{M} \end{bmatrix} \quad (7)$$

In this expression $\hat{\mathbf{R}}$ is a *rotation-of-axes transformation matrix* containing two identical rotation matrices in diagonal positions. The reverse transformation is

$$\mathbf{A} = \hat{\mathbf{R}}^{-1}\mathbf{A'} = \hat{\mathbf{R}}^{\mathrm{T}}\mathbf{A'} = \begin{bmatrix} \mathbf{R}^{\mathrm{T}} & \mathbf{0} \\ \mathbf{0} & \mathbf{R}^{\mathrm{T}} \end{bmatrix} \begin{bmatrix} \mathbf{F'} \\ \mathbf{M'} \end{bmatrix} \quad (8)$$

As mentioned before, displacements and other types of vectors also can be transformed to and from local and global directions. Therefore, we shall convert the equations of motion for a finite element from local axes to global axes. For this purpose, let us rewrite Eq. (3.3-22) for inclined axes, as follows:

$$\mathbf{M'}\ddot{\mathbf{q}}' + \mathbf{K'}\mathbf{q'} = \mathbf{p'}(t) + \mathbf{p}'_b(t) \quad (9)$$

In accordance with Eqs. (4) and (8), premultiply Eq. (9) with $\hat{\mathbf{R}}^T$, and substitute $\mathbf{q}' = \hat{\mathbf{R}} \mathbf{q}$ and $\ddot{\mathbf{q}}' = \hat{\mathbf{R}} \ddot{\mathbf{q}}$ to obtain

$$\mathbf{M} \ddot{\mathbf{q}} + \mathbf{K} \mathbf{q} = \mathbf{p}(t) + \mathbf{p}_b(t) \qquad (3.3\text{-}22)$$

in which

$$\mathbf{K} = \hat{\mathbf{R}}^T \mathbf{K}' \hat{\mathbf{R}} \qquad (10)$$

and

$$\mathbf{M} = \hat{\mathbf{R}}^T \mathbf{M}' \hat{\mathbf{R}} \qquad (11)$$

Also,

$$\mathbf{p}(t) = \hat{\mathbf{R}}^T \mathbf{p}'(t) \qquad (12)$$

and

$$\mathbf{p}_b(t) = \hat{\mathbf{R}}^T \mathbf{p}'_b(t) \qquad (13)$$

Here the matrix $\hat{\mathbf{R}}$ contains rotation submatrices for all the nodes of the element.

After stiffnesses, masses, and nodal loads for individual elements have been transformed to global directions, we can assemble them by the *direct stiffness method* [5]. With this approach we need only add the contributions from all the elements to obtain stiffnesses, masses, and nodal loads for the whole structure. Thus, by summation* we have

$$\mathbf{S}_s = \sum_{i=1}^{n_e} \mathbf{K}_i \qquad \mathbf{M}_s = \sum_{i=1}^{n_e} \mathbf{M}_i \qquad (14)$$

and

$$\mathbf{A}_s(t) = \sum_{i=1}^{n_e} \mathbf{p}_i(t) \qquad \mathbf{A}_{sb}(t) = \sum_{i=1}^{n_e} \mathbf{p}_{bi}(t) \qquad (15)$$

where $n_e$ is the number of elements. In Eqs. (14) the symbols $\mathbf{S}_s$ and $\mathbf{M}_s$ represent the *structural stiffness matrix* and the *structural mass matrix* for all the nodes. Similarly, the action vectors $\mathbf{A}_s(t)$ and $\mathbf{A}_{sb}(t)$ in Eqs. (15) are *actual* and *equivalent nodal loads* for the whole structure. Then the undamped equations of motion for the assembled structure become

$$\mathbf{M}_s \ddot{\mathbf{D}}_s + \mathbf{S}_s \mathbf{D}_s = \mathbf{A}_s(t) + \mathbf{A}_{sb}(t) \qquad (16)$$

in which $\mathbf{D}_s$ and $\ddot{\mathbf{D}}_s$ are vectors of structural displacements and accelerations. Equation (16) gives the *structural equations of motion* for all nodal displacements, regardless of whether they are free or restrained.

In preparation for solving Eq. (16), we can rearrange and partition it, as follows:

---

* For the operations in Eqs. (14) and (15), the matrix or vector on the right must be expanded with zeros to make it the same size as the matrix or vector on the left.

## Sec. 3.5  Transformation and Assemblage of Elements

$$\begin{bmatrix} \mathbf{M}_{FF} & \mathbf{M}_{FR} \\ \mathbf{M}_{RF} & \mathbf{M}_{RR} \end{bmatrix} \begin{bmatrix} \ddot{\mathbf{D}}_F \\ \ddot{\mathbf{D}}_R \end{bmatrix} + \begin{bmatrix} \mathbf{S}_{FF} & \mathbf{S}_{FR} \\ \mathbf{S}_{RF} & \mathbf{S}_{RR} \end{bmatrix} \begin{bmatrix} \mathbf{D}_F \\ \mathbf{D}_R \end{bmatrix} = \begin{bmatrix} \mathbf{A}_F(t) \\ \mathbf{A}_R(t) \end{bmatrix} \quad (17)$$

In this equation actual and equivalent nodal loads have been combined into a single action vector. The subscript $F$ refers to *free nodal displacements*, while the subscript $R$ denotes *restrained nodal displacements*. Writing Eq. (17) in two parts gives

$$\mathbf{M}_{FF}\ddot{\mathbf{D}}_F + \mathbf{M}_{FR}\ddot{\mathbf{D}}_R + \mathbf{S}_{FF}\mathbf{D}_F + \mathbf{S}_{FR}\mathbf{D}_R = \mathbf{A}_F(t) \quad (18a)$$

and

$$\mathbf{M}_{RF}\ddot{\mathbf{D}}_F + \mathbf{M}_{RR}\ddot{\mathbf{D}}_R + \mathbf{S}_{RF}\mathbf{D}_F + \mathbf{S}_{RR}\mathbf{D}_R = \mathbf{A}_R(t) \quad (18b)$$

If support motions (at restraints) are zero, Eqs. (18) simplify to

$$\mathbf{M}_{FF}\ddot{\mathbf{D}}_F + \mathbf{S}_{FF}\mathbf{D}_F = \mathbf{A}_F(t) \quad (19a)$$

and

$$\mathbf{M}_{RF}\ddot{\mathbf{D}}_F + \mathbf{S}_{RF}\mathbf{D}_F = \mathbf{A}_R(t) \quad (19b)$$

These equations will be used in subsequent work for calculating free displacements $\mathbf{D}_F$ and *support reactions* $\mathbf{A}_R(t)$.

In many problems it is sufficiently accurate merely to lump tributary masses at the nodes of a discretized continuum [8]. When doing so, we form the *lumped mass matrix* $\mathbf{M}_\ell$ for the whole structure as

$$\mathbf{M}_\ell = \begin{bmatrix} \mathbf{M}_1 & 0 & \ldots & 0 & \ldots & 0 \\ 0 & \mathbf{M}_2 & \ldots & 0 & \ldots & 0 \\ \ldots & \ldots & \ldots & \ldots & \ldots & \ldots \\ 0 & 0 & \ldots & \mathbf{M}_j & \ldots & 0 \\ \ldots & \ldots & \ldots & \ldots & \ldots & \ldots \\ 0 & 0 & \ldots & 0 & \ldots & \mathbf{M}_{n_n} \end{bmatrix} \quad (20)$$

where $n_n$ is the number of nodes. The typical submatrix $\mathbf{M}_j$ in Eq. (20) signifies a small diagonal array defined to be

$$\mathbf{M}_j = M_j \mathbf{I}_0 \quad (21)$$

In this expression $M_j$ is the tributary mass lumped at node $j$, and $\mathbf{I}_0$ is an identity matrix with 1 replaced by 0 wherever a nontranslational displacement occurs. Thus, the lumped-mass approach has the advantage that the mass matrix $\mathbf{M}_\ell$ is always diagonal, although not always positive-definite.

As an example of transformation and assemblage of element properties, we shall consider the plane truss in Fig. 3.11(a). For this structure let us find the stiffness matrix $\mathbf{S}_s$ and the consistent-mass matrix $\mathbf{M}_s$ in rearranged and partitioned forms. Assume that the cross-sectional area of member 1 is equal to $A$, that of member 2 is equal to $0.6A$, and that of member 3 is equal to $0.8A$.

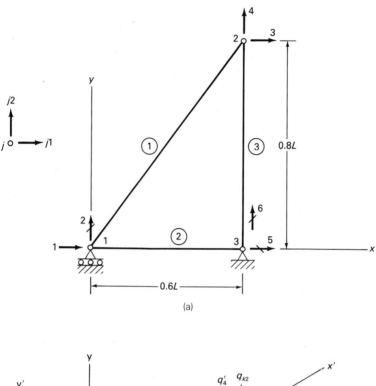

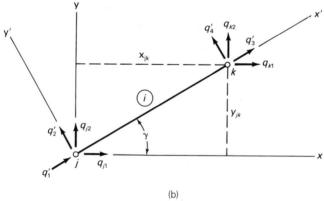

**Figure 3.11** Plane truss: (a) structure; (b) member.

An arbitrary system for numbering the members and joints of the truss appears in Fig. 3.11(a). Member numbers are enclosed in circles adjacent to the members, and joint numbers are placed adjacent to the joints. The numbering system for joint displacements is shown by numbered arrows that depict the positive directions of the possible displacements, including those at restrained points. These displacement numbers are obtained from the joint numbers, as follows:

## Sec. 3.5  Transformation and Assemblage of Elements

$$j1 = 2j - 1 \qquad j2 = 2j \qquad (22)$$

Those for a typical joint $j$ are illustrated at the left of Fig. 3.11(a). For this structure we have three degrees of freedom (numbered 1, 3, and 4) and three restrained displacements (numbered 2, 5, and 6). Note that the arrows for restrained displacements have small diagonal cross lines to distinguish them from free displacements.

Figure 3.11(b) shows a typical member $i$ in the plane truss. Its ends are numbered $j$ and $k$, and local axes $x'$ and $y'$ have their origin at point $j$. Axis $x'$ lies along the member, and its positive direction is from $j$ to $k$, whereas the direction of $y'$ is perpendicular to the member. These axes are inclined at the angle $\gamma$ from the global axes $x$ and $y$. Using the length of the member as

$$L = \sqrt{x_{jk}^2 + y_{jk}^2} \qquad (23)$$

we find the *direction cosines* of the inclined axes to be

$$\lambda_{11} = \cos \gamma = \frac{x_{jk}}{L} = c_x \qquad \lambda_{12} = \sin \gamma = \frac{y_{jk}}{L} = c_y$$
$$\lambda_{21} = -\sin \gamma = -c_y \qquad \lambda_{22} = \cos \gamma = c_x \qquad (24)$$

In these expressions the symbols $c_x$ and $c_y$ denote direction cosines of the member axis itself, with respect to axes $x$ and $y$. Table 3.1 summarizes member information for the truss, including the arbitrarily chosen joint numbers $j$ and $k$, the cross-sectional areas, the lengths, and the direction cosines $c_x$ and $c_y$.

TABLE 3.1  Member Information for Truss of Fig. 3.11(a)

| Member $i$ | Joint $j$ | Joint $k$ | Area | Length | Direction Cosines $c_x$ | $c_y$ |
|---|---|---|---|---|---|---|
| 1 | 1 | 2 | $A$ | $L$ | 0.6 | 0.8 |
| 2 | 1 | 3 | $0.6A$ | $0.6L$ | 1.0 | 0 |
| 3 | 2 | 3 | $0.8A$ | $0.8L$ | 0 | $-1.0$ |

By considering two nodal displacements in the $x'$ and $y'$ directions at each end of the member, we can write the *element stiffness matrix* $\mathbf{K}'$ for local axes as follows:

$$\mathbf{K}' = \frac{EA}{L} \begin{bmatrix} 1 & 0 & -1 & 0 \\ 0 & 0 & 0 & 0 \\ -1 & 0 & 1 & 0 \\ 0 & 0 & 0 & 0 \end{bmatrix} \qquad (25)$$

This matrix is the same as that for the axial element in Eq. (3.4-4), except that the zeros are inserted for stiffnesses in the $y'$ direction. Thus, the size of the matrix is expanded from 2 × 2 to 4 × 4. Equation (10) enables us to transform the stiffness matrix from local to global axes. Hence,

$$\mathbf{K} = \hat{\mathbf{R}}^T \mathbf{K}' \hat{\mathbf{R}} = \frac{EA}{L} \begin{bmatrix} c_x^2 & c_x c_y & -c_x^2 & -c_x c_y \\ c_x c_y & c_y^2 & -c_x c_y & -c_y^2 \\ -c_x^2 & -c_x c_y & c_x^2 & c_x c_y \\ -c_x c_y & -c_y^2 & c_x c_y & c_y^2 \end{bmatrix} \quad (26)$$

For this type of element the 4 × 4 rotation-of-axes transformation matrix $\hat{\mathbf{R}}$ is

$$\hat{\mathbf{R}} = \begin{bmatrix} \mathbf{R} & \mathbf{0} \\ \mathbf{0} & \mathbf{R} \end{bmatrix} \quad (27)$$

where the submatrix $\mathbf{R}$ can be written as

$$\mathbf{R} = \begin{bmatrix} c_x & c_y \\ -c_y & c_x \end{bmatrix} \quad (28)$$

The latter matrix contains only the direction cosines of the member axis.

When Eq. (26) and the data from Table 3.1 are applied to each of the members, we obtain the following stiffness matrices for global axes:

$$\mathbf{K}_1 = \frac{EA}{L} \begin{bmatrix} 0.36 & 0.48 & -0.36 & -0.48 \\ 0.48 & 0.64 & -0.48 & -0.64 \\ -0.36 & -0.48 & 0.36 & 0.48 \\ -0.48 & -0.64 & 0.48 & 0.64 \end{bmatrix} \begin{matrix} 1 \\ 2 \\ 3 \\ 4 \end{matrix}$$
$$\quad\quad\quad\quad\quad\; 1 \quad\quad 2 \quad\quad 3 \quad\quad 4$$

$$\mathbf{K}_2 = \frac{EA}{L} \begin{bmatrix} 1 & 0 & -1 & 0 \\ 0 & 0 & 0 & 0 \\ -1 & 0 & 1 & 0 \\ 0 & 0 & 0 & 0 \end{bmatrix} \begin{matrix} 1 \\ 2 \\ 5 \\ 6 \end{matrix} \quad (29)$$
$$\quad\quad\quad\quad\; 1 \quad 2 \quad 5 \quad 6$$

$$\mathbf{K}_3 = \frac{EA}{L} \begin{bmatrix} 0 & 0 & 0 & 0 \\ 0 & 1 & 0 & -1 \\ 0 & 0 & 0 & 0 \\ 0 & -1 & 0 & 1 \end{bmatrix} \begin{matrix} 3 \\ 4 \\ 5 \\ 6 \end{matrix}$$
$$\quad\quad\quad\quad\; 3 \quad 4 \quad 5 \quad 6$$

The terms in each of these matrices may be transferred to the appropriate locations in the structural stiffness matrix $\mathbf{S}_s$ by calculating the following joint displacement indexes:

$$\begin{matrix} j1 = 2j - 1 & j2 = 2j \\ k1 = 2k - 1 & k2 = 2k \end{matrix} \quad (30)$$

Sec. 3.5  Transformation and Assemblage of Elements    103

Note that these indexes appear as subscripts for the $x$ and $y$ translations at joints $j$ and $k$ in Fig. 3.11(b). As an aid in the transferring process, the numerical values of $j1$ through $k2$ are listed down the right-hand side and across the bottom of each matrix in Eqs. (29). After assembling the structural stiffness matrix $\mathbf{S}_s$ using the first of Eqs. (14), we obtain

$$\mathbf{S}_s = \frac{EA}{L} \begin{bmatrix} 1.36 & 0.48 & -0.36 & -0.48 & -1 & 0 \\ 0.48 & 0.64 & -0.48 & -0.64 & 0 & 0 \\ -0.36 & -0.48 & 0.36 & 0.48 & 0 & 0 \\ -0.48 & -0.64 & 0.48 & 1.64 & 0 & -1 \\ -1 & 0 & 0 & 0 & 1 & 0 \\ 0 & 0 & 0 & -1 & 0 & 1 \end{bmatrix} \begin{matrix} 1 \\ 2 \\ 3 \\ 4 \\ 5 \\ 6 \end{matrix} \quad (31a)$$

$$\phantom{\mathbf{S}_s = \frac{EA}{L}} \; 1 \quad\; 2 \quad\;\; 3 \quad\;\; 4 \quad\;\; 5 \quad 6$$

Rearranging and partitioning this matrix in the form shown by Eq. (17) produces

$$\mathbf{S}_s = \begin{bmatrix} \mathbf{S}_{FF} & \mathbf{S}_{FR} \\ \mathbf{S}_{RF} & \mathbf{S}_{RR} \end{bmatrix}$$

$$= \frac{EA}{L} \left[ \begin{array}{ccc|ccc} 1.36 & -0.36 & -0.48 & 0.48 & -1 & 0 \\ -0.36 & 0.36 & 0.48 & -0.48 & 0 & 0 \\ -0.48 & 0.48 & 1.64 & -0.64 & 0 & -1 \\ \hline 0.48 & -0.48 & -0.64 & 0.64 & 0 & 0 \\ -1 & 0 & 0 & 0 & 1 & 0 \\ 0 & 0 & -1 & 0 & 0 & 1 \end{array} \right] \begin{matrix} 1 \\ 3 \\ 4 \\ 2 \\ 5 \\ 6 \end{matrix} \quad (31b)$$

$$\phantom{= \frac{EA}{L}} \;\; 1 \quad\; 3 \quad\;\; 4 \quad\;\; 2 \quad\; 5 \quad 6$$

where the indexes down the right side and across the bottom show the rearrangement.

As with stiffnesses, we can generate the *consistent-mass matrix* $\mathbf{M}'$ for local axes of a typical member to be

$$\mathbf{M}' = \frac{\rho AL}{6} \begin{bmatrix} 2 & 0 & 1 & 0 \\ 0 & 2 & 0 & 1 \\ 1 & 0 & 2 & 0 \\ 0 & 1 & 0 & 2 \end{bmatrix} \quad (32)$$

In this instance, the terms in Eq. (3.4-5) for the axial element are repeated for the $y'$ direction, because accelerations in that sense also give rise to inertial actions. Thus, the consistent-mass matrix for inclined axes is of size $4 \times 4$. It may be transformed to global axes using Eq. (11). However, when that equation is applied, we find the resulting matrix $\mathbf{M}$ to be the same as matrix $\mathbf{M}'$ in Eq. (32). Therefore, we conclude that for a plane truss element the consistent-mass matrix is invariant with rotation of axes.

Applying Eq. (32) and the data from Table 3.1, we find the consistent-mass matrices for the members to be

$$\mathbf{M}_1 = \mathbf{M} \qquad \mathbf{M}_2 = 0.36\mathbf{M} \qquad \mathbf{M}_3 = 0.64\mathbf{M} \tag{33}$$

In these expressions the matrix $\mathbf{M}$ is the same as $\mathbf{M}'$ in Eq. (32). Terms in matrices $\mathbf{M}_1$, $\mathbf{M}_2$, and $\mathbf{M}_3$ may be transferred to the consistent-mass matrix $\mathbf{M}_s$ for the whole structure using again the displacement indexes $j1$ through $k2$. Thus, when $\mathbf{M}_s$ is assembled in accordance with the second of Eqs. (14), we find that

$$\mathbf{M}_s = \frac{\rho AL}{6} \begin{bmatrix} 2.72 & 0 & 1 & 0 & 0.36 & 0 \\ 0 & 2.72 & 0 & 1 & 0 & 0.36 \\ 1 & 0 & 3.28 & 0 & 0.64 & 0 \\ 0 & 1 & 0 & 3.28 & 0 & 0.64 \\ 0.36 & 0 & 0.64 & 0 & 2 & 0 \\ 0 & 0.36 & 0 & 0.64 & 0 & 2 \end{bmatrix} \begin{matrix} 1 \\ 2 \\ 3 \\ 4 \\ 5 \\ 6 \end{matrix} \tag{34a}$$

$$\phantom{\mathbf{M}_s = \frac{\rho AL}{6}} \;\; 1 \;\;\;\; 2 \;\;\;\; 3 \;\;\;\; 4 \;\;\;\; 5 \;\;\;\; 6$$

Rearranging and partitioning this matrix as shown in Eq. (17) gives

$$\mathbf{M}_s = \begin{bmatrix} \mathbf{M}_{FF} & \mathbf{M}_{FR} \\ \mathbf{M}_{RF} & \mathbf{M}_{RR} \end{bmatrix}$$

$$= \frac{\rho AL}{6} \left[ \begin{array}{ccc|ccc} 2.72 & 1 & 0 & 0 & 0.36 & 0 \\ 1 & 3.28 & 0 & 0 & 0.64 & 0 \\ 0 & 0 & 3.28 & 1 & 0 & 0.64 \\ \hline 0 & 0 & 1 & 2.72 & 0 & 0.36 \\ 0.36 & 0.64 & 0 & 0 & 2 & 0 \\ 0 & 0 & 0.64 & 0.36 & 0 & 2 \end{array} \right] \begin{matrix} 1 \\ 3 \\ 4 \\ 2 \\ 5 \\ 6 \end{matrix} \tag{34b}$$

$$\phantom{= \frac{\rho AL}{6}} \;\; 1 \;\;\;\; 3 \;\;\;\; 4 \;\;\;\; 2 \;\;\;\; 5 \;\;\;\; 6$$

where the indexes at the right and bottom again show the pattern of rearrangement.

Alternatively, the consistent mass matrix $\mathbf{M}_s$ in Eq. (34b) could be replaced by a lumped-mass matrix $\mathbf{M}_\ell$, as follows:

$$\mathbf{M}_\ell = \rho AL \left[ \begin{array}{ccc|ccc} 0.68 & 0 & 0 & 0 & 0 & 0 \\ 0 & 0.82 & 0 & 0 & 0 & 0 \\ 0 & 0 & 0.82 & 0 & 0 & 0 \\ \hline 0 & 0 & 0 & 0.68 & 0 & 0 \\ 0 & 0 & 0 & 0 & 0.5 & 0 \\ 0 & 0 & 0 & 0 & 0 & 0.5 \end{array} \right] \begin{matrix} 1 \\ 3 \\ 4 \\ 2 \\ 5 \\ 6 \end{matrix} \tag{35}$$

$$\phantom{\mathbf{M}_\ell = \rho AL} \;\; 1 \;\;\;\; 3 \;\;\;\; 4 \;\;\;\; 2 \;\;\;\; 5 \;\;\;\; 6$$

For each mass matrix $\mathbf{M}_s$ and $\mathbf{M}_\ell$, the total mass of the truss is $2\rho AL$ for both the $x$ and $y$ directions, as it should be. If nonstructural concentrated masses also exist at the joints of a truss, they may be added directly to the diagonal terms in either $\mathbf{M}_s$ or $\mathbf{M}_\ell$.

At this point we should observe that considering the masses for the three-member truss to be associated only with joint translations is a poor approximation to the truth. In fact, there often will be flexure in the members that significantly influences the dynamic characteristics of the structure. For example, if a time-varying force is applied in the $x$ direction at the middle of member 3 [see Fig. 3.11(a)], the flexural deformation in that member will dominate the response. We shall deal with this type of problem for trusses by the component-mode method in Sec. 10.6. In the meantime, we will use $\mathbf{M}_s$ or $\mathbf{M}_\ell$ in the forms shown here, even though the numbers of members and joints are small.

## 3.6 VIBRATIONAL ANALYSIS

In Sec. 3.5 we derived undamped equations of motion for free nodal displacements of a discretized structure [see Eq. (3.5-19a)]. If there are no applied actions, these equations can be written in homogeneous form as

$$\mathbf{M}\ddot{\mathbf{D}} + \mathbf{S}\mathbf{D} = \mathbf{0} \tag{1}$$

Here the subscript $F$ (for free displacements) is omitted to simplify the notation.

Equation (1) has a known solution [9] that may be stated as follows:

$$\mathbf{D}_i = \mathbf{\Phi}_i \sin(\omega_i t + \alpha_i) \qquad (i = 1, 2, \ldots, n) \tag{2}$$

where $n$ is the *number of degrees of freedom*. In this harmonic expression, $\mathbf{\Phi}_i$ is a vector of nodal amplitudes (the *mode shape*) for the $i$th mode of vibration. The symbol $\omega_i$ represents the *angular frequency* of mode $i$, and $\alpha_i$ denotes the *phase angle*. By differentiating Eq. (2) twice with respect to time $t$, we also find that

$$\ddot{\mathbf{D}}_i = -\omega_i^2 \mathbf{\Phi}_i \sin(\omega_i t + \alpha_i) \tag{3}$$

Substitution of Eqs. (2) and (3) into Eq. (1) allows cancellation of the term $\sin(\omega_i t + \alpha_i)$, which leaves

$$(\mathbf{S} - \omega_i^2 \mathbf{M})\mathbf{\Phi}_i = \mathbf{0} \tag{4}$$

This manipulation has the effect of separating the variable time from those of space, and we are left with a set of $n$ homogeneous algebraic equations.

Equation (4) has the form of the *algebraic eigenvalue problem*. From the theory of homogeneous equations [10], nontrivial solutions exist only if the determinant of the coefficient matrix is equal to zero. Thus,

$$|\mathbf{S} - \omega_i^2 \mathbf{M}| = 0 \tag{5}$$

Expansion of this determinant yields a polynomial of order $n$ called the *characteristic equation*. The $n$ roots $\omega_i^2$ of this polynomial are the *characteristic values*, or *eigenvalues*. Substitution of these roots (one at a time) into the homogeneous equations [Eq. (4)] produces the *characteristic vectors*, or *eigenvectors* $\mathbf{\Phi}_i$, within arbitrary constants. Alternatively [9], each eigenvector may be found as any column of the *adjoint matrix* $\mathbf{H}_i^a$ of the *characteristic matrix* $\mathbf{H}_i$, obtained from Eq. (4), as follows:

$$\mathbf{H}_i \mathbf{\Phi}_i = \mathbf{0} \tag{6}$$

where

$$\mathbf{H}_i = \mathbf{S} - \omega_i^2 \mathbf{M} \tag{7}$$

The methods implied by Eqs. (5), (6), and (7) are conducive to hand calculations for problems having small numbers of degrees of freedom. Examples at the end of this section demonstrate such calculations. However, a structure with a large number of degrees of freedom must be handled by a computer program, as described in Sec. 3.8. Such a program would include a subprogram (or subroutine) for calculating eigenvalues and eigenvectors. The most efficient type of subroutine for structural vibrations accepts the eigenvalue problem only in the following *standard, symmetric form*:

$$(\mathbf{A} - \lambda_i \mathbf{I}) \mathbf{X}_i = \mathbf{0} \tag{8}$$

in which $\mathbf{A}$ is a symmetric matrix and $\mathbf{I}$ is an identity matrix. The symbol $\lambda_i$ denotes the $i$th eigenvalue, and $\mathbf{X}_i$ is the corresponding eigenvector for a new system of $n$ homogeneous equations. We can put Eq. (4) into the form of Eq. (8) by factoring either matrix $\mathbf{S}$ or matrix $\mathbf{M}$, using the *Cholesky square-root method* [1, 5]. We choose to factor $\mathbf{S}$ for an important reason that will soon be apparent. Thus,

$$\mathbf{S} = \mathbf{U}^T \mathbf{U} \tag{9}$$

where the factor $\mathbf{U}$ is an upper triangular matrix. Substitute Eq. (9) into Eq. (4) to obtain

$$(\mathbf{U}^T \mathbf{U} - \omega_i^2 \mathbf{M}) \mathbf{\Phi}_i = \mathbf{0}$$

Then premultiply this equation by $\mathbf{U}^{-T}$ and insert $\mathbf{I} = \mathbf{U}^{-1} \mathbf{U}$ after matrix $\mathbf{M}$, which yields

$$\mathbf{U}^{-T}(\mathbf{U}^T \mathbf{U} - \omega_i^2 \mathbf{M} \mathbf{U}^{-1} \mathbf{U}) \mathbf{\Phi}_i = \mathbf{0}$$

Rewriting terms in reverse order, we find that

$$(\mathbf{M}_U - \lambda_i \mathbf{I}) \mathbf{\Phi}_{Ui} = \mathbf{0} \tag{10}$$

where

$$\mathbf{M}_U = \mathbf{U}^{-T} \mathbf{M} \mathbf{U}^{-1} \qquad \lambda_i = \frac{1}{\omega_i^2} \qquad \mathbf{\Phi}_{Ui} = \mathbf{U} \mathbf{\Phi}_i \tag{11}$$

Equation (10) is now in the standard, symmetric form of the eigenvalue problem given by Eq. (8). The matrix $\mathbf{A}$ is represented by $\mathbf{M}_U$ [the first of

## Sec. 3.6  Vibrational Analysis

Eqs. (11)], which is guaranteed to be symmetric. In addition, we see that the eigenvalue $\lambda_i$ is equal to the reciprocal of the square of the angular frequency. This is the consequence of choosing to factor matrix **S** and is numerically advantageous because the highest eigenvalue (corresponding to the lowest frequency) has the greatest accuracy. The eigenvector $\mathbf{\Phi}_{Ui}$ in Eq. (10) is related to $\mathbf{\Phi}_i$ by the last of Eqs. (11). This expression constitutes a change of coordinates to a new set where the stiffness matrix is equal to **I**. After the eigenvalues and eigenvectors have been found from Eq. (10), the angular frequencies and mode shapes (in the original coordinates) can be determined as

$$\omega_i = \frac{1}{\sqrt{\lambda_i}} \qquad \mathbf{\Phi}_i = \mathbf{U}^{-1}\mathbf{\Phi}_{Ui} \tag{12}$$

Because the matrix $\mathbf{M}_U$ in the new coordinates is symmetric, all of its eigenvectors are linearly independent [10]. In addition, two eigenvectors $\mathbf{\Phi}_{Ui}$ and $\mathbf{\Phi}_{Uj}$ corresponding to distinct eigenvalues $\lambda_i$ and $\lambda_j$ will be orthogonal with respect to each other. Thus,

$$\mathbf{\Phi}_{Ui}^T \mathbf{\Phi}_{Uj} = 0 \qquad (i \neq j) \tag{13}$$

However, the back-transformation given as the second expression in Eqs. (12) does not necessarily preserve orthogonality among the eigenvectors. Instead, the eigenvectors $\mathbf{\Phi}_i$ and $\mathbf{\Phi}_j$ in the original coordinates are orthogonal with respect to matrix **S**, as follows:

$$\mathbf{\Phi}_i^T \mathbf{S}\, \mathbf{\Phi}_j = \mathbf{\Phi}_{Ui}^T \mathbf{U}^{-T} \mathbf{S}\, \mathbf{U}^{-1} \mathbf{\Phi}_{Uj} = \mathbf{\Phi}_{Ui}^T \mathbf{\Phi}_{Uj} = 0 \tag{14}$$

In Sec. 4.2 we will show that the eigenvectors are also orthogonal with respect to matrix **M**.

From vibrational analysis of structures, we may find *repeated frequencies* (or *repeated eigenvalues*) as roots of the characteristic equation. If an eigenvalue is repeated $m$ times, it is said to be of *multiplicity m*. We can find $m$ linearly independent eigenvectors corresponding to the repeated eigenvalue, but such vectors are not unique. A new set of eigenvectors can always be formed as linear combinations of the original set and still satisfy the eigenvalue problem. That is,

$$\mathbf{M}_U \sum_{i=1}^{m} c_i \mathbf{\Phi}_{Ui} = \lambda_m \sum_{i=1}^{m} c_i \mathbf{\Phi}_{Ui} \tag{15}$$

where $c_i$ is a scalar multiplier of the $i$th modal vector and $\lambda_m$ is the repeated eigenvalue. It is always advantageous to form a new set of eigenvectors that are orthogonal with respect to each other. They will automatically be orthogonal to eigenvectors corresponding to distinct eigenvalues. But there is still an infinity of choices for nonunique orthogonal vectors corresponding to a repeated eigenvalue. The *Gram–Schmidt orthogonalization procedure* [11] is a formal mathematical approach commonly used to construct an orthogonal set of eigenvectors from a linearly independent set.

If the stiffness matrix is semidefinite, it cannot be factored as described here, because at least one rigid-body mode (with $\omega_i = 0$) is present. In such a case it would be possible to factor the mass matrix instead, using the reduction technique in Sec. 6.6 if necessary to achieve a positive-definite matrix. If the mass matrix is diagonal (as in the lumped-mass method), its factorization becomes simply

$$\mathbf{M} = \mathbf{M}^{1/2}\mathbf{M}^{1/2} \tag{16}$$

where $\mathbf{M}^{1/2}$ contains diagonal terms equal to the square roots of those in $\mathbf{M}$. In this case the transformation to standard, symmetric form yields

$$(\mathbf{S}_M - \lambda_i \mathbf{I})\mathbf{\Phi}_{Mi} = \mathbf{0} \tag{17}$$

where

$$\mathbf{S}_M = \mathbf{M}^{-1/2}\mathbf{S}\,\mathbf{M}^{-1/2} \qquad \lambda_i = \omega_i^2 \qquad \mathbf{\Phi}_{Mi} = \mathbf{M}^{1/2}\mathbf{\Phi}_i \tag{18}$$

In the first of Eqs. (18) the matrix $\mathbf{M}^{-1/2}$ contains the reciprocals of the diagonal terms in $\mathbf{M}^{1/2}$. After finding the eigenvalues and eigenvectors from Eq. (17), we can obtain the angular frequencies and mode shapes from

$$\omega_i = \sqrt{\lambda_i} \qquad \mathbf{\Phi}_i = \mathbf{M}^{-1/2}\mathbf{\Phi}_{Mi} \tag{19}$$

Note that the numerical advantage mentioned before is not present in this approach.

In general, the process of directly extracting the roots of the characteristic equation for Eq. (8) must be done iteratively and is not efficient for large problems. If the eigenvalues and eigenvectors of all the modes are to be found, it is best to use Householder transformations (see Sec. B.2) and convert matrix $\mathbf{A}$ to tridiagonal form. Then the final values of $\lambda_i$ and the vectors $\mathbf{X}_i$ can be determined by iteration with the QR algorithm. <u>On the other hand, if only a few modes are desired, the method of inverse iteration with spectral shifting is more efficient</u> (see Sec. B.1).

**Example 3.1**

Figure 3.12(a) shows a plane truss with two degrees of freedom at joint 1. Assume that the cross-sectional areas of members 1 and 2 are equal to $0.8A$ and $A$. Using the consistent-mass approach, find the angular frequencies and mode shapes for this structure.

The $2 \times 2$ stiffness matrix $\mathbf{S}$ for the free displacements in this problem is

$$\mathbf{S} = \frac{EA}{L}\begin{bmatrix} 0.36 & -0.48 \\ -0.48 & 1.64 \end{bmatrix} \tag{a}$$

and the consistent-mass matrix has the diagonal form

$$\mathbf{M} = \frac{\rho AL}{6}\begin{bmatrix} 3.28 & 0 \\ 0 & 3.28 \end{bmatrix} \tag{b}$$

If we let $s = EA/L$ and $m = 3.28\rho AL/6$, the homogeneous equations in Eq. (4) become

## Sec. 3.6 Vibrational Analysis

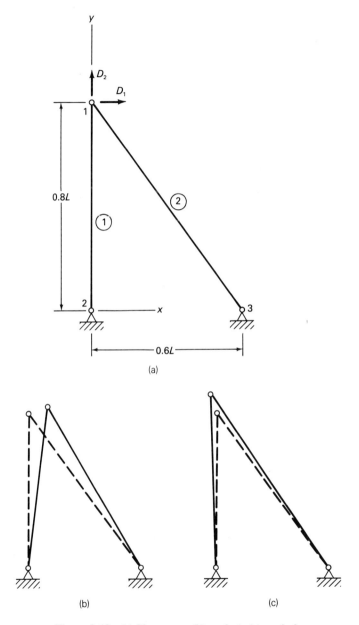

**Figure 3.12** (a) Plane truss; (b) mode 1; (c) mode 2.

$$\begin{bmatrix} 0.36s - m\omega_i^2 & -0.48s \\ -0.48s & 1.64s - m\omega_i^2 \end{bmatrix} \begin{bmatrix} \Phi_{1i} \\ \Phi_{2i} \end{bmatrix} = \begin{bmatrix} 0 \\ 0 \end{bmatrix} \tag{c}$$

in which the 2 × 2 array of coefficients is the characteristic matrix $\mathbf{H}_i$. Setting the determinant of $\mathbf{H}_i$ equal to zero in accordance with Eq. (5) produces the characteristic equation, as follows:

$$m^2\omega_i^4 - 2ms\omega_i^2 + 0.36s^2 = 0 \tag{d}$$

The roots of this quadratic equation are

$$\omega_1^2 = 0.2\frac{s}{m} \qquad \omega_2^2 = 1.8\frac{s}{m} \tag{e}$$

from which we find the angular frequencies to be

$$\omega_1 = \sqrt{\frac{0.2s}{m}} = \frac{0.6049}{L}\sqrt{\frac{E}{\rho}} \qquad \omega_2 = \sqrt{\frac{1.8s}{m}} = \frac{1.815}{L}\sqrt{\frac{E}{\rho}} \tag{f}$$

Substitution of these values (one at a time) into the homogeneous equations (c) produces the mode shapes

$$\mathbf{\Phi}_1 = \begin{bmatrix} 3 \\ 1 \end{bmatrix} \qquad \mathbf{\Phi}_2 = \begin{bmatrix} -1 \\ 3 \end{bmatrix} \tag{g}$$

which are scaled arbitrarily. These shapes are depicted in Fig. 3.12(b) and (c).

### Example 3.2

A cantilever beam consisting of one flexural element appears in Fig. 3.13(a). It is fixed at node 1 but has two degrees of freedom at node 2. Assuming that the beam is prismatic, determine the angular frequencies and mode shapes, using translational consistent mass terms.

From Eq. (3.4-24), we find the 2 × 2 stiffness matrix for the free displacements to be

$$\mathbf{S} = \frac{2EI}{L^3}\begin{bmatrix} 6 & -3L \\ -3L & 2L^2 \end{bmatrix} \tag{h}$$

Also, Eq. (3.4-26) gives the consistent-mass matrix as

$$\mathbf{M}_t = \frac{\rho AL}{210}\begin{bmatrix} 78 & -11L \\ -11L & 2L^2 \end{bmatrix} \tag{i}$$

Letting $s = 2EI/L^3$ and $m = \rho AL/210$, we substitute matrices $\mathbf{S}$ and $\mathbf{M}_t$ into Eq. (4) to obtain the homogeneous equations

$$\begin{bmatrix} 6(s - 13m\omega_i^2) & -L(3s - 11m\omega_i^2) \\ -L(3s - 11m\omega_i^2) & 2L^2(s - m\omega_i^2) \end{bmatrix}\begin{bmatrix} \Phi_{1i} \\ \Phi_{2i} \end{bmatrix} = \begin{bmatrix} 0 \\ 0 \end{bmatrix} \tag{j}$$

where the 2 × 2 coefficient matrix is the characteristic matrix $\mathbf{H}_i$. As indicated by Eq. (5), we set the determinant of $\mathbf{H}_i$ equal to zero, producing

$$35m^2\omega_i^4 - 102ms\omega_i^2 + 3s^2 = 0 \tag{k}$$

which is the characteristic equation. Then the roots of Eq. (k) are found to be

## Sec. 3.6  Vibrational Analysis

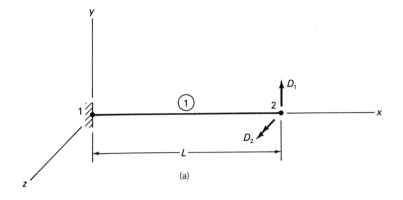

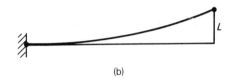

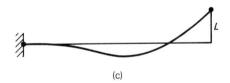

**Figure 3.13** (a) Cantilever beam; (b) mode 1; (c) mode 2.

$$\omega_1^2 = 0.02971 \frac{s}{m} \qquad \omega_2^2 = 2.885 \frac{s}{m} \tag{$\ell$}$$

Substituting the known values of $s$ and $m$ and taking square roots, we find the angular frequencies, as follows:

$$\omega_1 = \frac{3.533}{L^2} \sqrt{\frac{EI}{\rho A}} \qquad \omega_2 = \frac{34.81}{L^2} \sqrt{\frac{EI}{\rho A}} \tag{m}$$

When these formulas are compared with exact values [9], the errors are found to be $e_1 = +0.48\%$ and $e_2 = +58\%$. Thus, the first-mode frequency is a good approximation, but the second-mode frequency is very poor.

In this example we obtain the mode shapes by using the first column of the adjoint matrix $\mathbf{H}_i^a$, which is

$$\mathbf{H}_{1i}^a = \begin{bmatrix} 2L^2(s - m\omega_i^2) \\ L(3s - 11m\omega_i^2) \end{bmatrix} \tag{n}$$

Substitution of $\omega_1^2$ and $\omega_2^2$ from Eqs. ($\ell$) into this column yields

$$\boldsymbol{\Phi}_1 = \begin{bmatrix} L \\ 1.378 \end{bmatrix} \qquad \boldsymbol{\Phi}_2 = \begin{bmatrix} L \\ 7.622 \end{bmatrix} \tag{o}$$

Of course, the second column of $\mathbf{H}_i^a$ would serve equally well. Figures 3.13(b) and (c) show the mode shapes in Eqs. (o), which are both scaled so that the translation is numerically equal to $L$.

## 3.7 SYMMETRIC AND ANTISYMMETRIC MODES

Figure 3.14 shows two examples of symmetric structures. The plane frame in part (a) of the figure has one plane of symmetry, while the discretized plate in part (b) has two such planes, as indicated by the centerlines. When a structure has one or more planes of symmetry, the natural mode shapes for vibrations all will be either symmetric or antisymmetric with respect to those planes [12]. In problems of this type we need only analyze a portion of the original structure. The reduction to a smaller sized problem may be accomplished by introducing artificial restraints at joints located on planes of symmetry. In addition, the properties of members that lie in those planes must be altered. These changes may be incorporated into the structural data for a computer program and do not require any additional coding. If there is one plane of symmetry [as in Fig. 3.14(a)], only half of the structure need be analyzed. If two planes of symmetry exist [as in Fig. 3.14(b)], only a quarter need be analyzed, and so on.

When a vibrational mode is *symmetric* with respect to a plane of structural symmetry, the nodal displacements, strains, stresses, and reactions will also be symmetric with respect to the same plane. Therefore, nodes located on a plane of symmetry must be restrained in such a manner that the structure deforms symmetrically with respect to that plane. Figure 3.15(a) illustrates schematically a typical node $j$ located on a plane of symmetry that is normal to the $x$ axis. The figure also shows nodes $k$ and $k'$ that are symmetrically located on opposite sides of the plane. Displacement vectors at each of these nodes indicate a symmetric pattern of deformation. Note that translations in the $y$ and $z$ directions and rotations in the $x$ sense are all in positive directions at both points $k$ and $k'$. Therefore, we conclude that the same displacements on the plane of symmetry must be free to occur. These displacement vectors at point $j$ are labeled $j2$, $j3$, and $j4$. On the other hand, the translations in the $x$ direction and the rotations in the $y$ and $z$ senses are in opposite directions at points $k$ and $k'$. Thus, the same displacements on the plane of symmetry must be set equal to zero. Hence, the vectors labeled $j1$, $j5$, and $j6$ at point $j$ need to be restrained, as indicated by the small slashes on their arrows. In general, the component of nodal translation normal to a plane of symmetry and the components of rotation in the plane must be prevented in order to enforce a symmetric pattern of distortion.

## Sec. 3.7 Symmetric and Antisymmetric Modes

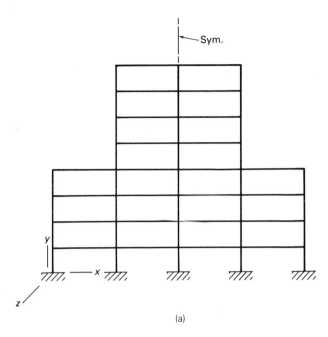

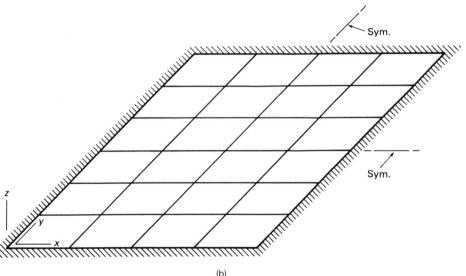

**Figure 3.14** Symmetric structures: (a) frame; (b) plate.

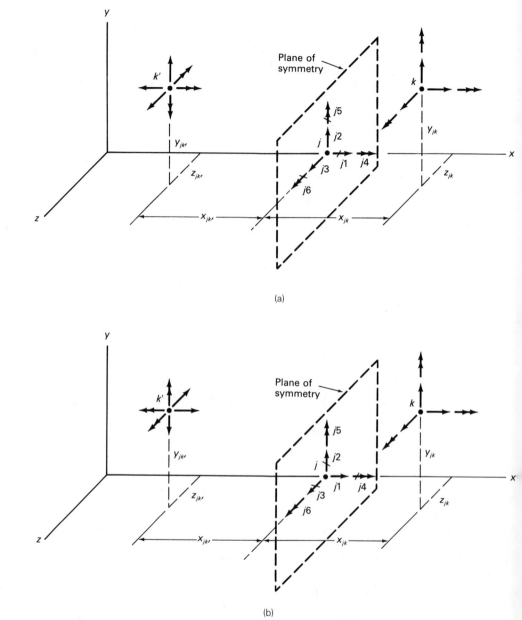

**Figure 3.15** Restraints on planes of symmetry: (a) symmetric modes; (b) antisymmetric modes.

## Sec. 3.7  Symmetric and Antisymmetric Modes

If a vibrational mode is *antisymmetric* with respect to a plane of structural symmetry, the nodal displacements, strains, stresses, and reactions will also be antisymmetric with respect to the same plane. For this case Fig. 3.15(b) depicts displacements at points $k$ and $k'$ that represent an antisymmetric pattern of deformation. That is, the translations in the $x$ direction and the rotations in the $y$ and $z$ senses are all in positive directions at both points $k$ and $k'$. From this we conclude that the same displacements on the plane of symmetry must be allowed to occur freely. They are the displacements labeled $j1$, $j5$, and $j6$ at point $j$. On the other hand, the translations in the $y$ and $z$ directions and the rotations in the $x$ sense are in opposite directions at point $k$ and $k'$. Therefore, the same displacements on the plane of symmetry need to be set equal to zero. This may be accomplished by introducing restraints corresponding to $j2$, $j3$, and $j4$, as indicated by the slashes on their vectors. In summary, the components of nodal translation in a plane of symmetry and the component of rotation normal to the plane must be prevented to give a pattern of distortion that is antisymmetric with respect to the plane.

If a member of a framed structure or an element in a discretized continuum lies in a plane of symmetry, we must divide its rigidities by two in order to cut the structure into equal parts. In the case where a member lies in two planes of symmetry, we need to divide its rigidities by four, and so on. If a member or a finite element is normal to and bisected by a plane of symmetry, we must divide it into two equal parts and introduce new nodes on the bisecting plane that are restrained as described above.

**Example 3.3**

Figure 3.16(a) shows a simply supported beam composed of two flexural elements. This beam has four degrees of freedom and is symmetric with respect to its centerline, as indicated in the figure. To take advantage of symmetry, we shall analyze only the right-hand half, using restraints at node 2 for symmetric and antisymmetric deformations. Figure 3.16(b) illustrates the symmetric case, for which the rotation $D_3$ in the plane of symmetry is restrained. In addition, the antisymmetric case is given in Fig. 3.16(c), where the translation $D_2$ in the plane of symmetry is restrained. For each of these cases we have only two degrees of freedom instead of the four degrees of freedom in the original problem. Now let us find two angular frequencies and mode shapes from each of the subsidiary problems, using translational consistent mass terms.

For the symmetric case in Fig. 3.16(b), the $2 \times 2$ stiffness matrix for the free displacements ($D_2$ and $D_4$) is

$$\mathbf{S} = \frac{2EI}{\ell^3} \begin{bmatrix} 6 & 3\ell \\ 3\ell & 2\ell^2 \end{bmatrix} \tag{a}$$

which is drawn from Eq. (3.4-24). Also, the $2 \times 2$ consistent-mass matrix becomes

$$\mathbf{M}_t = \frac{\rho A \ell}{420} \begin{bmatrix} 156 & -13\ell \\ -13\ell & 4\ell^2 \end{bmatrix} \tag{b}$$

as given by Eq. (3.4-26). The characteristic matrix $\mathbf{H}_i$ for this case has the form

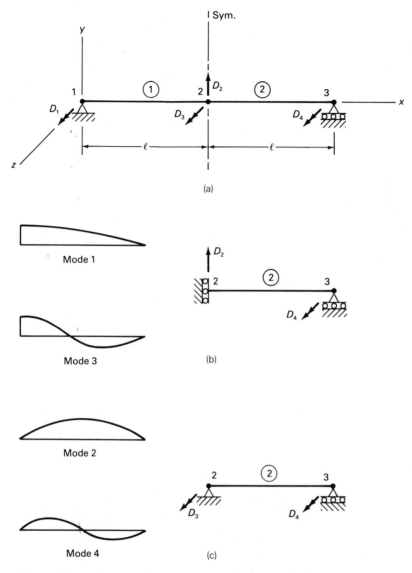

**Figure 3.16** (a) Symmetric beam; (b) symmetric modes; (c) antisymmetric modes.

$$\mathbf{H}_i = \mathbf{S} - \omega_i^2 \mathbf{M}_t = \begin{bmatrix} 6(s - 26m\omega_i^2) & \ell(3s + 13m\omega_i^2) \\ \ell(3s + 13m\omega_i^2) & 2\ell^2(s - 2m\omega_i^2) \end{bmatrix} \quad (c)$$

where $s = 2EI/\ell^3$ and $m = \rho A \ell/420$. Setting the determinant of matrix $\mathbf{H}_i$ equal to zero gives

$$455m^2\omega_i^4 - 414ms\omega_i^2 + 3s^2 = 0 \quad (d)$$

which is the characteristic equation. The roots of Eq. (d) are

### Sec. 3.7 Symmetric and Antisymmetric Modes

$$\omega_1^2 = 0.007305 \frac{s}{m} \qquad \omega_3^2 = 0.9026 \frac{s}{m} \qquad (e)$$

Substituting the known values of $s$ and $m$ and taking square roots, we find that

$$\omega_1 = \frac{9.909}{L^2} \sqrt{\frac{EI}{\rho A}} \qquad \omega_3 = \frac{110.1}{L^2} \sqrt{\frac{EI}{\rho A}} \qquad (f)$$

where $L = 2\ell$. When these formulas for the angular frequencies are compared with exact values [9], the errors are found to be $e_1 = +0.40\%$ and $e_3 = +24\%$.

We obtain mode shapes corresponding to $\omega_1$ and $\omega_3$ using the first column of the adjoint matrix $\mathbf{H}_i^a$, as follows:

$$\mathbf{H}_{1i}^a = \begin{bmatrix} 2\ell^2(s - 2m\omega_i^2) \\ -\ell(3s + 13m\omega_i^2) \end{bmatrix} \qquad (g)$$

Substitution of $\omega_1^2$ and $\omega_3^2$ from Eqs. (e) into this column produces

$$\mathbf{\Phi}_1 = \begin{bmatrix} \ell \\ -1.570 \end{bmatrix} \qquad \mathbf{\Phi}_3 = \begin{bmatrix} \ell \\ 9.149 \end{bmatrix} \qquad (h)$$

These mode shapes appear in the left-hand portion of Fig. 3.16(b). Of course, each of them represents half of a symmetric mode shape for the whole beam.

Considering now the antisymmetric case in Fig. 3.16(c), we form the stiffness matrix for the free displacements $D_3$ and $D_4$ as

$$\mathbf{S} = \frac{2EI}{\ell} \begin{bmatrix} 2 & 1 \\ 1 & 2 \end{bmatrix} \qquad (i)$$

and the consistent mass matrix is

$$\mathbf{M}_t = \frac{\rho A \ell^3}{420} \begin{bmatrix} 4 & -3 \\ -3 & 4 \end{bmatrix} \qquad (j)$$

Then the characteristic matrix becomes

$$\mathbf{H}_i = \begin{bmatrix} 2(s - 2m\omega_i^2) & s + 3m\omega_i^2 \\ s + 3m\omega_i^2 & 2(s - 2m\omega_i^2) \end{bmatrix} \qquad (k)$$

where $s$ and $m$ are the same as before. Expanding the determinant of $\mathbf{H}_i$ and setting it equal to zero gives the characteristic equation

$$7m^2\omega_i^4 - 22ms\omega_i^2 + 3s^2 = 0 \qquad (\ell)$$

from which the roots are

$$\omega_2^2 = \frac{1}{7}\frac{s}{m} \qquad \omega_4^2 = 3\frac{s}{m} \qquad (m)$$

Proceeding as before, we find the angular frequencies to be

$$\omega_2 = \frac{43.82}{L^2} \sqrt{\frac{EI}{\rho A}} \qquad \omega_4 = \frac{200.8}{L^2} \sqrt{\frac{EI}{\rho A}} \qquad (n)$$

for which the errors are $+11\%$ and $+27\%$.

Mode shapes are given by the first column of $\mathbf{H}_i^a$, which is

$$\mathbf{H}_{1i}^a = \begin{bmatrix} 2(s - 2m\omega_i^2) \\ -(s + 3m\omega_i^2) \end{bmatrix} \tag{o}$$

Substituting $\omega_2^2$ and $\omega_4^2$ from Eqs. (m) into this vector yields

$$\mathbf{\Phi}_2 = \begin{bmatrix} 1 \\ -1 \end{bmatrix} \qquad \mathbf{\Phi}_4 = \begin{bmatrix} 1 \\ 1 \end{bmatrix} \tag{p}$$

These mode shapes are displayed in the left-hand part of Fig. 3.16(c), where each of them depicts half of an antisymmetric mode for the whole beam.

The error calculated for the first angular frequency $\omega_1$ is acceptable, while those for the other modes are not. Better accuracy for these modes could be obtained by using more finite elements with more nodal degrees of freedom.

## 3.8 PROGRAM VIB FOR VIBRATIONAL ANALYSIS

In this section we discuss a computer program named VIB for vibrational analysis of any type of linearly elastic framed structure or discretized continuum. Steps in the main program appear in Flowchart 3.1, which calls seven subprograms indicated by the names in double boxes. Subprogram SDAT reads and writes input data for a particular type of structure and calculates nodal displacement indexes. Subprogram STIF generates the structural stiffness matrix (for free nodal displacements only) by assembling contributions from element stiffnesses, as indicated by the first of Eqs. (3.5-14). Next, the consistent mass matrix for free nodal accelerations in the structure is assembled by Subprogram CMAS, using contributions from individual elements [see the second of Eqs. (3.5-14)]. The subprogram named STASYM then converts the eigenvalue problem to standard, symmetric form by factoring the structural stiffness matrix, as shown in Eq. (3.6-9). If the stiffness matrix is found not to be positive definite, the mass matrix is factored instead. However, if the mass matrix is also found not to be positive definite, an error message is written and calculations stop. Otherwise, the subprogram EIGEN2 solves the eigenvalue problem using Householder transformations and the QR algorithm (see Sec. B.2). Then the eigenvectors are transformed back to the original coordinates with Subprogram TRAVEC, using the second of Eqs. (3.6-12). Finally, the subprogram named RES1 writes the resulting angular frequencies and mode shapes obtained from solution of the eigenvalue problem. As shown by the flowchart, several structures of the same type may be processed in one run of the program.

Program VIB may be specialized to become VIBCB for continuous beams, VIBPT for plane trusses, and so on. The main program for each specialization has four subprograms that are different for each type of structure, as indicated by the second footnote below Flowchart 3.1. For example, the subprogram named SDAT becomes SDATCB for a continuous beam, SDATPT for a plane

Sec. 3.8  Program VIB for Vibrational Analysis

**Flowchart 3.1  Main program for VIB***

101
- SDAT † → 1. Read and write structural data.
- STIF † → 2. Generate structural stiffness matrix.
- CMAS † → 3. Generate structural consistent mass matrix.
- STASYM → 4. Convert eigenvalue problem to standard, symmetric form.
- EIGEN2 → 5. Solve eigenvalue problem by Sec. B.2.
- TRAVEC → 6. Transform eigenvectors to original coordinates.
- RES1 † → 7. Write angular frequencies and mode shapes.
- (101) Go to 101 and process another structure.

END

---

*Applies to any type of linearly elastic structure.
† Subprograms that differ for every type of structure.

truss, and so on. Notation for this and other programs is given as Part 5 in the list of notation near the end of the book. Detailed steps in the logic for various subprograms are shown in the flowchart for Program DYNAPT, which appears in Appendix C.

Table 3.2 shows preparation of structural data for plane trusses. In the second line of the table are the number of nodes NN, the number of elements NE, the number of restrained nodes NRN, the modulus of elasticity E, and the mass density RHO. Each line of the data for nodal coordinates (NN lines total) contains a node number J, the $x$ coordinate $X(J)$ of the node, and the $y$ coordinate

TABLE 3.2  Structural Data for Plane Trusses

| Type of Data | No. of Lines | Items on Data Lines |
|---|---|---|
| Problem identification | 1 | Descriptive title |
| Structural parameters | 1 | NN, NE, NRN, E, RHO |
| Plane truss data | | |
| (a) Nodal coordinates | NN | J, X(J), Y(J) |
| (b) Element information | NE | I, JN(I), KN(I), AX(I) |
| (c) Nodal restraints | NRN | J, NRL(2J–1), NRL(2J) |

Y(J). The element information (NE lines) consists of the element number I, the $j$ node JN(I) at one end, the $k$ node KN(I) at the other end, and the cross-sectional area AX(I).

Each of the NRN lines in the last block of data contains a node number J and two code numbers which indicate the conditions of restraint at that node. The symbol NRL(2J-1) denotes the condition of restraint against translation in the $x$ direction at node J, and the term NRL(2J) gives the restraint against translation in the $y$ direction. The convention adopted in this book is that if the restraint exists, the integer 1 is assigned as the value of NRL; but if there is no restraint, a value of zero is assigned. Of course, the vector NRL must initially contain only zeros (by clearing it) before the restraint information is read.

If desired, we could include data for external masses that are idealized to be concentrated at the nodes of the structure. Such extra masses may be conveniently added to the data lines for nodal coordinates. When the program assembles the consistent mass matrix for the structure, these concentrated masses would be added to diagonal terms for translational accelerations. Such a procedure for handling superimposed masses could be applied to any type of framed structure or discretized continuum.

### Example 3.4

We will now use the specialized program VIBPT to find frequencies and mode shapes for the plane truss with three members examined previously in Sec. 3.5. This truss is reproduced in Fig. 3.17(a), where the free and restrained nodal translations are indicated by numbered arrows. Structural data for this problem are listed in Table 3.3, in which the following numerical values are assigned to the parameters $E$, $\rho$, $L$, and $A$:

$$E = 3.0 \times 10^4 \text{ k/in.}^2 \qquad \rho = 7.35 \times 10^{-7} \text{ k-s}^2/\text{in.}^4$$
$$L = 250 \text{ in.} \qquad A = 10 \text{ in.}^2$$

where the material is steel and US units are used. (See Appendix A for a discussion of systems of units and material properties.)

Table 3.4 contains the computer results for this example. In the first part of the table, we see an "echo" print of the data read by the computer. Also computed and printed are the element lengths EL, the direction cosine CX and CY, the number of degrees of

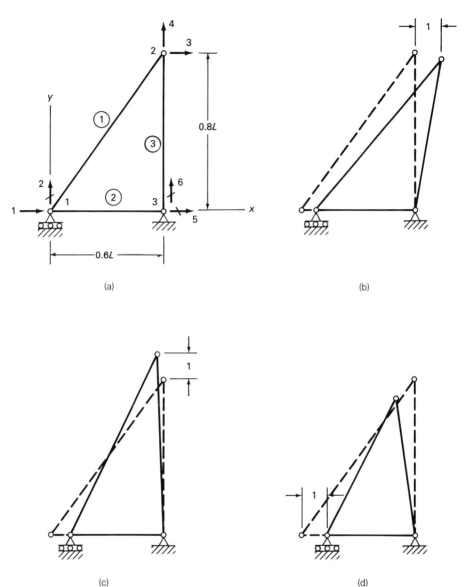

**Figure 3.17** (a) Plane truss; (b) mode 1; (c) mode 2; (d) mode 3.

freedom NDF, and the number of nodal restraints NNR. This is followed by the angular frequencies and mode shapes for each of the natural modes of vibration, which may be confirmed by hand calculations. The angular frequency for mode 1 is $\omega_1 = 420.0$ rad/sec (or s$^{-1}$), that for mode 2 is $\omega_2 = 1168$ s$^{-1}$, and that for mode 3 is $\omega_3 = 1862$ s$^{-1}$. Each modal vector has been normalized with respect to its largest term, and the mode shapes are depicted in Fig. 3.17(b)–(d).

**TABLE 3.3  Structural Data for Example 3.4**

| Type of Data | Alphanumerical Values |
|---|---|
| Problem identification<br>Structural parameters | Example 3.4: Three-member plane truss<br>3  3  2  30000.0  0.000000735 |
| Plane truss data<br>(a) Nodal coordinates | 1  0.0  0.0<br>2  150.0  200.0<br>3  150.0  0.0 |
| (b) Element information | 1  1  2  10.0<br>2  1  3  6.0<br>3  2  3  8.0 |
| (c) Nodal restraints | 1  0  1<br>3  1  1 |

**TABLE 3.4  Computer Output for Example 3.4**

```
PROGRAM VIBPT

*** EXAMPLE 3.4: THREE-MEMBER PLANE TRUSS ***

STRUCTURAL PARAMETERS
   NN   NE  NRN           E         RHO
    3    3    2   3.0000E+04   7.3500E-07

NODAL COORDINATES
  NODE          X              Y
    1        0.000          0.000
    2      150.000        200.000
    3      150.000          0.000

ELEMENT INFORMATION
ELEM.    J    K           AX         EL        CX        CY
  1      1    2       10.0000   250.0000    0.6000    0.8000
  2      1    3        6.0000   150.0000    1.0000    0.0000
  3      2    3        8.0000   200.0000    0.0000   -1.0000

NODAL RESTRAINTS
 NODE   NR1  NR2
   1     0    1
   3     1    1

NUMBER OF DEGREES OF FREEDOM:  NDF =    3
NUMBER OF NODAL RESTRAINTS:    NNR =    3

MODE     1
ANGULAR FREQUENCY   4.1995E+02
  NODE        DJ1            DJ2
    1     2.3137E-01     0.0000E+00
    2     1.0000E+00    -2.4722E-01
    3     0.0000E+00     0.0000E+00
```

### Sec. 3.8  Program VIB for Vibrational Analysis

TABLE 3.4 (Continued)

```
MODE       2
ANGULAR FREQUENCY   1.1677E+03
   NODE         DJ1              DJ2
      1   8.6725E-01       0.0000E+00
      2  -1.7149E-01       1.0000E+00
      3   0.0000E+00       0.0000E+00

MODE       3
ANGULAR FREQUENCY   1.8618E+03
   NODE         DJ1              DJ2
      1   1.0000E+00       0.0000E+00
      2  -6.0504E-01      -6.1068E-01
      3   0.0000E+00       0.0000E+00
```

**Example 3.5**

Figure 3.18(a) shows a plane truss with a larger number of members. The cross-sectional areas of diagonal members are equal to $1.5A$, and those of other members are equal to $A$. This truss happens to be symmetric with respect to its centerline, so we need only analyze half the structure. For this purpose, we impose restraints in the plane of symmetry for symmetric and antisymmetric deformations, as indicated in Figs. 3.18(b) and (c). In addition, the cross-sectional area of member 12 is divided by 2. Assuming that

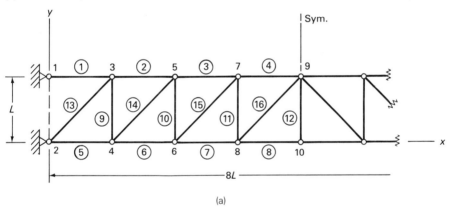

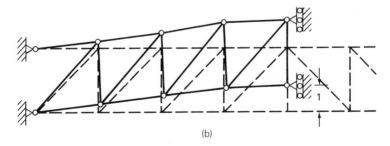

**Figure 3.18** (a) Plane truss; (b) mode 1 (symmetric); (c) mode 2 (antisymmetric).

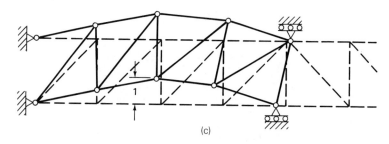

**Figure 3.18** (*cont.*)

the truss is aluminum, we give parameters the following numerical values:

$$E = 6.9 \times 10^7 \text{ kPa} \qquad \rho = 2.62 \text{ Mg/m}^3$$
$$L = 5 \text{ m} \qquad A = 6 \times 10^{-3} \text{ m}^2$$

where SI units are implied (again, see Appendix A).

To process this truss with Program VIBPT, we must analyze half the structure twice. In the first analysis, restraints on the plane of symmetry allow only symmetric modes of vibration; and the second analysis uses restraint data for only antisymmetric modes. Figures 3.18(b) and (c) illustrate the mode shapes corresponding to the first and second angular frequencies $\omega_1 = 79.55 \text{ s}^{-1}$ and $\omega_2 = 168.9 \text{ s}^{-1}$. We see that the first mode is symmetric with respect to the plane of symmetry, while the second mode is antisymmetric.

## REFERENCES

1. Weaver, W., Jr., and Johnston, P. R., *Finite Elements for Structural Analysis*, Prentice-Hall, Englewood Cliffs, N.J., 1984.
2. Zienkiewicz, O. C., *The Finite Element Method*, 4th ed., McGraw-Hill, Maidenhead, Berkshire, England, 1987.
3. Cook, R. D., *Concepts and Applications of Finite Element Analysis*, 2nd ed., Wiley, New York, 1981.
4. Timoshenko, S. P., and Goodier, J. N., *Theory of Elasticity*, 3rd ed., McGraw-Hill, New York, 1970.
5. Weaver, W., Jr., and Gere, J. M., *Matrix Analysis of Framed Structures*, 2nd ed., Van Nostrand Reinhold, New York, 1980.
6. Oden, J. T., *Mechanics of Elastic Structures*, McGraw-Hill, New York, 1967.
7. Archer, J. S., "Consistent Matrix Formulations for Structural Analysis Using Finite-Element Techniques," *AIAA J.*, Vol. 3, No. 10, 1965, pp. 1910–1918.
8. Clough, R. W., "Analysis of Structural Vibrations and Dynamic Response," *Rec. Adv. Mat. Methods Struct. Anal. Des.*, ed. R. H. Gallagher, Y. Yamada, and J. T. Oden, University of Alabama Press, Huntsville, Ala., 1971, pp. 25–45.

9. Timoshenko, S. P., Young, D. H., and Weaver, W., Jr., *Vibration Problems in Engineering,* 4th ed., Wiley, New York, 1974.
10. Gere, J. M., and Weaver, W., Jr., *Matrix Algebra for Engineers,* 2nd ed., Brooks/Cole, Monterey, Calif., 1983.
11. Hohn, F. E., *Elementary Matrix Algebra,* 3rd ed., Macmillan, New York, 1973.
12. Glockner, P. G., "Symmetry in Structural Mechanics," *ASCE, J. Struct. Div.,* Vol. 99, No. ST1, 1973, pp. 71–89.

## PROBLEMS

**3.4-1.** Figure P3.4-1 shows an axial element with a linearly distributed load (force per unit length) given by the formula $b_x = b_1 + (b_2 - b_1)x/L$. Find the equivalent nodal loads $\mathbf{p}_b(t) = \{p_{b1}, p_{b2}\}$ due to this influence.

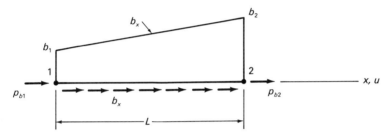

**Figure P3.4-1**

**3.4-2.** A parabolically distributed load (force per unit length) has the formula $b_x = b_2(x/L)^2$, as illustrated in Fig. P3.4-2. Determine the equivalent nodal loads $\mathbf{p}_b(t) = \{p_{b1}, p_{b2}\}$ resulting from this body force.

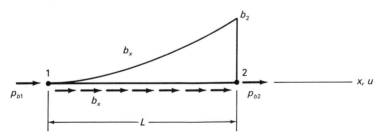

**Figure P3.4-2**

**3.4-3.** Assume that an axial element has three nodes, as shown in Fig. P3.4-3(a). In terms of the coordinate $x$ measured from node 2, the quadratic displacement shape functions in parts (b)–(d) of the figure are: $f_1 = (2x - L)x/L^2$, $f_2 = (L^2 - 4x^2)/L^2$, and $f_3 = (2x + L)x/L^2$. Derive the 3 × 3 stiffness matrix $\mathbf{K}$ for this element if the axial rigidity $EA$ is constant along the length.

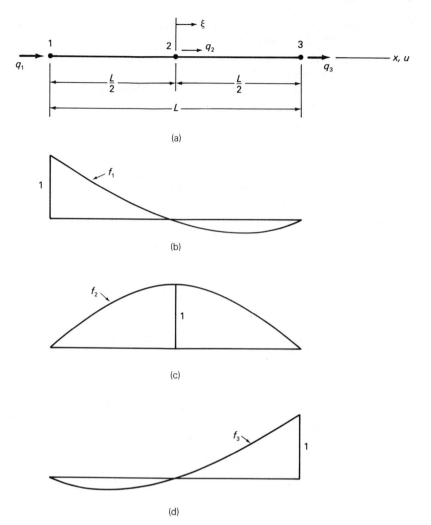

**Figure P3.4-3**

**3.4-4.** For the axial element with three nodes [Fig. P3.4-3(a)], derive the 3 × 3 consistent-mass matrix **M**, assuming that $\rho$ and $A$ are constant along the length.

**3.4-5.** Let a uniformly distributed load $b_x$ (force per unit length) be applied to the axial element with three nodes [Fig. P3.4-3(a)]. Find the equivalent nodal loads $\mathbf{p}_b(t) = \{p_{b1}, p_{b2}, p_{b3}\}$ due to this body force.

**3.4-6.** For the torsional element shown in Fig. P3.4-6, obtain the equivalent nodal loads $\mathbf{p}_b(t) = \{p_{b1}, p_{b2}\}$ caused by a parabolically distributed moment (per unit length) given as $m_x = m_{x1}[1 - (x/L)^2]$.

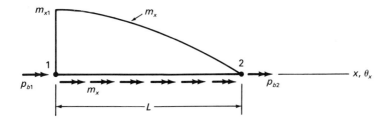

**Figure P3.4-6**

**3.4-7.** Suppose that a concentrated moment $M_x$ is applied to the torsional element at the distance $x$ from node 1, as shown in Fig. P3.4-7. Determine the equivalent nodal loads $\mathbf{p}_b(t) = \{p_{b1}, p_{b2}\}$ caused by this moment.

**Figure P3.4-7**

**3.4-8.** The flexural element shown in Fig. P3.4-8 is subjected to a triangular load $b_y = b_2 x/L$ (force per unit length). Derive the equivalent nodal loads $\mathbf{p}_b(t) = \{p_{b1}, p_{b2}, p_{b3}, p_{b4}\}$ indicated at points 1 and 2.

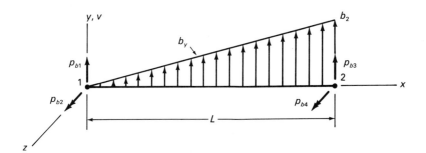

**Figure P3.4-8**

**3.4-9.** Figure P3.4-9 depicts a concentrated force $P_y$ and a concentrated moment $M_z$ applied to a flexural element at the distance $x$ from node 1. Obtain the equivalent nodal loads $\mathbf{p}_b(t) = \{p_{b1}, p_{b2}, p_{b3}, p_{b4}\}$ for each of these actions.

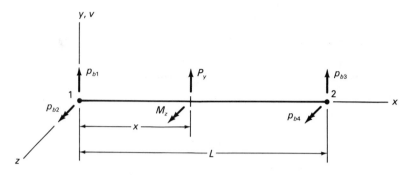

**Figure P3.4-9**

**3.4-10.** In Fig. P3.4-10 a linearly distributed load $b_y = b_1 + (b_2 - b_1)x/L$ (force per unit length) acts on a flexural element. Find the equivalent nodal loads $\mathbf{p}_b(t) = \{p_{b1}, p_{b2}, p_{b3}, p_{b4}\}$ due to this influence.

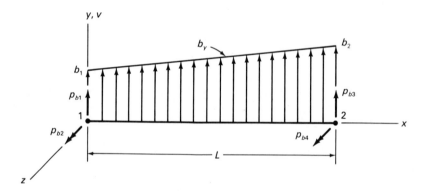

**Figure P3.4-10**

**3.4-11.** Rederive the $2 \times 2$ stiffness matrix $\mathbf{K}$ for a torsional element, using moment $M_x$ and twist $\psi$ as generalized stress and strain. Assume that the torsional rigidity $GJ$ is constant along the length.

**3.4-12.** Derive again the $4 \times 4$ stiffness matrix $\mathbf{K}$ for a flexural element, with moment $M_z$ and curvature $\phi$ as generalized stress and strain. Let the flexural rigidity $EI$ be constant along the length.

**3.5-1.** The plane truss shown in Fig. P3.5-1 has cross-sectional areas of $0.6A$ and $A$ for members 1 and 2. For this structure find the stiffness matrix $\mathbf{S}_s$ and the consistent mass matrix $\mathbf{M}_s$ in rearranged and partitioned forms.

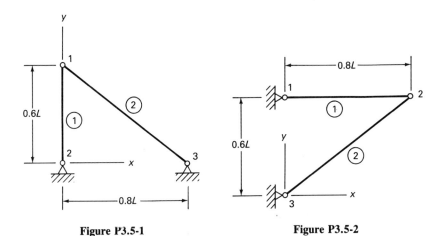

Figure P3.5-1        Figure P3.5-2

**3.5-2.** Repeat Prob. 3.5-1 for the plane truss shown in Fig. P3.5-2, assuming that cross-sectional areas of members 1 and 2 are $0.8A$ and $A$.

**3.5-3.** For the plane truss shown in Fig. P3.5-3, repeat Prob. 3.5-1. In this case assume that the cross-sectional areas of members 1 through 4 are equal to $A$ and that the area for member 5 is $\sqrt{2}\,A$.

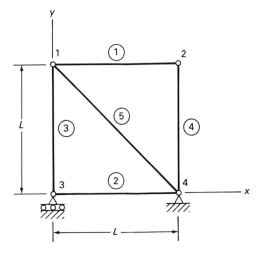

Figure P3.5-3

**3.5-4.** Repeat Prob. 3.5-1 for the plane truss shown in Fig. P3.5-4. Assume that the cross-sectional areas of members 1 and 2 are $0.8A$, those of members 3 and 4 are $0.6A$, and that of member 5 is equal to $A$.

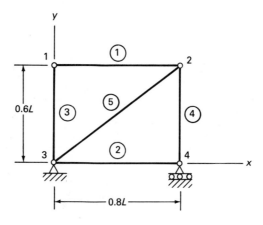

**Figure P3.5-4**

**3.5-5.** For the plane truss shown in Fig. P3.5-5, repeat Prob. 3.5-1. Assume that the cross-sectional areas of members 1 and 2 are $0.8A$, those for members 3 and 4 are $0.6A$, and those for members 5 and 6 are equal to $A$.

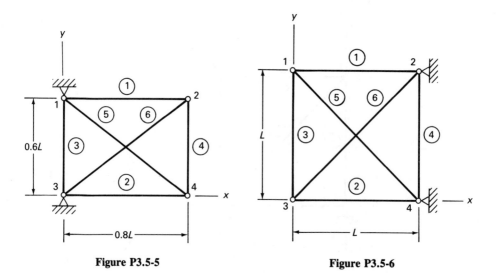

**Figure P3.5-5**     **Figure P3.5-6**

**3.5-6.** Repeat Prob. 3.5-1 for the plane truss shown in Fig. P3.5-6. In this case let the cross-sectional areas for members 1 through 4 be equal to $A$, while those for members 5 and 6 are equal to $\sqrt{2}\,A$.

**3.5-7.** Figure P3.5-7 shows a two-element beam for which the parameters $E$, $I$, $A$, and $\rho$ are constant along the length. Assemble the stiffness matrix $\mathbf{S}_s$ and the consistent-mass matrix $\mathbf{M}_s$ (for translational inertias) in rearranged and partitioned forms.

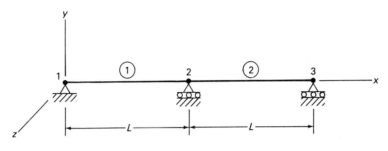

**Figure P3.5-7**

**3.5-8.** Repeat Prob. 3.5-7 for the two-element beam shown in Fig. P3.5-8.

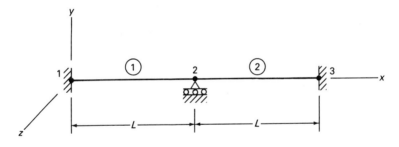

**Figure P3.5-8**

**3.5-9.** For the two-element beam shown in Fig. P3.5-9, repeat Prob. 3.5-7.

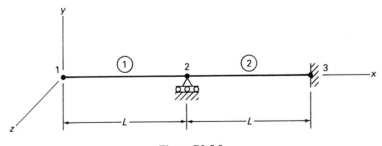

**Figure P3.5-9**

**3.5-10.** Repeat Prob. 3.5-7 for the two-element beam shown in Fig. P3.5-10.

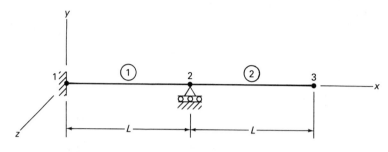

**Figure P3.5-10**

**3.6-1.** Assuming that the mass matrix **M** is positive-definite and not diagonal, factor it into $\mathbf{M} = \mathbf{V}^T\mathbf{V}$. Then transform Eq. (3.6-4) to the standard, symmetric form of Eq. (3.6-8). Also, show the back-transformation of eigenvectors to the original coordinates.

**3.6-2.** Figure P3.6-2 shows a plane truss with two degrees of freedom at joint 1. Assume that the cross-sectional areas of members 1 and 2 are equal to $A$ and $A\sqrt{2}$. By the consistent-mass method, find the angular frequencies and mode shapes for this structure.

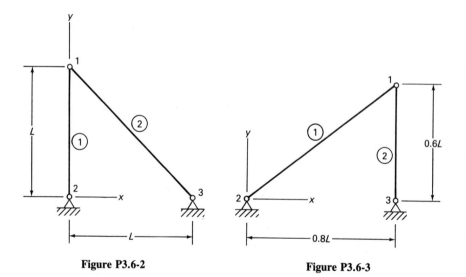

**Figure P3.6-2**  **Figure P3.6-3**

**3.6-3.** Repeat Prob. 3.6-2 for the plane truss shown in Fig. P3.6-3, but let the cross-sectional areas of members 1 and 2 be equal to $A$ and $0.6A$.

**3.6-4.** Repeat Prob. 3.6-2 for the plane truss shown in Fig. P3.6-4, but let the cross-sectional areas of members 1 and 2 be equal to $0.8A$ and $A$.

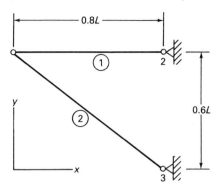
**Figure P3.6-4**

**3.6-5.** The beam shown in Fig. P3.6-5 consists of two prismatic flexural elements with two degrees of freedom at node 2. Member 1 has moment of inertia and cross-sectional area equal to $I$ and $A$, but member 2 has $2I$ and $2A$ for its properties. Determine the angular frequencies and mode shapes for this beam using translational consistent-mass terms.

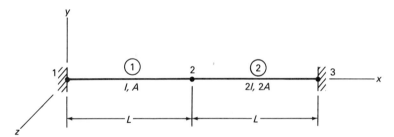

**Figure P3.6-5**

**3.6-6.** Repeat Prob. 3.6-5 for the continuous beam shown in Fig. P3.6-6. In this case the beam has constant values of $I$ and $A$ along its length and has two rotational degrees of freedom (at points 2 and 3).

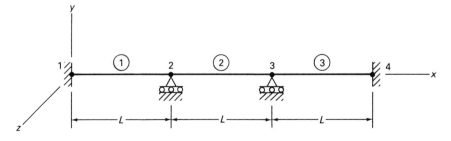

**Figure P3.6-6**

**3.6-7.** Repeat Prob. 3.6-5 for the two-element continuous beam shown in Fig. P3.6-7. Cross-sectional properties $I$ and $A$ are constant, and the free displacements at joints 1 and 2 are both rotational.

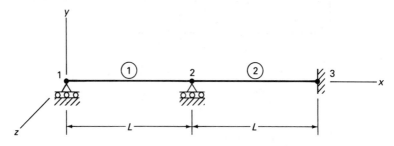

**Figure P3.6-7**

**3.7-1.** The fixed-end beam shown in Fig. P3.7-1 consists of two prismatic flexural elements having the same values of $I$ and $A$. Using only half the beam, find the angular frequencies for the **(a)** symmetric and **(b)** antisymmetric modes.

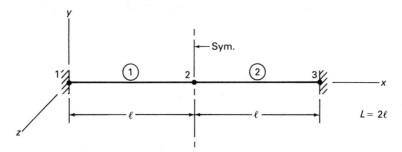

**Figure P3.7-1**

**3.7-2.** Figure P3.7-2 shows a prismatic continuous beam composed of four flexural elements. Determine the angular frequencies and mode shapes for **(a)** symmetric and **(b)** antisymmetric deformations, using only half the beam.

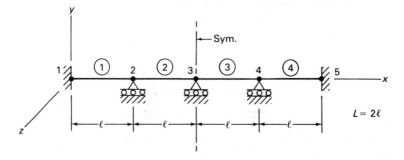

**Figure P3.7-2**

**3.7-3.** For the symmetric continuous beam shown in Fig. P3.7-3, find the angular frequencies and mode shapes for **(a)** symmetric and **(b)** antisymmetric distortions. Use only half the structure, assuming that each of the four elements has the same values of $I$ and $A$.

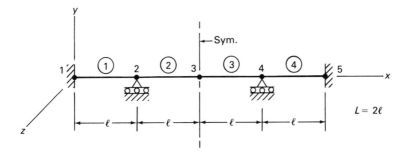

**Figure P3.7-3**

**3.7-4.** The symmetric plane truss shown in Fig. P3.7-4 has cross-sectional areas for members 1 and 2 equal to $\sqrt{2}\,A$, whereas those for members 3, 4, and 5 are equal to $A$. Using only half the structure, calculate the angular frequencies and the **(a)** symmetric and **(b)** antisymmetric mode shapes.

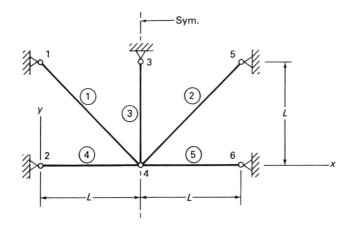

**Figure P3.7-4**

**3.7-5.** For the symmetric plane truss shown in Fig. P3.7-5, determine the angular frequencies and the **(a)** symmetric and **(b)** antisymmetric mode shapes. Use only half the structure, and assume that the cross-sectional areas of members 1 and 2 are $0.6A$, that for member 3 is $0.8A$, and those for members 4 and 5 are equal to $A$.

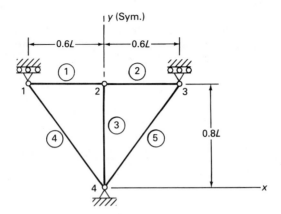

**Figure P3.7-5**

**3.7-6.** Figure P3.7-6 shows a plane truss having cross-sectional areas equal to $A$ for members 1 and 2, $0.6A$ for member 3, and $0.8A$ for members 4 and 5. Find the angular frequencies and the **(a)** symmetric and **(b)** antisymmetric mode shapes, using only half the truss.

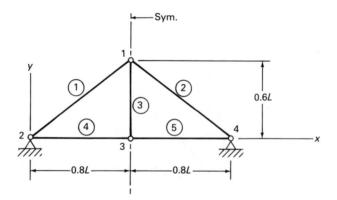

**Figure P3.7-6**

**3.7-7.** The symmetric plane truss shown in Fig. P3.7-7(a) has cross-sectional areas of all members equal to $A$. Using only half the structure, calculate the angular frequencies and the **(a)** symmetric and **(b)** antisymmetric mode shapes. For this purpose, part (b) of the figure shows symmetry restraints, and part (c) shows antisymmetry restraints. (*Note:* For the antisymmetric case, nodes 2 and 3 translate equally in the $x$ direction; so member 3 may be treated as a rigid body.)

Chap. 3 Problems

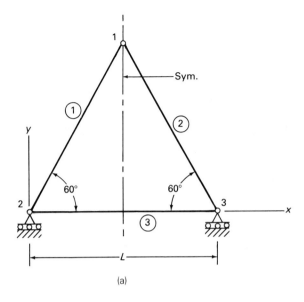

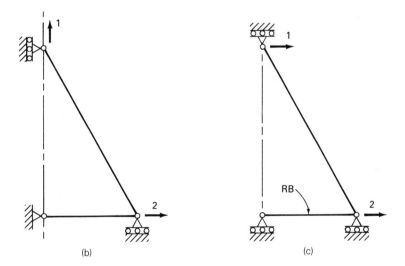

**Figure P3.7-7**

# 4

# Normal-Mode Method of Dynamic Analysis

## 4.1 INTRODUCTION

In Chapter 3 we discussed the formulation of action equations of motion for a MDOF structure using finite elements. Then we showed how a vibrational analysis can be performed by solving the algebraic eigenvalue problem associated with the homogeneous equations. Extracting the angular frequencies and mode shapes in this manner sets the stage for the *normal-mode method* of dynamic analysis [1], described in the present chapter.

A structure subjected to impulsive (time-varying) loads responds with a combination of rigid-body and flexible-body motions. If a structure is restrained (immobile), the response will involve only flexible-body motions. However, if the structure is unrestrained or partially restrained, certain rigid-body motions can occur as well. Within the linear theory of the normal-mode method, such motions may consist of small or large translations but only small rotations. Thus, problems involving large rigid-body rotations are beyond the scope of this theory.

Regarding the flexible-body part of the response, a linearly elastic solid or structure has an infinite number of degrees of freedom and an infinite number of natural modes of vibration. If the structure were analyzed as an elastic continuum, its flexible-body response to dynamic loads would consist of the sum of an infinite number of vibrational motions. However, if the structure is discretized by the finite-element method, the resulting analytical model will have only a finite number of nodal degrees of freedom and a finite number of natural modes of vibration. Therefore, such a model has only a finite number of vibrational motions contributing to its dynamic response.

Systems that are subjected to arbitrary dynamic loads become extremely difficult to analyze rigorously in their original physical coordinates. We can avoid these difficulties by using natural modes of vibration as generalized coordinates. When this path is followed, the equations of undamped motion become uncoupled. In these coordinates each equation may be solved as if it pertained to a system with only one degree of freedom. Superposition of these SDOF results is accomplished through a transformation back to the original coordinates. By this means we can evaluate time-varying nodal displacements, internal stresses, and support reactions for the analytical model.

In this chapter we develop the normal-mode method and apply it to simple structures, such as beams and plane trusses. In later chapters we will use it for more complicated framed structures and other continua discretized by finite elements. Undamped systems are treated first, and special considerations required for damped systems are discussed in the latter parts of the chapter.

An important advantage of the normal-mode method is that only the significant modal responses need be included in a dynamic analysis. The other modal responses may often be omitted without much loss of accuracy. This technique, known as *modal truncation*, can make the normal-mode method more efficient than the numerical integration methods to be described in Chapter 5.

## 4.2 PRINCIPAL AND NORMAL COORDINATES

In order to study relationships among the natural modes of vibration, let us consider modes $i$ and $j$ of the eigenvalue problem for action equations, as follows:

$$\mathbf{S} \, \boldsymbol{\Phi}_i = \omega_i^2 \mathbf{M} \, \boldsymbol{\Phi}_i \tag{1a}$$

$$\mathbf{S} \, \boldsymbol{\Phi}_j = \omega_j^2 \mathbf{M} \, \boldsymbol{\Phi}_j \tag{1b}$$

These expressions are modified versions of the algebraic eigenvalue problem given previously in Eq. (3.6-4). Premultiplication of Eq. (1a) by $\boldsymbol{\Phi}_j^T$ and postmultiplication of the transpose of Eq. (1b) by $\boldsymbol{\Phi}_i$ yields

$$\boldsymbol{\Phi}_j^T \mathbf{S} \, \boldsymbol{\Phi}_i = \omega_i^2 \boldsymbol{\Phi}_j^T \mathbf{M} \, \boldsymbol{\Phi}_i \tag{2a}$$

$$\boldsymbol{\Phi}_j^T \mathbf{S} \, \boldsymbol{\Phi}_i = \omega_j^2 \boldsymbol{\Phi}_j^T \mathbf{M} \, \boldsymbol{\Phi}_i \tag{2b}$$

The left-hand sides of Eqs. (2) are equal, so that subtraction of the second equation from the first produces the relationship

$$(\omega_i^2 - \omega_j^2) \, \boldsymbol{\Phi}_j^T \mathbf{M} \, \boldsymbol{\Phi}_i = 0 \tag{3}$$

On the other hand, if we divide both sides of Eq. (2a) by $\omega_i^2$ and both sides of

Eq. (2b) by $\omega_j^2$, the right-hand sides become equal. Then subtraction gives

$$\left(\frac{1}{\omega_i^2} - \frac{1}{\omega_j^2}\right) \mathbf{\Phi}_j^T \mathbf{S} \, \mathbf{\Phi}_i = 0 \tag{4}$$

To satisfy Eqs. (3) and (4) when $i \neq j$ and the eigenvalues are distinct $(\omega_i^2 \neq \omega_j^2)$, the following relationships must hold:

$$\mathbf{\Phi}_j^T \mathbf{M} \, \mathbf{\Phi}_i = \mathbf{\Phi}_i^T \mathbf{M} \, \mathbf{\Phi}_j = 0 \tag{5}$$

and

$$\mathbf{\Phi}_j^T \mathbf{S} \, \mathbf{\Phi}_i = \mathbf{\Phi}_i^T \mathbf{S} \, \mathbf{\Phi}_j = 0 \tag{6}$$

These expressions represent *orthogonality relationships* between the modal vectors $\mathbf{\Phi}_i$ and $\mathbf{\Phi}_j$. From Eq. (5) we see that the eigenvectors are orthogonal with respect to the mass matrix $\mathbf{M}$. Equation (6) also shows that they are orthogonal with respect to the stiffness matrix $\mathbf{S}$, as demonstrated in Sec. 3.6.

For the case when $i = j$, Eqs. (3) and (4) yield

$$\mathbf{\Phi}_i^T \mathbf{M} \, \mathbf{\Phi}_i = M_{Pi} \tag{7}$$

and

$$\mathbf{\Phi}_i^T \mathbf{S} \, \mathbf{\Phi}_i = S_{Pi} \tag{8}$$

in which $M_{Pi}$ and $S_{Pi}$ are constants that depend on how the eigenvector $\mathbf{\Phi}_i$ is normalized.

For operational efficiency, we place all of the eigenvectors columnwise into an $n \times n$ *modal matrix* of the form

$$\mathbf{\Phi} = [\mathbf{\Phi}_1 \quad \mathbf{\Phi}_2 \quad \mathbf{\Phi}_3 \quad \ldots \quad \mathbf{\Phi}_n] \tag{9}$$

where $n$ is the number of degrees of freedom. Then we can state Eqs. (5) and (7) collectively as

$$\boxed{\mathbf{\Phi}^T \mathbf{M} \, \mathbf{\Phi} = \mathbf{M}_P} \tag{10}$$

in which $\mathbf{M}_P$ is a diagonal array that will be referred to as a *principal mass matrix*. Similarly, Eqs. (6) and (8) are combined into

$$\boxed{\mathbf{\Phi}^T \mathbf{S} \, \mathbf{\Phi} = \mathbf{S}_P} \tag{11}$$

where $\mathbf{S}_P$ is another diagonal array that will be called a *principal stiffness matrix*. Equations (10) and (11) represent *diagonalization* of matrices $\mathbf{M}$ and $\mathbf{S}$. If either of them is already diagonal, the operations merely scale the values on the diagonal.

To take advantage of the diagonalization process, let us reconsider the *action equations of motion* for free vibrations of an undamped MDOF system, as follows:

$$\mathbf{M} \ddot{\mathbf{D}} + \mathbf{S} \mathbf{D} = \mathbf{0} \tag{12}$$

## Sec. 4.2  Principal and Normal Coordinates

Premultiplication of this equation by $\mathbf{\Phi}^T$ and insertion of $\mathbf{I} = \mathbf{\Phi}\,\mathbf{\Phi}^{-1}$ before $\ddot{\mathbf{D}}$ and $\mathbf{D}$ produces

$$\underbrace{\mathbf{\Phi}^T \mathbf{M}\, \mathbf{\Phi}}\, \mathbf{\Phi}^{-1}\ddot{\mathbf{D}} + \underbrace{\mathbf{\Phi}^T \mathbf{S}\, \mathbf{\Phi}}\, \mathbf{\Phi}^{-1}\mathbf{D} = \mathbf{0}$$

which can be restated as

$$\boxed{\mathbf{M}_P \ddot{\mathbf{D}}_P + \mathbf{S}_P \mathbf{D}_P = \mathbf{0}} \tag{13}$$

By virtue of Eqs. (10) and (11), the generalized mass and stiffness matrices in Eq. (13) are <u>both diagonal</u>. Also, the displacement and acceleration vectors in the latter equation are defined to be

$$\mathbf{D}_P = \mathbf{\Phi}^{-1}\mathbf{D} \qquad \ddot{\mathbf{D}}_P = \mathbf{\Phi}^{-1}\ddot{\mathbf{D}} \tag{14}$$

The generalized displacements $\mathbf{D}_P$ given by the first of Eqs. (14) are called *principal coordinates*, for which the equations of motion [Eq. (13)] have neither inertial nor elasticity coupling. From Eqs. (14) we find that the displacements and accelerations in the original coordinates are related to those in principal coordinates as follows:

$$\boxed{\mathbf{D} = \mathbf{\Phi}\,\mathbf{D}_P \qquad \ddot{\mathbf{D}} = \mathbf{\Phi}\,\ddot{\mathbf{D}}_P} \tag{15}$$

Here we see that the generalized displacements in vector $\mathbf{D}_P$ operate as multipliers of the modal columns in $\mathbf{\Phi}$ to produce values of the actual displacements in vector $\mathbf{D}$. Thus, the shape functions for the principal coordinates of a MDOF system are its natural modes of vibration.

We now restate the eigenvalue problem in Eq. (1a) more comprehensively as

$$\mathbf{S}\,\mathbf{\Phi} = \mathbf{M}\,\mathbf{\Phi}\,\boldsymbol{\omega}^2 \tag{16}$$

in which the modal matrix $\mathbf{\Phi}$ is given by Eq. (9). The symbol $\boldsymbol{\omega}^2$ in Eq. (16) represents a diagonal matrix with values of $\omega_i^2$ in diagonal positions, as follows:

$$\boldsymbol{\omega}^2 = \begin{bmatrix} \omega_1^2 & 0 & 0 & \cdots & 0 \\ 0 & \omega_2^2 & 0 & \cdots & 0 \\ 0 & 0 & \omega_3^2 & \cdots & 0 \\ \cdots & \cdots & \cdots & \cdots & \cdots \\ 0 & 0 & 0 & \cdots & \omega_n^2 \end{bmatrix} \tag{17}$$

This array, sometimes called the *spectral matrix*, will be referred to as the *eigenvalue matrix*, or *matrix of characteristic values*. It postmultiplies the matrix $\mathbf{\Phi}$ in Eq. (16), so that a typical modal column $\mathbf{\Phi}_i$ is scaled by the corresponding eigenvalue $\omega_i^2$. Premultiplying Eq. (16) by $\mathbf{\Phi}^T$ and using the relationships (10) and (11), we obtain

$$\mathbf{S}_P = \mathbf{M}_P \boldsymbol{\omega}^2 \tag{18a}$$

Hence,

$$S_{Pi} = M_{Pi}\,\omega_i^2 \tag{18b}$$

Thus, in principal coordinates the $i$th principal stiffness is equal to the $i$th principal mass multiplied by the $i$th eigenvalue.

Because the modal vectors may be scaled arbitrarily, the principal coordinates are not unique. In fact, there is an infinite number of sets of such generalized displacements, but the most common choice is that for which the mass matrix is transformed to the identity matrix. We state this condition by specifying that $M_{Pi}$ in Eq. (7) must be equal to unity, as follows:

$$\Phi_{Ni}^T M \Phi_{Ni} = M_{Ni} = 1 \tag{19}$$

where

$$\Phi_{Ni} = \frac{\Phi_i}{C_i} \tag{20}$$

Under this condition, the scaled eigenvector $\Phi_{Ni}$ is said to be *normalized with respect to the mass matrix*. The constant $C_i$ in Eq. (20) is computed as

$$C_i = \pm\sqrt{\Phi_i^T M \Phi_i} = \pm\sqrt{\sum_{j=1}^{n} \Phi_{ji}\left(\sum_{k=1}^{n} M_{jk}\Phi_{ki}\right)} \tag{21a}$$

If the mass matrix is diagonal, this expression simplifies to

$$C_i = \pm\sqrt{\sum_{j=1}^{n} M_j \Phi_{ji}^2} \tag{21b}$$

When all of the vectors in the modal matrix are normalized in this manner, we use the subscript $N$ and revise Eq. (10) to become

$$\Phi_N^T M \Phi_N = M_N = I \tag{22}$$

Thus, the principal mass matrix is now the identity matrix. Furthermore, the principal stiffness matrix, from Eqs. (11) and (18a), is seen to be

$$S_N = \Phi_N^T S \Phi_N = \omega^2 \tag{23a}$$

Or, for the $i$th mode,

$$S_{Ni} = \Phi_{Ni}^T S \Phi_{Ni} = \omega_i^2 \tag{23b}$$

Therefore, when the eigenvectors are normalized with respect to $M$, the stiffnesses in principal coordinates are equal to the eigenvalues. This particular set of principal coordinates is known as *normal coordinates*. Rewriting Eq. (13) in normal coordinates, we have

$$M_N \ddot{D}_N + S_N D_N = 0 \tag{24a}$$

or

$$\ddot{D}_N + \omega^2 D_N = 0 \tag{24b}$$

The vectors $D_N$ and $\ddot{D}_N$ in Eqs. (24) contain displacements and accelerations in

## Sec. 4.2  Principal and Normal Coordinates

normal coordinates. From Eqs. (14), these vectors are related to their counterparts in physical coordinates by

$$\mathbf{D}_N = \mathbf{\Phi}_N^{-1} \mathbf{D} \qquad \ddot{\mathbf{D}}_N = \mathbf{\Phi}_N^{-1} \ddot{\mathbf{D}} \qquad (25)$$

and the reverse transformations are

$$\mathbf{D} = \mathbf{\Phi}_N \mathbf{D}_N \qquad \ddot{\mathbf{D}} = \mathbf{\Phi}_N \ddot{\mathbf{D}}_N \qquad (26)$$

The *inverse of the normalized modal matrix* required in Eqs. (25) may be easily found. We need only postmultiply Eq. (22) by $\mathbf{\Phi}_N^{-1}$ to obtain the relationship

$$\mathbf{\Phi}_N^{-1} = \mathbf{\Phi}_N^T \mathbf{M} \qquad (27)$$

Thus, the desired inverse may always be calculated by this simple matrix multiplication. If we wish to include only a selected number of modes $m$ in our analysis, the modal matrix will be of size $n \times m$. Then Eqs. (25) must be restated as

$$\mathbf{D}_N = \mathbf{\Phi}_N^T \mathbf{M} \mathbf{D} \qquad \ddot{\mathbf{D}}_N = \mathbf{\Phi}_N^T \mathbf{M} \ddot{\mathbf{D}} \qquad (28)$$

This technique for using only a limited number of modes is called *modal truncation*.

### Example 4.1

To demonstrate the use of normal coordinates for action equations, let us reconsider the three-member plane truss in Fig. 4.1(a), which we studied previously in Chapter 3. From Eq. (3.5-31b) the $3 \times 3$ stiffness matrix for the free nodal displacements indicated in Fig. 4.1(a) is

$$\mathbf{S} = \frac{EA}{L} \begin{bmatrix} 1.36 & -0.36 & -0.48 \\ -0.36 & 0.36 & 0.48 \\ -0.48 & 0.48 & 1.64 \end{bmatrix} \qquad (a)$$

In addition, the consistent mass matrix from Eq. (3.5-34b) is

$$\mathbf{M} = \frac{\rho A L}{6} \begin{bmatrix} 2.72 & 1 & 0 \\ 1 & 3.28 & 0 \\ 0 & 0 & 3.28 \end{bmatrix} \qquad (b)$$

For this structure, three eigenvectors were calculated in Example 3.4 (see Table 3.4). These vectors become the columns of the modal matrix, as follows:

$$\mathbf{\Phi} = \begin{bmatrix} 0.2314 & 0.8673 & 1.0000 \\ 1.0000 & -0.1715 & -0.6050 \\ -0.2472 & 1.0000 & -0.6107 \end{bmatrix} \qquad (c)$$

To normalize this array with respect to the mass matrix, we compute from Eq. (21a) the

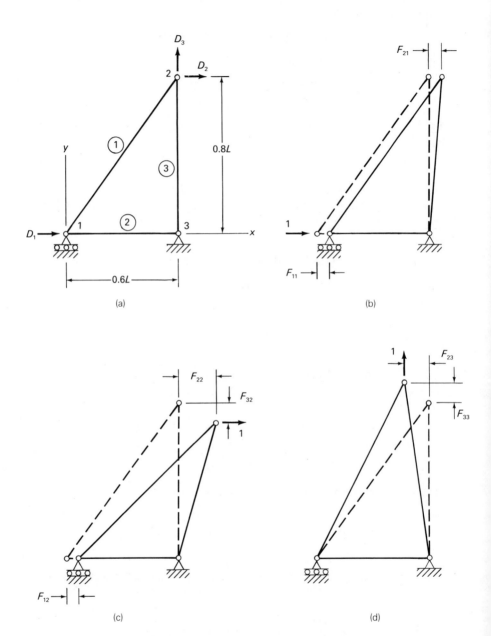

**Figure 4.1** (a) Plane truss; (b) condition $A_1 = 1$; (c) condition $A_2 = 1$; (d) condition $A_3 = 1$.

Sec. 4.2    Principal and Normal Coordinates    145

constants

$$C_1 = 0.8255\sqrt{\rho AL} \quad C_2 = 0.9242\sqrt{\rho AL} \quad C_3 = 0.8097\sqrt{\rho AL} \quad (d)$$

Dividing the columns of $\Phi$ by these values produces

$$\Phi_N = \frac{1}{\sqrt{\rho AL}} \begin{bmatrix} 0.2803 & 0.9384 & 1.2350 \\ 1.2114 & -0.1856 & -0.7472 \\ -0.2995 & 1.0820 & -0.7542 \end{bmatrix} \quad (e)$$

Substitution of Eqs. (a) and (e) into Eq. (23a) yields the result

$$S_N = \Phi_N^T S \, \Phi_N = \frac{E}{\rho L^2} \begin{bmatrix} 0.2701 & 0 & 0 \\ 0 & 2.088 & 0 \\ 0 & 0 & 5.308 \end{bmatrix} = \omega^2 \quad (f)$$

which contains the values of $\omega_1^2$, $\omega_2^2$, and $\omega_3^2$ on the diagonal. Their square roots are the same as the angular frequencies given in Table 3.4. Of course, the eigenvalues in Eq. (f) were already available; so the advantages of transforming stiffnesses to normal coordinates are not obvious. These advantages will become apparent for the response calculations in later sections.

If a structure is *statically determinate* and *immobile* [2], flexibility coefficients are not difficult to calculate. In such a case, the action equations of motion expressed by Eq. (12) may be replaced by *displacement equations of motion*, as follows:

$$\mathbf{F M \ddot{D} + D = 0} \quad (29)$$

in which

$$\mathbf{F = S^{-1}} \quad (30)$$

The *flexibility matrix* $\mathbf{F}$ contains values of free nodal displacements due to unit values of the corresponding actions. We transform Eq. (29) to principal coordinates by substituting Eqs. (15) for $\mathbf{D}$ and $\ddot{\mathbf{D}}$. Then premultiplication by $\Phi^{-1}$ and insertion of the identity matrix $\mathbf{I} = \Phi^{-T}\Phi^T$ before $\mathbf{M}$ produces

$$\Phi^{-1} \mathbf{F} \, \Phi^{-T} \Phi^T \mathbf{M} \, \Phi \, \ddot{\mathbf{D}}_P + \mathbf{D}_P = 0$$

which can be rewritten as

$$\mathbf{F}_P \mathbf{M}_P \ddot{\mathbf{D}}_P + \mathbf{D}_P = 0 \quad (31)$$

The symbol $\mathbf{F}_P$ in Eq. (31) represents a *principal flexibility matrix*, corresponding to $\mathbf{S}_P$, and is defined as

$$\mathbf{F}_P = \Phi^{-1} \mathbf{F} \, \Phi^{-T} = \mathbf{S}_P^{-1} \quad (32)$$

Of course, this definition applies only when $\mathbf{S}$ (and hence $\mathbf{S}_P$) is positive-definite. Furthermore, the expanded form of the eigenvalue problem in Eq. (16) is replaced by

$$\mathbf{F}\,\mathbf{M}\,\mathbf{\Phi} = \mathbf{\Phi}\,\boldsymbol{\lambda} \qquad (33)$$

The eigenvalue matrix $\boldsymbol{\lambda}$ in this expression consists of a diagonal array containing values of $\lambda_i = 1/\omega_i^2$ in diagonal positions, as follows:

$$\boldsymbol{\lambda} = \begin{bmatrix} \lambda_1 & 0 & 0 & \cdots & 0 \\ 0 & \lambda_2 & 0 & \cdots & 0 \\ 0 & 0 & \lambda_3 & \cdots & 0 \\ \cdots & \cdots & \cdots & \cdots & \cdots \\ 0 & 0 & 0 & \cdots & \lambda_n \end{bmatrix} = \boldsymbol{\omega}^{-2} \qquad (34)$$

Transforming Eq. (33) to principal coordinates as before, we find that

$$\mathbf{F}_P\,\mathbf{M}_P = \boldsymbol{\lambda} \qquad (35)$$

When the modal matrix is normalized with respect to the mass matrix, the principal flexibility matrix from Eqs. (32) and (35) takes the form

$$\mathbf{F}_N = \mathbf{\Phi}_N^{-1}\,\mathbf{F}\,\mathbf{\Phi}_N^{-T} = \boldsymbol{\lambda} = \boldsymbol{\omega}^{-2} = \mathbf{S}_N^{-1} \qquad (36)$$

Thus, the flexibility matrix in normal coordinates becomes the eigenvalue matrix $\boldsymbol{\lambda}$, which is also equal to the inverse of $\boldsymbol{\omega}^2$. From this we conclude that Eq. (24b) gives the equations of motion in normal coordinates, regardless of the method of formulation in the original coordinates.

### Example 4.2

As an example of the use of normal coordinates for displacement equations, we again consider the three-member truss in Fig. 4.1(a). Because this structure is statically determinate and immobile, we can find terms in the flexibility matrix by the *unit-load method* [2]. Figures 4.1(b)–(d) show the flexibilities obtained from applying unit values of $A_1$, $A_2$, and $A_3$ (all forces). The resulting flexibility matrix for this structure is

$$\mathbf{F} = \frac{L}{EA}\begin{bmatrix} 1.000 & 1.000 & 0 \\ 1.000 & 5.556 & -1.333 \\ 0 & -1.333 & 1.000 \end{bmatrix} \qquad (g)$$

Substituting $\mathbf{\Phi}_N^T$ and $\mathbf{M}$ from Example 4.1 into Eq. (27), we obtain the inverse of $\mathbf{\Phi}_N$ as

$$\mathbf{\Phi}_N^{-1} = \mathbf{\Phi}_N^T\,\mathbf{M} = \sqrt{\rho A L}\begin{bmatrix} 0.3290 & 0.7089 & -0.1637 \\ 0.3945 & 0.05494 & 0.5915 \\ 0.4353 & -0.2026 & -0.4123 \end{bmatrix} \qquad (h)$$

With this matrix and the flexibility matrix from Eq. (g), we find $\mathbf{F}_N$ using Eq. (36), as follows:

$$\mathbf{F}_N = \mathbf{\Phi}_N^{-1}\,\mathbf{F}\,\mathbf{\Phi}_N^{-T} = \frac{\rho L^2}{E}\begin{bmatrix} 3.702 & 0 & 0 \\ 0 & 0.4789 & 0 \\ 0 & 0 & 0.1884 \end{bmatrix} = \boldsymbol{\lambda} \qquad (i)$$

### Sec. 4.3 Normal-Mode Response to Initial Conditions

This diagonal array contains the values of $\lambda_1 = 1/\omega_1^2$, $\lambda_2 = 1/\omega_2^2$, and $\lambda_3 = 1/\omega_3^2$, as confirmed by Eq. (f) in Example 4.1.

At this point we should remember that working with displacement equations and flexibilities is feasible only for structures that are statically determinate and restrained against mobilities. Upon reflection, we realize that this is a very limited class of structures that would be encountered rather infrequently. In general, the approach using action equations and stiffnesses is much more suitable for dynamic analysis.

## 4.3 NORMAL-MODE RESPONSE TO INITIAL CONDITIONS

For a MDOF system, suppose that we know the initial conditions (at time $t = 0$) of displacements $\mathbf{D}_0$ and velocities $\dot{\mathbf{D}}_0$, as follows:

$$\mathbf{D}_0 = \begin{bmatrix} D_{01} \\ D_{02} \\ D_{03} \\ \ldots \\ D_{0n} \end{bmatrix} \quad \dot{\mathbf{D}}_0 = \begin{bmatrix} \dot{D}_{01} \\ \dot{D}_{02} \\ \dot{D}_{03} \\ \ldots \\ \dot{D}_{0n} \end{bmatrix} \quad (1)$$

In accordance with Eqs. (4.2-25), these initial values may be transformed to normal coordinates by premultiplying them with the inverse of the normalized modal matrix. Thus, we have

$$\mathbf{D}_{N0} = \mathbf{\Phi}_N^{-1} \mathbf{D}_0 \quad \dot{\mathbf{D}}_{N0} = \mathbf{\Phi}_N^{-1} \dot{\mathbf{D}}_0 \quad (2)$$

The second relationship in Eqs. (2) is obtained by differentiation of the first with respect to time. The forms of the displacement and velocity vectors in normal coordinates are the same as those in physical coordinates given by Eqs. (1).

From Eq. (4.2-24b) we see that a typical equation of motion for undamped free vibrations in normal coordinates is

$$\ddot{D}_{Ni} + \omega_i^2 D_{Ni} = 0 \quad (i = 1, 2, \ldots, n) \quad (3)$$

Each equation of this type is uncoupled from all of the others, and we will treat the expression as if it pertained to a SDOF system [see Eq. (2.2-4)]. Knowing the conditions of initial displacement $D_{N0i}$ and velocity $\dot{D}_{N0i}$, we find the response of the $i$th normal coordinate as

$$D_{Ni} = D_{N0i} \cos \omega_i t + \frac{\dot{D}_{N0i}}{\omega_i} \sin \omega_i t \quad (i = 1, 2, \ldots, n) \quad (4)$$

This expression is drawn from Eq. (2.2-9) for an undamped one-degree system. We can apply Eq. (4) repetitively to calculate the terms in the vector of normal-mode displacements $\mathbf{D}_N = \{D_{Ni}\}$. These results are then transformed back to the original coordinates, using the operation given by the first of Eqs. (4.2-26).

Thus,
$$\mathbf{D} = \mathbf{\Phi}_N \mathbf{D}_N \tag{5}$$

This sequence of operations is the same regardless of whether the original equations of motion are written as action equations or displacement equations.

However, for the action-equation approach, there exists the possibility of one or more *rigid-body modes*. For such a principal mode the eigenvalue $\omega_i^2$ is zero, and Eq. (3) becomes

$$\ddot{D}_{Ni} = 0 \tag{6}$$

Integration of this equation twice with respect to time yields

$$D_{Ni} = D_{N0i} + \dot{D}_{N0i} t \tag{7}$$

This formula is used in place of Eq. (4) to evaluate the response of a rigid-body mode in normal coordinates, assuming that rotations are small.

### Example 4.3

Suppose that the plane truss in Examples 4.1 and 4.2 is subjected to an initial force $P_0$ in the $x$ direction at joint 2, as illustrated in Fig. 4.2. Let us find the free-vibrational response of this structure due to suddenly releasing the load.

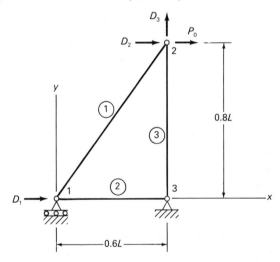

**Figure 4.2** Plane truss with initial load.

The initial displacements of the joints caused by the force may be calculated as the product of $P_0$ and the second column of the flexibility matrix $\mathbf{F}$ in Eq. (4.2-g). Thus,

$$\mathbf{D}_0 = \{1.000, 5.556, -1.333\} \frac{P_0 L}{EA} \tag{a}$$

Also, at time $t = 0$ the initial velocities are $\dot{\mathbf{D}}_0 = \mathbf{0}$. Using the first of Eqs. (2), we transform the initial displacements to normal coordinates, as follows:

Sec. 4.3   Normal-Mode Response to Initial Conditions

$$\mathbf{D}_{N0} = \mathbf{\Phi}_N^{-1}\mathbf{D}_0 = \sqrt{\rho A L}\,\{4.486,\ -0.08872,\ -0.1407\}\frac{P_0 L}{EA} \tag{b}$$

For this purpose the inverse of the normalized matrix $\mathbf{\Phi}_N^{-1}$ is available from Eq. (4.2-h). According to Eq. (4), we find the terms in the vector of normal mode displacements to be

$$\mathbf{D}_N = \sqrt{\rho A L}\begin{bmatrix} 4.486\cos\omega_1 t \\ -0.08872\cos\omega_2 t \\ -0.1407\cos\omega_3 t \end{bmatrix}\frac{P_0 L}{EA} \tag{c}$$

Then with the normalized modal matrix $\mathbf{\Phi}_N$ from Eq. (4.2-e), we transform these results back to the original coordinates, using Eq. (5) as follows:

$$\mathbf{D} = \mathbf{\Phi}_N \mathbf{D}_N$$

$$= \frac{P_0 L}{EA}\begin{bmatrix} 1.257\cos\omega_1 t - 0.08325\cos\omega_2 t - 0.1738\cos\omega_3 t \\ 5.434\cos\omega_1 t + 0.01647\cos\omega_2 t + 0.1051\cos\omega_3 t \\ -1.343\cos\omega_1 t - 0.09600\cos\omega_2 t + 0.1061\cos\omega_3 t \end{bmatrix} \tag{d}$$

In this example the response of the first mode of vibration is about one order of magnitude greater than each of the other two.

**Example 4.4**

Assume that the flexural element in Fig. 4.3(a) is initially at rest when node 2 is struck in such a manner that it suddenly acquires a velocity $\dot{D}_{03}$. Determine the small-displacement response of the element to this impact.

Nodes of the flexural element have no restraints against either translations in the $y$ direction or rotations in the $z$ sense. Therefore, the stiffness and mass matrices are

$$\mathbf{K} = \frac{2EI}{L^3}\begin{bmatrix} 6 & 3L & -6 & 3L \\ 3L & 2L^2 & -3L & L^2 \\ -6 & -3L & 6 & -3L \\ 3L & L^2 & -3L & 2L^2 \end{bmatrix} \tag{e}$$

and

$$\mathbf{M} = \frac{\rho A L}{420}\begin{bmatrix} 156 & 22L & 54 & -13L \\ 22L & 4L^2 & 13L & -3L^2 \\ 54 & 13L & 156 & -22L \\ -13L & -3L^2 & -22L & 4L^2 \end{bmatrix} \tag{f}$$

as given by Eqs. (3.4-24) and (3.4-26). In this case the stiffness matrix is positive-semidefinite, so we should expect to find two repeated zero roots when solving the eigenvalue problem. Thus, solution of Eq. (4.2-1a) yields the eigenvalues

$$\omega_1^2 = \omega_2^2 = 0 \qquad \omega_3^2 = 720\frac{EI}{\rho A L^4} \qquad \omega_4^2 = 8400\frac{EI}{\rho A L^4} \tag{g}$$

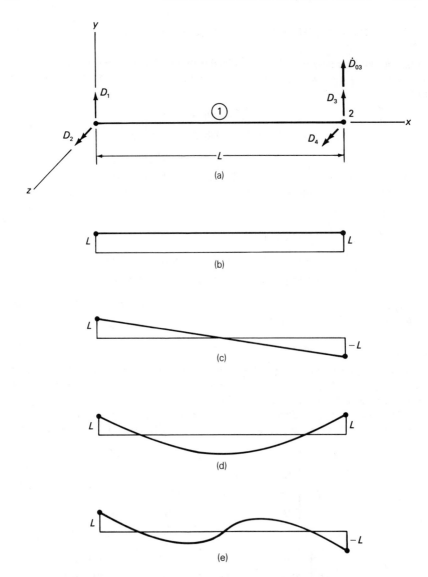

**Figure 4.3** (a) Flexural element with initial velocity; (b) mode 1; (c) mode 2; (d) mode 3; (e) mode 4.

as well as the modal matrix

$$\boldsymbol{\Phi} = \begin{bmatrix} L & L & L & L \\ 0 & -2 & -6 & -12 \\ L & -L & L & -L \\ 0 & -2 & 6 & -12 \end{bmatrix} \quad \text{(h)}$$

## Sec. 4.3  Normal-Mode Response to Initial Conditions

These eigenvalues and eigenvectors were found using a computer program called VIBCB for vibrational analysis of continuous beams (see Sec. 3.8). Each of the four column vectors in Eq. (h) is scaled so that the first translational displacement is numerically equal to $L$. Figures 4.3(b) and (c) depict the rigid-body modes, and Figs. 4.3(d) and (e) show the vibrational modes. For the purpose of normalizing columns in the modal matrix with respect to $\mathbf{M}$, we use Eq. (4.2-21a) to find the constants

$$C_1 = L\sqrt{m} \qquad C_2 = L\sqrt{\frac{m}{3}} \qquad C_3 = L\sqrt{\frac{m}{5}} \qquad C_4 = L\sqrt{\frac{m}{7}} \qquad (i)$$

in which $m = \rho A L$. Dividing the columns of Eq. (h) by these values, we obtain the normalized modal matrix as

$$\Phi_N = \frac{1}{L\sqrt{m}} \begin{bmatrix} L & L\sqrt{3} & L\sqrt{5} & L\sqrt{7} \\ 0 & -2\sqrt{3} & -6\sqrt{5} & -12\sqrt{7} \\ L & -L\sqrt{3} & L\sqrt{5} & -L\sqrt{7} \\ 0 & -2\sqrt{3} & 6\sqrt{5} & -12\sqrt{7} \end{bmatrix} \qquad (j)$$

Initial velocities in normal coordinates now may be calculated using the second expression in Eqs. (2). Hence,

$$\dot{\mathbf{D}}_{N0} = \Phi_N^{-1} \dot{\mathbf{D}}_0 = \Phi_N^T \mathbf{M} \begin{bmatrix} 0 \\ 0 \\ 1 \\ 0 \end{bmatrix} \dot{D}_{03} \quad \text{for dof 3}$$

$$= \Phi_N^T \begin{bmatrix} 54 \\ 13L \\ 156 \\ -22L \end{bmatrix} \frac{m\dot{D}_{03}}{420} = \begin{bmatrix} 35 \\ -14\sqrt{3} \\ 0 \\ \sqrt{7} \end{bmatrix} \frac{\sqrt{m}\,\dot{D}_{03}}{70} \qquad (k)$$

Equations (4) and (7) give the vibrational and rigid-body responses in normal coordinates as

$$\mathbf{D}_N = \{35t,\ -14\sqrt{3}\,t,\ 0,\ (\sqrt{7}\sin\omega_4 t)/\omega_4\}\sqrt{m}\,\frac{\dot{D}_{03}}{70} \qquad (\ell)$$

Note that there is no response of the third normal mode, which has a symmetric vibrational shape. Finally, with Eq. (5) we transform the results in Eq. ($\ell$) back to the original coordinates to obtain

$$\mathbf{D} = \Phi_N \mathbf{D}_N = \begin{bmatrix} -tL + (L\sin\omega_4 t)/\omega_4 \\ 12t - (12\sin\omega_4 t)/\omega_4 \\ 11tL - (L\sin\omega_4 t)/\omega_4 \\ 12t - (12\sin\omega_4 t)/\omega_4 \end{bmatrix} \frac{\dot{D}_{03}}{10L} \qquad (m)$$

These responses are valid only for small rigid-body rotations.

## 4.4 NORMAL-MODE RESPONSE TO APPLIED ACTIONS

Now we shall consider the case of an undamped MDOF system that is subjected to applied actions corresponding to the nodal displacements. For this situation, the action equations of motion are

$$\mathbf{M}\ddot{\mathbf{D}} + \mathbf{S}\mathbf{D} = \mathbf{A} \tag{1}$$

where the symbol $\mathbf{A}$ denotes a column vector of time-varying applied actions, as follows:

$$\mathbf{A} = \{A_1(t), A_2(t), \ldots, A_n(t)\} \tag{2}$$

and $n$ is the number of degrees of freedom. Equation (1) is transformed to normal coordinates by premultiplying both sides with $\mathbf{\Phi}_N^T$ and substituting Eqs. (4.2-26) for $\mathbf{D}$ and $\ddot{\mathbf{D}}$ to produce

$$\mathbf{\Phi}_N^T \mathbf{M} \mathbf{\Phi}_N \ddot{\mathbf{D}}_N + \mathbf{\Phi}_N^T \mathbf{S} \mathbf{\Phi}_N \mathbf{D}_N = \mathbf{\Phi}_N^T \mathbf{A}$$

This equation may also be written as

$$\ddot{\mathbf{D}}_N + \omega^2 \mathbf{D}_N = \mathbf{A}_N \tag{3}$$

The symbol $\mathbf{A}_N$ on the right-hand side of Eq. (3) represents a vector of applied actions in normal coordinates, computed by the operation

$$\mathbf{A}_N = \mathbf{\Phi}_N^T \mathbf{A} \tag{4}$$

In expanded form, the results of this multiplication are

$$\begin{bmatrix} A_{N1} \\ A_{N2} \\ \ldots \\ A_{Nn} \end{bmatrix} = \begin{bmatrix} \Phi_{N11} A_1 + \Phi_{N21} A_2 + \ldots + \Phi_{Nn1} A_n \\ \Phi_{N12} A_1 + \Phi_{N22} A_2 + \ldots + \Phi_{Nn2} A_n \\ \ldots \ldots \ldots \ldots \ldots \ldots \ldots \ldots \ldots \ldots \\ \Phi_{N1n} A_1 + \Phi_{N2n} A_2 + \ldots + \Phi_{Nnn} A_n \end{bmatrix} \tag{5}$$

The $i$th equation of motion in normal coordinates is

$$\ddot{D}_{Ni} + \omega_i^2 D_{Ni} = A_{Ni} \qquad (i = 1, 2, \ldots, n) \tag{6}$$

where the *normal-mode load* $A_{Ni}$ is taken from the $i$th row in Eq. (5).

Each of the normal-mode equations of motion given by Eq. (6) is uncoupled from all of the others and has the same form as a SDOF system with unit mass. Therefore, we can calculate the response of a typical *vibrational motion* using *Duhamel's integral*, as follows:

$$D_{Ni} = \frac{1}{\omega_i} \int_0^t A_{Ni} \sin \omega_i(t - t') \, dt' \tag{7}$$

This expression is drawn from Eq. (2.6-6) and was derived for an undamped one-degree system that is initially at rest. We apply it repetitively to calculate

(2.4.27)

### Sec. 4.4  Normal-Mode Response to Applied Actions

the terms in the vector of normal-mode displacements $\mathbf{D}_N = \{D_{Ni}\}$. Then the results are transformed back to the original coordinates using Eq. (4.3-5).

For a normal mode corresponding to a *rigid-body motion*, the eigenvalue $\omega_i^2$ is zero. Then Eq. (6) becomes

$$\ddot{D}_{Ni} = A_{Ni} \tag{8}$$

In this instance, the normal-mode response (with the system initially at rest) is

$$D_{Ni} = \int_0^t \int_0^{t'} A_{Ni} \, dt'' \, dt' \tag{9}$$

This equation replaces Eq. (7) whenever a rigid-body mode is encountered.

In summary, we calculate the dynamic response of an undamped MDOF system to applied actions by first transforming those actions to normal coordinates using Eq. (4). Then the response of each vibrational mode is obtained from the integral in Eq. (7), and that for each rigid-body mode is determined by the double integral in Eq. (9). Finally, the values of the actual displacements in physical coordinates are found with the back-transformation operation of Eq. (4.3-5). If applied actions do not correspond to the nodal displacements, the appropriate *equivalent nodal loads* discussed in Sec. 3.3 can always be found as a preliminary step (see Example 4.6 at the end of this section).

At this point, let us examine the effect of a dynamic load $A_j(t)$, corresponding to the $j$th nodal displacement, on the response of the $k$th nodal displacement $D_k$. From Eq. (5) the $i$th normal-mode load due to $A_j$ is

$$A_{Ni} = \Phi_{Nji} A_j \tag{10}$$

If the system has only vibrational modes, the response of the $i$th mode is found from Eq. (7) to be

$$D_{Ni} = \frac{\Phi_{Nji}}{\omega_i} \int_0^t A_j \sin \omega_i (t - t') \, dt' \tag{11}$$

Transformation of this result back to the original coordinates by Eq. (4.3-5) yields the response of the $k$th nodal displacement as

$$(D_k)_{A_j} = \sum_{i=1}^n \left[ \frac{\Phi_{Nki} \Phi_{Nji}}{\omega_i} \int_0^t A_j \sin \omega_i (t - t') \, dt' \right] \tag{12}$$

Similarly, the response of the $j$th nodal displacement caused by a dynamic load $A_k(t)$, corresponding to the $k$th nodal displacement, may be written

$$(D_j)_{A_k} = \sum_{i=1}^n \left[ \frac{\Phi_{Nji} \Phi_{Nki}}{\omega_i} \int_0^t A_k \sin \omega_i (t - t') \, dt' \right] \tag{13}$$

If $A_j = A_k$, the right-hand sides of Eqs. (12) and (13) are equal, and we may equate the left-hand sides to obtain

$$(D_k)_{A_j} = (D_j)_{A_k} \tag{14}$$

This relationship constitutes a *reciprocal theorem for dynamic loads* [3] that is similar to Maxwell's reciprocal theorem for static loads [2]. It states that the dynamic response of the $k$th nodal displacement due to any time-varying action corresponding to the $j$th displacement is equal to the response of the $j$th displacement due to the same action applied at the $k$th displacement. The theorem holds for systems with rigid-body modes as well as vibrational modes, as can be seen by using Eq. (9) in place of Eq. (7) in Eq. (11).

**Example 4.5**

We shall consider again the plane truss of Examples 4.1, 4.2, and 4.3. Figure 4.4(a) shows a step force of magnitude $P = P_2$ applied in the $x$ direction at joint 2. Let us determine the response of the structure to this suddenly applied load, starting from rest.

The vector of applied actions for this case is

$$\mathbf{A} = \{0, P_2, 0\} \tag{a}$$

As required by Eq. (4), we transform vector $\mathbf{A}$ to normal coordinates with the operation:

$$\mathbf{A}_N = \mathbf{\Phi}_N^T \mathbf{A} = \begin{bmatrix} 1.2114 \\ -0.1856 \\ -0.7472 \end{bmatrix} \frac{P_2}{\sqrt{m}} \tag{b}$$

where $m = \rho AL$. For this purpose the normalized modal matrix $\mathbf{\Phi}_N$ is available from Eq. (4.2-e). From the result of the Duhamel integral for a step function, given as Eq. (2.6-8),

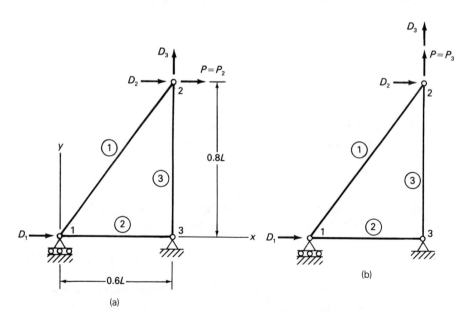

**Figure 4.4** Plane truss: (a) with step force $P_2$; (b) with step force $P_3$.

Sec. 4.4　Normal-Mode Response to Applied Actions　　155

we obtain the vector of normal-mode displacements, as follows:

$$\mathbf{D}_N = \begin{bmatrix} 1.2114(1 - \cos \omega_1 t)/\omega_1^2 \\ -0.1856(1 - \cos \omega_2 t)/\omega_2^2 \\ -0.7472(1 - \cos \omega_3 t)/\omega_3^2 \end{bmatrix} \frac{P_2}{\sqrt{m}} \quad (c)$$

Substitution of $\omega_1^2 = 0.2701 E/\rho L^2$, $\omega_2^2 = 2.088 E/\rho L^2$, and $\omega_3^2 = 5.308 E/\rho L^2$ into Eq. (c) yields the simpler form

$$\mathbf{D}_N = \begin{bmatrix} 4.485(1 - \cos \omega_1 t) \\ -0.08889(1 - \cos \omega_2 t) \\ -0.1408(1 - \cos \omega_3 t) \end{bmatrix} \frac{P_2 \rho L^2}{E\sqrt{m}} \quad (d)$$

Transforming this solution back to the original coordinates using Eq. (4.3-5), we find that

$$\mathbf{D} = \mathbf{\Phi}_N \mathbf{D}_N = \begin{bmatrix} 1.000 - 1.257 c_1 + 0.08341 c_2 + 0.1739 c_3 \\ 5.556 - 5.433 c_1 - 0.01650 c_2 - 0.1052 c_3 \\ -1.333 + 1.343 c_1 + 0.09618 c_2 - 0.1062 c_3 \end{bmatrix} \frac{P_2 L}{EA} \quad (e)$$

where $c_1 = \cos \omega_1 t$, $c_2 = \cos \omega_2 t$, and $c_3 = \cos \omega_3 t$. Inspection of these results shows that the joints of the truss vibrate about the displaced positions:

$$\mathbf{D}_{st} = \{1.000, 5.556, -1.333\} \frac{P_2 L}{EA} \quad (f)$$

due to the force $P_2$ applied statically.

Proceeding in a similar manner, we can also calculate the response of the truss to a step force of magnitude $P = P_3$ applied in the $y$ direction at joint 2, as indicated in Fig. 4.4(b). In this case, the results are

$$\mathbf{D} = \begin{bmatrix} 0 + 0.3108 c_1 - 0.4863 c_2 + 0.1755 c_3 \\ -1.333 + 1.343 c_1 + 0.09618 c_2 - 0.1062 c_3 \\ 1.000 - 0.3321 c_1 - 0.5607 c_2 - 0.1072 c_3 \end{bmatrix} \frac{P_3 L}{EA} \quad (g)$$

Equation (g) shows that the joints of the truss vibrate about the displaced positions:

$$\mathbf{D}_{st} = \{0, -1.333, 1.000\} \frac{P_3 L}{EA} \quad (h)$$

due to applying the force $P_3$ statically. If we let the step force $P_2$ be equal to $P_3$, then the response $D_3$ caused by $P_2$ in Eq. (e) will be equal to the response $D_2$ caused by $P_3$ in Eq. (g). This equality confirms the reciprocal theorem for dynamic loads in Eq. (14).

**Example 4.6**

Assume that the unrestrained flexural element in Example 4.4 is subjected to a ramp force $P = P_1 t/t_1$ applied in the $y$ direction at its center, as indicated in Fig. 4.5. We shall calculate the response at the nodes due to this influence, beginning with the element at rest.

Using $x = L/2$ in the displacement shape functions $\mathbf{f}$ from Eq. (3.4-17) and applying Eq. (3.3-25) without integration, we find the equivalent nodal actions to be

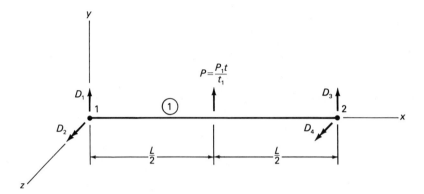

**Figure 4.5** Flexural element with ramp force.

$$\mathbf{A}_P = \mathbf{f}^T P = \{4, L, 4, -L\}\frac{P_1 t}{8t_1} \tag{i}$$

which consists of a force and a moment at each end. Premultiplication of this vector with the transpose of the normalized modal matrix from Eq. (4.3-j) yields

$$\mathbf{A}_N = \mathbf{\Phi}_N^T \mathbf{A}_P = \{2, 0, -\sqrt{5}, 0\}\frac{P_1 t}{2t_1\sqrt{m}} \tag{j}$$

where $m = \rho A L$. The vector in Eq. (j) contains normal-mode loads of types 1 and 3 for this example. Now we integrate the first term with Eq. (9) to obtain

$$D_{N1} = \frac{P_1 t^3}{6t_1\sqrt{m}} \tag{k}$$

which is a symmetric rigid-body motion [see Fig. 4.3(b)]. Similarly, evaluation of the third term in Eq. (j) in accordance with Eq. (2.6-c) for a ramp function produces

$$D_{N3} = -\frac{P_1\sqrt{5}}{2t_1\omega_3^2\sqrt{m}}\left(t - \frac{1}{\omega_3}\sin\omega_3 t\right) \tag{$\ell$}$$

which is a symmetric vibrational response [see Fig. 4.3(d)]. The antisymmetric rigid-body and flexural modes 2 and 4 do not respond to the centrally-placed load. Altogether, the vector of normal-mode responses is

$$\mathbf{D}_N = \frac{P_1}{6t_1\sqrt{m}}\begin{bmatrix} t^3 \\ 0 \\ -3\sqrt{5}\left(t - \frac{1}{\omega_3}\sin\omega_3 t\right)\Big/\omega_3^2 \\ 0 \end{bmatrix} \tag{m}$$

Transformation of these displacements back to physical coordinates using Eq. (4.3-5) gives

Sec. 4.5  Normal-Mode Response to Support Motions

$$\mathbf{D} = \mathbf{\Phi}_N \mathbf{D}_N = \frac{P_1}{6mLt_1} \begin{bmatrix} Lt^3 - 15L\left(t - \frac{1}{\omega_3}\sin\omega_3 t\right)\Big/\omega_3^2 \\ 90\left(t - \frac{1}{\omega_3}\sin\omega_3 t\right)\Big/\omega_3^2 \\ Lt^3 - 15L\left(t - \frac{1}{\omega_3}\sin\omega_3 t\right)\Big/\omega_3^2 \\ -90\left(t - \frac{1}{\omega_3}\sin\omega_3 t\right)\Big/\omega_3^2 \end{bmatrix} \qquad (n)$$

Here we see that the first rigid-body mode contributes translations equal to $P_1 t^3/6mt_1$ at both nodes 1 and 2.

## 4.5 NORMAL-MODE RESPONSE TO SUPPORT MOTIONS

We are often interested in the response of structures to support motions instead of applied actions. In this section we discuss problems in which either rigid-body ground displacements or accelerations are specified. In addition, independent motions of multiple restraints will be treated.

Figure 4.6 shows six possible displacement components $D_{g1}, D_{g2}, \ldots, D_{g6}$ for a point $g$ that is assumed to be a reference point on ground. The figure also depicts a typical joint (or node) $j$ on a structure that is connected to ground. The six possible displacement components at point $j$ are labeled $D_{j1}, D_{j2}, \ldots, D_{j6}$. A location vector $\mathbf{r}_{gj}$ is directed from point $g$ to point $j$ and has scalar components $x_{gj}$, $y_{gj}$, and $z_{gj}$. We may calculate the displacements at $j$ due to *rigid-body displacements of the ground* at point $g$ using the concept of *translation of axes* [2, 4]. For this purpose, the rotational components of the ground displacements must be small. Under this condition, the displacements at $j$ in terms of those at $g$ are

$$\mathbf{D}_j = \mathbf{T}_{gj}^T \mathbf{D}_g \qquad (1)$$

where

$$\mathbf{D}_j = \{D_{j1}, D_{j2}, \ldots, D_{j6}\} \qquad (2)$$

and

$$\mathbf{D}_g = \{D_{g1}, D_{g2}, \ldots, D_{g6}\} \qquad (3)$$

In Eq. (1) the transformation matrix is

$$\mathbf{T}_{gj}^T = \begin{bmatrix} \mathbf{I}_3 & \mathbf{c}_{gj}^T \\ \mathbf{0} & \mathbf{I}_3 \end{bmatrix} \qquad (4)$$

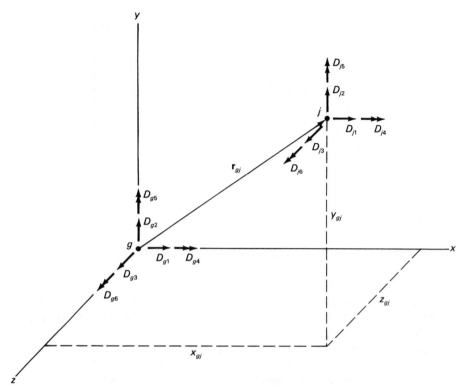

**Figure 4.6** Rigid-body ground displacements.

in which $\mathbf{I}_3$ is an identity matrix of order 3 and

$$\mathbf{c}_{gj}^T = -\mathbf{c}_{gj} = \mathbf{c}_{jg} = \begin{bmatrix} 0 & z_{gj} & -y_{gj} \\ -z_{gj} & 0 & x_{gj} \\ y_{gj} & -x_{gj} & 0 \end{bmatrix} \quad (5)$$

This skew-symmetric submatrix contains positive and negative values of the components of vector $\mathbf{r}_{gj}$. These components are arranged in a manner that produces the cross product of a small rotation vector at $g$ and the location vector $\mathbf{r}_{gj}$. Then the time-varying displacements $\mathbf{\Delta}_g$ at all free joints may be written as

$$\mathbf{\Delta}_g = \mathbf{T}_g^T \mathbf{D}_g \quad (6)$$

where

$$\mathbf{T}_g^T = \begin{bmatrix} \mathbf{T}_{g1}^T \\ \mathbf{T}_{g2}^T \\ \ldots \\ \mathbf{T}_{gn_j}^T \end{bmatrix} \quad (7)$$

and $n_j$ is the number of such joints.

## Sec. 4.5  Normal-Mode Response to Support Motions

To include the effects of ground displacements in the action equations of motion [see Eq. (4.2-12)], we write them as

$$\mathbf{M}\ddot{\mathbf{D}} + \mathbf{S}(\mathbf{D} - \boldsymbol{\Delta}_g) = \mathbf{0} \tag{8}$$

This matrix equation has the same form as Eq. (2.3-11) and can be restated

$$\mathbf{M}\ddot{\mathbf{D}} + \mathbf{S}\mathbf{D} = \mathbf{A}_g \tag{9}$$

in which

$$\mathbf{A}_g = \mathbf{S}\,\boldsymbol{\Delta}_g = \mathbf{S}\,\mathbf{T}_g^\mathrm{T}\mathbf{D}_g \tag{10}$$

Thus, the vector $\mathbf{A}_g$ on the right-hand side of Eq. (9) contains equivalent nodal actions due to rigid-body ground displacements.

Similarly, the accelerations at a typical joint $j$ may be expressed in terms of *rigid-body ground accelerations* at point $g$, as follows:

$$\ddot{\mathbf{D}}_j = \mathbf{T}_{gj}^\mathrm{T}\ddot{\mathbf{D}}_g \tag{11}$$

where

$$\ddot{\mathbf{D}}_j = \{\ddot{D}_{j1}, \ddot{D}_{j2}, \ldots, \ddot{D}_{j6}\} \tag{12}$$

and

$$\ddot{\mathbf{D}}_g = \{\ddot{D}_{g1}, \ddot{D}_{g2}, \ldots, \ddot{D}_{g6}\} \tag{13}$$

Then the vector of accelerations $\ddot{\boldsymbol{\Delta}}_g$ at all free joints becomes

$$\ddot{\boldsymbol{\Delta}}_g = \mathbf{T}_g^\mathrm{T}\ddot{\mathbf{D}}_g \tag{14}$$

In order to use this vector in the equations of motion, we must change to the *relative coordinates*:

$$\mathbf{D}^* = \mathbf{D} - \boldsymbol{\Delta}_g \qquad \ddot{\mathbf{D}}^* = \ddot{\mathbf{D}} - \ddot{\boldsymbol{\Delta}}_g \tag{15}$$

In these expressions the symbol $\mathbf{D}^*$ denotes a vector of displacements relative to the ground, and the vector $\ddot{\mathbf{D}}^*$ contains the corresponding relative accelerations. Substituting $\mathbf{D} - \boldsymbol{\Delta}_g$ and $\ddot{\mathbf{D}}$ from Eqs. (15) into Eq. (8) and rearranging, we find that

$$\mathbf{M}\ddot{\mathbf{D}}^* + \mathbf{S}\mathbf{D}^* = \mathbf{A}_g^* \tag{16}$$

in which

$$\mathbf{A}_g^* = -\mathbf{M}\ddot{\boldsymbol{\Delta}}_g = -\mathbf{M}\mathbf{T}_g^\mathrm{T}\ddot{\mathbf{D}}_g \tag{17}$$

Therefore, the vector $\mathbf{A}_g^*$ on the right-hand side of Eq. (16) consists of equivalent nodal actions caused by rigid-body ground accelerations. If we compare Eq. (16) to its SDOF counterpart in Eq. (2.3-18), we see that both are of the same form.

After the equivalent nodal actions $\mathbf{A}_g$ or $\mathbf{A}_g^*$ have been found, the response calculations in absolute or relative coordinates proceed as described in Sec. 4.4 for applied actions. For the latter case, the absolute displacements at free nodes may be calculated from the first of Eqs. (15) as

$$\mathbf{D} = \mathbf{D}^* + \mathbf{\Delta}_g \tag{18}$$

which is the sum of the relative displacements and the effects of the rigid-body ground displacements.

**Example 4.7**

Figure 4.7 shows the plane truss used previously in Example 3.1. In this truss the cross-sectional areas of members 1 and 2 are equal to $0.8A$ and $A$. Now let us calculate the response of the structure to a small rigid-body step-rotation $D_{g6} = \theta_z$ about point 2.

From Example 3.1, the stiffness and mass matrices for the free displacements at point 1 are

$$\mathbf{S} = s \begin{bmatrix} 0.36 & -0.48 \\ -0.48 & 1.64 \end{bmatrix} \quad \mathbf{M} = m \begin{bmatrix} 1 & 0 \\ 0 & 1 \end{bmatrix} \tag{a}$$

in which $s = EA/L$ and $m = 3.28\, \rho AL/6$. We also know that

$$\omega_1^2 = 0.2 \frac{s}{m} \quad \omega_2^2 = 1.8 \frac{s}{m} \quad \mathbf{\Phi}_N = \frac{1}{\sqrt{10m}} \begin{bmatrix} 3 & -1 \\ 1 & 3 \end{bmatrix} \tag{b}$$

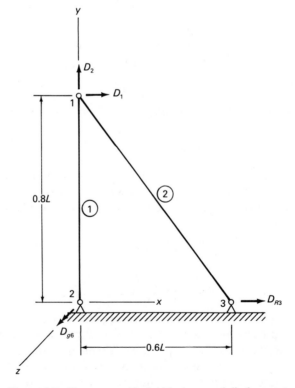

**Figure 4.7** Plane truss with rigid-body ground displacement.

## Sec. 4.5  Normal-Mode Response to Support Motions

The rigid-body rotation $\theta_z$, centered at point 2, causes the following step-translations at point 1:

$$\mathbf{\Delta}_g = \mathbf{T}_g^T \mathbf{D}_g = \begin{bmatrix} -0.8L \\ 0 \end{bmatrix} \theta_z \quad \text{(c)}$$

as given by Eq. (6). The $2 \times 1$ operator $\mathbf{T}_g^T$ in Eq. (c) contains only terms from the first and second rows and the sixth column of the general operator in Eq. (4). Equivalent nodal loads (forces) at joint 1 are

$$\mathbf{A}_g = \mathbf{S}\,\mathbf{\Delta}_g = \begin{bmatrix} -0.288 \\ 0.384 \end{bmatrix} sL\theta_z \quad \text{(d)}$$

which result from applying Eq. (10). From this point we may proceed to calculate the response by the method shown in Sec. 4.4. Thus, Eq. (4.4-4) gives the normal-mode loads as

$$\mathbf{A}_N = \mathbf{\Phi}_N^T \mathbf{A}_g = \begin{bmatrix} -1 \\ 3 \end{bmatrix} \frac{0.48 sL\theta_z}{\sqrt{10m}} \quad \text{(e)}$$

Then the normal-mode responses to these step loads become

$$\mathbf{D}_N = \begin{bmatrix} -(1 - \cos \omega_1 t)/\omega_1^2 \\ 3(1 - \cos \omega_2 t)/\omega_2^2 \end{bmatrix} \frac{0.48 sL\theta_z}{\sqrt{10m}} \quad \text{(f)}$$

These expressions may be simplified by substituting $\omega_1^2$ and $\omega_2^2$ from Eqs. (b), as follows:

$$\mathbf{D}_N = \begin{bmatrix} -3(1 - \cos \omega_1 t) \\ 1 - \cos \omega_2 t \end{bmatrix} \frac{0.8 mL\theta_z}{\sqrt{10m}} \quad \text{(g)}$$

Transformation of this vector back to physical coordinates with Eq. (4.3-5) produces

$$\mathbf{D} = \mathbf{\Phi}_N \mathbf{D}_N = \begin{bmatrix} -10 + 9c_1 + c_2 \\ 3c_1 - 3c_2 \end{bmatrix} \frac{0.8L\theta_z}{10} \quad \text{(h)}$$

where $c_1 = \cos \omega_1 t$ and $c_2 = \cos \omega_2 t$. Here we see that the truss vibrates about the displaced position given by Eq. (c).

### Example 4.8

The prismatic cantilever beam analyzed in Example 3.2 is shown again in Fig. 4.8. We shall determine the steady-state response of node 2 caused by a rigid-body rotational acceleration $\ddot{D}_{g6} = \ddot{\theta}_z \sin \Omega t$ of ground at node 1.

Stiffness and mass matrices for node 2 are

$$\mathbf{S} = s \begin{bmatrix} 6 & -3L \\ -3L & 2L^2 \end{bmatrix} \qquad \mathbf{M} = m \begin{bmatrix} 78 & -11L \\ -11L & 2L^2 \end{bmatrix} \quad \text{(i)}$$

where $s = 2EI/L^3$ and $m = \rho AL/210$. Normalization of the modal matrix (from Example 3.2) with respect to the mass matrix yields

$$\mathbf{\Phi}_N = \frac{1}{L\sqrt{m}} \begin{bmatrix} 0.1394L & 0.1943L \\ 0.1921 & 1.481 \end{bmatrix} \quad \text{(j)}$$

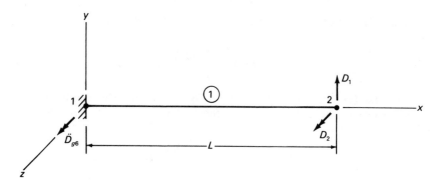

**Figure 4.8** Cantilever beam with rigid-body ground acceleration.

Due to the ground acceleration at node 1, the accelerations at node 2 are

$$\ddot{\mathbf{A}}_g = \mathbf{T}_g^T \ddot{\mathbf{D}}_g = \begin{bmatrix} L \\ 1 \end{bmatrix} \ddot{\theta}_z \sin \Omega t \tag{k}$$

as given by Eq. (14). In this case, the 2 × 1 operator $\mathbf{T}_g^T$ in Eq. (k) has only terms from rows 2 and 6 and column 6 of the general operator in Eq. (4). Equivalent nodal loads (a force and a moment) at node 2 become

$$\mathbf{A}_g^* = -\mathbf{M}\ddot{\mathbf{A}}_g = -\begin{bmatrix} 67 \\ -9L \end{bmatrix} mL\ddot{\theta}_z \sin \Omega t \tag{\ell}$$

which is dictated by Eq. (17). Then the normal-mode loads are

$$\mathbf{A}_N^* = \mathbf{\Phi}_N^T \mathbf{A}_g^* = -\begin{bmatrix} 7.613 \\ -0.3109 \end{bmatrix} L\sqrt{m}\,\ddot{\theta}_z \sin \Omega t \tag{m}$$

From these actions we find the steady-state normal-mode responses to be

$$\mathbf{D}_N^* = -\begin{bmatrix} 7.613\beta_1/\omega_1^2 \\ -0.3109\beta_2/\omega_2^2 \end{bmatrix} L\sqrt{m}\,\ddot{\theta}_z \sin \Omega t \tag{n}$$

Upon transforming this vector back to physical coordinates, we find the relative displacements:

$$\mathbf{D}^* = \mathbf{\Phi}_N \mathbf{D}_N^* = -\begin{bmatrix} (1.061 b_1 - 0.06041 b_2)L \\ 1.461 b_1 - 0.4604 b_2 \end{bmatrix} \ddot{\theta}_z \sin \Omega t \tag{o}$$

where $b_1 = \beta_1/\omega_1^2$ and $b_2 = \beta_2/\omega_2^2$.

When a structure has multiple connections to ground, it is also possible to calculate the response to *independent motions of support restraints* by generating the appropriate stiffness and mass coefficients [1, 5]. In such a case, the relative displacements of the supports must be small in order to retain linear behavior. Let us rewrite the undamped equations of motion for all possible nodal displacements in rearranged and partitioned form, as follows:

## Sec. 4.5  Normal-Mode Response to Support Motions

$$\begin{bmatrix} \mathbf{M}_{FF} & \mathbf{M}_{FR} \\ \mathbf{M}_{RF} & \mathbf{M}_{RR} \end{bmatrix} \begin{bmatrix} \ddot{\mathbf{D}}_F \\ \ddot{\mathbf{D}}_R \end{bmatrix} + \begin{bmatrix} \mathbf{S}_{FF} & \mathbf{S}_{FR} \\ \mathbf{S}_{RF} & \mathbf{S}_{RR} \end{bmatrix} \begin{bmatrix} \mathbf{D}_F \\ \mathbf{D}_R \end{bmatrix} = \begin{bmatrix} \mathbf{A}_F \\ \mathbf{A}_R \end{bmatrix} \quad (3.5\text{-}17)$$

As in Sec. 3.5, the subscript $F$ denotes free displacements, while the subscript $R$ pertains to restraint displacements. Writing these equations in two parts produces

$$\mathbf{M}_{FF}\ddot{\mathbf{D}}_F + \mathbf{M}_{FR}\ddot{\mathbf{D}}_R + \mathbf{S}_{FF}\mathbf{D}_F + \mathbf{S}_{FR}\mathbf{D}_R = \mathbf{A}_F \quad (3.5\text{-}18\text{a})$$

and

$$\mathbf{M}_{RF}\ddot{\mathbf{D}}_F + \mathbf{M}_{RR}\ddot{\mathbf{D}}_R + \mathbf{S}_{RF}\mathbf{D}_F + \mathbf{S}_{RR}\mathbf{D}_R = \mathbf{A}_R \quad (3.5\text{-}18\text{b})$$

Rearranging Eq. (3.5-18a) gives

$$\mathbf{M}_{FF}\ddot{\mathbf{D}}_F + \mathbf{S}_{FF}\mathbf{D}_F = \mathbf{A}_F - \mathbf{S}_{FR}\mathbf{D}_R - \mathbf{M}_{FR}\ddot{\mathbf{D}}_R \quad (19)$$

In this form we can see that the terms

$$\mathbf{A}_{FR} = -\mathbf{S}_{FR}\mathbf{D}_R - \mathbf{M}_{FR}\ddot{\mathbf{D}}_R \quad (20)$$

on the right-hand side of Eq. (19) are equivalent nodal loads. They are caused by the independent restraint displacements in the vector $\mathbf{D}_R$ and their corresponding accelerations in $\ddot{\mathbf{D}}_R$. After the displacements $\mathbf{D}_F$ and the accelerations $\ddot{\mathbf{D}}_F$ are found by normal-mode analysis, the reactions $\mathbf{A}_R$ at support points may be obtained from Eq. (3.5-18b) if desired. Of course, any restraints without motions may be represented with zeros in vectors $\mathbf{D}_R$ and $\ddot{\mathbf{D}}_R$ (or omitted altogether).

### Example 4.9

We shall now reconsider the plane truss in Example 4.7, which appears in Fig. 4.7. Let the support point 3 have a sudden independent step translation $D_{R3} = d$ in the $x$ direction. Then determine the response of the free displacements at joint 1.

By including terms for the restraint displacement at point 3, we can extend the stiffness matrix in Eqs. (a) to become

$$\mathbf{S} = \begin{bmatrix} \mathbf{S}_{FF} & \mathbf{S}_{FR} \\ \mathbf{S}_{RF} & \mathbf{S}_{RR} \end{bmatrix} = s \begin{bmatrix} 0.36 & -0.48 & -0.36 \\ -0.48 & 1.64 & 0.48 \\ -0.36 & 0.48 & 0.36 \end{bmatrix} \quad (p)$$

Stiffnesses for unmoving restraints are omitted from Eq. (p). Similarly, the mass matrix from Eqs. (a) is extended to

$$\mathbf{M} = \begin{bmatrix} \mathbf{M}_{FF} & \mathbf{M}_{FR} \\ \mathbf{M}_{RF} & \mathbf{M}_{RR} \end{bmatrix} = m_1 \begin{bmatrix} 3.28 & 0 & 1 \\ 0 & 3.28 & 0 \\ 1 & 0 & 2 \end{bmatrix} \quad (q)$$

where $m_1 = \rho AL/6$. Then from Eq. (20), the equivalent nodal loads due to the step displacement $D_{R3} = d$ are

$$\mathbf{A}_{FR} = -\mathbf{S}_{FR}\mathbf{D}_R - \mathbf{M}_{FR}\ddot{\mathbf{D}}_R = -s\begin{bmatrix} -0.36 \\ 0.48 \end{bmatrix} d = \begin{bmatrix} 0.36 \\ -0.48 \end{bmatrix} sd \quad (r)$$

Next, the normal-mode loads may be calculated as

$$\mathbf{A}_N = \mathbf{\Phi}_N^T \mathbf{A}_{FR} = \begin{bmatrix} 0.6 \\ -1.8 \end{bmatrix} \frac{sd}{\sqrt{10m}} \tag{s}$$

From these actions, we find the normal-mode responses to be

$$\mathbf{D}_N = \begin{bmatrix} 0.6(1 - \cos \omega_1 t)/\omega_1^2 \\ -1.8(1 - \cos \omega_2 t)/\omega_2^2 \end{bmatrix} \frac{sd}{\sqrt{10m}} = \begin{bmatrix} 3(1 - \cos \omega_1 t) \\ -(1 - \cos \omega_2 t) \end{bmatrix} \frac{md}{\sqrt{10m}} \tag{t}$$

Finally, the transformation back to physical coordinates yields

$$\mathbf{D} = \mathbf{\Phi}_N \mathbf{D}_N = \begin{bmatrix} 10 - 9c_1 - c_2 \\ -3c_1 + 3c_2 \end{bmatrix} \frac{d}{10} \tag{u}$$

In this case, joint 1 vibrates about the displaced position:

$$\mathbf{D}_{st} = \begin{bmatrix} d \\ 0 \end{bmatrix} \tag{v}$$

obtained by static analysis.

## 4.6 DAMPING IN MDOF SYSTEMS

Damping in solids and structures is not understood as well as stiffness and mass properties. Often the effects of damping upon the response of a vibratory system can be ignored, as has been done in Secs. 4.2 through 4.6. For example, the influence of a small amount of damping on the response of a structure during an excitation of short duration is not likely to be significant. In addition, damping plays a minor role in the steady-state response of a system to a periodic forcing function when the frequency of the excitation is not near a resonance. However, for a periodic function with a frequency at or near a natural frequency, damping is of primary importance and must be taken into account. Because its effects are usually not known in advance, damping should ordinarily be included in a vibrational analysis until its importance is ascertained.

When a discretized solid or structure is assumed to have *viscous damping*, the action equations of motion may be written as

$$\mathbf{M}\ddot{\mathbf{D}} + \mathbf{C}\dot{\mathbf{D}} + \mathbf{S}\mathbf{D} = \mathbf{A} \tag{1}$$

which applies only to free nodal displacements. The *damping matrix* $\mathbf{C}$ premultiplying the velocity vector $\dot{\mathbf{D}}$ in Eq. (1) has the general form

$$\mathbf{C} = \begin{bmatrix} C_{11} & C_{12} & C_{13} & \cdots & C_{1n} \\ C_{21} & C_{22} & C_{23} & \cdots & C_{2n} \\ C_{31} & C_{32} & C_{33} & \cdots & C_{3n} \\ \cdots & \cdots & \cdots & \cdots & \cdots \\ C_{n1} & C_{n2} & C_{n3} & \cdots & C_{nn} \end{bmatrix} \tag{2}$$

## Sec. 4.6 Damping in MDOF Systems

This matrix contains *damping coefficients* that are defined as actions required for unit velocities. That is, any term $C_{jk}$ in an array of viscous damping coefficients is an action of type $j$ equilibrating damping actions associated with a unit velocity of type $k$. This definition is similar to those for stiffness and mass terms and implies that the damping matrix is also symmetric.

To form the damping matrix, we consider first the systems for which this array is assumed to be linearly related to the mass and stiffness matrices. That is, we take

$$\mathbf{C} = a\mathbf{M} + b\mathbf{S} \qquad (3)$$

where $a$ and $b$ are constants. The formula in Eq. (3), attributed to Rayleigh [3], is called *proportional damping* because the matrix $\mathbf{C}$ is proportional to a linear combination of $\mathbf{S}$ and $\mathbf{M}$. In such a case the equations of motion [Eq. (1)] are uncoupled by the same transformation as that for the undamped system. Thus, in principal coordinates we have

$$\mathbf{M}_P \ddot{\mathbf{D}}_P + \mathbf{C}_P \dot{\mathbf{D}}_P + \mathbf{S}_P \mathbf{D}_P = \mathbf{A}_P \qquad (4)$$

where

$$\mathbf{C}_P = \mathbf{\Phi}^T \mathbf{C} \mathbf{\Phi} = a\mathbf{M}_P + b\mathbf{S}_P \qquad (5)$$

The symbol $\mathbf{C}_P$ represents a diagonal array that will be referred to as a *principal damping matrix*, and it consists of a linear combination of $\mathbf{M}_P$ and $\mathbf{S}_P$. When the modal matrix is normalized with respect to $\mathbf{M}$, the damping matrix in normal coordinates becomes

$$\mathbf{C}_N = \mathbf{\Phi}_N^T \mathbf{C} \mathbf{\Phi}_N = a\mathbf{I} + b\boldsymbol{\omega}^2 \qquad (6)$$

The diagonal matrix $\boldsymbol{\omega}^2$ in this expression contains the characteristic values $\omega_i^2$ for the undamped case [see Eq. (4.2-17)]. Therefore, the $i$th equation of motion in normal coordinates is

$$\ddot{D}_{Ni} + (a + b\omega_i^2)\dot{D}_{Ni} + \omega_i^2 D_{Ni} = A_{Ni} \qquad (i = 1, 2, \ldots, n) \qquad (7)$$

To make this expression analogous to that for a SDOF system (see Chapter 2), we introduce the notations

$$C_{Ni} = 2n_i = a + b\omega_i^2 \qquad \gamma_i = \frac{n_i}{\omega_i} \qquad (8)$$

In these relationships the term $C_{Ni} = 2n_i$ is defined as the *modal damping constant* for the $i$th normal mode, and $\gamma_i$ represents the corresponding *modal damping ratio*. Using the first of these definitions in Eq. (7), we obtain

$$\ddot{D}_{Ni} + 2n_i \dot{D}_{Ni} + \omega_i^2 D_{Ni} = A_{Ni} \qquad (i = 1, 2, \ldots, n) \qquad (9)$$

Each of the $n$ equations represented by this expression is uncoupled from all of the others. Therefore, we can determine the response of the $i$th mode in the same manner as that for a SDOF system with viscous damping.

From the definitions in Eqs. (8), we may express the modal damping ratio $\gamma_i$ in terms of the constants $a$ and $b$, as follows:

$$\gamma_i = \frac{a + b\omega_i^2}{2\omega_i} \tag{10}$$

This formula is useful for studying the effects upon the modal damping of varying the constants $a$ and $b$ in Eq. (3). For example, setting the constant $a$ equal to zero (while $b$ is nonzero) implies that the damping matrix is proportional to the stiffness matrix. This type of damping is sometimes referred to as *relative damping* because it is associated with relative velocities of displacement coordinates. Thus, under the condition that $a = 0$, Eq. (10) becomes

$$\gamma_i = \frac{b\omega_i}{2} \tag{11}$$

which means that the damping ratio in each principal mode is proportional to the undamped angular frequency of that mode. Therefore, the responses of the higher modes of a system will be damped out more quickly than those of the lower modes.

On the other hand, setting $b$ equal to zero (while $a$ is nonzero) implies that the damping matrix is proportional to the mass matrix. This type of damping is sometimes called *absolute damping* because it is associated with absolute velocities of displacement coordinates. In this case Eq. (10) simplifies to

$$\gamma_i = \frac{a}{2\omega_i} \tag{12}$$

so that the damping ratio in each mode is inversely proportional to the undamped angular frequency. Under this condition the lower modes of a system will be suppressed more strongly than the higher modes.

It has been shown by Caughy [6] that the criterion given by Eq. (3) is sufficient but not necessary for the existence of principal modes in damped systems. The essential condition is that the transformation which diagonalizes the damping matrix also uncouples the equations of motion. This criterion is less restrictive than that in Eq. (3) and encompasses more possibilities.

However, in the most general case, the damping coefficients in matrix $\mathbf{C}$ are such that the damping matrix cannot be diagonalized simultaneously with the mass and stiffness matrices. In this instance, the natural modes that do exist have phase relationships that complicate the analysis. The eigenvalues for this type of system are either real and negative or complex with negative real parts. The complex eigenvalues occur as conjugate pairs, and the corresponding eigenvectors also consist of complex conjugate pairs. In highly damped systems, where the imaginary terms due to dissipative forces are significant, the method of Foss [7] may be used. This approach involves transformation of the $n$ second-order equations of motion into $2n$ uncoupled first-order equations.

## Sec. 4.6  Damping in MDOF Systems

Lightly damped structures need not be treated in such a complicated manner, especially in view of the fact that the nature of damping in physical systems is not well understood. The simplest approach consists of assuming that the equations of motion are uncoupled by use of the modal matrix obtained for the structure without damping. In other words, the matrix $\boldsymbol{\Phi}$ is assumed to be orthogonal with respect to not only $\mathbf{M}$ and $\mathbf{S}$ but also $\mathbf{C}$, as follows:

$$\boldsymbol{\Phi}_j^T \mathbf{C} \, \boldsymbol{\Phi}_i = \boldsymbol{\Phi}_i^T \mathbf{C} \, \boldsymbol{\Phi}_j = 0 \qquad (i \neq j) \tag{13}$$

This expression implies that any off-diagonal terms resulting from the operation $\mathbf{C}_P = \boldsymbol{\Phi}^T \mathbf{C} \, \boldsymbol{\Phi}$ are small and can be neglected. In addition, it is more convenient to obtain experimentally (or to assume) the damping ratio $\gamma_i$ for the natural modes of vibration than to determine the damping coefficients in matrix $\mathbf{C}$ directly. We can usually find the damping ratio $\gamma_1$ for the first mode of vibration by field testing a structure or by previous experience. As mentioned in Sec. 2.4, the range of this constant for metal structures is approximately 0.01 to 0.05, while that for reinforced concrete is about 0.05 to 0.10. With the value of $\gamma_1$ on hand, we can extrapolate to other values of $\gamma_i$ using the approximate formula:

$$\gamma_i \approx \gamma_1 \left(\frac{\omega_i}{\omega_1}\right)^{e_1} \qquad (0.5 \leq e_1 \leq 0.7) \tag{14}$$

This expression suppresses the higher modes in accordance with damping experiments [8], but not as severely as in Eq. (11). Alternatively, we can simply determine $\gamma_1$ and then let $\gamma_i = \gamma_1$ for all other modes.

Now we rewrite Eq. (9) in terms of $\gamma_i$ as

$$\ddot{D}_{Ni} + 2\gamma_i \omega_i \dot{D}_{Ni} + \omega_i^2 D_{Ni} = A_{Ni} \qquad (i = 1, 2, \ldots, n) \tag{15}$$

where $C_{Ni} = 2\gamma_i \omega_i$. In order that this equation may pertain to a lightly damped structure, let us also specify that $0 \leq \gamma_i \leq 0.20$ for all modes. The type of damping associated with this set of assumptions is of great practical value, and it will be referred to simply as *modal damping*. It should be remembered that this concept is based on the normal coordinates for the undamped system and that damping ratios are specified in those coordinates.

When modal damping is assumed in the normal coordinates for a structure, it may also be of interest to determine the damping matrix $\mathbf{C}$ in the original (or physical) coordinates. This array can be found by means of the reverse transformation

$$\mathbf{C} = \boldsymbol{\Phi}_N^{-T} \mathbf{C}_N \boldsymbol{\Phi}_N^{-1} \tag{16}$$

Instead of attempting to invert $\boldsymbol{\Phi}_N$, however, we use the relationship $\boldsymbol{\Phi}_N^{-1} = \boldsymbol{\Phi}_N^T \mathbf{M}$ [see Eq. (4.2-27)] and rewrite Eq. (16) as

$$\mathbf{C} = \mathbf{M} \, \boldsymbol{\Phi}_N \mathbf{C}_N \boldsymbol{\Phi}_N^T \mathbf{M} \tag{17}$$

This form of the transformation is especially appropriate when not all of the natural modes are included in the analysis (modal truncation).

**Example 4.10**

As an example of modal and physical damping, we shall reconsider the 3-DOF plane truss analyzed in Examples 4.1 and 4.2. For this case, let us assume that the structure is made of steel and that the damping ratio for the first mode of vibration is $\gamma_1 = 0.02$. From Eq. (4.2-f) the three angular frequencies are

$$\omega_1 = 0.5197 c_1 \qquad \omega_2 = 1.445 c_1 \qquad \omega_3 = 2.304 c_1 \qquad (a)$$

where $c_1 = (\sqrt{E/\rho})/L$. Applying Eq. (14) with $e_1 = 0.6$., we find that

$$\gamma_2 = 0.02 \left(\frac{1.445}{0.5197}\right)^{0.6} = 0.03694$$

$$\gamma_3 = 0.02 \left(\frac{2.304}{0.5197}\right)^{0.6} = 0.04887 \qquad (b)$$

Then the normal-mode damping constants are

$$C_{N1} = 2(0.02)(0.5197)c_1 = 0.02079 c_1$$
$$C_{N2} = 2(0.03694)(1.445)c_1 = 0.1068 c_1 \qquad (c)$$
$$C_{N3} = 2(0.04887)(2.304)c_1 = 0.2252 c_1$$

Using these values in Eq. (16) along with $\Phi_N^{-1}$ from Eq. (4.2-h), we obtain

$$\mathbf{C} = \Phi_N^{-T} \mathbf{C}_N \Phi_N^{-1}$$

$$= A\sqrt{\rho E} \begin{bmatrix} 0.06154 & -0.01270 & -0.01662 \\ -0.01270 & 0.02002 & 0.01987 \\ -0.01662 & 0.01987 & 0.07620 \end{bmatrix} \qquad (d)$$

which is the symmetrical damping matrix in physical coordinates.

## 4.7 DAMPED RESPONSE TO PERIODIC FORCING FUNCTIONS

As mentioned in Sec. 4.6, damping is of greatest importance when a periodic excitation has a frequency that is close to one of the natural frequencies of a MDOF system. In this section we consider the normal-mode approach for calculating steady-state responses of discretized structures to periodic forcing functions. Knowing the imposed frequency of such a function and the natural frequencies of the system, we can obtain in a direct manner the steady-state responses of the modes having frequencies in the vicinity of the imposed frequency. Both simple harmonic and general periodic forcing functions will be discussed, and modal damping will be assumed, as described in Sec. 4.6.

If a lightly damped structure is subjected to a set of actions that are all proportional to the simple harmonic function $\cos \Omega t$, the action vector $\mathbf{A}$ may be written as

$$\mathbf{A} = \mathbf{P} \cos \Omega t \qquad (1)$$

### Sec. 4.7  Damped Response to Periodic Forcing Functions

where

$$\mathbf{P} = \{P_1, P_2, P_3, \ldots, P_n\} \tag{2}$$

In Eq. (1) the terms in $\mathbf{P}$ act as scale factors on the function $\cos \Omega t$. Transformation of the action equations of motion to normal coordinates produces the typical modal equation

$$\ddot{D}_{Ni} + 2n_i \dot{D}_{Ni} + \omega_i^2 D_{Ni} = p_{mi} \cos \Omega t \qquad (i = 1, 2, \ldots, n) \tag{3}$$

in which $p_{mi}$ is a constant. This equation has the same form as Eq. (2.4-23), so we can take the damped steady-state response of the $i$th mode to be

$$D_{Ni} = \frac{p_{mi}}{\omega_i^2} \beta_i \cos(\Omega t - \theta_i) \tag{4}$$

The magnification factor $\beta_i$ in this expression is

$$\beta_i = \frac{1}{\sqrt{[1 - (\Omega/\omega_i)^2]^2 + (2\gamma_i \Omega/\omega_i)^2}} \tag{5}$$

and the phase angle $\theta_i$ is

$$\theta_i = \tan^{-1}\left[\frac{2\gamma_i \Omega/\omega_i}{1 - (\Omega/\omega_i)^2}\right] \tag{6}$$

Equations (4), (5), and (6) are drawn from Eqs. (2.4-31), (2.4-32), and (2.4-33), respectively. The response given by Eq. (4) may then be transformed back to the original coordinates in the usual manner, using Eq. (4.3-5).

To determine the response of the mode having its angular frequency $\omega_i$ closest to the impressed angular frequency, we need only use the modal column $\mathbf{\Phi}_{Ni}$ in the transformations to and from normal coordinates. That is, Eq. (4.4-4) is specialized to

$$p_{mi} = \mathbf{\Phi}_{Ni}^T \mathbf{P} \tag{7}$$

and the back-transformation in Eq. (4.3-5) becomes

$$\mathbf{D} = \mathbf{\Phi}_{Ni} D_{Ni} \tag{8}$$

If desired, this process can be repeated for other modes with frequencies in the vicinity of $\Omega$.

Now we shall consider a lightly damped structure subjected to a set of actions that are all proportional to the general periodic function $f(t)$. In this case the applied action vector $\mathbf{A}$ may be written as

$$\mathbf{A} = \mathbf{F}(t) = \mathbf{P} f(t) \tag{9}$$

where the vector $\mathbf{P}$ is given by Eq. (2). Proceeding as described in Sec. 2.5, we express $f(t)$ in the form of a Fourier series, as follows:

$$f(t) = a_0 + \sum_{j=1}^{\infty} (a_j \cos j\Omega t + b_j \sin j\Omega t) \tag{10}$$

which is drawn from Eq. (2.5-1) with $i$ replaced by $j$. The coefficients $a_j$, $b_j$, and $a_0$ in Eq. (10) may be evaluated as indicated by Eqs. (2.5-3).

Transformation of the action equations of motion to normal coordinates produces the typical modal equation

$$\ddot{D}_{Ni} + 2n_i \dot{D}_{Ni} + \omega_i^2 D_{Ni} = p_{mi} f(t) \qquad (i = 1, 2, \ldots, n) \qquad (11)$$

where $p_{mi}$ is again the constant given by Eq. (7). From the solution of Prob. 2.5-5, we take the damped steady-state response of the $i$th mode to be

$$D_{Ni} = \frac{p_{mi}}{\omega_i^2} \left\{ a_0 + \sum_{j=1}^{\infty} \beta_{ij} [a_j \cos(j\Omega t - \theta_{ij}) + b_j \sin(j\Omega t - \theta_{ij})] \right\} \qquad (12)$$

in which the magnification factor $\beta_{ij}$ is

$$\beta_{ij} = \frac{1}{\sqrt{[1 - (j\Omega/\omega_i)^2]^2 + (2\gamma_i j\Omega/\omega_i)^2}} \qquad (13)$$

and the phase angle $\theta_{ij}$ is

$$\theta_{ij} = \tan^{-1}\left[\frac{2\gamma_i j\Omega/\omega_i}{1 - (j\Omega/\omega_i)^2}\right] \qquad (14)$$

Because a multiplicity of terms contribute to the $i$th mode in Eq. (12), the possibility of resonance ($j\Omega \approx \omega_i$) is much greater for a general periodic function than for a simple harmonic function. Therefore, it becomes more difficult to predict in advance which of the natural modes will be strongly affected. However, after the forcing function has been expressed as a Fourier series, each of the $j\Omega$ frequencies can be compared with the $\omega_i$ frequencies for the purpose of predicting large-amplitude forced vibrations.

### Example 4.11

Let the plane truss in Fig. 4.1(a) be subjected to a simple harmonic forcing function $P_2 \sin \Omega t$, applied in the $x$ direction at joint 2 (corresponding to displacement $D_2$). The angular frequency of the forcing function is $\Omega = (0.5/L)\sqrt{E/\rho}$. Calculate the steady-state response of the structure, assuming that the modal damping ratios are $\gamma_1 = 0.02$, $\gamma_2 = 0.035$, and $\gamma_3 = 0.05$ (see Example 4.10).

The square of the impressed angular frequency is

$$\Omega^2 = \frac{0.25E}{L^2 \rho} \qquad (a)$$

This value is fairly close to the square of the first angular frequency

$$\omega_1^2 = \frac{0.2701E}{L^2 \rho} \qquad (b)$$

as given in Table 3.4. Therefore, we should expect the first mode of the structure to be the primary contributor to the response. Using Eq. (7), we determine the first normal-mode load scale factor to be

Sec. 4.7  Damped Response to Periodic Forcing Functions    171

$$p_{m1} = \mathbf{\Phi}_{N1}^T \mathbf{P} = \frac{1}{\sqrt{\rho AL}}[0.2803 \quad 1.2114 \quad -0.2995]\begin{bmatrix} 0 \\ P_2 \\ 0 \end{bmatrix}$$

$$= \frac{1.2114 P_2}{\sqrt{\rho AL}}$$
(c)

The magnification factor for the first mode is obtained from Eq. (5) as

$$\beta_1 = \frac{1}{\sqrt{\left(1 - \frac{0.25}{0.2701}\right)^2 + (0.04)^2\left(\frac{0.25}{0.2701}\right)}} = 11.94 \tag{d}$$

From Eq. (4) we find the damped steady-state response of the first mode to be

$$D_{N1} = \frac{1.2114 P_2 L^2 \rho}{0.2701 E \sqrt{\rho AL}}(11.94)\sin(\Omega t - \theta_1)$$

$$= \frac{53.55 P_2 L^2 \rho}{E\sqrt{\rho AL}} \sin(\Omega t - \theta_1) \tag{e}$$

where

$$\theta_1 = \tan^{-1}\left[\frac{\frac{(0.04)(0.5)}{(0.5197)}}{1 - \frac{0.25}{0.2701}}\right] = 27° \; 21' \tag{f}$$

as given by Eq. (6). Transforming the response of the first mode back to the original coordinates with Eq. (8), we obtain

$$\mathbf{D} = \mathbf{\Phi}_{N1} D_{N1} = \{15.01, \; 64.87, \; -16.04\}\frac{P_2 L}{EA}\sin(\Omega t - \theta_1) \tag{g}$$

Proceeding in a similar manner, we can determine the response contributed by the second mode as

$$\mathbf{D} = \{-0.09478, \; 0.01875, \; -0.1093\}\frac{P_2 L}{EA}\sin(\Omega t - \theta_2) \tag{h}$$

and that due to the third mode is found to be

$$\mathbf{D} = \{-0.1824, \; 0.1104, \; 0.1114\}\frac{P_2 L}{EA}\sin(\Omega t - \theta_3) \tag{i}$$

The amplitudes in both of these vectors are small compared to those in Eq. (g). Furthermore, the influence of damping is significant for the first mode but negligible for the other two.

**Example 4.12**

Figure 4.9 shows a periodic forcing function in the shape of a square wave. If this function is applied in the $x$ direction at joint 1 of the truss in Fig. 4.1(a), find the damped steady-state response for each of the normal modes.

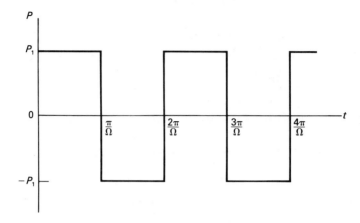

**Figure 4.9** Periodic forcing function.

Expanding the square wave as a Fourier series (see Prob. 2.5-1), we obtain

$$F(t) = P_1 f(t) = \frac{4P_1}{\pi}\left(\sin \Omega t + \frac{1}{3}\sin 3\Omega t + \ldots\right) \quad (j)$$

Transformation of the load vector to normal coordinates produces

$$\mathbf{A}_N = \mathbf{\Phi}_N^T \mathbf{A} = \mathbf{\Phi}_N^T \begin{bmatrix} F(t) \\ 0 \\ 0 \end{bmatrix} = \begin{bmatrix} \Phi_{N11} \\ \Phi_{N12} \\ \Phi_{N13} \end{bmatrix} F(t) \quad (k)$$

in which the load corresponds to displacement $D_1$. In accordance with Eq. (12), the normal-mode responses become

$$\mathbf{D}_N = \frac{4P_1}{\pi} \begin{bmatrix} \Phi_{N11}\left\{\beta_{11}\sin(\Omega t - \theta_{11}) + \dfrac{\beta_{13}}{3}\sin(3\Omega t - \theta_{13}) + \ldots\right\}\big/\omega_1^2 \\ \Phi_{N12}\left\{\beta_{21}\sin(\Omega t - \theta_{21}) + \dfrac{\beta_{23}}{3}\sin(3\Omega t - \theta_{23}) + \ldots\right\}\big/\omega_2^2 \\ \Phi_{N13}\left\{\beta_{31}\sin(\Omega t - \theta_{31}) + \dfrac{\beta_{33}}{3}\sin(3\Omega t - \theta_{33}) + \ldots\right\}\big/\omega_3^2 \end{bmatrix} \quad (\ell)$$

where the magnification factors and phase angles are given by Eqs. (13) and (14), respectively.

## 4.8 DAMPED RESPONSE TO ARBITRARY FORCING FUNCTIONS

Damping should be included in transient response calculations whenever it might be significant. For example, if the duration of a forcing function is relatively long compared to the natural periods of a structure, damping could be important. Also, when the time of interest is short but the modal damping ratios

## Sec. 4.8  Damped Response to Arbitrary Forcing Functions

are relatively high ($\gamma_i > 0.05$), the effects of damping could have some consequence. Therefore, we shall now modify the formulations in Secs. 4.3, 4.4, and 4.5 to account for the influence of damping on vibrational response in normal coordinates. Action equations and modal damping will be assumed throughout the discussion.

In Sec. 4.3 we formulated the normal-mode responses of a MDOF structure to initial conditions of displacement and velocity at time $t = 0$. In the presence of damping, the free-vibrational response of the $i$th mode, given by Eq. (4.3-4), must be changed to

$$D_{Ni} = e^{-n_i t}\left(D_{N0i} \cos \omega_{di} t + \frac{\dot{D}_{N0i} + n_i D_{N0i}}{\omega_{di}} \sin \omega_{di} t\right) \quad (1)$$

which is drawn from Eq. (2.4-6). The angular frequency of damped vibration in Eq. (1) is

$$\omega_{di} = \sqrt{\omega_i^2 - n_i^2} = \omega_i \sqrt{1 - \gamma_i^2} \quad (2)$$

in which $\omega_i$ is the undamped angular frequency. Transformation of the initial-condition vectors $\mathbf{D}_0$ and $\dot{\mathbf{D}}_0$ to normal coordinates remains the same as in Eqs. (4.3-2), and back-transformation of the response is still given by Eq. (4.3-5).

Similarly, the calculation of normal-mode responses to applied actions, as described in Sec. 4.4, requires only a few modifications associated with modal damping. Transformation of applied actions to normal coordinates is the same as in Eq. (4.4-4), but Duhamel's integral in Eq. (4.4-7) must now be written as

$$D_{Ni} = \frac{e^{-n_i t}}{\omega_{di}} \int_0^t e^{n_i t'} A_{Ni} \sin \omega_{di}(t - t') \, dt' \quad (3)$$

which is taken from Eq. (2.6-4).

Normal-mode responses to support motions, covered in Sec. 4.5, may also be altered to include the effects of modal damping. For rigid-body ground accelerations, there is neither displacement coupling nor velocity coupling between the masses and the ground in relative coordinates. There exists only inertial coupling with ground, which is the same as that for the structure without damping. To determine nodal responses relative to the ground, we first calculate the equivalent nodal actions in the vector $\mathbf{A}_g^*$, as given by Eq. (4.5-17). Transformation of these actions to normal coordinates yields the equivalent modal loads

$$\mathbf{A}_N^* = \mathbf{\Phi}_N^T \mathbf{A}_g^* \quad (4)$$

Then the relative response $D_{Ni}^*$ in each normal coordinate is obtained from Eq. (3), with $A_{Ni}^*$ replacing $A_{Ni}$. Finally, these displacements are back-transformed using

$$\mathbf{D}^* = \mathbf{\Phi}_N \mathbf{D}_N^* \quad (5)$$

which gives the relative responses in physical coordinates. As before, the absolute displacements at free nodes may be found with Eq. (4.5-18).

In certain cases where support displacements are specified, there is velocity coupling between free displacement coordinates and support restraints. This situation can arise for either rigid-body ground displacements or independent motions of multiple restraints. Methods for handling such circumstances are described in Ref. 1.

**Example 4.13**

We shall now repeat the first part of Example 4.5, including the effects of modal damping. Recall that the three-member truss in Fig. 4.4(a) is subjected to a step force $P_2$, corresponding to displacement $D_2$.

Symbolically transforming the vector of applied actions to normal coordinates produces

$$\mathbf{A}_N = \mathbf{\Phi}_N^T \mathbf{A} = \mathbf{\Phi}_N^T \begin{bmatrix} 0 \\ P_2 \\ 0 \end{bmatrix} = \begin{bmatrix} \Phi_{N21} \\ \Phi_{N22} \\ \Phi_{N23} \end{bmatrix} P_2 \qquad (a)$$

Due to the step function, the damped normal-mode responses are

$$\mathbf{D}_N = P_2 \begin{bmatrix} \Phi_{N21}\left\{1 - e^{-n_1 t}\left(\cos \omega_{d1} t + \dfrac{n_1}{\omega_{d1}} \sin \omega_{d1} t\right)\right\}\Big/\omega_1^2 \\ \Phi_{N22}\left\{1 - e^{-n_2 t}\left(\cos \omega_{d2} t + \dfrac{n_2}{\omega_{d2}} \sin \omega_{d2} t\right)\right\}\Big/\omega_2^2 \\ \Phi_{N23}\left\{1 - e^{-n_3 t}\left(\cos \omega_{d3} t + \dfrac{n_3}{\omega_{d3}} \sin \omega_{d3} t\right)\right\}\Big/\omega_3^2 \end{bmatrix} \qquad (b)$$

These expressions are drawn from Eq. (2.7-2b).

**Example 4.14**

Suppose that the ground in Fig. 4.4(a) accelerates in the $y$ direction in accordance with the ramp function $\ddot{D}_{g2} = a_2 t/t_2$. Formulate the damped responses of the normal modes, starting from rest.

For this problem we work in relative coordinates, where the equivalent nodal load vector is

$$\mathbf{A}_g^* = -\mathbf{M}\,\ddot{\mathbf{\Delta}}_g = -\mathbf{M} \begin{bmatrix} 0 \\ \dfrac{a_2 t}{t_2} \\ 1 \end{bmatrix} = -\begin{bmatrix} 0 \\ 0 \\ 1 \end{bmatrix} M_{33} \dfrac{a_2 t}{t_2} \qquad (c)$$

as given by Eq. (4.5-17). Then the equivalent normal-mode loads are

$$\mathbf{A}_N^* = \mathbf{\Phi}_N^T \mathbf{A}_g^* = -\begin{bmatrix} \Phi_{N31} \\ \Phi_{N32} \\ \Phi_{N33} \end{bmatrix} M_{33} \dfrac{a_2 t}{t_2} \qquad (d)$$

For the ramp function, the damped normal-mode responses become

$$D_{Ni} = -\Phi_{N3i}\frac{M_{33}a_2}{\omega_i^4 t_2}\left[\omega_i^2 t - 2n_i + e^{-n_i t}\left(2n_i \cos\omega_{di}t - \frac{\omega_{di}^2 - n_i^2}{\omega_{di}}\sin\omega_{di}t\right)\right] \quad \text{(e)}$$

where $i = 1, 2, 3$. The expression in Eq. (e) is a modified version of Eq. (2.7-2c).

## 4.9 STEP-BY-STEP RESPONSE CALCULATIONS

In Sec. 2.7 we examined step-by-step solutions for SDOF structures, where the forcing functions are not necessarily analytical expressions. The basic approach in that section was to approximate the forcing function (or data points) using *piecewise-linear* interpolation and then to use the Duhamel integral within small time steps. We shall now incorporate this technique into the normal-mode method for calculating transient responses of MDOF structures. As in the preceding sections, modal damping will be assumed throughout. Because of the extensive calculations required, it is implied that the method of this section is to be programmed for a digital computer. Such a program is described in Sec. 4.10, where numerical examples are also presented.

Let us consider again the piecewise-linear type of interpolation illustrated by Fig. 2.18. Without loss of generality, only one such forcing function $f_\ell(\Delta t_j)$ will be handled at a time, and the piecewise-linear action vector $\mathbf{A}_{\ell j}$ (or $\mathbf{A}_{\ell j}^*$) may be expressed as

$$\mathbf{A}_{\ell j} = \mathbf{F}_\ell(\Delta t_j) = \mathbf{P}f_\ell(\Delta t_j) \quad (j+1 = 1, 2, \ldots, n_1) \quad (1)$$

where $\Delta t_j$ represents a small but finite time step, and $n_1$ is the number of steps. In this form the values of $\mathbf{P}$ act as scale factors on the common function $f_\ell(\Delta t_j)$. If more than one such function is applied simultaneously, the responses for each of them handled separately can be superimposed.

Transformation of the action equations of motion to normal coordinates produces the typical modal equation

$$\ddot{D}_{Ni} + 2n_i\dot{D}_{Ni} + \omega_i^2 D_{Ni} = A_{Ni,j} + \Delta A_{Ni,j}\frac{t'}{\Delta t_j} \quad (2)$$

$(i = 1, 2, \ldots, n)$ and $j + 1 = 1, 2, \ldots, n_1)$

where $t' = t - t_j$. The symbol $A_{Ni,j}$ in Eq. (2) represents the $i$th normal-mode load at time $t_j$. Thus,

$$A_{Ni,j} = \mathbf{\Phi}_{Ni}^T \mathbf{A}_{\ell j} = \mathbf{\Phi}_{Ni}^T \mathbf{P}f_\ell(\Delta t_j) \quad (3)$$

In addition, we have the change in the $i$th modal load during the time step $\Delta t_j$, defined as follows:

$$\Delta A_{Ni,j} = A_{Ni,j+1} - A_{Ni,j} \quad (4)$$

where the symbol $A_{ni,j+1}$ denotes the action at time $t_{j+1}$.

In a manner similar to that in Sec. 2.7, we can express the response of the $i$th mode at the end of the $j$th time step as the sum of three parts, which are

$$D_{Ni,j+1} = (D_{N1} + D_{N2} + D_{N3})_{i,j+1} \tag{5}$$

The first part of the response consists of the $i$th free-vibrational motion caused by the conditions of displacement and velocity at time $t_j$ (the beginning of the interval). Therefore, we have

$$(D_{N1})_{i,j+1} = e^{-n_i \Delta t_j} \left( D_{Ni,j} \cos \omega_{di} \Delta t_j + \frac{\dot{D}_{Ni,j} + n_i D_{Ni,j}}{\omega_{di}} \sin \omega_{di} \Delta t_j \right) \tag{6a}$$

This formula represents an extension of Eq. (2.7-5a).

The other two parts of the response in Eq. (5) are due to the linear forcing function within the time step. The rectangular portion of this impulse yields

$$(D_{N2})_{i,j+1} = \frac{A_{Ni,j}}{\omega_i^2} \left[ 1 - e^{-n_i \Delta t_j} \left( \cos \omega_{di} \Delta t_j + \frac{n_i}{\omega_{di}} \sin \omega_{di} \Delta t_j \right) \right] \tag{6b}$$

which is taken from Eq. (2.7-5b); and that associated with the triangular portion becomes

$$(D_{N3})_{i,j+1} = \frac{\Delta A_{Ni,j}}{\omega_i^4 \Delta t_j} \left[ \omega_i^2 \Delta t_j - 2n_i + e^{-n_i \Delta t_j} \left( 2n_i \cos \omega_{di} \Delta t_j - \frac{\omega_{di}^2 - n_i^2}{\omega_{di}} \sin \omega_{di} \Delta t_j \right) \right] \tag{6c}$$

which is drawn from Eq. (2.7-5c).

We can also write the velocity of the $i$th mode at the end of the $j$th time step in three parts, as follows:

$$\dot{D}_{Ni,j+1} = (\dot{D}_{N1} + \dot{D}_{N2} + \dot{D}_{N3})_{i,j+1} \tag{7}$$

These three contributions may be formed by extending the notation in Eqs. (2.7-6) to obtain

$$(\dot{D}_{N1})_{i,j+1} = e^{-n_i \Delta t_j} \left[ -\left( D_{Ni,j} \omega_{di} + n_i \frac{\dot{D}_{Ni,j} + n_i D_{Ni,j}}{\omega_{di}} \right) \sin \omega_{di} \Delta t_j + \dot{D}_{Ni,j} \cos \omega_{di} \Delta t_j \right] \tag{8a}$$

$$(\dot{D}_{N2})_{i,j+1} = \frac{A_{Ni,j}}{\omega_{di}} e^{-n_i \Delta t_j} \sin \omega_{di} \Delta t_j \tag{8b}$$

$$(\dot{D}_{N3})_{i,j+1} = \frac{\Delta A_{Ni,j}}{\omega_i^2 \Delta t_j} \left[ 1 - e^{-n_i \Delta t_j} \left( \cos \omega_{di} \Delta t_j + \frac{n_i}{\omega_{di}} \sin \omega_{di} \Delta t_j \right) \right] \tag{8c}$$

Equations (5) through (8) constitute recurrence formulas for calculating the damped response of each normal mode at the end of the $j$th time step. They also

Sec. 4.10  Program NOMO for Normal-Mode Response            177

provide the initial conditions of displacement and velocity at the beginning of step $j + 1$. These expressions may be applied repetitively to obtain the time history of response for each of the normal modes. Then the results for each time station are transformed back to the original coordinates in the usual manner.

If the $i$th mode of a structure is a rigid-body motion, appropriate expressions for rigid-body response must be used instead of the recurrence formulas given above. That is, the displacements in Eqs. (5) and (6) are replaced by

$$D_{Ni,j+1} = D_{Ni,j} + \dot{D}_{Ni,j}\Delta t_j + \tfrac{1}{2}(A_{Ni,j} + \tfrac{1}{3}\Delta A_{Ni,j})(\Delta t_j)^2 \tag{9}$$

and the velocities in Eqs. (7) and (8) are supplanted by

$$\dot{D}_{Ni,j+1} = \dot{D}_{Ni,j} + (A_{Ni,j} + \tfrac{1}{2}\Delta A_{Ni,j})\Delta t_j \tag{10}$$

Equations (9) and (10) pertain to rigid-body motions with no absolute damping.

## 4.10 PROGRAM NOMO FOR NORMAL-MODE RESPONSE

The normal-mode method for calculating dynamic responses of structures will now be applied in a program named NOMO. This program can be used to analyze any type of linearly elastic framed structure or discretized continuum. The main program for NOMO calls six subprograms, as shown by the double boxes in Flowchart 4.1. The first subprogram appearing in the flow chart is VIB, which is the program for vibrational analysis described previously in Sec. 3.8. Here it is treated as a subprogram that calls the seven other subprograms in Flowchart 3.1. However, as the last step in Subprogram RES1, the eigenvectors are normalized with respect to the mass matrix. Then the *number of loading systems* NLS is read, the *loading number* LN is initialized to zero, and LN is incremented by one. Next, Subprogram DYLO reads and writes dynamic load data, and the output includes the loading number LN as well as the number of loading systems NLS. This is followed by Subprogram TRANOR, which transforms initial conditions and actual or equivalent nodal loads to normal coordinates, using Eqs. (4.3-2) and (4.4-4). At its beginning, TRANOR reads and writes the number of modes NMODES for the purpose of using modal truncation.

Subprogram TIHIST calculates time histories of normal-mode displacements and velocities with the step-by-step method described in Sec. 4.9. Because uniform time steps are to be used, the coefficients of $D_{Ni,j}$, $\dot{D}_{Ni,j}$, $A_{Ni,j}$, and $\Delta A_{Ni,j}$ in Eqs. (4.9-5) through (4.9-10) become constants that need be determined only once at the beginning of the analysis. For vibrational motions, the responses in Eqs. (4.9-5) through (4.9-8) can be written in eight parts, as follows:

$$D_{Ni,j+1} = C_1 D_{Ni,j} + C_2 \dot{D}_{Ni,j} + C_3 A_{Ni,j} + C_4 \Delta A_{Ni,j} \tag{1}$$

$$\dot{D}_{Ni,j+1} = C_5 D_{Ni,j} + C_6 \dot{D}_{Ni,j} + C_7 A_{Ni,j} + C_8 \Delta A_{Ni,j} \tag{2}$$

**Flowchart 4.1    Main program for NOMO***

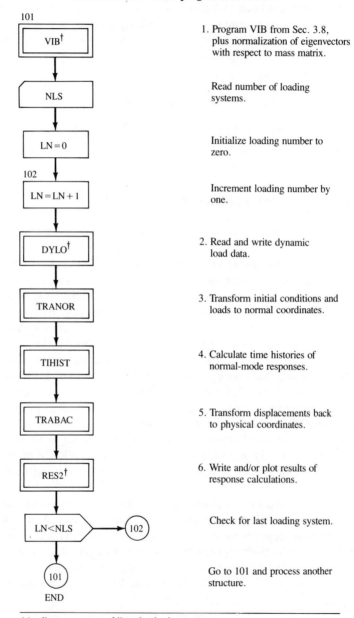

*Applies to any type of linearly elastic structure.
†Subprograms that differ for every type of structure.

## Sec. 4.10  Program NOMO for Normal-Mode Response

The constant coefficients $C_1$ through $C_8$ appearing in Eqs. (1) and (2) have the definitions

$$C_1 = e^{-n_i \Delta t}\left(\cos \omega_{di} \Delta t + \frac{n_i}{\omega_{di}} \sin \omega_{di} \Delta t\right)$$

$$C_2 = \frac{1}{\omega_{di}} e^{-n_i \Delta t} \sin \omega_{di} \Delta t \qquad C_3 = \frac{1}{\omega_i^2}(1 - C_1) \tag{3}$$

$$C_4 = \frac{1}{\omega_i^2 \Delta t}(\Delta t - C_2 - 2n_i C_3) \qquad C_5 = -\omega_i^2 C_2$$

$$C_6 = C_1 - 2n_i C_2 \qquad C_7 = C_2 \qquad C_8 = \frac{1}{\Delta t} C_3$$

These coefficients, as well as simpler constants for rigid-body motions [see Eqs. (4.9-9) and (4.9-10)], are coded in Subprogram TIHIST.

After the response calculations have been completed, the normal-mode displacements are transformed back to physical coordinates by Subprogram TRABAC, using Eq. (4.3-5). The last subprogram, named RES2, optionally writes and/or plots resulting time histories of nodal displacements and axial forces in members. After writing, the maximum and minimum values of these quantities and the times of occurrence are written as well. At the end of the flowchart the test of LN against NLS determines whether to return for another loading system or another structure.

The logic in Program NOMO implies that the time histories of nodal displacements are stored in a matrix of size NDF × NTS, where NDF is the number of degrees of freedom and NTS is the number of time steps. Although this procedure is conducive to plotting and calculation of internal actions or stresses, the use of such a large block of storage is not efficient. We could transfer blocks of these displacements to auxiliary storage if desired. However, this approach would require more data, more intricate logic, and more detailed explanations.

Program NOMO may be specialized to become NOMOCB for continuous beams, NOMOPT for plane trusses, and so on. The main program for each specialization has certain subprograms that are different for each type of structure, as indicated by the second footnote below Flowchart 4.1. That is, the subprogram named DYLO becomes DYLOCB for continuous beams, DYLOPT for plane trusses, and so on. As for Program VIB, notation for Program NOMO appears in Part 5 of the list of notation near the back of the book. Also, the flowchart for Program DYNAPT in Appendix C contains detailed steps for the logic in the subprograms of Program NOMO.

Table 4.1 shows preparation of *dynamic load data* for plane trusses. In the first line of the table are the number of time steps NTS, the duration of the uniform time step DT, and the damping ratio DAMPR, pertaining to all modes.

**TABLE 4.1 Dynamic Load Data for Plane Trusses**

| Type of Data | No. of Lines | Items on Data Lines |
|---|---|---|
| Dynamic parameters | 1 | NTS, DT, DAMPR |
| Initial conditions<br>(a) Condition parameters<br>(b) Displacements<br>(c) Velocities | <br>1<br>NNID<br>NNIV | <br>NNID, NNIV<br>J, D0(2J-1), D0(2J)<br>J, V0(2J-1), V0(2J) |
| Applied actions<br>(a) Load parameters<br>(b) Nodal loads<br>(c) Line loads | <br>1<br>NLN<br>NEL | <br>NLN, NEL<br>J, AS(2J-1), AS(2J)<br>I, BL1, BL2, BL3, BL4 |
| Ground accelerations<br>(a) Acceleration parameter<br>(b) Acceleration factors[a] | <br>1<br>1 | <br>IGA<br>GAX, GAY |
| Forcing function<br>(a) Function parameter<br>(b) Function ordinates | <br>1<br>NFO | <br>NFO<br>K, T(K), F0(K) |

[a]Omit when IGA = 0.

Next, the initial-condition parameters NNID and NNIV give the number of nodes with initial displacements and initial velocities, respectively. Each line of the data for initial displacements (NNID lines total) contains a node number J, the $x$-component of displacement D0(2J-1), and the $y$-component D0(2J). Similarly, the NNIV lines of data for initial velocities indicate terms that are analogous to those for initial displacements.

Applied actions and ground accelerations carry appropriate dimensions and are to be multiplied by a dimensionless forcing function given at the end of the table. In the data for applied actions, the load parameters are the number of loaded nodes NLN and the number of elements with line loads NEL. Data for nodal loads (NLN lines) consist of a node number J and scale factors for the $x$-component of force AS(2J-1) and the $y$-component AS(2J). Figure 4.10(a) depicts these components of the applied force acting at node $j$ of a plane truss. Also, Fig. 4.10(b) shows linearly varying line loads (force per unit length) applied in the $x$ and $y$ directions along the length of a typical plane-truss element $i$. Data for this condition of loading appear as part (c) under applied actions, where each of NEL lines contains an element number I and scale factors for the load intensities BL1 through BL4 shown in Fig. 4.10(b). It can easily be shown that the equivalent nodal loads in structural directions at joints $j$ and $k$ are

## Sec. 4.10 Program NOMO for Normal-Mode Response

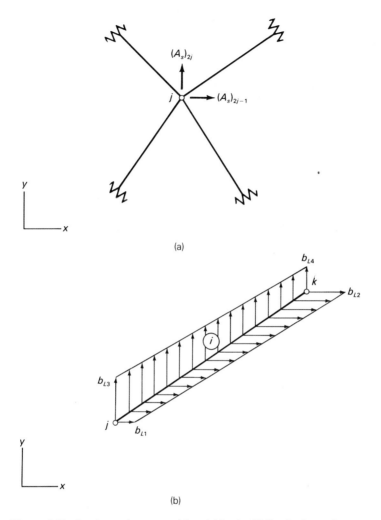

**Figure 4.10** Loads on plane truss: (a) nodal loads; (b) line loads on element.

$$AS(J1) = (2BL1 + BL3)L/6$$
$$AS(J2) = (2BL2 + BL4)L/6$$
$$AS(K1) = (BL1 + 2BL3)L/6$$
$$AS(K2) = (BL2 + 2BL4)L/6$$
(4)

where subscripts J1 through K2 are obtained from Eqs. (3.5-30). Of course, an infinity of other load sets could be applied to the element and their equivalent nodal loads derived.

Data for ground accelerations consist of two lines, the first of which gives a parameter IGA for ground accelerations. If this number is nonzero, ground accelerations exist; otherwise, a zero indicates nonexistence. On the second line are scale factors GAX and GAY for ground accelerations in the $x$ and $y$ directions. The possibility of rotational ground acceleration is omitted from this program.

For either applied actions or ground accelerations, the load data must include information defining a dimensionless *piecewise-linear forcing function*. Such a function is given as the last block of data in Table 4.1. On line (a) we have the number of function ordinates NFO. Ordinates of the forcing function are given in NFO lines, each of which contains a subscript K, the time T(K) when the function ordinate occurs, and the value of the function ordinate FO(K) at that time. For simplicity, we restrict the time T(K) to be equal to an even number of time steps DT. If the forcing function has a discontinuity at time T(K), two lines are required to define FO(K) on both sides of the discontinuity (for the same time). Note again that the function ordinates will receive dimensions only when they are multiplied within the program by either applied-action or ground-acceleration scale factors, thus creating time-varying *proportional loads*.

In all computer programs for dynamic analysis we write and plot selectively to limit the volume of output. Table 4.2 shows our method for selecting nodal displacements and element stresses for writing and/or plotting in Subprogram RES2. Line (a) of the table contains output parameters that have the following meanings:

$$\text{IWR} = \text{indicator for writing (0 or 1)}$$
$$\text{IPL} = \text{indicator for plotting* (0 or 1)}$$
$$\text{NNO} = \text{number of nodes for output}$$
$$\text{NEO} = \text{number of elements for output}$$

In line (b) the node numbers JNO( ) for output of displacements are listed, and element numbers IEO( ) for output of stresses appear in line (c) of the table. For framed structures, this output usually consists of generalized (or integrated) stresses at the $j$ and $k$ ends of members (see Chapter 6). However, for plane

**TABLE 4.2 Output Selection Data**

| Type of Data | No. of Lines | Items on Data Lines |
|---|---|---|
| Output Selection | | |
| (a) Output parameters | 1 | IWR, IPL, NNO, NEO |
| (b) Nodal displacements | NNO | JNO(1), JNO(2), . . . , JNO(NNO) |
| (c) Element stresses | NEO | IEO(1), IEO(2), . . . , IEO(NEO) |

*Because plotting capability varies among program users, the parameter IPL may be set equal to zero in all cases if desired.

Sec. 4.10  Program NOMO for Normal-Mode Response                                    183

trusses we calculate and write only the axial force at the $k$ end, which has the same sign as the axial force in an unloaded member. On the other hand, with two- and three-dimensional finite elements, the stresses are calculated at sampling points for numerical integration (see Chapters 7 and 8). If desired, we could specialize the variables JNO( ) and IEO( ) to have a second subscript denoting a particular type of nodal displacement or element stress.

**Example 4.15**

To illustrate using Program NOMOPT, we shall analyze the three-member plane truss used previously in Example 3.4. Dynamic responses of this structure will be obtained for the following influences:

1. Initial displacements of 0.1 in. at all nodal degrees of freedom, with DAMPR = 0.0
2. Piecewise-linear force in $x$ direction at node 2, with DAMPR = 0.0
3. Same as case 1, but with DAMPR = 0.1
4. Same as case 2, but with DAMPR = 0.1

In all cases we take the values of $E$, $\rho$, $L$, and $A$ as given in Example 3.4, where the material is steel and US units apply.

Table 4.3 contains a partial listing of the output from Program NOMOPT for the four analyses considered. With the formats coded in the program, this output becomes

**TABLE 4.3  Computer Output for Example 4.15**

```
PROGRAM NOMOPT

*** EXAMPLE 4.15: THREE-MEMBER PLANE TRUSS ***

STRUCTURAL PARAMETERS
  NN    NE   NRN         E           RHO
   3     3    2      3.0000E+04   7.3500E-07

NODAL COORDINATES
  NODE         X              Y
    1       0.000          0.000
    2     150.000        200.000
    3     150.000          0.000

ELEMENT INFORMATION
  ELEM.    J    K        AX          EL         CX        CY
    1      1    2      10.0000     250.0000    0.6000    0.8000
    2      1    3       6.0000     150.0000    1.0000    0.0000
    3      2    3       8.0000     200.0000    0.0000   -1.0000

NODAL RESTRAINTS
  NODE   NR1   NR2
    1     0     1
    3     1     1

NUMBER OF DEGREES OF FREEDOM:  NDF =    3
NUMBER OF NODAL RESTRAINTS:    NNR =    3

STIFFNESS MATRIX DECOMPOSED
```

**TABLE 4.3** (Continued)

```
MODE     1
ANGULAR FREQUENCY   4.1995E+02
   NODE         DJ1          DJ2
      1    2.3137E-01   0.0000E+00
      2    1.0000E+00  -2.4722E-01
      3    0.0000E+00   0.0000E+00

MODE     2
ANGULAR FREQUENCY   1.1677E+03
   NODE         DJ1          DJ2
      1    8.6725E-01   0.0000E+00
      2   -1.7149E-01   1.0000E+00
      3    0.0000E+00   0.0000E+00

MODE     3
ANGULAR FREQUENCY   1.8618E+03
   NODE         DJ1          DJ2
      1    1.0000E+00   0.0000E+00
      2   -6.0504E-01  -6.1068E-01
      3    0.0000E+00   0.0000E+00

*** LOADING NUMBER   1 OF   4 ***

DYNAMIC PARAMETERS
   NTS         DT         DAMPR
    20    1.0000E-03   0.0000E+00

INITIAL CONDITIONS
NNID NNIV
  2    0

INITIAL DISPLACEMENTS
   NODE         D01          D02
      1    1.0000E-01   0.0000E+00
      2    1.0000E-01   1.0000E-01

APPLIED ACTIONS
  NLN  NEL
   0    0

GROUND ACCELERATIONS
  IGA
   0

NORMAL MODE SOLUTION
NMODES =    3

OUTPUT SELECTION
  IWR  IPL  NNO  NEO
   1    1    2    1

NODES:       1    2

ELEMENTS:    1

DISPLACEMENT TIME HISTORY FOR NODE    1
  STEP        TIME          DJ1          DJ2
     0    0.0000E+00   1.0000E-01   0.0000E+00
     1    1.0000E-03   6.7053E-02   0.0000E+00
     2    2.0000E-03  -3.2735E-02   0.0000E+00
     3    3.0000E-03  -1.0087E-01   0.0000E+00
     4    4.0000E-03  -1.5500E-02   0.0000E+00
```

Sec. 4.10    Program NOMO for Normal-Mode Response                               **185**

**TABLE 4.3**  (Continued)

```
         5   5.0000E-03   9.7853E-02   0.0000E+00
         6   6.0000E-03   4.9458E-02   0.0000E+00
         7   7.0000E-03  -7.4544E-02   0.0000E+00
         8   8.0000E-03  -1.0603E-01   0.0000E+00
         9   9.0000E-03  -5.4252E-02   0.0000E+00
        10   1.0000E-02   2.7924E-02   0.0000E+00
        11   1.1000E-02   9.2962E-02   0.0000E+00
        12   1.2000E-02   4.0849E-02   0.0000E+00
        13   1.3000E-02  -8.1017E-02   0.0000E+00
        14   1.4000E-02  -6.9040E-02   0.0000E+00
        15   1.5000E-02   6.8279E-02   0.0000E+00
        16   1.6000E-02   1.1979E-01   0.0000E+00
        17   1.7000E-02   4.7138E-02   0.0000E+00
        18   1.8000E-02  -3.6778E-02   0.0000E+00
        19   1.9000E-02  -8.3692E-02   0.0000E+00
        20   2.0000E-02  -5.2695E-02   0.0000E+00

MAXIMUM                   1.1979E-01   0.0000E+00
TIME OF MAXIMUM           1.6000E-02   2.0000E-02
MINIMUM                  -1.0603E-01   0.0000E+00
TIME OF MINIMUM           8.0000E-03   2.0000E-02

DISPLACEMENT TIME HISTORY FOR NODE    2
    STEP        TIME         DJ1          DJ2
         0   0.0000E+00   1.0000E-01   1.0000E-01
         1   1.0000E-03   8.5266E-02   1.6390E-02
         2   2.0000E-03   7.2848E-02  -1.0676E-01
         3   3.0000E-03   6.0750E-02  -1.0298E-01
         4   4.0000E-03  -5.4078E-03   3.5295E-03
         5   5.0000E-03  -8.4207E-02   1.0144E-01
         6   6.0000E-03  -9.8212E-02   1.0809E-01
         7   7.0000E-03  -8.5678E-02   2.3188E-03
         8   8.0000E-03  -9.3358E-02  -9.5989E-02
         9   9.0000E-03  -8.2739E-02  -3.8360E-02
        10   1.0000E-02  -5.1067E-02   9.6986E-02
        11   1.1000E-02  -2.9195E-02   1.0992E-01
        12   1.2000E-02   1.9009E-02  -7.1293E-03
        13   1.3000E-02   9.6665E-02  -1.0699E-01
        14   1.4000E-02   1.2087E-01  -1.0634E-01
        15   1.5000E-02   8.8728E-02  -1.2480E-02
        16   1.6000E-02   7.6180E-02   8.6587E-02
        17   1.7000E-02   7.2053E-02   5.6728E-02
        18   1.8000E-02   3.4921E-02  -7.7855E-02
        19   1.9000E-02  -3.4530E-03  -1.1650E-01
        20   2.0000E-02  -3.8910E-02   2.4600E-03

MAXIMUM                   1.2087E-01   1.0992E-01
TIME OF MAXIMUM           1.4000E-02   1.1000E-02
MINIMUM                  -9.8212E-02  -1.1650E-01
TIME OF MINIMUM           6.0000E-03   1.9000E-02

MEMBER FORCE TIME HISTORY FOR ELEMENT    1
    STEP        TIME         AM1
         0   0.0000E+00   9.6000E+01
         1   1.0000E-03   2.8848E+01
         2   2.0000E-03  -2.6472E+01
         3   3.0000E-03   1.7503E+01
         4   4.0000E-03   1.0655E+01
         5   5.0000E-03  -3.3704E+01
         6   6.0000E-03  -2.5581E+00
```

**TABLE 4.3** (Continued)

```
 7  7.0000E-03  -5.7902E+00
 8  8.0000E-03  -8.3027E+01
 9  9.0000E-03  -5.7337E+01
10  1.0000E-02   3.6233E+01
11  1.1000E-02   1.7570E+01
12  1.2000E-02  -2.2569E+01
13  1.3000E-02   2.5220E+01
14  1.4000E-02   3.4642E+01
15  1.5000E-02   2.7426E+00
16  1.6000E-02   5.1722E+01
17  1.7000E-02   7.2398E+01
18  1.8000E-02  -2.3117E+01
19  1.9000E-02  -5.4068E+01
20  2.0000E-02   1.2287E+01

MAXIMUM              9.6000E+01
TIME OF MAXIMUM      0.0000E+00
MINIMUM             -8.3027E+01
TIME OF MINIMUM      8.0000E-03

*** LOADING NUMBER  2 OF  4 ***

DYNAMIC PARAMETERS
  NTS        DT         DAMPR
   20   1.0000E-03   0.0000E+00

INITIAL CONDITIONS
 NNID NNIV
   0    0

APPLIED ACTIONS
  NLN  NEL
   1    0

NODAL LOADS
 NODE        AJ1           AJ2
   2    2.0000E+01    0.0000E+00

GROUND ACCELERATIONS
  IGA
   0

FORCING FUNCTION
  NFO
   7

FUNCTION ORDINATES
   K       TIME        FACTOR
   1   0.0000E+00   0.0000E+00
   2   1.0000E-03   1.5000E-01
   3   3.0000E-03   8.5000E-01
   4   4.0000E-03   1.0000E+00
   5   5.0000E-03   8.5000E-01
   6   7.0000E-03   1.5000E-01
   7   8.0000E-03   0.0000E+00

NORMAL MODE SOLUTION
NMODES =    3

OUTPUT SELECTION
  IWR  IPL  NNO  NEO
   1    1    1    1
```

**TABLE 4.3**  (Continued)

NODES:       2

ELEMENTS:    1

```
DISPLACEMENT TIME HISTORY FOR NODE     2
   STEP      TIME          DJ1           DJ2
      0  0.0000E+00    0.0000E+00    0.0000E+00
      1  1.0000E-03    5.3221E-04   -2.0013E-05
      2  2.0000E-03    4.4538E-03   -5.1199E-04
      3  3.0000E-03    1.5965E-02   -3.0147E-03
      4  4.0000E-03    3.7705E-02   -8.9069E-03
      5  5.0000E-03    6.4468E-02   -1.6887E-02
      6  6.0000E-03    8.7960E-02   -2.2416E-02
      7  7.0000E-03    9.9425E-02   -2.3223E-02
      8  8.0000E-03    9.4028E-02   -2.1405E-02
      9  9.0000E-03    7.1941E-02   -1.8146E-02
     10  1.0000E-02    3.7520E-02   -1.1247E-02
     11  1.1000E-02   -3.2319E-03   -2.5762E-05
     12  1.2000E-02   -4.4130E-02    1.1875E-02
     13  1.3000E-02   -7.7419E-02    2.0566E-02
     14  1.4000E-02   -9.6296E-02    2.4408E-02
     15  1.5000E-02   -9.8483E-02    2.3290E-02
     16  1.6000E-02   -8.4354E-02    1.9033E-02
     17  1.7000E-02   -5.5403E-02    1.3650E-02
     18  1.8000E-02   -1.6520E-02    6.0714E-03
     19  1.9000E-02    2.4923E-02   -4.9980E-03
     20  2.0000E-02    6.2330E-02   -1.6433E-02

MAXIMUM                 9.9425E-02    2.4408E-02
TIME OF MAXIMUM         7.0000E-03    1.4000E-02
MINIMUM                -9.8483E-02   -2.3223E-02
TIME OF MINIMUM         1.5000E-02    7.0000E-03


MEMBER FORCE TIME HISTORY FOR ELEMENT     1
   STEP      TIME          AM1
      0  0.0000E+00    0.0000E+00
      1  1.0000E-03    4.8184E-01
      2  2.0000E-03    3.2844E+00
      3  3.0000E-03    9.0341E+00
      4  4.0000E-03    1.6478E+01
      5  5.0000E-03    2.1947E+01
      6  6.0000E-03    2.7113E+01
      7  7.0000E-03    3.1667E+01
      8  8.0000E-03    3.0814E+01
      9  9.0000E-03    2.2442E+01
     10  1.0000E-02    1.0842E+01
     11  1.1000E-02   -7.8519E-01
     12  1.2000E-02   -1.3479E+01
     13  1.3000E-02   -2.4655E+01
     14  1.4000E-02   -3.0138E+01
     15  1.5000E-02   -3.0917E+01
     16  1.6000E-02   -2.7591E+01
     17  1.7000E-02   -1.7788E+01
     18  1.8000E-02   -4.0957E+00
     19  1.9000E-02    8.1356E+00
     20  2.0000E-02    1.8881E+01

MAXIMUM                 3.1667E+01
TIME OF MAXIMUM         7.0000E-03
MINIMUM                -3.0917E+01
TIME OF MINIMUM         1.5000E-02
```

self-explanatory. Note that the first page of the listing repeats the vibrational analysis given previously in Table 3.4. For each load case we use number of time steps NTS = 20 and duration of time steps DT = 1 ms. Partial results for the four load cases are also plotted (by computer) in Figs. 4.11 and 4.12. Figure 4.11 shows that the initial displacements excite all three modes of vibration. However, the responses in Fig. 4.12(b) due to the applied force in Fig. 4.12(a) demonstrate mostly first-mode contributions.

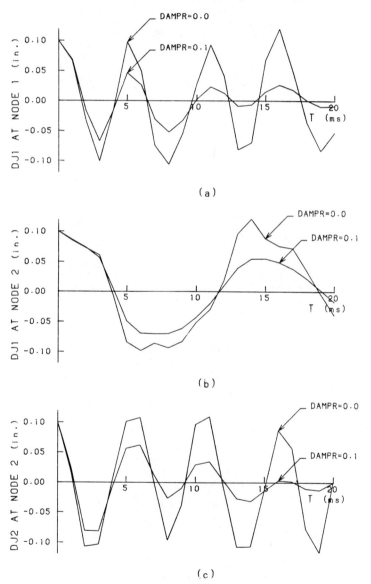

**Figure 4.11** Three-member plane truss: responses to initial displacements.

## Sec. 4.10 Program NOMO for Normal-Mode Response

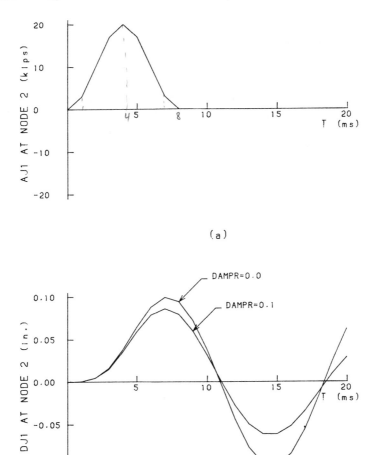

**Figure 4.12** Three-member plane truss: (a) applied force; (b) responses.

### Example 4.16

As a second example using Program NOMOPT, let us reconsider the symmetric plane truss from Sec. 3.8 (see Example 3.5). For this structure dynamic responses due to the following influences are desired:

1. Step force of magnitude 100 kN applied in the $y$ direction at node 3, with DAMPR = 0.0
2. Piecewise-linear ground acceleration in the $y$ direction, with DAMPR = 0.0
3. Same as case 2, but with DAMPR = 0.1

For these cases we use the values of $E$, $\rho$, $L$, and $A$ stated in Example 3.5, where the material is aluminum and units are SI.

To take advantage of symmetry, we must decompose the load for case 1 into symmetric and antisymmetric components and then analyze half the structure twice. Thus, half the force must be applied using symmetric restraints on the plane of symmetry; and the other half of the force must be applied with antisymmetric restraints. Of course, the results of these two analyses must be added to find the total solution for the left-hand part of the truss.

Figure 4.13 consists of computer plots showing the total step force of 100 kN in part (a), symmetric responses to half the force in part (b), and antisymmetric responses

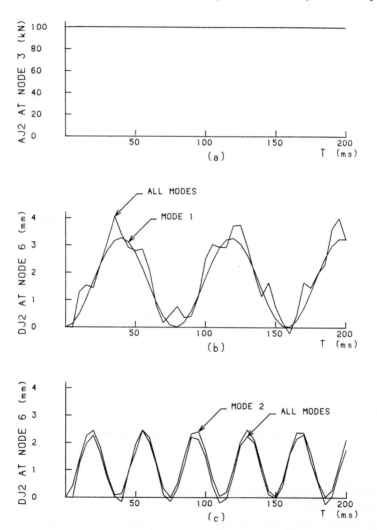

**Figure 4.13** Plane truss: (a) applied force; (b) symmetric responses; (c) antisymmetric responses.

## Sec. 4.10  Program NOMO for Normal-Mode Response

to half the force in part (c). For each load case the number of time steps NTS = 40, and the duration of time steps DT = 5 ms. The responses in Fig. 4.13(b) consist of the y-translation at node 6 for all symmetric modes as well as truncation to mode 1. Similarly, Fig. 4.13(c) shows the responses from all antisymmetric modes and truncation to mode 2 (for the same nodal translation). We see that the first (symmetric) and second (antisymmetric) modes produce the major contributions to the responses.

Figure 4.14(a) is a computer plot of the piecewise-linear ground acceleration in the y direction. Because this influence induces symmetric inertial loads, we need only consider symmetric responses of the structure. Time histories of the resulting y-translation at node 10 appear in Fig. 4.14(b), with and without damping. Evidently, the first mode is the primary contributor to both responses.

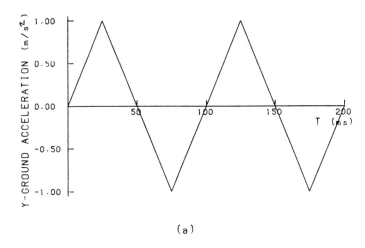

(a)

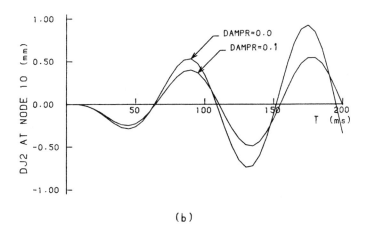

(b)

**Figure 4.14** Plane truss: (a) ground acceleration; (b) responses.

## REFERENCES

1. Timoshenko, S. P., Young, D. H., and Weaver, W., Jr., *Vibration Problems in Engineering,* 4th ed., Wiley, New York, 1974.
2. Weaver, W., Jr., and Gere, J. M., *Matrix Analysis of Framed Structures,* 2nd ed., Van Nostrand Reinhold, New York, 1980.
3. Rayleigh, J. W. S., *Theory of Sound,* Vol. 1, Dover, New York, 1945.
4. Weaver, W., Jr., and Johnston, P. R., *Finite Elements for Structural Analysis,* Prentice-Hall, Englewood Cliffs, N.J., 1984.
5. Weaver, W., Jr., "Dynamics of Discrete-Parameter Structures," in *Developments in Theoretical and Applied Mechanics,* Vol. 2, ed. W. A. Shaw, Pergamon Press, New York, 1965, pp. 629–651.
6. Caughy, T. K., "Classical Normal Modes in Damped Linear Systems," *J. Appl. Mech.,* Vol. 27, 1960, pp. 269–271; also Vol. 32, 1965, pp. 583–588.
7. Foss, K. A., "Coordinates which Uncouple the Equations of Motion of Damped Linear Dynamic Systems," *J. Appl. Mech.,* Vol. 25, 1958, pp. 361–364.
8. Louie, J. J. C., "Damping in Structures—a Review," Engineer thesis, Department of Civil Engineering, Stanford University, June 1976.

## PROBLEMS

**4.3-1.** Assume that the plane truss in Fig. 3.12(a) is at rest when joint 2 is suddenly struck so that it acquires an initial velocity $\dot{D}_{01}$ in the $x$ direction. Determine the free-vibrational response of the structure due to this impact. The properties, angular frequencies, and mode shapes for this problem are all given in Example 3.1.

**4.3-2.** Calculate the free-vibrational response of the cantilever beam in Fig. 3.13(a) to the sudden release of a static force $P_0$ in the $y$ direction at node 2. Example 3.2 gives the properties, angular frequencies, and mode shapes for this problem.

**4.3-3.** For the plane truss in Prob. 3.6-2, find the response caused by initial displacements $D_{01} = D_{02} = d$ at joint 1.

**4.3-4.** Obtain the response of the plane truss in Prob. 3.6-3 to an initial velocity $\dot{D}_{02}$ in the $y$ direction at joint 1.

**4.3-5.** For the plane truss in Prob. 3.6-4, determine the response to the sudden release of a force $P_0$ in the negative $y$ direction at joint 1.

**4.3-6.** Suppose that node 2 of the beam in Prob. 3.6-5 has an initial velocity $\dot{D}_{01}$ in the $y$ direction. Find the response of the structure due to this influence.

**4.3-7.** Let the beam in Prob. 3.6-6 initially have a small positive rotation $\theta_{z0}$ at node 2 and a negative rotation $-\theta_{z0}$ at node 3. Calculate the response of the structure that results.

**4.3-8.** Assume that an initial positive moment $M_{z0}$ at node 1 of the beam in Prob. 3.6-7 is suddenly released. Obtain the response of the structure caused by this condition.

**4.3-9.** If joint 1 of the symmetric plane truss in Prob. 3.7-7 has an initial velocity $\dot{D}_{01}$ in the $x$ direction, determine the antisymmetric response of the structure.

**4.4-1.** (a) Calculate the response of the plane truss in Fig. 3.12(a) caused by a step force of magnitude $P = P_1$ applied in the $x$ direction at joint 1. (b) Confirm the reciprocal theorem for dynamic loads by obtaining the response due to the same forcing function applied in the $y$ direction at joint 1. Example 3.1 gives the properties, angular frequencies, and mode shapes for this truss.

**4.4-2.** Assume that a harmonically-varying force $P = P_1 \sin \Omega t$ acts in the $y$ direction at node 2 of the cantilever beam in Fig. 3.13(a). (a) Find the steady-state response at the free end of the beam due to this loading. (b) Confirm the reciprocal theorem for dynamic loads by calculating the response caused by a moment $M = M_2 \sin \Omega t$ in the $z$ sense at node 2, where $M_2$ is numerically equal to $P_1$. The properties, angular frequencies and mode shapes for this problem are all given in Example 3.2.

**4.4-3.** For the plane truss in Prob. 3.6-2, determine the response caused by a ramp force $P = P_1 t/t_1$ applied in the negative $y$ direction at joint 1.

**4.4-4.** Obtain the response of the plane truss in Prob. 3.6-3 to a step force of magnitude $P = P_1$ applied in the $y$ direction at the middle of element 1.

**4.4-5.** Assume that a harmonically-varying force $P = P_1 \cos \Omega t$ is applied in the $x$ direction at joint 1 of the plane truss in Prob. 3.6-4. Calculate the steady-state response of the structure due to this influence.

**4.4-6.** Apply a step moment of magnitude $M = M_2$ in the $z$ sense at node 2 of the beam in Prob. 3.6-5, and find the response of the structure.

**4.4-7.** Suppose that a harmonically-varying force $P = P_1 \sin \Omega t$ is applied in the $y$ direction at the midpoint of element 2 in Prob. 3.6-6. Determine the steady-state nodal responses of the beam caused by this loading.

**4.4-8.** Let the beam in Prob. 3.6-7 be subjected to a ramp moment $M = M_2 t/t_1$ in the $z$ sense at node 2, and calculate the response of the structure.

**4.4-9.** For the symmetric plane truss in Prob. 3.7-7, obtain the antisymmetric response to a ramp force $P = P_1 t/t_1$ applied in the negative $x$ direction at joint 1.

**4.5-1.** For the plane truss in Fig. 4.7, calculate the relative response of joint 1 to a rigid-body ramp acceleration $\ddot{D}_{g1} = a_1 t/t_1$ of ground in the $x$ direction.

**4.5-2.** Considering the cantilever beam in Fig. 4.8, let the ground at point 1 have a sudden translation $D_{g2} = d$ in the $y$ direction. Find the response at node 2 due to this influence.

**4.5-3.** Suppose that the ground at point 3 in Prob. 3.6-2 has a small rigid-body step rotation $D_{g6} = \theta_z$. Determine the response of joint 1 in the plane truss caused by this motion.

**4.5-4.** For the plane truss in Prob. 3.6-3, evaluate the relative steady-state response at joint 1 due to a rigid-body ground acceleration $\ddot{D}_{g2} = a \cos \Omega t$ in the $y$ direction.

**4.5-5.** Assume that the support at point 3 of the plane truss in Prob. 3.6-4 translates independently in the $y$ direction according to the function $D_{R3} = d \sin \Omega t$. Calculate the resulting steady-state response at joint 1.

**4.5-6.** If the ground in Prob. 3.6-5 has a sudden small rigid-body step rotation $D_{g6} = -\theta_z$ in the negative $z$ sense at point 1, find the response of the beam at node 2.

**4.5-7.** For the continuous beam in Prob. 3.6-6, determine the relative steady-state response due to a rigid-body rotational acceleration $\ddot{D}_{g6} = \ddot{\theta}_z \sin \Omega t$ of ground at node 3.

**4.5-8.** Let the support at point 3 in Prob. 3.6-7 have a small harmonic rotation $D_{R3} = \theta_z \sin \Omega t$ in the $z$ sense. Calculate the steady-state responses at nodes 1 and 2 of the beam caused by this influence.

# 5

# Direct Numerical Integration Methods

## 5.1 INTRODUCTION

In this chapter we shall discuss various numerical integration methods for calculating dynamic responses of structures. These techniques are usually more efficient than the normal-mode method, unless modal truncation is used. Thus, either approach involves approximations of different types. While the normal-mode method applies only to linearly elastic structures, direct numerical integration can be used for either linear or nonlinear systems.

Any method for direct numerical integration of second-order differential equations of motion may be visualized as some type of *finite-difference formulation*. Although we use finite elements for discretization in space, we find that the discretization in time is more conveniently handled by finite differences. This approach is most apparent in the next section, where a second central difference formula is converted to an explicit extrapolator for dynamic response. Although it is not always evident, the other approximation formulas used in subsequent sections are also various types of finite-difference expressions.

For a SDOF analytical model, the damped equation of motion is

$$m\ddot{u} + c\dot{u} + ku = P(t) \tag{1}$$

in which the terms were all defined previously in Chapter 2. On the other hand, the damped equations of motion for a MDOF structure have the matrix form

$$\mathbf{M}\ddot{\mathbf{D}} + \mathbf{C}\dot{\mathbf{D}} + \mathbf{S}\mathbf{D} = \mathbf{A}(t) \tag{2}$$

as described in Chapter 4. To begin any numerical integration procedure, we must have available the initial conditions of displacements, velocities, and loads at time $t_0 = 0$. From these quantities we may calculate the initial acceleration for a SDOF system as

$$\ddot{u}_0 = \frac{1}{m}(P_0 - ku_0 - c\dot{u}_0) \tag{3}$$

which is obtained from Eq. (1) at time $t_0$. Similarly, for a MDOF structure, Eq. (2) yields

$$\ddot{\mathbf{D}}_0 = \mathbf{M}^{-1}(\mathbf{A}_0 - \mathbf{S}\,\mathbf{D}_0 - \mathbf{C}\,\dot{\mathbf{D}}_0) \tag{4}$$

If the mass matrix $\mathbf{M}$ is singular, the vector $\ddot{\mathbf{D}}_0$ in Eq. (4) may be taken as a null vector.

Figure 5.1 shows a graph of the numerical solution for the response of a SDOF system. This plot is represented as a smooth curve in the $u$-$t$ plane, even though it may actually have slight discontinuities. The symbols $u_0, u_1, u_2, \ldots, u_{j-1}, u_j, u_{j+1}, \ldots$ denote values of $u$ at the time stations $t_0, t_1, t_2, \ldots, t_{j-1}, t_j, t_{j+1}, \ldots$, and so on. The time interval $\Delta t_j$ between times $t_j$ and $t_{j+1}$ is usually taken to be of uniform duration $\Delta t$, although such a specialization is not necessary. The objective of the numerical integration process is to calculate the

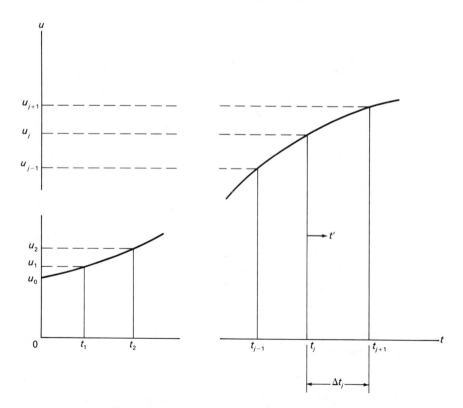

**Figure 5.1** Solution by direct numerical integration.

## 5.2 EXTRAPOLATION WITH EXPLICIT FORMULAS

An explicit extrapolation procedure consists of expressing the displacement at time $t_{j+1}$ in terms of the displacements, velocities, or accelerations at previous time stations. Referring to Fig. 5.1, we can write the *second central difference* of the displacement at time $t_j$ as the acceleration:

$$\ddot{u}_j \approx \frac{1}{(\Delta t)^2}(u_{j-1} - 2u_j + u_{j+1}) \tag{1}$$

which implies a SDOF system with a uniform time step $\Delta t$. Similarly, the velocity at time $t_j$ may be approximated as

$$\dot{u}_j \approx \frac{1}{2\,\Delta t}(u_{j+1} - u_{j-1}) \tag{2}$$

Substituting $\ddot{u}_j$ and $\dot{u}_j$ from Eqs. (1) and (2) into Eq. (5.1-1) and rearranging the result produces

$$\left[\frac{m}{(\Delta t)^2} + \frac{c}{2\,\Delta t}\right] u_{j+1} = P^*(t_j) \tag{3}$$

where

$$P^*(t_j) = P(t_j) - \left[k - \frac{2m}{(\Delta t)^2}\right] u_j - \left[\frac{m}{(\Delta t)^2} - \frac{c}{2\,\Delta t}\right] u_{j-1} \tag{4}$$

Equation (3) is known as the *central-difference predictor*, which can be applied repetitively to obtain $u_{j+1}$ for each time step. Then the acceleration $\ddot{u}_j$ and the velocity $\dot{u}_j$ at time $t_j$ may also be found using Eqs. (1) and (2) if desired.

The central-difference predictor in Eq. (3) is a two-step formula, so it cannot be applied directly in the first time step. In order to derive a starting procedure, we apply Eqs. (1) and (2) at time $t_0$ and solve for $u_{-1}$, as follows:

$$u_{-1} = u_0 - \dot{u}_0\,\Delta t + \tfrac{1}{2}\ddot{u}_0(\Delta t)^2 \tag{5}$$

This result can be used in Eq. (4) to evaluate $P^*(t_1)$ for the first time step.

For a MDOF structure, the expressions given above can be generalized in a matrix format. Thus, Eqs. (1) through (4) become

$$\ddot{\mathbf{D}}_j \approx \frac{1}{(\Delta t)^2}(\mathbf{D}_{j-1} - 2\mathbf{D}_j + \mathbf{D}_{j+1}) \tag{6}$$

$$\dot{\mathbf{D}}_j \approx \frac{1}{2\,\Delta t}(\mathbf{D}_{j+1} - \mathbf{D}_{j-1}) \tag{7}$$

$$\left[\frac{1}{(\Delta t)^2}\mathbf{M} + \frac{1}{2\,\Delta t}\mathbf{C}\right]\mathbf{D}_{j+1} = \mathbf{A}^*(t_j) \tag{8}$$

$$\mathbf{A}^*(t_j) = \mathbf{A}(t_j) - \left[\mathbf{S} - \frac{2}{(\Delta t)^2}\mathbf{M}\right]\mathbf{D}_j - \left[\frac{1}{(\Delta t)^2}\mathbf{M} - \frac{1}{2\,\Delta t}\mathbf{C}\right]\mathbf{D}_{j-1} \tag{9}$$

In addition, the generalized form of Eq. (5) is

$$\mathbf{D}_{-1} = \mathbf{D}_0 - \dot{\mathbf{D}}_0\,\Delta t + \tfrac{1}{2}\ddot{\mathbf{D}}_0(\Delta t)^2 \tag{10}$$

which may be used to start the procedure. If there is no damping and the mass matrix is diagonal, a so-called *nodewise solution* may be devised. That is, the solution for a particular displacement $D_{i,j+1}$ in Eq. (8) becomes

$$D_{i,j+1} = \frac{(\Delta t)^2}{M_i}A_i^*(t_j) \qquad (i = 1, 2, \ldots, n) \tag{11}$$

which is simply a multiplication of the effective action $A_i^*(t_j)$ by the scalar $(\Delta t)^2/M_i$. Therefore, high-speed computer core storage need only contain information for one displacement (or more conveniently, one node) at a time.

The central-difference predictor is probably the most widely used explicit formula for solving structural dynamics problems. However, all expressions of this type have a *critical time step*, above which the solution becomes numerically unstable and diverges [1]. Nevertheless, among all of the known second-order predictors, the central-difference method has the largest stable time step [2]. The value of the critical time step for this technique is

$$(\Delta t)_{\text{cr}} = \frac{2}{\omega_n} = \frac{T_n}{\pi} \tag{12}$$

The symbol $\omega_n$ in this expression denotes the largest angular frequency in the analytical model, and $T_n$ is the smallest period.

Key and Beisinger [3] applied the central-difference predictor to the linear dynamic analysis of thin shells. In addition, Krieg and Key [4] showed that using a diagonal mass matrix improves the accuracy of the procedure. Successful use of this method in nonlinear analysis has also been reported by Key [5], who used artificial damping to control the inherent instability. Morino et al. [6], searched for the optimal predictor for systems of second-order differential equations and concluded that the central-difference formula is best. The main disadvantage of this method is that for a fine network of elements a very small time step is required to obtain stable results without damping. The topics of numerical stability and accuracy will be discussed more thoroughly in Sec. 5.6.

**Example 5.1**

Suppose that an undamped SDOF linear system is subjected to a step force of magnitude $P_1$, starting from rest. Let us calculate the approximate response using the central-difference predictor with twenty uniform time steps of duration $\Delta t = T/20$.

Sec. 5.2    Extrapolation with Explicit Formulas    199

In this example we can express the mass $m$ in terms of the stiffness $k$ and the period $T$ as

$$m = \frac{k}{\omega^2} = k\left(\frac{T}{2\pi}\right)^2 \tag{a}$$

At time $t_0 = 0$, we have $u_0 = \dot{u}_0 = 0$, so the initial acceleration is

$$\ddot{u}_0 = \frac{P_0}{m} = \frac{P_1}{m} \tag{b}$$

which is drawn from Eq. (5.1-3) with $P_0 = P_1$. To start the solution, we apply Eq. (5) to obtain

$$u_{-1} = \frac{1}{2}\ddot{u}_0(\Delta t)^2 = \frac{P_1}{2m}\left(\frac{T}{20}\right)^2 = \frac{P_1}{2k}\left(\frac{2\pi}{T}\right)^2\left(\frac{T}{20}\right)^2 = 0.04935\frac{P_1}{k} \tag{c}$$

Then the central-difference predictor in Eq. (3) gives the displacement at time $t_1$ as

$$u_1 = \frac{(\Delta t)^2}{m}\left[P_1 - 0 - \frac{m}{(\Delta t)^2}u_{-1}\right] = \left(\frac{T}{20}\right)^2\left(\frac{2\pi}{T}\right)^2\frac{P_1}{k} - u_{-1}$$

$$= (0.09870 - 0.04935)\frac{P_1}{k} = 0.04935\frac{P_1}{k} \tag{d}$$

which is the same as $u_{-1}$ in Eq. (c). A second application of the predictor at time $t_2$ yields

$$u_2 = \frac{(\Delta t)^2}{m}\left\{P_1 - \left[k - \frac{2m}{(\Delta t)^2}\right]u_1 - 0\right\}$$

$$= \left[\left(\frac{T}{20}\right)^2\left(\frac{2\pi}{T}\right)^2(1 - 0.04935) + (2)(0.04935)\right]\frac{P_1}{k}$$

$$= 0.1925\frac{P_1}{k} \tag{e}$$

In each subsequent time step, Eq. (3) is applied in the same manner, and Eqs. (1) and (2) could be employed to find the accelerations and velocities as well.

The approximate values of $u_j$ obtained by this procedure are listed in Table 5.1, along with the exact values. In addition, Fig. 5.2 shows plots of the approximate results for both zero damping and a damping ratio of $\gamma = 0.1$. For the scale used in Fig. 5.2, the plotted curves are indistinguishable from exact responses.

### Example 5.2

Figure 5.3(a) shows the plane truss analyzed previously in Example 3.1. Recall that the cross-sectional areas of members 1 and 2 are equal to $0.8A$ and $A$. Now we shall determine the undamped response of this structure to the force $P(t)$, applied in the $x$ direction at joint 1. The time history of the dynamic load appears in Fig. 5.3(b). Let us calculate the displacements $D_1$ and $D_2$ using the central-difference predictor with 20 uniform time steps of duration $\Delta t = T_1/20$, starting from rest.

From Example 3.1 the stiffness and mass matrices for the free displacements at

**TABLE 5.1  Response for Example 5.1
Using Central-Difference Predictor**[a]

| j | Approx. u | Exact u | j | Approx. u | Exact u |
|---|---|---|---|---|---|
| 1 | 0.04935 | 0.04894 | 11 | 1.947 | 1.951 |
| 2 | 0.1925 | 0.1910 | 12 | 1.800 | 1.809 |
| 3 | 0.4154 | 0.4122 | 13 | 1.574 | 1.588 |
| 4 | 0.6960 | 0.6910 | 14 | 1.292 | 1.309 |
| 5 | 1.007 | 1.000 | 15 | 0.9804 | 1.000 |
| 6 | 1.317 | 1.309 | 16 | 0.6712 | 0.6910 |
| 7 | 1.595 | 1.588 | 17 | 0.3944 | 0.4122 |
| 8 | 1.815 | 1.809 | 18 | 0.1774 | 0.1910 |
| 9 | 1.955 | 1.951 | 19 | 0.04157 | 0.04894 |
| 10 | 2.000 | 2.000 | 20 | 0.00034 | 0 |

[a] Tabulated values to be multiplied by $P_1/k$.

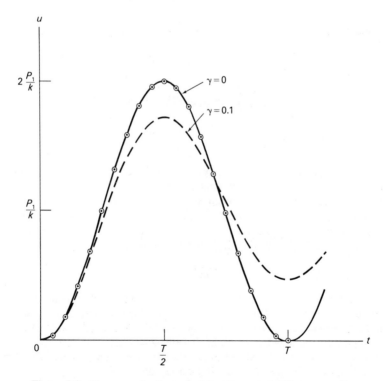

**Figure 5.2**  Responses for Example 5.1 using central-difference predictor.

## Sec. 5.2  Extrapolation with Explicit Formulas  201

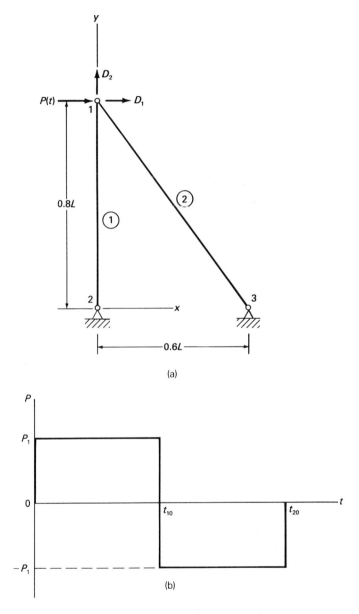

**Figure 5.3**  (a) Plane truss; (b) applied force.

joint 1 are

$$\mathbf{S} = s \begin{bmatrix} 0.36 & -0.48 \\ -0.48 & 1.64 \end{bmatrix} \qquad \mathbf{M} = m \begin{bmatrix} 1 & 0 \\ 0 & 1 \end{bmatrix} \qquad (f)$$

in which $s = EA/L$ and $m = 3.28\, \rho AL/6$. In addition, we have

$$\omega_1^2 = 0.2 \frac{s}{m} \qquad \omega_2^2 = 1.8 \frac{s}{m} \qquad (g)$$

Using the first expression in Eqs. (g), we can relate the mass constant $m$ to the stiffness constant $s$ and the fundamental period $T_1$, as follows:

$$m = 0.2s \left( \frac{T_1}{2\pi} \right)^2 \qquad (h)$$

At time $t_0 = 0$ the initial displacements and velocities are $(D_1)_0 = (D_2)_0 = (\dot{D}_1)_0 = (\dot{D}_2)_0 = 0$. Therefore, Eq. (5.1-4) gives the initial accelerations as

$$\ddot{\mathbf{D}}_0 = \mathbf{M}^{-1}\mathbf{A}_0 = \frac{1}{m}\begin{bmatrix} 1 & 0 \\ 0 & 1 \end{bmatrix} \begin{bmatrix} P_1 \\ 0 \end{bmatrix} = \begin{bmatrix} 1 \\ 0 \end{bmatrix} \frac{P_1}{m} \qquad (i)$$

Before beginning the step-by-step procedure, we apply Eq. (10) to find

$$\mathbf{D}_{-1} = \frac{1}{2} \ddot{\mathbf{D}}_0 (\Delta t)^2 = \frac{1}{2} \begin{bmatrix} 1 \\ 0 \end{bmatrix} \frac{P_1}{m} \left( \frac{T_1}{20} \right)^2$$

$$= \frac{1}{2} \begin{bmatrix} 1 \\ 0 \end{bmatrix} \frac{P_1}{0.2s} \left( \frac{2\pi}{T_1} \right)^2 \left( \frac{T_1}{20} \right)^2 = \begin{bmatrix} 0.2467 \\ 0 \end{bmatrix} \frac{P_1}{s} \qquad (j)$$

Using this result in the matrix form of the predictor [Eq. (8)] at time $t_1$ produces

$$\mathbf{D}_1 = \frac{(\Delta t)^2}{m} \begin{bmatrix} 1 & 0 \\ 0 & 1 \end{bmatrix} \left\{ \begin{bmatrix} P_1 \\ 0 \end{bmatrix} - \frac{m}{(\Delta t)^2} \begin{bmatrix} 1 & 0 \\ 0 & 1 \end{bmatrix} \mathbf{D}_{-1} \right\}$$

$$= \left\{ \begin{bmatrix} 0.4934 \\ 0 \end{bmatrix} - \begin{bmatrix} 0.2467 \\ 0 \end{bmatrix} \right\} \frac{P_1}{s} = \begin{bmatrix} 0.2467 \\ 0 \end{bmatrix} \frac{P_1}{s} \qquad (k)$$

which is the same as $\mathbf{D}_{-1}$ in Eq. (j). Applying Eq. (8) again at time $t_2$ yields

$$\mathbf{D}_2 = \frac{(\Delta t)^2}{m} \begin{bmatrix} 1 & 0 \\ 0 & 1 \end{bmatrix} \left\{ \begin{bmatrix} P_1 \\ 0 \end{bmatrix} - s \begin{bmatrix} 0.36 & -0.48 \\ -0.48 & 1.64 \end{bmatrix} \mathbf{D}_1 + \frac{2m}{(\Delta t)^2} \begin{bmatrix} 1 & 0 \\ 0 & 1 \end{bmatrix} \mathbf{D}_1 \right\}$$

$$= \left( \frac{T_1}{20} \right)^2 \frac{P_1}{0.2s} \left( \frac{2\pi}{T_1} \right)^2 \begin{bmatrix} 1 - 0.08883 \\ 0 + 0.1184 \end{bmatrix} + 2 \begin{bmatrix} 0.2467 \\ 0 \end{bmatrix} \frac{P_1}{s}$$

$$= \left\{ \begin{bmatrix} 0.4496 \\ 0.05845 \end{bmatrix} + \begin{bmatrix} 0.4935 \\ 0 \end{bmatrix} \right\} \frac{P_1}{s} = \begin{bmatrix} 0.9431 \\ 0.05845 \end{bmatrix} \frac{P_1}{s} \qquad (\ell)$$

In subsequent calculations we apply Eq. (8) repetitively to find the response at each of the remaining time stations. Note that the mass matrix in this example is diagonal, so Eq. (8) does not require solving simultaneous equations.

Results from this approximate analysis for $D_1$ and $D_2$ are given in Table 5.2, where

## Sec. 5.3  Iteration with Implicit Formulas

**TABLE 5.2  Responses for Example 5.2 Using Central-Difference Predictor**[a]

| j | Approx. $D_1$ | Exact $D_1$ | Approx. $D_2$ | Exact $D_2$ |
|---|---|---|---|---|
| 1  | 0.2467  | 0.2432  | 0       | 0.00471 |
| 2  | 0.9431  | 0.9322  | 0.05845 | 0.06831 |
| 3  | 1.979   | 1.963   | 0.2930  | 0.2932  |
| 4  | 3.227   | 3.210   | 0.7593  | 0.7350  |
| 5  | 4.574   | 4.556   | 1.375   | 1.333   |
| 6  | 5.928   | 5.901   | 1.962   | 1.932   |
| 7  | 7.188   | 7.148   | 2.365   | 2.374   |
| 8  | 8.223   | 8.179   | 2.557   | 2.598   |
| 9  | 8.898   | 8.868   | 2.627   | 2.662   |
| 10 | 9.107   | 9.111   | 2.679   | 2.667   |
| 11 | 7.839   | 8.382   | 2.720   | 2.653   |
| 12 | 5.330   | 6.315   | 2.416   | 2.462   |
| 13 | 1.952   | 3.221   | 1.420   | 1.787   |
| 14 | −1.929  | −0.5187 | −0.2636 | 0.4618  |
| 15 | −6.024  | −4.556  | −2.191  | −1.333  |
| 16 | −10.06  | −8.592  | −3.772  | −3.128  |
| 17 | −13.70  | −12.33  | −4.683  | −4.454  |
| 18 | −16.50  | −15.43  | −5.049  | −5.128  |
| 19 | −18.07  | −17.49  | −5.238  | −5.319  |
| 20 | −18.15  | −18.22  | −5.467  | −5.333  |

[a] Tabulated values to be multiplied by $P_1/s$.

exact values are also listed. Figure 5.4 depicts plots of these responses for both the approximate and the exact analyses. Here we see that the curves for the approximate responses deviate slightly from those for the exact responses.

## 5.3 ITERATION WITH IMPLICIT FORMULAS

The technique to be explained in this section is called the *predictor-corrector method* [6, 7]. In each time step an *explicit* formula (a *predictor*) is first used to estimate the response at the end of the step. This is followed by one or more applications of an *implicit* formula (a *corrector*) to improve the results. Although such an iterative procedure is not required for linear analysis, it is included here for use in nonlinear problems, where physical properties can change in each cycle of iteration.

By an approach that we shall refer to as the *average-acceleration method*, the velocity $\dot{u}_{j+1}$ for a SDOF system at time $t_{j+1}$ is approximated as

$$\dot{u}_{j+1} = \dot{u}_j + \tfrac{1}{2}(\ddot{u}_j + \ddot{u}_{j+1})\,\Delta t_j \tag{1}$$

in which $\dot{u}_j$ is the velocity at the preceding time station $t_j$ (see Fig. 5.1). This formula, known to numerical analysts as the *trapezoidal rule*, implies that the

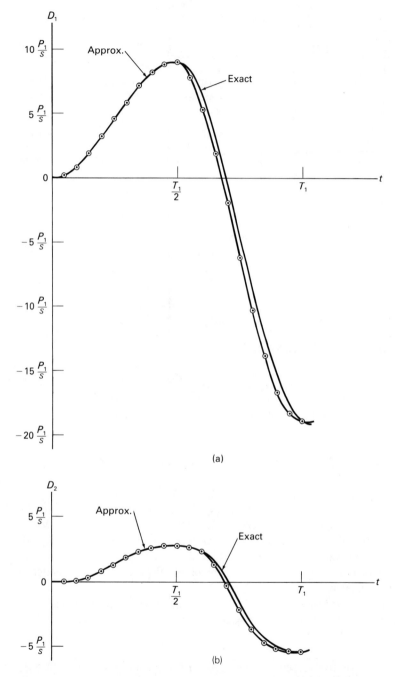

**Figure 5.4** Responses for Example 5.2 using central-difference predictor: (a) $D_1$; (b) $D_2$.

### Sec. 5.3  Iteration with Implicit Formulas

acceleration in the step is taken to be the average of $\ddot{u}_j$ and $\ddot{u}_{j+1}$. Similarly, the displacement $u_{j+1}$ at the end of the step is approximated by the trapezoidal rule with the expression

$$u_{j+1} = u_j + \tfrac{1}{2}(\dot{u}_j + \dot{u}_{j+1})\,\Delta t_j \tag{2}$$

where the velocity in the step is taken to be the average of $\dot{u}_j$ and $\dot{u}_{j+1}$. Substitution of Eq. (1) into Eq. (2) yields

$$u_{j+1} = u_j + \dot{u}_j\,\Delta t_j + \tfrac{1}{4}(\ddot{u}_j + \ddot{u}_{j+1})(\Delta t_j)^2 \tag{3}$$

When applying this method, we do not use Eq. (3) directly; but Eqs. (1) and (2) are used in succession. Because the value of $\ddot{u}_{j+1}$ is not known in advance, the approximation is said to be implicit, so the solution must be iterative within each step. The following recurrence equations represent the $i$th iteration of the $j$th step:

$$(\dot{u}_{j+1})_i = Q_j + \tfrac{1}{2}(\ddot{u}_{j+1})_{i-1}\Delta t_j \qquad (i > 1) \tag{4}$$

$$(u_{j+1})_i = R_j + \tfrac{1}{2}(\dot{u}_{j+1})_i \Delta t_j \qquad (i \geq 1) \tag{5}$$

$$(\ddot{u}_{j+1})_i = \frac{1}{m}(P_{j+1} - ku_{j+1} - c\dot{u}_{j+1})_i \qquad (i \geq 1) \tag{6}$$

where

$$Q_j = \dot{u}_j + \tfrac{1}{2}\ddot{u}_j\,\Delta t_j \tag{7}$$

and

$$R_j = u_j + \tfrac{1}{2}\dot{u}_j\,\Delta t_j \tag{8}$$

This iterative procedure is not self-starting because it requires a supplementary formula for determining the first estimate of $\dot{u}_{j+1}$ in each time step. After evaluating the initial acceleration from Eq. (5.1-3), we may start the iteration for the first step by approximating $\dot{u}_1$ with *Euler's extrapolation formula*, as follows:

$$(\dot{u}_1)_1 = \dot{u}_0 + \ddot{u}_0\,\Delta t_0 \qquad (j = 0;\ i = 1) \tag{9}$$

Then the first approximations for $u_1$ and $\ddot{u}_1$ are obtained from Eqs. (5) and (6). All subsequent iterations for the first time step involve the repetitive use of Eqs. (4), (5), and (6).

To start the iteration in the $j$th time step, we may again apply Euler's formula to determine a first estimate of $\dot{u}_{j+1}$ as

$$(\dot{u}_{j+1})_1 = \dot{u}_j + \ddot{u}_j\,\Delta t_j \qquad (i = 1) \tag{10}$$

Both Eqs. (9) and (10) imply constant values of the accelerations within the steps. To improve the accuracy of the results for the first iteration of the $j$th step, we can use the slightly more elaborate formula that is valid only for uniform time steps:

$$(\dot{u}_{j+1})_1 = \dot{u}_{j-1} + 2\ddot{u}_j \, \Delta t \qquad (i = 1) \tag{11}$$

This expression spans two equal time steps from $t_{j-1}$ to $t_{j+1}$ (see Fig. 5.1) and utilizes the midpoint acceleration at time $t_j$.

Equations (10) and (11) are called *explicit predictors* because they provide estimates of $\dot{u}_{j+1}$ in terms of previous values of $\dot{u}$ and $\ddot{u}$. On the other hand, Eq. (1) is referred to as an *implicit corrector* that yields an improved value of $\dot{u}_{j+1}$ after an estimation of $\ddot{u}_{j+1}$ has been obtained. The method described here involves one application of a predictor, followed by repetitive applications of the corrector.

An iterative type of solution requires some criterion for stopping or changing the step size, such as a limit on the number of iterations. A convenient method for measuring the *rate of convergence* is to control the number of significant figures in $u_{j+1}$, as follows:

$$|(u_{j+1})_i - (u_{j+1})_{i-1}| < \epsilon_u |(u_{j+1})_i| \tag{12}$$

where $\epsilon_u$ is some small number selected by the analyst. For example, an accuracy of approximately four digits may be specified by taking $\epsilon_u = 0.0001$. That level of accuracy is used in the numerical examples of this section. For a MDOF structure, we use the length of a translational or rotational vector in Eq. (12), which is equal to the square root of the sum of the squares of its components.

Another implicit approach for approximating responses is known as the *linear-acceleration method*. As indicated by its name, this technique has the assumption that the acceleration varies linearly within each time step. Thus, an expression for $\ddot{u}$ during the step $\Delta t_j$ may be written

$$\ddot{u}(t') = \ddot{u}_j + (\ddot{u}_{j+1} - \ddot{u}_j) \frac{t'}{\Delta t_j} \tag{13}$$

where $t'$ is measured from the beginning of the step (see Fig. 5.1). If the acceleration varies linearly, the corresponding velocity and displacement will vary quadratically and cubically with time. Therefore,

$$\dot{u}(t') = \dot{u}_j + \ddot{u}_j t' + (\ddot{u}_{j+1} - \ddot{u}_j) \frac{(t')^2}{2 \, \Delta t_j} \tag{14}$$

and

$$u(t') = u_j + \dot{u}_j t' + \ddot{u}_j \frac{(t')^2}{2} + (\ddot{u}_{j+1} - \ddot{u}_j) \frac{(t')^3}{6 \, \Delta t_j} \tag{15}$$

At the end of the step the velocity and displacement become

$$\dot{u}_{j+1} = \dot{u}_j + \tfrac{1}{2}(\ddot{u}_j + \ddot{u}_{j+1}) \, \Delta t_j \tag{16}$$

and

$$u_{j+1} = u_j + \dot{u}_j \Delta t_j + \tfrac{1}{6}(2\ddot{u}_j + \ddot{u}_{j+1})(\Delta t_j)^2 \tag{17}$$

## Sec. 5.3  Iteration with Implicit Formulas

Equation (16) is the same as Eq. (1) of the average-acceleration method, but Eq. (17) is slightly different from its counterpart in Eq. (3).

We will apply the linear-acceleration method in a manner analogous to that for the average-acceleration approach. Because Eq. (16) is the same as Eq. (1), the recurrence expression for the $i$th iteration of $\dot{u}_{j+1}$ is the same as that in Eq. (4). To obtain a direct relationship between $u_{j+1}$ and $\dot{u}_{j+1}$, we solve for $\ddot{u}_{j+1}$ in Eq. (16) and substitute the result into Eq. (17), which yields

$$u_{j+1} = u_j + \tfrac{1}{3}(2\dot{u}_j + \dot{u}_{j+1})\,\Delta t_j + \tfrac{1}{6}\ddot{u}_j(\Delta t_j)^2 \tag{18}$$

Thus, we form the recurrence equation for the $i$th iteration of $u_{j+1}$ as:

$$(u_{j+1})_i = R_j^* + \tfrac{1}{3}(\dot{u}_{j+1})_i\,\Delta t_j \qquad (i \geq 1) \tag{19}$$

where

$$R_j^* = u_j + \tfrac{2}{3}\dot{u}_j\,\Delta t_j + \tfrac{1}{6}\ddot{u}_j(\Delta t_j)^2 \tag{20}$$

The formulas given earlier [see Eqs. (9) and (10) or (11)] may be used again to start the iteration in each step.

For MDOF structures, we can generalize the recurrence equations for iteration into matrix formats. Considering first the average-acceleration method, we replace Eqs. (4) through (8) by

$$(\dot{\mathbf{D}}_{j+1})_i = \mathbf{Q}_j + \tfrac{1}{2}(\ddot{\mathbf{D}}_{j+1})_{i-1}\,\Delta t_j \qquad (i > 1) \tag{21}$$

$$(\mathbf{D}_{j+1})_i = \mathbf{R}_j + \tfrac{1}{2}(\dot{\mathbf{D}}_{j+1})_i\,\Delta t_j \qquad (i \geq 1) \tag{22}$$

$$(\ddot{\mathbf{D}}_{j+1})_i = \mathbf{M}^{-1}(\mathbf{A}_{j+1} - \mathbf{S}\,\mathbf{D}_{j+1} - \mathbf{C}\,\dot{\mathbf{D}}_{j+1})_i \qquad (i \geq 1) \tag{23}$$

where

$$\mathbf{Q}_j = \dot{\mathbf{D}}_j + \tfrac{1}{2}\ddot{\mathbf{D}}_j\,\Delta t_j \tag{24}$$

and

$$\mathbf{R}_j = \mathbf{D}_j + \tfrac{1}{2}\dot{\mathbf{D}}_j\,\Delta t_j \tag{25}$$

Also, Eqs. (9), (10), and (11) are supplanted by the matrix expressions

$$(\dot{\mathbf{D}}_1)_1 = \dot{\mathbf{D}}_0 + \ddot{\mathbf{D}}_0\,\Delta t_0 \qquad (j = 0;\ i = 1) \tag{26}$$

$$(\dot{\mathbf{D}}_{j+1})_1 = \dot{\mathbf{D}}_j + \ddot{\mathbf{D}}_j\,\Delta t_j \qquad (i = 1) \tag{27}$$

$$(\dot{\mathbf{D}}_{j+1})_1 = \dot{\mathbf{D}}_{j-1} + 2\ddot{\mathbf{D}}_j\,\Delta t \qquad (i = 1) \tag{28}$$

Similarly, for the linear-acceleration method, Eqs. (19) and (20) generalize to

$$(\mathbf{D}_{j+1})_i = \mathbf{R}_j^* + \tfrac{1}{3}(\dot{\mathbf{D}}_{j+1})_i\,\Delta t_j \qquad (i \geq 1) \tag{29}$$

where

$$\mathbf{R}_j^* = \mathbf{D}_j + \tfrac{2}{3}\dot{\mathbf{D}}_j\,\Delta t_j + \tfrac{1}{6}\ddot{\mathbf{D}}_j(\Delta t_j)^2 \tag{30}$$

Of course, Eqs. (21), (23), (24), (26), (27), and (28) apply to both methods.

It is well known that the linear-acceleration method is somewhat more accurate than the average-acceleration method [7]. However, it has been shown [8] that the former technique is only *conditionally stable*. Therefore, as in the central-difference procedure, the solution diverges if the time step is too large. On the other hand, the average-acceleration method is *unconditionally stable*, although less accurate. As with the central-difference predictor, a nodewise iterative solution of a MDOF problem is feasible if the mass matrix is diagonal.

**Example 5.3**

We shall now repeat Example 5.1 using the iteration methods described in this section. Recall that $m = k(T/2\pi)^2$, as given by Eq. (5.2-a), and that the initial acceleration at time $t_0 = 0$ becomes $\ddot{u}_0 = P_1/m$, in accordance with Eq. (5.2-b).

To apply the average-acceleration method, we start the first iteration in the first time step using Eq. (9) to estimate the velocity at time $t_1 = \Delta t = T/20$, as follows:

$$(\dot{u}_1)_1 = \dot{u}_0 + \ddot{u}_0 \Delta t = 0 + \left(\frac{P_1}{m}\right)\left(\frac{T}{20}\right) = 0.05 \frac{P_1 T}{m} \quad (a)$$

Then the displacement at time $t_1$ is found from Eq. (5) to be

$$(u_1)_1 = R_0 + \frac{1}{2}(\dot{u}_1)_1 \Delta t = 0 + \frac{0.05}{2k}\left(\frac{P_1 T^2}{20}\right)\left(\frac{2\pi}{T}\right)^2 = 0.04935 \frac{P_1}{k} \quad (b)$$

which is the same as $u_1$ in Eq. (5.2-d). Next, we obtain the acceleration at time $t_1$ from Eq. (6) as

$$(\ddot{u}_1)_1 = \frac{1}{m}(P_1 - ku_1)_1 = \frac{P_1}{m}(1 - 0.04935) = 0.9507 \frac{P_1}{m} \quad (c)$$

For the second iteration in the first time step, Eqs. (4), (5), and (6) yield

Eq. (4): $\quad (\dot{u}_1)_2 = (1 + 0.9507)\dfrac{P_1 T}{40m} = 0.04877\dfrac{P_1 T}{m}$

Eq. (5): $\quad (u_1)_2 = 0 + \dfrac{0.04877}{2k}\left(\dfrac{P_1 T^2}{20}\right)\left(\dfrac{2\pi}{T}\right)^2 = 0.04813\dfrac{P_1}{k}$

Eq. (6): $\quad (\ddot{u}_1)_2 = \dfrac{P_1}{m}(1 - 0.04813) = 0.9519\dfrac{P_1}{m}$

Third iteration:

Eq. (4): $\quad (\dot{u}_1)_3 = (1 + 0.9519)\dfrac{P_1 T}{40m} = 0.04880\dfrac{P_1 T}{m}$

Eq. (5): $\quad (u_1)_3 = 0 + \dfrac{0.04880}{2k}\left(\dfrac{P_1 T^2}{20}\right)\left(\dfrac{2\pi}{T}\right)^2 = 0.04816\dfrac{P_1}{k}$

Eq. (6): $\quad (\ddot{u}_1)_3 = \dfrac{P_1}{m}(1 - 0.04816) = 0.9518\dfrac{P_1}{m}$

Fourth iteration:

Sec. 5.3    Iteration with Implicit Formulas    209

Eq. (4):    $(\dot{u}_1)_4 = (1 + 0.9518)\dfrac{P_1 T}{40m} = 0.04880 \dfrac{P_1 T}{m}$

Eq. (5):    $(u_1)_4 = 0 + \dfrac{0.04880}{2k}\left(\dfrac{P_1 T^2}{20}\right)\left(\dfrac{2\pi}{T}\right)^2 = 0.04816 \dfrac{P_1}{k}$

Eq. (6):    $(\ddot{u}_1)_4 = \dfrac{P_1}{m}(1 - 0.04816) = 0.9518 \dfrac{P_1}{m}$

In the fourth iteration we see that the response has converged to within four significant digits. Results for 20 time steps are given in Table 5.3, along with the number of iterations $n_i$ required in each step.

TABLE 5.3    Response for Example 5.3 Using Iteration Methods[a]

| j | Average-Acceleration Method | | Linear-Acceleration Method | |
|---|---|---|---|---|
|   | $n_i$ | Approx. $u$ | $n_i$ | Approx. $u$ |
| 1  | 4 | 0.04816 | 4 | 0.04855 |
| 2  | 4 | 0.1880  | 4 | 0.1895  |
| 3  | 4 | 0.4061  | 4 | 0.4091  |
| 4  | 4 | 0.6813  | 4 | 0.6861  |
| 5  | 4 | 0.9873  | 3 | 0.9936  |
| 6  | 4 | 1.294   | 3 | 1.302   |
| 7  | 4 | 1.573   | 3 | 1.581   |
| 8  | 3 | 1.797   | 3 | 1.803   |
| 9  | 3 | 1.944   | 3 | 1.948   |
| 10 | 3 | 2.000   | 3 | 2.000   |
| 11 | 3 | 1.959   | 3 | 1.955   |
| 12 | 3 | 1.827   | 3 | 1.818   |
| 13 | 3 | 1.614   | 3 | 1.601   |
| 14 | 4 | 1.343   | 3 | 1.326   |
| 15 | 4 | 1.038   | 3 | 1.019   |
| 16 | 4 | 0.7300  | 4 | 0.7105  |
| 17 | 4 | 0.4478  | 4 | 0.4299  |
| 18 | 4 | 0.2187  | 4 | 0.2047  |
| 19 | 4 | 0.06497 | 4 | 0.05665 |
| 20 | 5 | 0.00126 | 5 | 0.00025 |

[a] Tabulated values to be multiplied by $P_1/k$.

Next, we apply the linear-acceleration method, using Eqs. (19) and (20) in place of Eqs. (5) and (8). In this case the approximate responses are calculated somewhat more accurately and with fewer iterations. Results for this second analysis also appear in Table 5.3.

**Example 5.4**

Let us analyze by iterative methods the plane truss in Example 5.2, which was shown in Fig. 5.3(a). For this problem we have the relationships $m = 0.2s(T_1/2\pi)^2$ and $\ddot{\mathbf{D}}_0 = \{1, 0\}P_1/m$, as given by Eqs. (5.2-h) and (5.2-i).

For the average-acceleration method, we begin the first iteration in the first time step by estimating the velocities at time $t_1$ with Eq. (26), as follows:

$$(\dot{\mathbf{D}}_1)_1 = \dot{\mathbf{D}}_0 + \ddot{\mathbf{D}}_0 \, \Delta t = 0 + \begin{bmatrix} 1 \\ 0 \end{bmatrix}\left(\frac{P_1}{m}\right)\left(\frac{T_1}{20}\right) = \begin{bmatrix} 0.05 \\ 0 \end{bmatrix}\frac{P_1 T_1}{m} \quad \text{(d)}$$

Next, we find the displacements at time $t_1$ from Eq. (22) to be

$$(\mathbf{D}_1)_1 = \mathbf{R}_0 + \frac{1}{2}(\dot{\mathbf{D}}_1)_1 \, \Delta t = 0 + \begin{bmatrix} 0.05 \\ 0 \end{bmatrix}\frac{P_1 T_1}{0.2 s}\left(\frac{T_1}{40}\right)\left(\frac{2\pi}{T_1}\right)^2 = \begin{bmatrix} 0.2467 \\ 0 \end{bmatrix}\frac{P_1}{s} \quad \text{(e)}$$

which is the same as $\mathbf{D}_1$ in Eq. (5.2-k). Then the acceleration at time $t_1$ is calculated from Eq. (23) as

$$(\ddot{\mathbf{D}}_1)_1 = \mathbf{M}^{-1}(\mathbf{A}_1 - \mathbf{S}\,\mathbf{D}_1)_1 = \frac{1}{m}\begin{bmatrix} 1 & 0 \\ 0 & 1 \end{bmatrix}\left\{\begin{bmatrix} P_1 \\ 0 \end{bmatrix} - s\begin{bmatrix} 0.36 & -0.48 \\ -0.48 & 1.64 \end{bmatrix}\begin{bmatrix} 0.2467 \\ 0 \end{bmatrix}\frac{P_1}{s}\right\}$$

$$= \begin{bmatrix} 0.9112 \\ 0.1184 \end{bmatrix}\frac{P_1}{m} \quad \text{(f)}$$

For the second iteration in the first time step, Eqs. (21), (22), and (23) produce

Eq. (21): $\quad (\dot{\mathbf{D}}_1)_2 = \mathbf{Q}_0 + \frac{1}{2}(\ddot{\mathbf{D}}_1)_1 \, \Delta t = \left\{\begin{bmatrix} 1 \\ 0 \end{bmatrix} + \begin{bmatrix} 0.9112 \\ 0 \end{bmatrix}\right\}\frac{P_1}{m}\left(\frac{\Delta t}{2}\right)$

$$= \begin{bmatrix} 1.9112 \\ 0.1184 \end{bmatrix}\frac{P_1}{m}\left(\frac{T_1}{40}\right) = \begin{bmatrix} 0.04778 \\ 0.002960 \end{bmatrix}\frac{P_1 T_1}{m}$$

Eq. (22): $\quad (\mathbf{D}_1)_2 = \mathbf{R}_0 + \frac{1}{2}(\dot{\mathbf{D}}_1)_2 \, \Delta t = 0 + \begin{bmatrix} 0.04778 \\ 0.002960 \end{bmatrix}\frac{P_1 T_1}{0.2 s}\left(\frac{T_1}{40}\right)\left(\frac{2\pi}{T_1}\right)^2$

$$= \begin{bmatrix} 0.2358 \\ 0.01461 \end{bmatrix}\frac{P_1}{s}$$

Eq. (23): $\quad (\ddot{\mathbf{D}}_1)_2 = \mathbf{M}^{-1}(\mathbf{A}_1 - \mathbf{S}\,\mathbf{D}_1)_2$

$$= \frac{1}{m}\begin{bmatrix} 1 & 0 \\ 0 & 1 \end{bmatrix}\left\{\begin{bmatrix} P_1 \\ 0 \end{bmatrix} - s\begin{bmatrix} 0.36 & -0.48 \\ -0.48 & 1.64 \end{bmatrix}\begin{bmatrix} 0.2358 \\ 0.01461 \end{bmatrix}\frac{P_1}{s}\right\}$$

$$= \begin{bmatrix} 0.9221 \\ 0.08922 \end{bmatrix}\frac{P_1}{m}$$

Such calculations are repeated until convergence of displacements is obtained to within four significant figures (in the sixth iteration). At that stage the values of the displacements calculated from Eq. (22) are

$$(\mathbf{D}_1)_6 = \begin{bmatrix} 0.2369 \\ 0.01168 \end{bmatrix}\frac{P_1}{s} \quad \text{(g)}$$

Table 5.4 contains the displacements $D_1$ and $D_2$ for twenty time steps, as well as the number of iterations $n_i$ for each step.

**TABLE 5.4 Responses for Example 5.4 Using Iteration Methods**[a]

| j | Average-Acceleration Method | | | Linear-Acceleration Method | | |
|---|---|---|---|---|---|---|
| | $n_i$ | Approx. $D_1$ | Approx. $D_2$ | $n_i$ | Approx. $D_1$ | Approx. $D_2$ |
| 1  | 6 | 0.2369  | 0.01168  | 5 | 0.2400  | 0.00834  |
| 2  | 6 | 0.9121  | 0.08378  | 5 | 0.9220  | 0.07618  |
| 3  | 5 | 1.932   | 0.2961   | 4 | 1.948   | 0.2941   |
| 4  | 4 | 3.173   | 0.7005   | 4 | 3.192   | 0.7165   |
| 5  | 4 | 4.515   | 1.263    | 3 | 4.535   | 1.298    |
| 6  | 4 | 5.851   | 1.864    | 3 | 5.875   | 1.902    |
| 7  | 4 | 7.081   | 2.357    | 3 | 7.112   | 2.373    |
| 8  | 3 | 8.102   | 2.648    | 3 | 8.139   | 2.629    |
| 9  | 3 | 8.807   | 2.734    | 3 | 8.838   | 2.697    |
| 10 | 3 | 9.101   | 2.693    | 3 | 9.110   | 2.670    |
| 11 | 4 | 8.691   | 2.596    | 4 | 8.742   | 2.618    |
| 12 | 4 | 7.150   | 2.409    | 3 | 7.146   | 2.482    |
| 13 | 4 | 4.451   | 1.950    | 3 | 4.394   | 2.031    |
| 14 | 6 | 0.9388  | 1.014    | 5 | 0.8498  | 1.024    |
| 15 | 4 | −3.005  | −0.4398  | 3 | −3.106  | −0.5508  |
| 16 | 4 | −7.028  | −2.193   | 4 | −7.142  | −2.386   |
| 17 | 4 | −10.82  | −3.841   | 4 | −10.97  | −4.002   |
| 18 | 4 | −14.09  | −5.009   | 3 | −14.27  | −5.034   |
| 19 | 3 | −16.53  | −5.537   | 3 | −16.73  | −5.425   |
| 20 | 3 | −17.87  | −5.533   | 3 | −18.02  | −5.395   |

[a]Tabulated values to be multiplied by $P_1/s$.

As a second analysis, we apply the linear-acceleration method, using Eqs. (29) and (30) in place of Eqs. (22) and (24). The results for this case are also listed in Table 5.4. Comparison of these responses with the exact results in Table 5.2 shows that they are more accurate than those for the average-acceleration method.

## 5.4 DIRECT LINEAR EXTRAPOLATION

If the equations of motion for a MDOF structure are linear, it is possible to avoid iteration of implicit formulas for numerical solutions. Instead, direct linear extrapolation procedures may be devised for both the average- and linear-acceleration methods. We can formulate either total-response algorithms for linear systems or incremental-response methods for nonlinear systems. Because the number of arithmetic operations is about the same for either approach, we choose to develop the incremental technique, which applies to both linear and nonlinear problems. Furthermore, it is feasible to set up simultaneous equations for incremental accelerations, velocities, or displacements. Here we will use incremental displacements as unknowns and solve a *pseudostatic problem* for each time step. To save space, only MDOF structures will be considered; and any SDOF system becomes merely a special case.

At time $t_j$ (see Fig. 5.1), the damped equations of motion for a MDOF linearly elastic structure are

$$\mathbf{M}\ddot{\mathbf{D}}_j + \mathbf{C}\dot{\mathbf{D}}_j + \mathbf{S}\mathbf{D}_j = \mathbf{A}_j \tag{1}$$

Similarly, at time $t_{j+1} = t_j + \Delta t_j$, the equations of motion become

$$\mathbf{M}(\ddot{\mathbf{D}}_j + \Delta\ddot{\mathbf{D}}_j) + \mathbf{C}(\dot{\mathbf{D}}_j + \Delta\dot{\mathbf{D}}_j) + \mathbf{S}(\mathbf{D}_j + \Delta\mathbf{D}_j) = \mathbf{A}_j + \Delta\mathbf{A}_j \tag{2}$$

Subtraction of Eq. (1) from Eq. (2) produces the *incremental equations of motion* as

$$\mathbf{M}\Delta\ddot{\mathbf{D}}_j + \mathbf{C}\Delta\dot{\mathbf{D}}_j + \mathbf{S}\Delta\mathbf{D}_j = \Delta\mathbf{A}_j \tag{3}$$

These equations will be used for both the average- and linear-acceleration algorithms developed in the following discussion.

For the average-acceleration method, the incremental velocities obtained by the trapezoidal rule at the end of time step $\Delta t_j$ are

$$\Delta\dot{\mathbf{D}}_j = \tfrac{1}{2}(\ddot{\mathbf{D}}_j + \ddot{\mathbf{D}}_{j+1})\,\Delta t_j = \ddot{\mathbf{D}}_j\,\Delta t_j + \tfrac{1}{2}\Delta\ddot{\mathbf{D}}_j\,\Delta t_j \tag{4}$$

Similarly, the incremental displacements at the end of the step become

$$\Delta\mathbf{D}_j = \tfrac{1}{2}(\dot{\mathbf{D}}_j + \dot{\mathbf{D}}_{j+1})\,\Delta t_j = \dot{\mathbf{D}}_j\,\Delta t_j + \tfrac{1}{2}\Delta\dot{\mathbf{D}}_j\,\Delta t_j \tag{5}$$

Substitution of Eq. (4) into Eq. (5) yields

$$\Delta\mathbf{D}_j = \dot{\mathbf{D}}_j\,\Delta t_j + \tfrac{1}{2}\ddot{\mathbf{D}}_j(\Delta t_j)^2 + \tfrac{1}{4}\Delta\ddot{\mathbf{D}}_j(\Delta t_j)^2 \tag{6}$$

Solving for the incremental accelerations in Eq. (6) gives

$$\Delta\ddot{\mathbf{D}}_j = \frac{4}{(\Delta t_j)^2}\Delta\mathbf{D}_j - \frac{4}{\Delta t_j}\dot{\mathbf{D}}_j - 2\ddot{\mathbf{D}}_j \tag{7}$$

Substituting this expression into Eq. (4) produces

$$\Delta\dot{\mathbf{D}}_j = \frac{2}{\Delta t_j}\Delta\mathbf{D}_j - 2\dot{\mathbf{D}}_j \tag{8}$$

Now we define vectors $\overline{\mathbf{Q}}_j$ and $\overline{\mathbf{R}}_j$ that contain only combinations of $\dot{\mathbf{D}}_j$ and $\ddot{\mathbf{D}}_j$, as follows:

$$\overline{\mathbf{Q}}_j = \frac{4}{\Delta t_j}\dot{\mathbf{D}}_j + 2\ddot{\mathbf{D}}_j \tag{9}$$

$$\overline{\mathbf{R}}_j = 2\dot{\mathbf{D}}_j \tag{10}$$

Using these definitions, we rewrite Eqs. (7) and (8) as

## Sec. 5.4 Direct Linear Extrapolation

$$\Delta \ddot{D}_j = \frac{4}{(\Delta t_j)^2} \Delta D_j - \overline{Q}_j \qquad (11)$$

$$\Delta \dot{D}_j = \frac{2}{\Delta t_j} \Delta D_j - \overline{R}_j \qquad (12)$$

Now substitute Eqs. (11) and (12) into Eq. (3) to obtain

$$M \left[ \frac{4}{(\Delta t_j)^2} \Delta D_j - \overline{Q}_j \right] + C \left[ \frac{2}{\Delta t_j} \Delta D_j - \overline{R}_j \right] + S \, \Delta D_j = \Delta A_j$$

Collecting terms, we rewrite this equation in the form

$$\overline{S} \, \Delta D_j = \Delta \overline{A}_j \qquad (13)$$

in which

$$\overline{S} = S + \frac{4}{(\Delta t_j)^2} M + \frac{2}{\Delta t_j} C \qquad (14)$$

and

$$\Delta \overline{A}_j = \Delta A_j + M \overline{Q}_j + C \overline{R}_j \qquad (15)$$

Thus, the pseudostatic equations represented by Eq. (13) are to be solved for the incremental displacements $\Delta D_j$ in each step. Then the incremental accelerations $\Delta \ddot{D}_j$ and velocities $\Delta \dot{D}_j$ may be found using Eqs. (11) and (12). Finally, the total values of $D_{j+1}$, $\dot{D}_{j+1}$, and $\ddot{D}_{j+1}$ are

$$D_{j+1} = D_j + \Delta D_j \qquad (16)$$

$$\dot{D}_{j+1} = \dot{D}_j + \Delta \dot{D}_j \qquad (17)$$

$$\ddot{D}_{j+1} = \ddot{D}_j + \Delta \ddot{D}_j \qquad (18)$$

In summary, the procedure for obtaining dynamic responses of a MDOF structure consists of the following calculations in each time step:

### Direct Linear Extrapolation by the Average-Acceleration Method

1. Determine $\overline{Q}_j$ and $\overline{R}_j$ from Eqs. (9) and (10).
2. Find $\overline{S}$ and $\Delta \overline{A}_j$ using Eqs. (14) and (15).
3. Solve Eq. (13) for $\Delta D_j$.
4. Calculate $\Delta \ddot{D}_j$ and $\Delta \dot{D}_j$ with Eqs. (11) and (12).
5. Add the incremental displacements, velocities, and accelerations to preceding values using Eqs. (16), (17), and (18).

Turning now to the linear-acceleration method, we may again derive the incremental velocity vector from the trapezoidal rule to be

$$\Delta \dot{D}_j = \ddot{D}_j \, \Delta t_j + \tfrac{1}{2} \Delta \ddot{D}_j \, \Delta t_j \qquad (4)$$

However, the incremental displacements in the step must be found from Eq. (5.3-17), as follows:

$$\Delta \mathbf{D}_j = \dot{\mathbf{D}}_j \Delta t_j + \tfrac{1}{6}(2\ddot{\mathbf{D}}_j + \ddot{\mathbf{D}}_{j+1})(\Delta t_j)^2$$
$$= \dot{\mathbf{D}}_j \Delta t_j + \tfrac{1}{2}\ddot{\mathbf{D}}_j (\Delta t_j)^2 + \tfrac{1}{6}\Delta \ddot{\mathbf{D}}_j (\Delta t_j)^2 \quad (19)$$

Solving for the incremental accelerations in Eq. (19) gives

$$\Delta \ddot{\mathbf{D}}_j = \frac{6}{(\Delta t_j)^2} \Delta \mathbf{D}_j - \frac{6}{\Delta t_j} \dot{\mathbf{D}}_j - 3\ddot{\mathbf{D}}_j \quad (20)$$

Now substitute Eq. (20) into Eq. (4) to obtain

$$\Delta \dot{\mathbf{D}}_j = \frac{3}{\Delta t_j} \Delta \mathbf{D}_j - 3\dot{\mathbf{D}}_j - \frac{\Delta t_j}{2} \ddot{\mathbf{D}}_j \quad (21)$$

Then define

$$\mathbf{Q}_j^* = \frac{6}{\Delta t_j} \dot{\mathbf{D}}_j + 3\ddot{\mathbf{D}}_j \quad (22)$$

$$\mathbf{R}_j^* = 3\dot{\mathbf{D}}_j + \frac{\Delta t_j}{2} \ddot{\mathbf{D}}_j \quad (23)$$

and rewrite Eqs. (20) and (21) as

$$\Delta \ddot{\mathbf{D}}_j = \frac{6}{(\Delta t_j)^2} \Delta \mathbf{D}_j - \mathbf{Q}_j^* \quad (24)$$

$$\Delta \dot{\mathbf{D}}_j = \frac{3}{\Delta t_j} \Delta \mathbf{D}_j - \mathbf{R}_j^* \quad (25)$$

Next, we substitute Eqs. (24) and (25) into Eq. (3) and collect terms to find

$$\mathbf{S}^* \Delta \mathbf{D}_j = \Delta \mathbf{A}_j^* \quad (26)$$

in which

$$\mathbf{S}^* = \mathbf{S} + \frac{6}{(\Delta t_j)^2} \mathbf{M} + \frac{3}{\Delta t_j} \mathbf{C} \quad (27)$$

and

$$\Delta \mathbf{A}_j^* = \Delta \mathbf{A}_j + \mathbf{M}\, \mathbf{Q}_j^* + \mathbf{C}\, \mathbf{R}_j^* \quad (28)$$

Equations (26), (27), and (28) are analogous to Eqs. (13), (14), and (15), derived previously for the average-acceleration method. Thus, the procedure for direct linear extrapolation by the linear-acceleration method follows the same steps as before, except that Eqs. (9) through (15) are replaced by Eqs. (22) through (28).

## Sec. 5.4  Direct Linear Extrapolation

### Example 5.5

Again, we shall repeat Example 5.1, using the direct linear extrapolation methods derived in this section. For that purpose, we need the relationships $m = k(T/2\pi)^2$ and $\ddot{u}_0 = P_1/m$ from Eqs. (5.2-a) and (5.2-b).

Starting with the average-acceleration method, we apply Eqs. (9) through (18) in the first time step. When doing so, we use notation for the undamped SDOF system, as follows:

Eq. (9): $\quad \overline{Q}_0 = \dfrac{4}{\Delta t}\dot{u}_0 + 2\ddot{u}_0 = 0 + 2\dfrac{P_1}{m} = 2\dfrac{P_1}{m}$  (a)

Eq. (10): $\quad \overline{R}_0 = 2\dot{u}_0 = 0$  (b)

Eq. (14): $\quad \bar{k} = k + \dfrac{4m}{(\Delta t)^2} = k + 4k\left(\dfrac{T}{2\pi}\right)^2\left(\dfrac{20}{T}\right)^2 = 41.53k$  (c)

Eq. (15): $\quad \Delta \overline{P}_0 = \Delta P_0 + m\overline{Q}_0 = 0 + 2P_1 = 2P_1$  (d)

Eq. (13): $\quad \Delta u_0 = \dfrac{\Delta \overline{P}_0}{\bar{k}} = \dfrac{2P_1}{41.53k} = 0.04816\dfrac{P_1}{k}$  (e)

Eq. (11): $\quad \Delta \ddot{u}_0 = \dfrac{4}{(\Delta t)^2}\Delta u_0 - \overline{Q}_0 = 4\left(\dfrac{20}{T}\right)^2\left(0.04816\dfrac{P_1}{k}\right) - 2\dfrac{P_1}{m}$

$\qquad = 4\left(\dfrac{20}{2\pi}\right)^2\dfrac{k}{m}\left(0.04816\dfrac{P_1}{k}\right) - 2\dfrac{P_1}{m} = -0.04815\dfrac{P_1}{m}$  (f)

Eq. (12): $\quad \Delta \dot{u}_0 = \dfrac{2}{\Delta t}\Delta u_0 - \overline{R}_0 = 2\left(\dfrac{20}{T}\right)\left(0.04816\dfrac{P_1}{k}\right) - 0$

$\qquad = \left(\dfrac{40}{T}\right)\left(0.04816\dfrac{P_1}{m}\right)\left(\dfrac{T}{2\pi}\right)^2 = 0.04880\dfrac{P_1 T}{m}$  (g)

Eq. (16): $\quad u_1 = u_0 + \Delta u_0 = 0 + 0.04816\dfrac{P_1}{k} = 0.04816\dfrac{P_1}{k}$  (h)

Eq. (17): $\quad \dot{u}_1 = \dot{u}_0 + \Delta \dot{u}_0 = 0 + 0.04880\dfrac{P_1 T}{m} = 0.04880\dfrac{P_1 T}{m}$  (i)

Eq. (18): $\quad \ddot{u}_1 = \ddot{u}_0 + \Delta \ddot{u}_0 = \dfrac{P_1}{m} - 0.04815\dfrac{P_1}{m} = 0.9518\dfrac{P_1}{m}$  (j)

Note that the values in Eqs. (h), (i), and (j) are the same as those obtained in Example 5.3 by the method of iteration. This procedure is repeated for each of 20 time steps.

Values of the response from direct linear extrapolation by the average-acceleration method are listed in Table 5.5. Also given in the table are slightly more accurate values computed by the linear-acceleration method.

**TABLE 5.5 Response for Example 5.5 Using Direct Linear Extrapolation**[a]

| j | Avg.-Accel. | Lin.-Accel. | j | Avg.-Accel. | Lin.-Accel. |
|---|---|---|---|---|---|
| 1 | 0.04816 | 0.04855 | 11 | 1.959 | 1.955 |
| 2 | 0.1880 | 0.1895 | 12 | 1.827 | 1.818 |
| 3 | 0.4061 | 0.4091 | 13 | 1.614 | 1.601 |
| 4 | 0.6813 | 0.6861 | 14 | 1.343 | 1.326 |
| 5 | 0.9873 | 0.9936 | 15 | 1.038 | 1.019 |
| 6 | 1.295 | 1.302 | 16 | 0.7300 | 0.7105 |
| 7 | 1.573 | 1.581 | 17 | 0.4478 | 0.4299 |
| 8 | 1.797 | 1.803 | 18 | 0.2188 | 0.2047 |
| 9 | 1.944 | 1.947 | 19 | 0.06500 | 0.05673 |
| 10 | 2.000 | 2.000 | 20 | 0.00130 | 0.00033 |

[a]Tabulated values to be multiplied by $P_1/k$.

## Example 5.6

Now we will calculate the responses of the plane truss in Example 5.2 by direct linear extrapolation methods. From Eqs. (5.2-h) and (5.2-i) we know that $m = 0.2s(T_1/2\pi)^2$ and $\ddot{\mathbf{D}}_0 = \{1, 0\}P_1/m$.

For the average-acceleration method in matrix form, we use Eqs. (9) through (18) as shown. Thus, for the first time step we have

Eq. (9): $\quad \overline{\mathbf{Q}}_0 = \dfrac{4}{\Delta t}\dot{\mathbf{D}}_0 + 2\ddot{\mathbf{D}}_0 = 0 + 2\begin{bmatrix}1\\0\end{bmatrix}\dfrac{P_1}{m} = \begin{bmatrix}2\\0\end{bmatrix}\dfrac{P_1}{m}$  (k)

Eq. (10): $\quad \overline{\mathbf{R}}_0 = 2\dot{\mathbf{D}}_0 = 0$  ($\ell$)

Eq. (14): $\quad \overline{\mathbf{S}} = \mathbf{S} + \dfrac{4}{(\Delta t)^2}\mathbf{M} = \mathbf{S} + \dfrac{(4)(20)^2}{T_1^2}m\begin{bmatrix}1 & 0\\0 & 1\end{bmatrix}$

$= s\begin{bmatrix}0.36 & -0.48\\-0.48 & 1.64\end{bmatrix} + \dfrac{(4)(20)^2}{T_1^2}\begin{bmatrix}1 & 0\\0 & 1\end{bmatrix}0.2s\left(\dfrac{T_1}{2\pi}\right)^2$

$= s\begin{bmatrix}8.466 & -0.48\\-0.48 & 9.746\end{bmatrix}$  (m)

Eq. (15): $\quad \Delta\overline{\mathbf{A}}_0 = \Delta\mathbf{A}_0 + \mathbf{M}\,\mathbf{Q}_0$

$= 0 + m\begin{bmatrix}1 & 0\\0 & 1\end{bmatrix}\begin{bmatrix}2\\0\end{bmatrix}\dfrac{P_1}{m} = \begin{bmatrix}2\\0\end{bmatrix}P_1$  (n)

Eq. (13): $\quad \Delta\mathbf{D}_0 = \overline{\mathbf{S}}^{-1}\,\Delta\overline{\mathbf{A}}_0 = \dfrac{1}{82.28}\begin{bmatrix}9.746 & 0.48\\0.48 & 8.466\end{bmatrix}\begin{bmatrix}2\\0\end{bmatrix}\dfrac{P_1}{s}$

$= \begin{bmatrix}0.2369\\0.01167\end{bmatrix}\dfrac{P_1}{s}$  (o)

Sec. 5.5  Newmark's Generalized Acceleration Method

Eq. (11): $\Delta \ddot{\mathbf{D}}_0 = \dfrac{4}{(\Delta t)^2} \Delta \mathbf{D}_0 - \overline{\mathbf{Q}}_0 = \dfrac{(4)(20)^2}{T_1^2} \begin{bmatrix} 0.2369 \\ 0.01167 \end{bmatrix} \dfrac{P_1}{s} - \begin{bmatrix} 2 \\ 0 \end{bmatrix} \dfrac{P_1}{m}$

$= \dfrac{(4)(20)^2(0.2s)}{(2\pi)^2 m} \begin{bmatrix} 0.2369 \\ 0.01167 \end{bmatrix} \dfrac{P_1}{s} - \begin{bmatrix} 2 \\ 0 \end{bmatrix} \dfrac{P_1}{m}$

$= \begin{bmatrix} -0.07976 \\ 0.09459 \end{bmatrix} \dfrac{P_1}{m}$  (p)

Eq. (12): $\Delta \dot{\mathbf{D}}_0 = \dfrac{2}{\Delta t} \Delta \mathbf{D}_0 - \overline{\mathbf{R}}_0 = \left(\dfrac{40}{T_1}\right) \begin{bmatrix} 0.2369 \\ 0.01167 \end{bmatrix} \dfrac{P_1}{s} - 0$

$= \left(\dfrac{40}{T_1}\right) \begin{bmatrix} 0.2369 \\ 0.01167 \end{bmatrix} \dfrac{0.2 P_1}{m} \left(\dfrac{T_1}{2\pi}\right)^2$

$= \begin{bmatrix} 0.04801 \\ 0.002365 \end{bmatrix} \dfrac{P_1 T_1}{m}$  (q)

Eq. (16): $\mathbf{D}_1 = \mathbf{D}_0 + \Delta \mathbf{D}_0 = 0 + \begin{bmatrix} 0.2369 \\ 0.01167 \end{bmatrix} \dfrac{P_1}{s}$

$= \begin{bmatrix} 0.2369 \\ 0.01167 \end{bmatrix} \dfrac{P_1}{s}$  (r)

Eq. (17): $\dot{\mathbf{D}}_1 = \dot{\mathbf{D}}_0 + \Delta \dot{\mathbf{D}}_0 = 0 + \begin{bmatrix} 0.04801 \\ 0.002365 \end{bmatrix} \dfrac{P_1 T_1}{m}$

$= \begin{bmatrix} 0.04801 \\ 0.002365 \end{bmatrix} \dfrac{P_1 T_1}{m}$  (s)

Eq. (18): $\ddot{\mathbf{D}}_1 = \ddot{\mathbf{D}}_0 + \Delta \ddot{\mathbf{D}}_0 = \begin{bmatrix} 1 \\ 0 \end{bmatrix} \dfrac{P_1}{m} + \begin{bmatrix} -0.07976 \\ 0.09459 \end{bmatrix} \dfrac{P_1}{m}$

$= \begin{bmatrix} 0.9202 \\ 0.09459 \end{bmatrix} \dfrac{P_1}{m}$  (t)

Here the values in Eqs. (r), (s), and (t) are the same as those found previously by iteration in Example 5.4. Following this time step, 19 other sets of calculations are performed in the same manner, and Table 5.6 gives the results.

For the linear-acceleration method, we apply Eqs. (22) through (28) in place of Eqs. (9) through (15). Responses for this second analysis are also listed in Table 5.6. Note that all of the values in the table are practically the same as those obtained by iteration in Table 5.4.

## 5.5 NEWMARK'S GENERALIZED ACCELERATION METHOD

In this section we study a family of methods that were described by Nathan M. Newmark in a renowned ASCE paper. Two other variants, derived by Wilson et al. and Hilber et al., will also be discussed.

**TABLE 5.6** Responses for Example 5.6 Using Direct Linear Extrapolation[a]

| j | Average-Acceleration Method | | Linear-Acceleration Method | |
|---|---|---|---|---|
| | Approx. $D_1$ | Approx. $D_2$ | Approx. $D_1$ | Approx. $D_2$ |
| 1  | 0.2369  | 0.01167 | 0.2400  | 0.00835 |
| 2  | 0.9121  | 0.08374 | 0.9220  | 0.07621 |
| 3  | 1.932   | 0.2961  | 1.948   | 0.2941  |
| 4  | 3.173   | 0.7006  | 3.192   | 0.7164  |
| 5  | 4.515   | 1.264   | 4.536   | 1.297   |
| 6  | 5.851   | 1.865   | 5.875   | 1.900   |
| 7  | 7.080   | 2.359   | 7.113   | 2.370   |
| 8  | 8.101   | 2.649   | 8.139   | 2.627   |
| 9  | 8.806   | 2.737   | 8.838   | 2.699   |
| 10 | 9.099   | 2.697   | 9.108   | 2.675   |
| 11 | 8.689   | 2.600   | 8.741   | 2.623   |
| 12 | 7.149   | 2.409   | 7.145   | 2.482   |
| 13 | 4.452   | 1.948   | 4.397   | 2.022   |
| 14 | 0.9392  | 1.013   | 0.8543  | 1.011   |
| 15 | −3.005  | −0.4404 | −3.104  | −0.5576 |
| 16 | −7.028  | −2.193  | −7.143  | −2.379  |
| 17 | −10.89  | −3.843  | −10.97  | −3.988  |
| 18 | −14.09  | −5.011  | −14.28  | −5.024  |
| 19 | −16.53  | −5.541  | −16.73  | −5.428  |
| 20 | −17.86  | −5.544  | −18.01  | −5.412  |

[a]Tabulated values to be multiplied by $P_1/s$.

## Newmark-$\beta$ Method

In his 1959 paper, Newmark [8] generalized certain direct numerical integration methods that had been in use up to that time. He presented equations for approximating the velocity and displacement of a SDOF system at time $t_{j+1}$, as follows:

$$\dot{u}_{j+1} = \dot{u}_j + [(1 - \gamma)\ddot{u}_j + \gamma \ddot{u}_{j+1}]\,\Delta t_j \tag{1}$$

$$u_{j+1} = u_j + \dot{u}_j\,\Delta t_j + [(\tfrac{1}{2} - \beta)\ddot{u}_j + \beta \ddot{u}_{j+1}](\Delta t_j)^2 \tag{2}$$

The parameter $\gamma$ in Eq. (1) produces *numerical* (or *algorithmic*) *damping* within the time step $\Delta t_j$. If $\gamma$ is taken to be less than $\tfrac{1}{2}$, an artificial negative damping results. On the other hand, if $\gamma$ is greater than $\tfrac{1}{2}$, such damping is positive. To avoid numerical damping altogether, the value of $\gamma$ must be equal to $\tfrac{1}{2}$; and Eq. (1) becomes the trapezoidal rule.

## Sec. 5.5  Newmark's Generalized Acceleration Method

The parameter $\beta$ in Eq. (2) controls the variation of acceleration within the time step. For this reason the technique is referred to as Newmark's generalized acceleration method (or Newmark's $\beta$ method). For example, if we take $\beta = 0$, Eq. (2) becomes

$$u_{j+1} = u_j + \dot{u}_j \, \Delta t_j + \tfrac{1}{2}\ddot{u}_j(\Delta t_j)^2 \tag{3}$$

This formula is known as the *constant-acceleration method*, because the acceleration $\ddot{u}_j$ at the beginning of the time step $\Delta t_j$ is taken to be constant within the step. Equation (3) also corresponds to a truncated Taylor series that results from using Euler's formula [see Eq. (5.3-10)] for velocity and the trapezoidal rule for displacement.

If we let $\beta = \tfrac{1}{4}$, Eq. (2) yields

$$u_{j+1} = u_j + \dot{u}_j \, \Delta t_j + \tfrac{1}{4}(\ddot{u}_j + \ddot{u}_{j+1})(\Delta t_j)^2 \tag{4}$$

This expression is the same as that in Eq. (5.3-3) for the *average-acceleration method*. When we take $\beta = \tfrac{1}{6}$, Eq. (2) produces

$$u_{j+1} = u_j + \dot{u}_j \, \Delta t_j + \tfrac{1}{6}(2\ddot{u}_j + \ddot{u}_{j+1})(\Delta t_j)^2 \tag{5}$$

In this case the formula is identical to Eq. (5.3-17) for the *linear-acceleration method*.

Now let us consider a MDOF structure and cast the Newark-$\beta$ method into matrix format. In each time step we shall solve a pseudostatic problem for incremental displacements by direct linear extrapolation, as in Sec. 5.4. For this purpose, Eq. (1) is restated in incremental matrix form, as follows:

$$\Delta \dot{\mathbf{D}}_j = [(1 - \gamma)\ddot{\mathbf{D}}_j + \gamma \ddot{\mathbf{D}}_{j+1}] \, \Delta t_j$$
$$= \ddot{\mathbf{D}}_j \, \Delta t_j + \gamma \, \Delta \ddot{\mathbf{D}}_j \, \Delta t_j \tag{6}$$

In addition, Eq. (2) is restated in incremental matrix form as

$$\Delta \mathbf{D}_j = \dot{\mathbf{D}}_j \, \Delta t_j + [(\tfrac{1}{2} - \beta)\ddot{\mathbf{D}}_j + \beta \ddot{\mathbf{D}}_{j+1}](\Delta t_j)^2$$
$$= \dot{\mathbf{D}}_j \, \Delta t_j + \tfrac{1}{2}\ddot{\mathbf{D}}_j(\Delta t_j)^2 + \beta \, \Delta \ddot{\mathbf{D}}_j(\Delta t_j)^2 \tag{7}$$

Solving for $\Delta \ddot{\mathbf{D}}_j$ in Eq. (7) produces

$$\Delta \ddot{\mathbf{D}}_j = \frac{1}{\beta(\Delta t_j)^2} \Delta \mathbf{D}_j - \frac{1}{\beta \, \Delta t_j} \dot{\mathbf{D}}_j - \frac{1}{2\beta} \ddot{\mathbf{D}}_j \tag{8}$$

Substitution of Eq. (8) into Eq. (6) yields

$$\Delta \dot{\mathbf{D}}_j = \frac{\gamma}{\beta \, \Delta t_j} \Delta \mathbf{D}_j - \frac{\gamma}{\beta} \dot{\mathbf{D}}_j - \left(\frac{\gamma}{2\beta} - 1\right) \Delta t_j \, \ddot{\mathbf{D}}_j \tag{9}$$

For convenience, we define the vectors

$$\hat{\mathbf{Q}}_j = \frac{1}{\beta \, \Delta t_j} \dot{\mathbf{D}}_j + \frac{1}{2\beta} \ddot{\mathbf{D}}_j \tag{10}$$

$$\hat{\mathbf{R}}_j = \frac{\gamma}{\beta} \dot{\mathbf{D}}_j + \left(\frac{\gamma}{2\beta} - 1\right) \Delta t_j \ddot{\mathbf{D}}_j \tag{11}$$

Now rewrite Eqs. (8) and (9) in the forms

$$\Delta \ddot{\mathbf{D}}_j = \frac{1}{\beta(\Delta t_j)^2} \Delta \mathbf{D}_j - \hat{\mathbf{Q}}_j \tag{12}$$

$$\Delta \dot{\mathbf{D}}_j = \frac{\gamma}{\beta \Delta t_j} \Delta \mathbf{D}_j - \hat{\mathbf{R}}_j \tag{13}$$

Then substitute Eqs. (12) and (13) into the incremental equations of motion, given previously as Eq. (5.4-3); and collect terms to obtain

$$\hat{\mathbf{S}} \, \Delta \mathbf{D}_j = \Delta \hat{\mathbf{A}}_j \tag{14}$$

in which

$$\hat{\mathbf{S}} = \mathbf{S} + \frac{1}{\beta(\Delta t_j)^2} \mathbf{M} + \frac{\gamma}{\beta \Delta t_j} \mathbf{C} \tag{15}$$

and

$$\Delta \hat{\mathbf{A}}_j = \Delta \mathbf{A}_j + \mathbf{M} \, \hat{\mathbf{Q}}_j + \mathbf{C} \, \hat{\mathbf{R}}_j \tag{16}$$

We solve the pseudostatic problem in Eq. (14) for the incremental displacements $\Delta \mathbf{D}_j$ and substitute them into Eqs. (12) and (13) to find the incremental accelerations and velocities $\Delta \ddot{\mathbf{D}}_j$ and $\Delta \dot{\mathbf{D}}_j$. Then the total values of displacements, velocities, and accelerations at time $t_{j+1}$ are determined using Eqs. (5.4-16), (5.4-17), and (5.4-18).

### Wilson-θ Method

Wilson et al. [9], extended the linear-acceleration method in a manner that makes it numerically stable. The basic assumption of the Wilson-$\theta$ approach is that the acceleration $\ddot{u}$ varies linearly over an extended time step $\Delta t_\theta = \theta \, \Delta t_j$, as depicted in Fig. 5.5. During that time step the incremental acceleration is $\Delta \ddot{u}_\theta = \theta \, \Delta \ddot{u}_j$. It has been shown [10] that the optimum value of the parameter $\theta$ is 1.420815, which can be rounded to 1.42.

As with the Newmark-$\beta$ method, we shall construct the Wilson-$\theta$ variant in an incremental matrix format applicable to a MDOF structure. From Eq. (5.4-4) for the linear-acceleration technique, we have the incremental velocities at the end of the extended time step $\Delta t_\theta$, as follows:

$$\Delta \dot{\mathbf{D}}_\theta = \ddot{\mathbf{D}}_j \, \Delta t_\theta + \tfrac{1}{2} \Delta \ddot{\mathbf{D}}_\theta \, \Delta t_\theta \tag{17}$$

where the symbol $\Delta \ddot{\mathbf{D}}_\theta$ denotes a vector of incremental accelerations. Similarly, Eq. (5.4-19) gives the incremental displacements as

$$\Delta \mathbf{D}_\theta = \dot{\mathbf{D}}_j \, \Delta t_\theta + \tfrac{1}{2} \ddot{\mathbf{D}}_j (\Delta t_\theta)^2 + \tfrac{1}{6} \Delta \ddot{\mathbf{D}}_\theta (\Delta t_\theta)^2 \tag{18}$$

## Sec. 5.5  Newmark's Generalized Acceleration Method

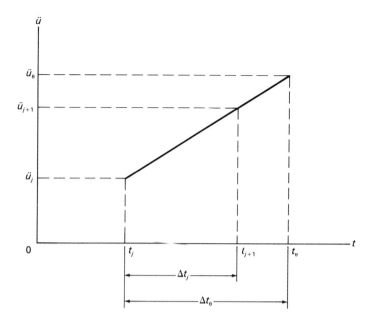

**Figure 5.5** Linear-acceleration method extended to Wilson-$\theta$ method.

We solve Eq. (18) for $\Delta\ddot{\mathbf{D}}_\theta$ to obtain

$$\Delta\ddot{\mathbf{D}}_\theta = \frac{6}{(\Delta t_\theta)^2}\Delta\mathbf{D}_\theta - \frac{6}{\Delta t_\theta}\dot{\mathbf{D}}_j - 3\ddot{\mathbf{D}}_j \tag{19}$$

Then substitute this expression into Eq. (17), which results in

$$\Delta\dot{\mathbf{D}}_\theta = \frac{3}{\Delta t_\theta}\Delta\mathbf{D}_\theta - 3\dot{\mathbf{D}}_j - \frac{\Delta t_\theta}{2}\ddot{\mathbf{D}}_j \tag{20}$$

Now define

$$\mathbf{Q}_\theta^* = \frac{6}{\Delta t_\theta}\dot{\mathbf{D}}_j + 3\ddot{\mathbf{D}}_j \tag{21}$$

$$\mathbf{R}_\theta^* = 3\dot{\mathbf{D}}_j + \frac{\Delta t_\theta}{2}\ddot{\mathbf{D}}_j \tag{22}$$

and restate Eqs. (19) and (20) as

$$\Delta\ddot{\mathbf{D}}_\theta = \frac{6}{(\Delta t_\theta)^2}\Delta\mathbf{D}_\theta - \mathbf{Q}_\theta^* \tag{23}$$

$$\Delta\dot{\mathbf{D}}_\theta = \frac{3}{\Delta t_\theta}\Delta\mathbf{D}_\theta - \mathbf{R}_\theta^* \tag{24}$$

Substitute Eqs. (23) and (24) into Eq. (5.4-3) and collect terms to find

$$\mathbf{S}_\theta^* \Delta\mathbf{D}_\theta = \Delta\mathbf{A}_\theta^* \tag{25}$$

The starred terms in this equation are

$$\mathbf{S}_\theta^* = \mathbf{S} + \frac{6}{(\Delta t_\theta)^2}\mathbf{M} + \frac{3}{\Delta t_\theta}\mathbf{C} \qquad (26)$$

and

$$\Delta \mathbf{A}_\theta^* = \Delta \mathbf{A}_\theta + \mathbf{M}\,\mathbf{Q}_\theta^* + \mathbf{C}\,\mathbf{R}_\theta^* \qquad (27)$$

where

$$\Delta \mathbf{A}_\theta = \theta\,\Delta \mathbf{A}_j \qquad (28)$$

The pseudostatic problem in Eq. (25) is solved for the incremental displacements $\Delta \mathbf{D}_\theta$. Then we can determine the incremental accelerations $\Delta \ddot{\mathbf{D}}_\theta$ from Eq. (23) and reduce them linearly by the formula

$$\Delta \ddot{\mathbf{D}}_j = \frac{1}{\theta}\,\Delta \ddot{\mathbf{D}}_\theta \qquad (29)$$

Next, the incremental velocities $\Delta \dot{\mathbf{D}}_j$ and displacements $\Delta \mathbf{D}_j$ are obtained from Eqs. (5.4-4) and (5.4-19) for the time step $\Delta t_j$. As before, Eqs. (5.4-16), (5.4-17), and (5.4-18) give total values of the displacements, velocities, and accelerations at time $t_{j+1}$.

### Hilber-α Method

To improve control of numerical damping, Hilber et al. [11] introduced a parameter $\alpha$ into the equations of motion at time $t_{j+1}$, as follows:

$$\mathbf{M}\,\ddot{\mathbf{D}}_{j+1} + \mathbf{C}\,\dot{\mathbf{D}}_{j+1} + (1 + \alpha)\mathbf{S}\,\mathbf{D}_{j+1} - \alpha \mathbf{S}\,\mathbf{D}_j = \mathbf{A}_{j+1} \qquad (30)$$

Subtracting similar equations of motion at time $t_j$ from Eq. (30) produces the incremental equations

$$\mathbf{M}\,\Delta\ddot{\mathbf{D}}_j + \mathbf{C}\,\Delta\dot{\mathbf{D}}_j + (1 + \alpha)\mathbf{S}\,\Delta\mathbf{D}_j - \alpha \mathbf{S}\,\Delta\mathbf{D}_{j-1} = \Delta \mathbf{A}_j \qquad (31)$$

Now substitute Eqs. (12) and (13) from Newmark's method into Eq. (31) and collect terms to find

$$\hat{\mathbf{S}}_\alpha\,\Delta \mathbf{D}_j = \Delta \hat{\mathbf{A}}_{\alpha j} \qquad (32)$$

in which

$$\hat{\mathbf{S}}_\alpha = \hat{\mathbf{S}} + \alpha \mathbf{S} \qquad (33)$$

and

$$\Delta \hat{\mathbf{A}}_{\alpha j} = \Delta \hat{\mathbf{A}}_j + \alpha \mathbf{S}\,\Delta \mathbf{D}_{j-1} \qquad (34)$$

Expressions for $\hat{\mathbf{S}}$ and $\Delta \hat{\mathbf{A}}_j$ were derived previously as Eqs. (15) and (16). For the first time step, we take $\Delta \mathbf{D}_{j-1} = \mathbf{0}$ in Eq. (34).

## 5.6 NUMERICAL STABILITY AND ACCURACY

To study the stability and accuracy of various one-step direct numerical integration procedures, we may cast them into *operator form* [12], as follows:

$$\mathbf{U}_{j+1} = \mathbf{A}\,\mathbf{U}_j + \mathbf{L}\,P_{j+1} \tag{1}$$

This expression pertains to a SDOF system that might equally well be considered as one natural mode of vibration for a MDOF structure. The symbol $\mathbf{U}_j$ in Eq. (1) represents a column vector containing the three response quantities $u_j$, $\dot{u}_j$, and $\ddot{u}_j$ at the time station $t_j$. That is,

$$\mathbf{U}_j = \{u_j,\ \dot{u}_j,\ \ddot{u}_j\} \tag{2}$$

and the vector $\mathbf{U}_{j+1}$ is similarly defined at time $t_{j+1}$ to be

$$\mathbf{U}_{j+1} = \{u_{j+1},\ \dot{u}_{j+1},\ \ddot{u}_{j+1}\} \tag{3}$$

The coefficient matrix $\mathbf{A}$ in Eq. (1) is a $3 \times 3$ array called the *amplification matrix* that we shall examine to answer questions about stability and accuracy. Finally, the symbol $\mathbf{L}$ denotes a column vector called the *load operator*, which is multiplied by the load $P_{j+1}$ at time $t_{j+1}$. If there is no loading, Eq. (1) simplifies to

$$\mathbf{U}_{j+1} = \mathbf{A}\,\mathbf{U}_j \tag{4}$$

for free-vibrational response.

To investigate the stability of a numerical algorithm, we apply *spectral decomposition* [13] to the amplification matrix $\mathbf{A}$, as follows:

$$\mathbf{A} = \boldsymbol{\Phi}\,\boldsymbol{\lambda}\,\boldsymbol{\Phi}^{-1} \tag{5}$$

In this equation $\boldsymbol{\lambda}$ is the *spectral matrix* of $\mathbf{A}$, containing eigenvalues $\lambda_1$, $\lambda_2$, and $\lambda_3$ in diagonal positions; and $\boldsymbol{\Phi}$ is the $3 \times 3$ *modal matrix* of $\mathbf{A}$, with eigenvectors $\boldsymbol{\Phi}_1$, $\boldsymbol{\Phi}_2$, and $\boldsymbol{\Phi}_3$ listed columnwise. If we start at time $t_0 = 0$ and take $n_j$ time steps using Eq. (4), we have

$$\mathbf{U}_{n_j} = \mathbf{A}^{n_j}\mathbf{U}_0 \tag{6}$$

where the vector $\mathbf{U}_0$ contains initial conditions, and vector $\mathbf{U}_{n_j}$ gives the response values at time $t_{n_j}$. Raising the decomposed form of matrix $\mathbf{A}$ in Eq. (5) to the power $n_j$ yields

$$\mathbf{A}^{n_j} = \boldsymbol{\Phi}\,\boldsymbol{\lambda}^{n_j}\boldsymbol{\Phi}^{-1} \tag{7}$$

Now let us define the *spectral radius* of matrix $\mathbf{A}$ as

$$(r)_\mathbf{A} = \max |\lambda_i| \qquad (i = 1, 2, 3) \tag{8}$$

Then Eq. (7) shows that we must have

$$(r)_\mathbf{A} \leq 1 \tag{9}$$

in order to keep the numerical solution from growing without bound. This condition is known as the *stability criterion* for a given method. By applying this criterion to the constant acceleration method (see Sec. 5.5), we find the critical time step for this conditionally stable approach to be

$$(\Delta t)_{cr} = \frac{T_n}{\pi} = 0.318 T_n \tag{10}$$

This value is the same as that given by Eq. (5.2-12) for the central-difference predictor. For the linear-acceleration method, the critical time step is

$$(\Delta t)_{cr} = \sqrt{3}\,\frac{T_n}{\pi} = 0.551 T_n \tag{11}$$

On the other hand, the spectral radius for the average-acceleration method is always unity. Therefore, it has no critical time step and is said to be unconditionally stable.

The matter of accuracy of a numerical integration procedure is closely related to that of stability. Figure 5.6 shows the undamped response of a SDOF system to an initial displacement $u_0$. The curve labeled 1 is the exact result, and those labeled 2, 3, and 4 represent various approximations. Curve 2 demonstrates an amplitude increase (AI) that implies an unstable algorithm. Curve 3 shows no amplitude change, and curve 4 depicts an amplitude decrease (AD). Because of the stability criterion in Eq. (9), only curves of types 3 and 4 are admissible approximations. Curve 3 may be considered to be the result for the average-acceleration method, which has a spectral radius of unity. Other admissible algorithms are represented by curve 4, which implies a spectral radius less

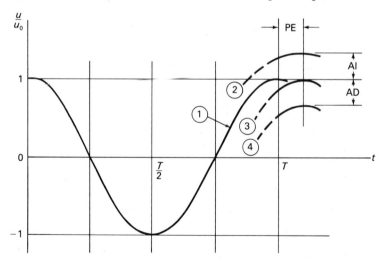

**Figure 5.6** Stability and accuracy of numerical integration methods.

than unity. Thus, one important type of error to be considered is *amplitude suppression*, as exhibited by curve 4.

All of the approximate responses in Fig. 5.6 also show *period elongation* (PE), which is a second type of error introduced by any numerical algorithm. Both the amplitude suppression and the period elongation may be made negligible by using sufficiently small time steps. Newmark [8] recommended a time step of duration equal to $\frac{1}{5}$ or $\frac{1}{6}$ of $T_n$, which is the smallest period of a MDOF structure. However, a more commonly used time step is $\Delta t = T_n/10$.

At first glance, the average-acceleration method appears to be the best choice among the implicit approaches, because it has no amplitude suppression and the least period elongation [12]. However, a small amount of amplitude suppression is desirable to reduce or eliminate unwanted responses of higher modes in an undamped MDOF structure. But if the spectral radius of an amplification matrix is too small, the response of the structure will be unduly suppressed, as in the Houbolt method [14]. Although the optimized Wilson-$\theta$ formulation produces a reasonable level of amplitude suppression, it also exhibits an undesirable tendency to overshoot the true response in the first few time steps [15]. Thus, we conclude that the best choice of algorithm is the Newmark-$\beta$ method with the Hilber-$\alpha$ modification, as described in Sec. 5.5. Probably, the optimum selection of parameters for this approach is to let $\alpha = -0.1$, $\beta = 0.3025$, and $\gamma = 0.6$.

## 5.7 PROGRAM DYNA FOR DYNAMIC RESPONSE

Now we shall describe a general-purpose program named DYNA that calculates dynamic responses of linearly elastic structures. This program includes vibrational analysis from Chapter 3, normal-mode responses from Chapter 4, and direct numerical integration from the present chapter. By virtue of a branch in the main program, responses may be obtained using either the normal-mode method or direct numerical integration. For the latter approach, we select the Newmark-$\beta$ method with the Hilber-$\alpha$ variant discussed in Secs. 5.5 and 5.6.

Flowchart 5.1 shows the main program for DYNA, which calls the five subprograms indicated in double boxes. As in Flowchart 4.1, Subprogram VIB calls the seven other subprograms given in Flowchart 3.1, including normalization of the eigenvectors with respect to the mass matrix. Then the program reads the number of loading systems NLS and checks it against zero to determine whether to process a loading system or another structure. Next, the loading number LN is initialized to zero, and then it is increased by one.

The second subprogram to be called is DYLO, which is almost the same as in Flowchart 4.1. However, the first dynamic parameter now becomes the indicator ISOLVE for type of solution. The two choices of solution are the normal-mode method (ISOLVE = 0) and direct numerical integration

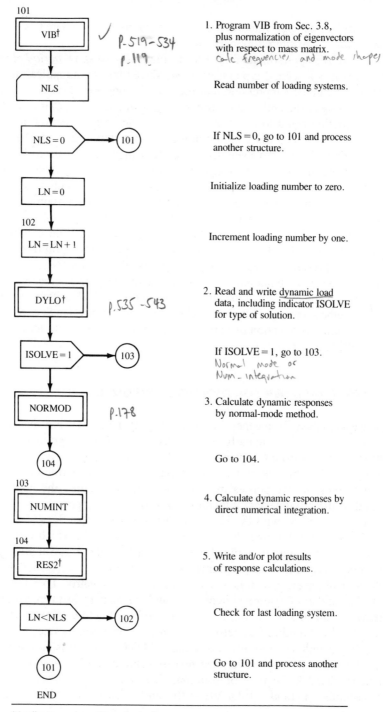

**Flowchart 5.1 Main program for DYNA***

*Applies to any type of linearly elastic structure.
†Subprograms that differ for every type of structure.

compares ISOLVE with unity to
to be used for dynamic response
NORMOD for the normal-mode
TIHIST, and TRABAC, which are
ernatively, if ISOLVE = 1, Sub-
nses by direct numerical integra-
e integration parameters ALPHA,
-α formulas.
necessary to generate the damping
ns. Toward this end, we substitute
from Subprogram VIB into Eq.
ysical coordinates.
in program is RES2, which writes
calculations as before. Testing LN
nines whether to return for another

loading system or another structure.

Program DYNA may be specialized to become DYNACB for continuous beams, <u>DYNAPT for plane trusses</u>, and so on. As before, the main program for each specialization has subprograms that are different for each type of application, as indicated by the second footnote below Flowchart 5.1. Note that Subprograms NORMOD and NUMINT are the same for all types of structures. As for Program NOMO, notation for Program DYNA is included in Part 5 of the list of notation near the back of the book. Details of the logic in Subprogram NUMINT appear in the flowchart for Program DYNAPT, given in Appendix C.

Thus, in several stages we have devised a program that will handle not only vibrational analysis but also two types of dynamic response calculations. Variants of the program apply to all linearly elastic framed structures and discretized continua discussed in this book.

**Example 5.7**

To show how Program DYNAPT is used, we shall repeat Example 4.15 (the three-member plane truss), using the same number and size of time steps. In this application we employ direct numerical integration instead of the normal-mode method by setting the parameter ISOLVE equal to unity. Also, the values of the integration parameters read by Subprogram NUMINT are taken to be ALPHA = $-0.1$, BETA = 0.3025, and GAMMA = 0.6.

Let us examine again the responses of the structure to the dynamic influences given in Example 4.15. Table 5.7 lists part of the output from Program DYNAPT for these four analyses. Figures 5.7 and 5.8 (on pages 233 and 234) also show computer plots of the results obtained. The responses in Fig. 5.7 due to initial conditions follow the same general trends as those in Fig. 4.11, but they differ because the time step is too large. On the other hand, the smoother responses in Fig. 5.8(b) caused by the applied force in Fig. 5.8(a) are practically indistinguishable from those in Fig. 4.12(b). This good correlation is due to the fact that the time step is short enough to model the slowly varying responses accurately.

**TABLE 5.7** Computer Output for Example 5.7

```
PROGRAM DYNAPT

*** EXAMPLE 5.7: THREE-MEMBER PLANE TRUSS ***


STRUCTURAL PARAMETERS
    NN   NE  NRN           E          RHO
     3    3    2   3.0000E+04   7.3500E-07

NODAL COORDINATES
  NODE          X                Y
     1      0.000            0.000
     2    150.000          200.000
     3    150.000            0.000

ELEMENT INFORMATION
  ELEM.   J    K        AX           EL         CX         CY
     1    1    2    10.0000     250.0000     0.6000     0.8000
     2    1    3     6.0000     150.0000     1.0000     0.0000
     3    2    3     8.0000     200.0000     0.0000    -1.0000

NODAL RESTRAINTS
  NODE   NR1  NR2
     1    0    1
     3    1    1

NUMBER OF DEGREES OF FREEDOM:  NDF =    3
NUMBER OF NODAL RESTRAINTS:    NNR =    3

STIFFNESS MATRIX DECOMPOSED

MODE     1
ANGULAR FREQUENCY   4.1995E+02
  NODE         DJ1             DJ2
     1   2.3137E-01      0.0000E+00
     2   1.0000E+00     -2.4722E-01
     3   0.0000E+00      0.0000E+00

MODE     2
ANGULAR FREQUENCY   1.1677E+03
  NODE         DJ1             DJ2
     1   8.6725E-01      0.0000E+00
     2  -1.7149E-01      1.0000E+00
     3   0.0000E+00      0.0000E+00

MODE     3
ANGULAR FREQUENCY   1.8618E+03
  NODE         DJ1             DJ2
     1   1.0000E+00      0.0000E+00
     2  -6.0504E-01     -6.1068E-01
     3   0.0000E+00      0.0000E+00

*** LOADING NUMBER   1 OF   4 ***

DYNAMIC PARAMETERS
 ISOLVE  NTS         DT        DAMPR
     1    20   1.0000E-03   0.0000E+00
```

**TABLE 5.7** (Continued)

```
INITIAL CONDITIONS
  NNID NNIV
    2    0

INITIAL DISPLACEMENTS
  NODE         D01            D02
    1    1.0000E-01     0.0000E+00
    2    1.0000E-01     1.0000E-01

APPLIED ACTIONS
  NLN  NEL
    0    0

GROUND ACCELERATIONS
  IGA
    0

DIRECT NUMERICAL INTEGRATION
ALPHA = -0.1000   BETA =  0.3025   GAMMA =  0.6000

OUTPUT SELECTION
  IWR  IPL  NNO  NEO
    1    1    2    1

NODES:       1    2

ELEMENTS:    1

DISPLACEMENT TIME HISTORY FOR NODE    1
 STEP       TIME          DJ1            DJ2
    0   0.0000E+00    1.0000E-01     0.0000E+00
    1   1.0000E-03    6.9151E-02     0.0000E+00
    2   2.0000E-03   -9.7574E-03     0.0000E+00
    3   3.0000E-03   -8.0921E-02     0.0000E+00
    4   4.0000E-03   -6.9989E-02     0.0000E+00
    5   5.0000E-03    2.1718E-02     0.0000E+00
    6   6.0000E-03    8.8292E-02     0.0000E+00
    7   7.0000E-03    3.9844E-02     0.0000E+00
    8   8.0000E-03   -7.5170E-02     0.0000E+00
    9   9.0000E-03   -1.2500E-01     0.0000E+00
   10   1.0000E-02   -5.7660E-02     0.0000E+00
   11   1.1000E-02    4.7566E-02     0.0000E+00
   12   1.2000E-02    8.9567E-02     0.0000E+00
   13   1.3000E-02    5.1576E-02     0.0000E+00
   14   1.4000E-02   -1.2272E-02     0.0000E+00
   15   1.5000E-02   -4.6102E-02     0.0000E+00
   16   1.6000E-02   -2.5622E-02     0.0000E+00
   17   1.7000E-02    3.6195E-02     0.0000E+00
   18   1.8000E-02    8.5913E-02     0.0000E+00
   19   1.9000E-02    6.2893E-02     0.0000E+00
   20   2.0000E-02   -2.9158E-02     0.0000E+00

MAXIMUM                 1.0000E-01     0.0000E+00
TIME OF MAXIMUM         0.0000E+00     2.0000E-02
MINIMUM                -1.2500E-01     0.0000E+00
TIME OF MINIMUM         9.0000E-03     2.0000E-02
```

**TABLE 5.7** (Continued)

```
DISPLACEMENT TIME HISTORY FOR NODE     2
STEP       TIME          DJ1          DJ2
  0    0.0000E+00    1.0000E-01    1.0000E-01
  1    1.0000E-03    8.8727E-02    3.4123E-02
  2    2.0000E-03    6.8467E-02   -8.5095E-02
  3    3.0000E-03    4.9368E-02   -1.2309E-01
  4    4.0000E-03    1.2394E-02   -4.4408E-02
  5    5.0000E-03   -5.2366E-02    6.8993E-02
  6    6.0000E-03   -1.1040E-01    1.1863E-01
  7    7.0000E-03   -1.2015E-01    7.5778E-02
  8    8.0000E-03   -8.9808E-02   -1.3763E-02
  9    9.0000E-03   -6.2312E-02   -7.2918E-02
 10    1.0000E-02   -5.2629E-02   -4.8878E-02
 11    1.1000E-02   -3.4703E-02    3.7619E-02
 12    1.2000E-02    9.6841E-03    9.6294E-02
 13    1.3000E-02    6.2708E-02    5.2826E-02
 14    1.4000E-02    9.8915E-02   -6.1248E-02
 15    1.5000E-02    1.1326E-01   -1.3161E-01
 16    1.6000E-02    1.0852E-01   -8.6674E-02
 17    1.7000E-02    8.0000E-02    2.3038E-02
 18    1.8000E-02    3.1105E-02    8.6592E-02
 19    1.9000E-02   -1.4868E-02    5.3559E-02
 20    2.0000E-02   -3.9867E-02   -2.6064E-02

MAXIMUM                1.1326E-01    1.1863E-01
TIME OF MAXIMUM        1.5000E-02    6.0000E-03
MINIMUM               -1.2015E-01   -1.3161E-01
TIME OF MINIMUM        7.0000E-03    1.5000E-02

MEMBER FORCE TIME HISTORY FOR ELEMENT     1
STEP       TIME          AM1
  0    0.0000E+00    9.6000E+01
  1    1.0000E-03    4.6852E+01
  2    2.0000E-03   -2.5370E+01
  3    3.0000E-03   -2.4355E+01
  4    4.0000E-03    1.6683E+01
  5    5.0000E-03    1.2893E+01
  6    6.0000E-03   -2.9174E+01
  7    7.0000E-03   -4.2452E+01
  8    8.0000E-03   -2.3751E+01
  9    9.0000E-03   -2.4867E+01
 10    1.0000E-02   -4.3301E+01
 11    1.1000E-02   -2.3119E+01
 12    1.2000E-02    3.4926E+01
 13    1.3000E-02    5.8727E+01
 14    1.4000E-02    2.1256E+01
 15    1.5000E-02   -1.1598E+01
 16    1.6000E-02    1.3378E+01
 17    1.7000E-02    5.3656E+01
 18    1.8000E-02    4.3667E+01
 19    1.9000E-02   -4.5712E+00
 20    2.0000E-02   -3.2732E+01

MAXIMUM                9.6000E+01
TIME OF MAXIMUM        0.0000E+00
MINIMUM               -4.3301E+01
TIME OF MINIMUM        1.0000E-02
```

TABLE 5.7 (Continued)

```
*** LOADING NUMBER    2 OF   4 ***

DYNAMIC PARAMETERS
ISOLVE NTS        DT       DAMPR
   1    20  1.0000E-03  0.0000E+00

INITIAL CONDITIONS
 NNID NNIV
   0    0

APPLIED ACTIONS
  NLN  NEL
   1    0

NODAL LOADS
 NODE        AJ1          AJ2
   2    2.0000E+01   0.0000E+00

GROUND ACCELERATIONS
  IGA
   0

FORCING FUNCTION
  NFO
   7

FUNCTION ORDINATES
    K       TIME        FACTOR
    1   0.0000E+00   0.0000E+00
    2   1.0000E-03   1.5000E-01
    3   3.0000E-03   8.5000E-01
    4   4.0000E-03   1.0000E+00
    5   5.0000E-03   8.5000E-01
    6   7.0000E-03   1.5000E-01
    7   8.0000E-03   0.0000E+00

DIRECT NUMERICAL INTEGRATION
ALPHA = -0.1000   BETA =  0.3025   GAMMA =  0.6000

OUTPUT SELECTION
  IWR  IPL  NNO  NEO
   1    1    1    1

NODES:       2

ELEMENTS:    1

DISPLACEMENT TIME HISTORY FOR NODE    2
  STEP      TIME          DJ1          DJ2
    0   0.0000E+00   0.0000E+00   0.0000E+00
    1   1.0000E-03   8.4579E-04  -1.0008E-04
    2   2.0000E-03   5.5055E-03  -8.5533E-04
    3   3.0000E-03   1.7942E-02  -3.4897E-03
    4   4.0000E-03   3.8969E-02  -8.9831E-03
    5   5.0000E-03   6.3694E-02  -1.6261E-02
    6   6.0000E-03   8.4309E-02  -2.1959E-02
    7   7.0000E-03   9.4437E-02  -2.3226E-02
    8   8.0000E-03   9.0374E-02  -2.0476E-02
    9   9.0000E-03   7.0970E-02  -1.6052E-02
   10   1.0000E-02   3.8775E-02  -1.0521E-02
```

**TABLE 5.7** (Continued)

```
11  1.1000E-02   1.3673E-04  -2.3257E-03
12  1.2000E-02  -3.7755E-02   8.5393E-03
13  1.3000E-02  -6.9338E-02   1.8614E-02
14  1.4000E-02  -9.0054E-02   2.3972E-02
15  1.5000E-02  -9.5688E-02   2.3788E-02
16  1.6000E-02  -8.4382E-02   1.9802E-02
17  1.7000E-02  -5.8635E-02   1.3542E-02
18  1.8000E-02  -2.3925E-02   5.5924E-03
19  1.9000E-02   1.4150E-02  -3.2004E-03
20  2.0000E-02   5.0421E-02  -1.1543E-02

MAXIMUM                 9.4437E-02   2.3972E-02
TIME OF MAXIMUM         7.0000E-03   1.4000E-02
MINIMUM                -9.5688E-02  -2.3226E-02
TIME OF MINIMUM         1.5000E-02   7.0000E-03

MEMBER FORCE TIME HISTORY FOR ELEMENT    1
STEP       TIME           AM1
   0   0.0000E+00    0.0000E+00
   1   1.0000E-03    6.1168E-01
   2   2.0000E-03    3.4981E+00
   3   3.0000E-03    9.6372E+00
   4   4.0000E-03    1.7047E+01
   5   5.0000E-03    2.2466E+01
   6   6.0000E-03    2.5649E+01
   7   7.0000E-03    2.8511E+01
   8   8.0000E-03    2.9596E+01
   9   9.0000E-03    2.4238E+01
  10   1.0000E-02    1.1821E+01
  11   1.1000E-02   -1.9340E+00
  12   1.2000E-02   -1.2285E+01
  13   1.3000E-02   -2.0281E+01
  14   1.4000E-02   -2.7599E+01
  15   1.5000E-02   -3.1270E+01
  16   1.6000E-02   -2.7564E+01
  17   1.7000E-02   -1.7935E+01
  18   1.8000E-02   -6.9027E+00
  19   1.9000E-02    3.9111E+00
  20   2.0000E-02    1.5351E+01

MAXIMUM                 2.9596E+01
TIME OF MAXIMUM         8.0000E-03
MINIMUM                -3.1270E+01
TIME OF MINIMUM         1.5000E-02
```

**Example 5.8**

Next, we repeat Example 4.16 (the symmetric plane truss), with the same number and size of time steps. However, the first load case in that example will be treated differently. Instead of using modal truncation, we shall compare results from the Newmark-$\beta$ method against those obtained by the Hilber-$\alpha$ technique. In the first instance, we have the integration parameters ALPHA = 0, BETA = 0.25, and GAMMA = 0.5; and in the second we take ALPHA = $-0.1$, BETA = 0.3025, and GAMMA = 0.6.

## Sec. 5.7  Program DYNA for Dynamic Response

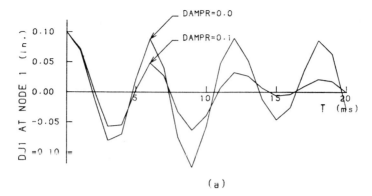

(a)

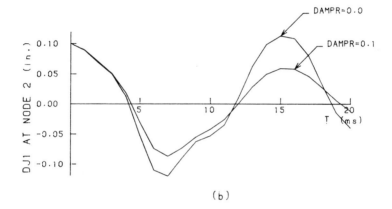

(b)

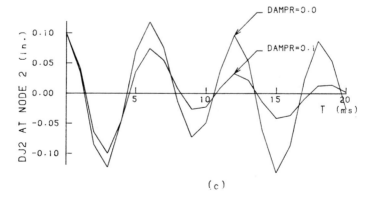

(c)

**Figure 5.7**  Three-member plane truss: responses to initial displacements.

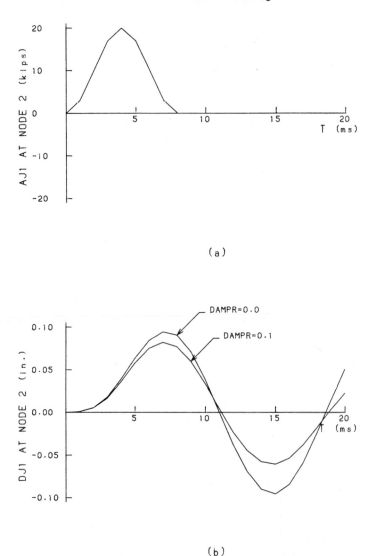

**Figure 5.8**  Three-member plane truss: (a) applied force; (b) responses.

Figures 5.9(b) and (c) show computer plots of the symmetric and anti-symmetric contributions to the *y* translation at node 6 for the Newmark and Hilber methods. These responses should be the same as those for all modes given previously in Figs. 4.13(b) and (c). They all match fairly well, even though the time step is rather long. Note that the approximate responses in Fig. 5.9(c) have a significant *period elongation* relative to that in Fig. 4.13(c). Also, the Hilber plot in Fig. 5.9(c) has a noticeable *amplitude suppression* relative to the Newmark plot, as it should.

### Sec. 5.7 Program DYNA for Dynamic Response

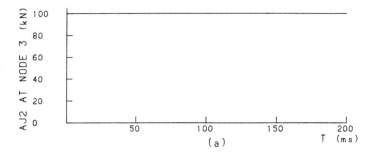

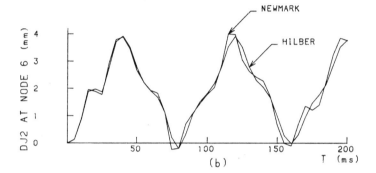

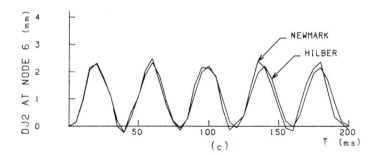

**Figure 5.9** Plane truss: (a) applied force; (b) symmetric responses; (c) antisymmetric responses.

Responses to the piecewise-linear ground acceleration in Fig. 5.10(a) are plotted in Fig. 5.10(b) for the two cases of DAMPR = 0.0 and 0.1. They were obtained by the Hilber-$\alpha$ approach, using the same integration parameters as before. These plots are indistinguishable from those in Fig. 4.14(b), which were found by the normal-mode method.

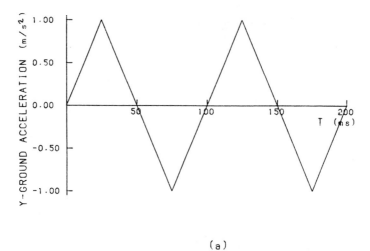

(a)

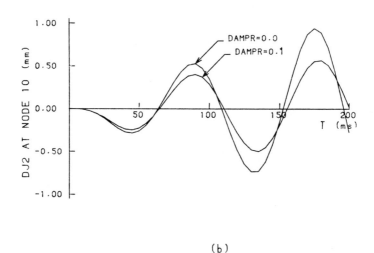

(b)

**Figure 5.10**  Plane truss: (a) ground acceleration; (b) responses.

# REFERENCES

1. Dahlquist, G. G., "A Special Stability Problem for Linear Multistep Methods," *Nord. Tidskr. Inf. Behandling*, Vol. 3, 1963, pp. 27–43.
2. Krieg, R. D., "Unconditional Stability in Numerical Time Integration Methods," *J. Appl. Mech.*, Vol. 40, No. 2, 1973, pp. 417–421.

3. Key, S. W., and Beisinger, Z. E., "Transient Dynamic Analysis of Thin Shells by the Finite Element Method," *Proc. 3rd Conf. Mat. Methods Struct. Mech.*, AFIT, Wright-Patterson AFB, Ohio, 1971, pp. 479–518.
4. Krieg, R. D., and Key, S. W., "Transient Shell Response by Numerical Time Integration," *Int. J. Numer. Methods Eng.*, Vol. 7, No. 3, 1973, pp. 273–286.
5. Key, S. W., "A Finite Element Procedure for the Large Deformation Dynamic Response of Axisymmetric Solids," *Comp. Methods Appl. Mech. Eng.*, Vol. 4, No. 2, 1974, pp. 195–218.
6. Morino, L., Leech, J. W., and Witmer, E. A., "Optimal Predictor-Corrector Method for Systems of Second-Order Differential Equations," *AIAA J.*, Vol. 12, No. 10, 1974, pp. 1343–1347.
7. Timoshenko, S. P., Young, D. H., and Weaver, W., Jr., *Vibration Problems in Engineering*, 4th ed., Wiley, New York, 1974.
8. Newmark, N. M., "A Method of Computation for Structural Dynamics," *ASCE J. Eng. Mech. Div.*, Vol. 85, No. EM3, 1959, pp. 67–94.
9. Wilson, E. L., Farhoomand, I., and Bathe, K. J., "Nonlinear Dynamic Analysis of Complex Structures," *Earthquake Eng. Struct. Dyn.*, Vol. 1, No. 3, 1973, pp. 241–252.
10. Bathe, K. J., and Wilson, E. L., "Stability and Accuracy Analysis of Direct Integration Methods," *Earthquake Eng. Struct. Dyn.*, Vol. 1, No. 3, 1973, pp. 283–291.
11. Hilber, H. M., Hughes, T. J. R., and Taylor, R. L., "Improved Numerical Dissipation for Time Integration Algorithms in Structural Mechanics," *Earthquake Eng. Struct. Dyn.*, Vol. 5, No. 3, 1977, pp. 283–292.
12. Bathe, K. J., *Finite Element Procedures in Engineering Analysis*, Prentice-Hall, Englewood Cliffs, N. J., 1982.
13. Gere, J. M., and Weaver, W., Jr., *Matrix Algebra for Engineers*, 2nd ed., Brooks/Cole, Monterey, Calif., 1983.
14. Houbolt, J. C., "A Recurrence Matrix Solution for the Dynamic Response of Elastic Aircraft," *J. Aero. Sci.*, Vol. 17, 1950, pp. 540–550.
15. Hilber, H. H., and Hughes, T. J. R., "Collocation, Dissipation, and 'Overshoot' for Time Integration Schemes in Structural Dynamics," *Earthquake Eng. Struct. Dyn.*, Vol. 6, No. 1, 1978, pp. 99–117.

## PROBLEMS*

**5.2-1.** Confirm the approximate results of Example 5.1 in Table 5.1.

**5.2-2.** Repeat Prob. 2.7-2 using the central-difference predictor.

**5.2-3.** Repeat Prob. 2.7-3 using the central-difference predictor.

**5.2-4.** Repeat Prob. 2.7-4 using the central-difference predictor.

**5.2-5.** Repeat Prob. 2.7-5 using the central-difference predictor.

**5.2-6.** Repeat Prob. 2.7-6 using the central-difference predictor.

*Solutions for problems in this chapter are rather tedious and should be handled using a personal computer.

**5.2-7.** Repeat Prob. 2.7-7 using the central-difference predictor.

**5.2-8.** Repeat Prob. 2.7-8 using the central-difference predictor.

**5.2-9.** Repeat Prob. 2.7-9 using the central-difference predictor.

**5.2-10.** Repeat Prob. 2.7-10 using the central-difference predictor.

**5.2-11.** Confirm the approximate results for the plane truss of Example 5.2 in Table 5.2.

**5.2-12.** Repeat Prob. 4.4-3 (plane truss) for 20 time steps, using the central-difference predictor with $\Delta t = T_1/20 = t_1$.

**5.2-13.** Repeat Prob. 4.4-4 (plane truss) for 20 time steps, using the central-difference predictor with $\Delta t = T_1/20$.

**5.3-1.** Confirm the iterative results of Example 5.3 in Table 5.3 for both the average- and linear-acceleration methods.

**5.3-2.** Repeat Prob. 2.7-2 using iteration by the average- and linear-acceleration methods.

**5.3-3.** Repeat Prob. 2.7-3 using iteration by the average- and linear-acceleration methods.

**5.3-4.** Repeat Prob. 2.7-4 using iteration by the average- and linear-acceleration methods.

**5.3-5.** Repeat Prob. 2.7-5 using iteration by the average- and linear-acceleration methods.

**5.3-6.** Repeat Prob. 2.7-6 using iteration by the average- and linear-acceleration methods.

**5.3-7.** Repeat Prob. 2.7-7 using iteration by the average- and linear-acceleration methods.

**5.3-8.** Repeat Prob. 2.7-8 using iteration by the average- and linear-acceleration methods.

**5.3-9.** Repeat Prob. 2.7-9 using iteration by the average- and linear-acceleration methods.

**5.3-10.** Repeat Prob. 2.7-10 using iteration by the average- and linear-acceleration methods.

**5.3-11.** Confirm the iterative results for the plane truss of Example 5.4 in Table 5.4, using both the average- and linear-acceleration methods.

**5.3-12.** Repeat Prob. 4.4-3 (plane truss) for 20 time steps, using iteration by the average- and linear-acceleration methods with $\Delta t = T_1/20 = t_1$.

**5.3-13.** Repeat Prob. 4.4-4 (plane truss) for 20 time steps, using iteration by the average- and linear-acceleration methods with $\Delta t = T_1/20$.

**5.4-1.** Confirm the approximate results of Example 5.5 in Table 5.5 for both the average- and linear-acceleration methods.

**5.4-2.** Repeat Prob. 2.7-2 using direct linear extrapolation by the average- and linear-acceleration methods.

**5.4-3.** Repeat Prob. 2.7-3 using direct linear extrapolation by the average- and linear-acceleration methods.

**5.4-4.** Repeat Prob. 2.7-4 using direct linear extrapolation by the average- and linear-acceleration methods.

**5.4-5.** Repeat Prob. 2.7-5 using direct linear extrapolation by the average- and linear-acceleration methods.

**5.4-6.** Repeat Prob. 2.7-6 using direct linear extrapolation by the average- and linear-acceleration methods.

**5.4-7.** Repeat Prob. 2.7-7 using direct linear extrapolation by the average- and linear-acceleration methods.

**5.4-8.** Repeat Prob. 2.7-8 using direct linear extrapolation by the average- and linear-acceleration methods.

**5.4-9.** Repeat Prob. 2.7-9 using direct linear extrapolation by the average- and linear-acceleration methods.

**5.4-10.** Repeat Prob. 2.7-10 using direct linear extrapolation by the average- and linear-acceleration methods.

**5.4-11.** Confirm the approximate results for the plane truss of Example 5.6 in Table 5.6, using both the average- and linear-acceleration methods.

**5.4-12.** Repeat Prob. 4.4-3 (plane truss) for 20 time steps, using direct linear extrapolation by the average- and linear-acceleration methods with $\Delta t = T_1/20 = t_1$.

**5.4-13.** Repeat Prob. 4.4-4 (plane truss) for 20 time steps, using direct linear extrapolation by the average- and linear-acceleration methods with $\Delta t = T_1/20$.

**5.4-14.** The equations of motion for the average-acceleration method may be converted to the form: $\overline{\mathbf{S}}\,\mathbf{D}_{j+1} = \overline{\mathbf{A}}_{j+1}$. Derive expressions for the matrices $\overline{\mathbf{S}}$ and $\overline{\mathbf{A}}_{j+1}$ to be used in this approach.

**5.4-15.** The equations of motion for the average-acceleration method may be converted to the form: $\overline{\mathbf{C}}\,\dot{\mathbf{D}}_{j+1} = \overline{\mathbf{A}}_{j+1}$. Derive expressions for the matrices $\overline{\mathbf{C}}$ and $\overline{\mathbf{A}}_{j+1}$ to be used in this approach.

**5.4-16.** The equations of motion for the average-acceleration method may be converted to the form: $\overline{\mathbf{M}}\,\ddot{\mathbf{D}}_{j+1} = \overline{\mathbf{A}}_{j+1}$. Derive expressions for the matrices $\overline{\mathbf{M}}$ and $\overline{\mathbf{A}}_{j+1}$ to be used in this approach.

**5.4-17.** The incremental equations of motion for the average-acceleration method may be converted to the form: $\overline{\mathbf{C}}\,\Delta\dot{\mathbf{D}}_j = \Delta\overline{\mathbf{A}}_j$. Derive expressions for the matrices $\overline{\mathbf{C}}$ and $\Delta\overline{\mathbf{A}}_j$ to be used in this approach.

**5.4-18.** The incremental equations of motion for the average-acceleration method may be converted to the form: $\overline{\mathbf{M}}\,\Delta\ddot{\mathbf{D}}_j = \Delta\overline{\mathbf{A}}_j$. Derive expressions for the matrices $\overline{\mathbf{M}}$ and $\Delta\overline{\mathbf{A}}_j$ to be used in this approach.

**5.4-19.** The equations of motion for the linear-acceleration method may be converted to the form: $\mathbf{S}^*\mathbf{D}_{j+1} = \mathbf{A}^*_{j+1}$. Derive expressions for the matrices $\mathbf{S}^*$ and $\mathbf{A}^*_{j+1}$ to be used in this approach.

**5.4-20.** The equations of motion for the linear-acceleration method may be converted to the form: $\mathbf{C}^*\dot{\mathbf{D}}_{j+1} = \mathbf{A}^*_{j+1}$. Derive expressions for the matrices $\mathbf{C}^*$ and $\mathbf{A}^*_{j+1}$ to be used in this approach.

**5.4-21.** The equations of motion for the linear-acceleration method may be converted to the form: $\mathbf{M}^*\ddot{\mathbf{D}}_j = \mathbf{A}^*_{j+1}$. Derive expressions for the matrices $\mathbf{M}^*$ and $\mathbf{A}^*_{j+1}$ to be used in this approach.

**5.4-22.** The incremental equations of motion for the linear-acceleration method may be converted to the form: $\mathbf{C}^* \Delta \dot{\mathbf{D}}_j = \Delta \mathbf{A}_j^*$. Derive expressions for the matrices $\mathbf{C}^*$ and $\Delta \mathbf{A}_j^*$ to be used in this approach.

**5.4-23.** The incremental equations of motion for the linear-acceleration method may be converted to the form: $\mathbf{M}^* \Delta \ddot{\mathbf{D}}_j = \Delta \mathbf{A}_j^*$. Derive expressions for the matrices $\mathbf{M}^*$ and $\Delta \mathbf{A}_j^*$ to be used in this approach.

**5.5-1.** The equations of motion for the Newmark-$\beta$ method may be converted to the form: $\hat{\mathbf{S}} \mathbf{D}_{j+1} = \hat{\mathbf{A}}_{j+1}$. Derive expressions for the matrices $\hat{\mathbf{S}}$ and $\hat{\mathbf{A}}_{j+1}$ to be used in this approach.

**5.5-2.** The equations of motion for the Newmark-$\beta$ method may be converted to the form: $\hat{\mathbf{C}} \dot{\mathbf{D}}_{j+1} = \hat{\mathbf{A}}_{j+1}$. Derive expressions for the matrices $\hat{\mathbf{C}}$ and $\hat{\mathbf{A}}_{j+1}$ to be used in this approach.

**5.5-3.** The equations of motion for the Newmark-$\beta$ method may be converted to the form: $\hat{\mathbf{M}} \ddot{\mathbf{D}}_{j+1} = \hat{\mathbf{A}}_{j+1}$. Derive expressions for the matrices $\hat{\mathbf{M}}$ and $\hat{\mathbf{A}}_{j+1}$ to be used in this approach.

**5.5-4.** The incremental equations of motion for the Newmark-$\beta$ method may be converted to the form: $\hat{\mathbf{C}} \Delta \dot{\mathbf{D}}_j = \Delta \hat{\mathbf{A}}_j$. Derive expressions for the matrices $\hat{\mathbf{C}}$ and $\Delta \hat{\mathbf{A}}_j$ to be used in this approach.

**5.5-5.** The incremental equations of motion for the Newmark-$\beta$ method may be converted to the form: $\hat{\mathbf{M}} \Delta \ddot{\mathbf{D}}_j = \Delta \hat{\mathbf{A}}_j$. Derive expressions for the matrices $\hat{\mathbf{M}}$ and $\Delta \hat{\mathbf{A}}_j$ to be used in this approach.

# 6

# Framed Structures

## 6.1 INTRODUCTION

A *framed structure* consists of *members* that are relatively long (or slender) compared to their cross-sectional dimensions. Points where members intersect, free ends of members, and points of support are called *joints* of the structure. In finite-element terminology, we refer to the members as linear (or perhaps curvilinear) elements, and we call the joints nodes. Thus, the matrix analysis of framed structures [1] becomes a subset of the more general theory of finite elements [2] for discretizing and analyzing continua. However, a framed structure inherently is divided into elements, unless it becomes necessary to subdivide members into smaller elements.

There are six distinct types of framed structures that designers use to resist commensurate sets of loading systems. Six commonly occurring *types of force systems* are: (a) parallel-coplanar, (b) concurrent-coplanar, (c) general-coplanar, (d) parallel in space, (e) concurrent in space, and (f) general in space. Each of the framed structures illustrated in Fig. 6.1 is specifically designed to resist one of these load sets at its joints. When the structures are loaded, their joints undergo translational displacements (corresponding to forces) and rotational displacements (corresponding to moments). In addition, certain types of internal actions (or stress resultants) arise in the members, depending on the type of structure. We will briefly describe the characteristics of each type of framed structure appearing in Fig. 6.1, even though beams and plane trusses already have been used for examples and problems in previous chapters.

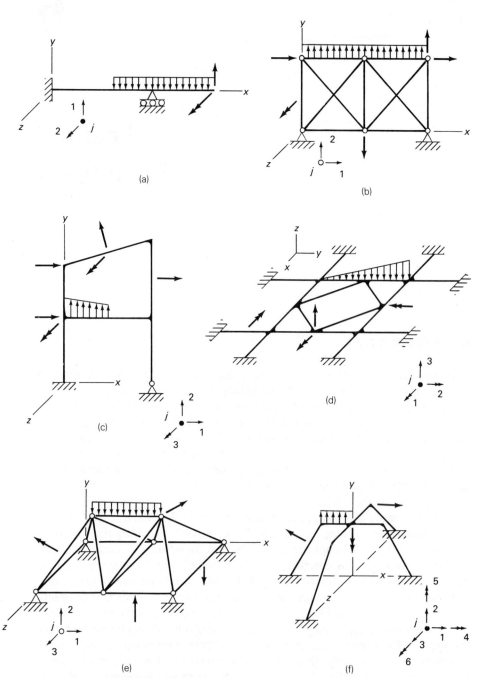

**Figure 6.1** Types of framed structures: (a) continuous beam; (b) plane truss; (c) plane frame; (d) grid; (e) space truss; (f) space frame.

## Sec. 6.1  Introduction

Figure 6.1(a) shows a straight *continuous beam*, subjected to a parallel-coplanar set of forces in the $x$-$y$ plane, which is a principal plane of bending. Any applied moments must act in the $z$ sense, as implied by the double-headed arrow at the right end of the beam. This type of loading system produces at a free (unrestrained) joint a translation in the $y$ direction and a rotation in the $z$ sense. These displacements are indicated in the figure by the arrows numbered 1 and 2 at a typical joint $j$. From flexural theory, internal member actions (generalized stresses) within the beam are a shearing force in the $y$ direction and a bending moment in the $z$ sense. If a force applied to a beam has a component in the $x$ direction, the resulting internal axial stress must be combined with the flexural stresses due to bending, even though the analyses are uncoupled.

Similarly, Fig. 6.1(b) depicts a *plane truss* in the $x$-$y$ plane with forces applied to its members as well as its hinged joints. As with a beam, if a moment is applied to a member, its direction must be parallel to the $z$ axis. Reactions at the pinned ends of members caused by loads become forces at the joints when their signs are reversed. Thus, the joints resist systems of concurrent-coplanar forces, which produce axial forces in the members. Due to the accompanying axial strains, joint displacements become translations in the $x$ and $y$ directions, as indicated by the numbered arrows in the figure. For a member with loads applied directly to it, the axial stress from the truss analysis must be combined with the flexural stresses arising from local bending, although the analyses are uncoupled.

The rigidly connected *plane frame* in Fig. 6.1(c) carries a general-coplanar set of forces in the $x$-$y$ plane, while applied moments act in the $z$ sense. Resulting displacements at a free joint are translations in the $x$ and $y$ directions and a rotation in the $z$ sense. Internal actions occurring in the members consist of an axial force, a shearing force, and a bending moment.

Figure 6.1(d) illustrates another type of planar structure, called a *grid*, which lies in the $x$-$y$ plane. This structure usually has rigid joints and is designed to resist forces in space that are all normal to its plane (or parallel to the $z$ axis). It follows that any applied moments have their vectors in the $x$-$y$ plane. Displacements at a typical free joint consist of rotations in the $x$ and $y$ senses and a translation in the $z$ direction. Internal member actions are a shearing force, a bending moment, and a torsional moment, or torque.

The *space truss* shown in Fig. 6.1(e) is similar to a plane truss, except that the members may have any directions in space. This type of structure carries forces at its hinged (or universal) joints that are concurrent in space. The forces acting on members and joints may be in arbitrary directions, but any moment applied to a member must have its vector normal to the axis of that member. The reason for this restriction is that a truss member is incapable of resisting torque. Displacements at a free joint are three components of translation in the $x$, $y$, and $z$ directions. Due to applied loads, a member of a space truss may have local flexure in two principal planes of bending as well as an axial force from the truss analysis.

Figure 6.1(f) shows a *space frame*, which is the most versatile and complicated type of framed structure. Locations of joints and directions of members are completely arbitrary, and the structure is designed to carry forces that are general in space. At a free joint there are three components of translation (as in a space truss) and three components of rotation as well. Internal member actions consist of an axial force, two shearing forces in principal planes of bending, two bending moments in the same principal planes, and a torsional moment.

A typical prismatic member in each type of framed structure has stiffnesses, masses, and equivalent nodal loads drawn from those of the axial, torsional and flexural elements described in Sec. 3.4. Such member properties are first expressed in local directions and then transformed to global directions by the rotation-of-axes technique explained in Sec. 3.5. Next, we can assemble member contributions to form stiffness, mass, and load matrices in the equations of motion for the whole structure [see Eqs. (3.5-14), (3.5-15), and (3.5-16)]. These equations are solved for joint displacements in structural directions, and the displacements pertaining to individual members are rotated to local directions. Using such local displacements, we can find internal actions at any point for axial, torsional, and flexural deformations with Eqs. (3.4-48), (3.4-50), and (3.4-52). Of course, internal actions due to static influences at time $t = 0$ must be added to those associated with dynamic response to obtain the total values. Support reactions can also be calculated from member end-actions if desired.

In this chapter we deal only with linearly elastic framed structures having small relative displacements and small absolute rotations due to dynamic loads. We also assume that there are no interactions among axial, torsional, and flexural deformations. Because continuous beams and plane trusses have been discussed before, no separate sections are devoted to them. However, Secs. 6.2 through 6.5 give further information about plane frames, grids, space trusses, and space frames. Then dynamic analysis programs are discussed in Sec. 6.6 for these types of framed structures, as well as for continuous beams. Next, Secs. 6.7 and 6.8 cover methods for reducing the number of degrees of freedom for beams, grids, plane frames, and space frames. Finally, Sec. 6.9 describes specialized computer programs that use these reduction methods in the dynamic analyses of plane and space frames.

## 6.2 PLANE FRAMES

Figure 6.2(a) shows a typical member $i$ of a plane frame with local (primed) and global (unprimed) axes. The three numbered displacements in local directions at each end consist of a translation in the $x'$ direction, a translation in the $y'$ direction, and a rotation in the $z'$ (or $z$) sense. Assuming the member is prismatic, we can write its $6 \times 6$ stiffness matrix for local directions as

## Sec. 6.2  Plane Frames

$$\mathbf{K}' = \begin{bmatrix} \mathbf{K}'_{jj} & \mathbf{K}'_{jk} \\ \mathbf{K}'_{kj} & \mathbf{K}'_{kk} \end{bmatrix} = \frac{EI_z}{L^3} \begin{bmatrix} r_1 & & & & & \\ 0 & 12 & & & Sym. & \\ 0 & 6L & 4L^2 & & & \\ -r_1 & 0 & 0 & r_1 & & \\ 0 & -12 & -6L & 0 & 12 & \\ 0 & 6L & 2L^2 & 0 & -6L & 4L^2 \end{bmatrix} \quad (1)$$

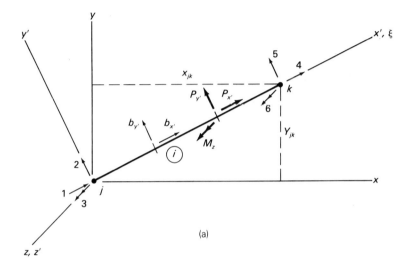

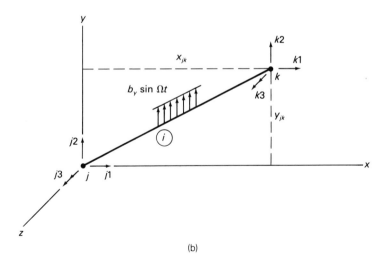

**Figure 6.2**  Plane frame member: (a) local directions; (b) global directions.

which is partitioned in accordance with joints $j$ and $k$. In matrix $\mathbf{K}'$ the dimensionless ratio $r_1$ is $AL^2/I_z$, where $I_z$ is the second moment of area of the cross section with respect to the $z$ axis. We form the stiffness matrix for a plane frame member by adding contributions from an axial element and a flexural element, which were derived in Sec. 3.4 [see Eqs. (3.4-4) and (3.4-24)]. Note that the terms from these two types of elements are uncoupled for local directions.

Similarly, we can form the $6 \times 6$ consistent-mass matrix in local directions, as follows:

$$\mathbf{M}' = \begin{bmatrix} \mathbf{M}'_{jj} & \mathbf{M}'_{jk} \\ \mathbf{M}'_{kj} & \mathbf{M}'_{kk} \end{bmatrix} = \frac{\rho AL}{420} \begin{bmatrix} 140 & & & & & \text{Sym.} \\ 0 & 156 & & & & \\ 0 & 22L & 4L^2 & & & \\ 70 & 0 & 0 & 140 & & \\ 0 & 54 & 13L & 0 & 156 & \\ 0 & -13L & -3L^2 & 0 & -22L & 4L^2 \end{bmatrix} \quad (2)$$

Again, this matrix is obtained by adding terms from those in Sec. 3.4 for an axial element and a flexural element [see Eqs. (3.4-5) and (3.4-26)].

In addition, equivalent nodal loads in local directions due to distributed forces on a plane frame member may be calculated from

$$\mathbf{p}'_b(t) = L \int_0^1 \mathbf{f}^T \mathbf{b}'(t) \, d\xi \quad (3)$$

In this expression, the vector of time-varying body forces $\mathbf{b}'(t)$ is

$$\mathbf{b}'(t) = \begin{bmatrix} b_{x'} \\ b_{y'} \end{bmatrix} \quad (4)$$

which contains forces (per unit length) in the $x'$ and $y'$ directions, as indicated in Fig. 6.2(a). The $2 \times 6$ matrix of displacement shape functions in Eq. (3) has the partitioned form

$$\mathbf{f} = [\mathbf{f}_j \quad \mathbf{f}_k] \quad (5)$$

where

$$\mathbf{f}_j = \begin{bmatrix} f_1 & 0 & 0 \\ 0 & f_2 & f_3 \end{bmatrix} \quad (6a)$$

The functions in this submatrix for the $j$ end are

$$f_1 = 1 - \xi \qquad f_2 = 2\xi^3 - 3\xi^2 + 1 \qquad f_3 = (\xi^3 - 2\xi^2 + \xi)L \quad (6b)$$

in which the dimensionless coordinate is $\xi = x'/L$. Also, for the $k$ end we have

$$\mathbf{f}_k = \begin{bmatrix} f_4 & 0 & 0 \\ 0 & f_5 & f_6 \end{bmatrix} \quad (7a)$$

where
$$f_4 = \xi \qquad f_5 = -2\xi^3 + 3\xi^2 \qquad f_6 = (\xi^3 - \xi^2)L \tag{7b}$$

Functions given in Eqs. (6b) and (7b) are drawn from Eqs. (3.4-1) and (3.4-17) for axial and flexural elements.

A plane frame member may also be subjected to time-varying concentrated forces $P_{x'}$ and $P_{y'}$ as well as a moment $M_z$ at any point [see Fig. 6.2(a)]. The concentrated forces may be handled in a manner similar to that for distributed forces $b_{x'}$ and $b_{y'}$, expect that no integrations are required. On the other hand, for the moment vector $M_z$ we need first derivatives with respect to $x'$ of $f_2, f_3, f_5$, and $f_6$ to determine the following equivalent nodal loads:

$$p'_{M2} = M_z \frac{df_2}{dx'} = \frac{M_z}{L} \frac{df_2}{d\xi} = \frac{6M_z}{L}(\xi^2 - \xi) \tag{8a}$$

$$p'_{M3} = M_z \frac{df_3}{dx'} = \frac{M_z}{L} \frac{df_3}{d\xi} = M_z(3\xi^2 - 4\xi + 1) \tag{8b}$$

$$p'_{M5} = M_z \frac{df_5}{dx'} = \frac{M_z}{L} \frac{df_5}{d\xi} = -\frac{6M_z}{L}(\xi^2 - \xi) = -p'_{M2} \tag{8c}$$

$$p'_{M6} = M_z \frac{df_6}{dx'} = \frac{M_z}{L} \frac{df_6}{d\xi} = M_z(3\xi^2 - 2\xi) \tag{8d}$$

in which the subscript $M$ replaces the subscript $b$. Note that no integrations are required for evaluating these formulas.

In order to convert stiffness, mass, and load matrices to global (or structural) directions, we compose a $6 \times 6$ rotation-of-axes transformation matrix $\hat{\mathbf{R}}$, as follows:

$$\hat{\mathbf{R}} = \begin{bmatrix} \mathbf{R} & \mathbf{0} \\ \mathbf{0} & \mathbf{R} \end{bmatrix} \tag{9}$$

in which

$$\mathbf{R} = \begin{bmatrix} c_x & c_y & 0 \\ -c_y & c_x & 0 \\ 0 & 0 & 1 \end{bmatrix} \tag{10}$$

Terms in the rows of this $3 \times 3$ rotation matrix $\mathbf{R}$ consist of direction cosines of axes $x'$, $y'$, and $z'$ with respect to $x, y$, and $z$. As in Sec. 3.5, we apply matrix $\hat{\mathbf{R}}$ and its transpose to obtain

$$\mathbf{K} = \hat{\mathbf{R}}^T \mathbf{K}' \hat{\mathbf{R}} = \begin{bmatrix} \mathbf{K}_{jj} & \mathbf{K}_{jk} \\ \mathbf{K}_{kj} & \mathbf{K}_{kk} \end{bmatrix} \tag{11}$$

and

$$\mathbf{M} = \hat{\mathbf{R}}^T \mathbf{M}' \hat{\mathbf{R}} = \begin{bmatrix} \mathbf{M}_{jj} & \mathbf{M}_{jk} \\ \mathbf{M}_{kj} & \mathbf{M}_{kk} \end{bmatrix} \tag{12}$$

and

$$\mathbf{p}_b(t) = \hat{\mathbf{R}}^T \mathbf{p}'_b(t) = \begin{bmatrix} \mathbf{p}_{bj}(t) \\ \mathbf{p}_{bk}(t) \end{bmatrix} \quad (13)$$

*local dir* (handwritten annotation)

Performing the multiplications indicated in Eq. (11) yields the stiffness submatrices

$$\mathbf{K}_{jj} = \frac{EI_z}{L^3} \begin{bmatrix} r_1 c_x^2 + 12 c_y^2 & & \text{Sym.} \\ (r_1 - 12) c_x c_y & r_1 c_y^2 + 12 c_x^2 & \\ -6L c_y & 6L c_x & 4L^2 \end{bmatrix} \quad (14a)$$

$$\mathbf{K}_{kj} = \frac{EI_z}{L^3} \begin{bmatrix} -r_1 c_x^2 - 12 c_y^2 & -(r_1 - 12) c_x c_y & 6L c_y \\ -(r_1 - 12) c_x c_y & -r_1 c_y^2 - 12 c_x^2 & -6L c_x \\ -6L c_y & 6L c_x & 2L^2 \end{bmatrix} \quad (14b)$$

$$\mathbf{K}_{kk} = \frac{EI_z}{L^3} \begin{bmatrix} r_1 c_x^2 + 12 c_y^2 & & \text{Sym.} \\ (r_1 - 12) c_x c_y & -r_1 c_y^2 + 12 c_x^2 & \\ 6L c_y & -6L c_x & 4L^2 \end{bmatrix} \quad (14c)$$

Also, multiplications given in Eq. (12) produce the consistent-mass submatrices

$$\mathbf{M}_{jj} = \frac{\rho AL}{420} \begin{bmatrix} 140 c_x^2 + 156 c_y^2 & & \text{Sym.} \\ -16 c_x c_y & 140 c_y^2 + 156 c_x^2 & \\ -22L c_y & 22L c_x & 4L^2 \end{bmatrix} \quad (15a)$$

$$\mathbf{M}_{kj} = \frac{\rho AL}{420} \begin{bmatrix} 70 c_x^2 + 54 c_y^2 & 16 c_x c_y & -13L c_y \\ 16 c_x c_y & 70 c_y^2 + 54 c_x^2 & 13L c_x \\ 13L c_y & -13L c_x & -3L^2 \end{bmatrix} \quad (15b)$$

$$\mathbf{M}_{kk} = \frac{\rho AL}{420} \begin{bmatrix} 140 c_x^2 + 156 c_y^2 & & \text{Sym.} \\ -16 c_x c_y & 140 c_y^2 + 156 c_x^2 & \\ 22L c_y & -22L c_x & 4L^2 \end{bmatrix} \quad (15c)$$

Of course, $\mathbf{K}_{jk} = \mathbf{K}_{kj}^T$ and $\mathbf{M}_{jk} = \mathbf{M}_{kj}^T$, because matrices $\mathbf{K}$ and $\mathbf{M}$ are symmetric. Finally, Eq. (13) results in the following subvectors of equivalent nodal loads:

$$\mathbf{p}_{bj}(t) = \begin{bmatrix} p'_{b1} c_x - p'_{b2} c_y \\ p'_{b1} c_y + p'_{b2} c_x \\ p'_{b3} \end{bmatrix} \quad (16a)$$

*AMLI $C_x$* (handwritten)
*$f_{b_1}' = 0$* (handwritten)
*$f_{b_4}' = 0$* (handwritten)

$$\mathbf{p}_{bk}(t) = \begin{bmatrix} p'_{b4} c_x - p'_{b5} c_y \\ p'_{b4} c_y + p'_{b5} c_x \\ p'_{b6} \end{bmatrix} \quad (16b)$$

Here we imply that body forces are given in the directions of local member axes.

Sec. 6.3  Grids

If such forces happen to be in structural directions instead, a conversion to member directions must precede use of the formulas in Eqs. (16).

After stiffnesses, masses, and equivalent nodal loads for individual members have been transformed to structural directions, we can assemble them in the manner given by Eqs. (3.5-14) and (3.5-15). For this purpose we must calculate the global indexes $j1$ through $k3$ shown in Fig. 6.2(b), as follows:

$$j1 = 3j - 2 \qquad j2 = 3j - 1 \qquad j3 = 3j$$
$$k1 = 3k - 2 \qquad k2 = 3k - 1 \qquad k3 = 3k \tag{17}$$

These indexes serve as subscripts for placing terms into the stiffness, mass, and load matrices for the whole structure.

**Example 6.1**

Figure 6.2(b) includes a uniformly distributed body force $b_y \sin \Omega t$, acting in the $y$ direction. Let us calculate the equivalent nodal loads in structural directions due to this influence.

First, we transform the body force to member directions, using a $2 \times 2$ rotation matrix, as follows:

$$\mathbf{b}'(t) = \mathbf{R}\,\mathbf{b}(t) = \begin{bmatrix} c_x & c_y \\ -c_y & c_x \end{bmatrix} \begin{bmatrix} 0 \\ 1 \end{bmatrix} b_y \sin \Omega t = \begin{bmatrix} c_y \\ c_x \end{bmatrix} b_y \sin \Omega t \tag{a}$$

Then we apply Eq. (3) to obtain

$$\mathbf{p}'_b(t) = L \int_0^1 \mathbf{f}^T \mathbf{b}'(t)\, d\xi$$

$$= \{6c_y,\ 6c_x,\ Lc_x,\ 6c_y,\ 6c_x,\ -Lc_x\} \frac{Lb_y}{12} \sin \Omega t \tag{b}$$

which are forces and moments in local directions. As the last step, Eqs. (16) give

$$\mathbf{p}_b(t) = \{0,\ 6,\ Lc_x,\ 0,\ 6,\ -Lc_x\} \frac{Lb_y}{12} \sin \Omega t \tag{c}$$

By inspection, we can see that these results are correct.

## 6.3 GRIDS

To make the analysis of a grid similar to that of a plane frame, we place the structure in the $x$-$y$ plane, as illustrated by the typical grid member in Fig. 6.3(a). The three numbered displacements in local directions at each end of the member are a rotation in the $x'$ sense, a rotation in the $y'$ sense, and a translation in the $z'$ (or $z$) direction. If the member is prismatic, its $6 \times 6$ stiffness matrix for local directions becomes

$$\mathbf{K}' = \begin{bmatrix} \mathbf{K}'_{jj} & \mathbf{K}'_{jk} \\ \mathbf{K}'_{kj} & \mathbf{K}'_{kk} \end{bmatrix} = \frac{EI_y}{L^3} \begin{bmatrix} r_2 L^2 & & & & & \\ 0 & 4L^2 & & & \text{Sym.} & \\ 0 & -6L & 12 & & & \\ -r_2 L^2 & 0 & 0 & r_2 L^2 & & \\ 0 & 2L^2 & -6L & 0 & 4L^2 & \\ 0 & 6L & -12 & 0 & 6L & 12 \end{bmatrix} \quad (1)$$

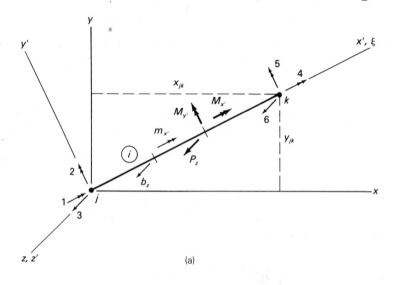

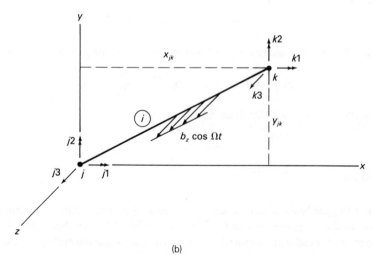

**Figure 6.3** Grid member: (a) local directions; (b) global directions.

## Sec. 6.3  Grids

In this matrix the dimensionless ratio $r_2$ is $GI_x/EI_y$, where $I_x$ is the *torsion constant* of the cross section and $I_y$ is its second moment of area about the $y'$ axis. Here we combine stiffnesses from a torsional element and a flexural element, as given by Eqs. (3.4-12) and (3.4-24). In this case the torsional and flexural terms are uncoupled for local directions.

We can also write the 6 × 6 consistent-mass matrix in local directions as

$$\mathbf{M}' = \begin{bmatrix} \mathbf{M}'_{jj} & \mathbf{M}'_{jk} \\ \mathbf{M}'_{kj} & \mathbf{M}'_{kk} \end{bmatrix} = \frac{\rho A L}{420} \begin{bmatrix} 140r_g^2 & & & & & \text{Sym.} \\ 0 & 4L^2 & & & & \\ 0 & -22L & 156 & & & \\ 70r_g^2 & 0 & 0 & 140r_g^2 & & \\ 0 & -3L^2 & 13L & 0 & 4L^2 & \\ 0 & -13L & 54 & 0 & 22L & 156 \end{bmatrix} \quad (2)$$

where $r_g^2 = J/A$ is the *radius of gyration* squared. Terms in matrix $\mathbf{M}'$ come from consistent mass matrices for torsional and flexural elements [see Eqs. (3.4-15) and (3.4-26)].

Furthermore, equivalent nodal loads in local directions for a grid member may be obtained using the previously stated expression

$$\mathbf{p}'_b(t) = L \int_0^1 \mathbf{f}^T \mathbf{b}'(t)\, d\xi \qquad (6.2\text{-}3)$$

However, in this case the vector of time-varying body forces is

$$\mathbf{b}'(t) = \begin{bmatrix} m_{x'} \\ b_z \end{bmatrix} \qquad (3)$$

This vector contains a distributed axial torque $m_{x'}$ (per unit length) in the $x'$ direction and a force $b_z$ (per unit length) in the $z$ direction, as indicated in Fig. 6.3(a). As before, the 2 × 6 matrix of displacement shape functions in Eq. (6.2-3) has the partitioned form

$$\mathbf{f} = [\mathbf{f}_j \quad \mathbf{f}_k] \qquad (6.2\text{-}5)$$

But now the functions $f_2$ and $f_3$ in the submatrix $\mathbf{f}_j$ are switched, and the sign of $f_3$ is reversed. Similarly, functions $f_5$ and $f_6$ in submatrix $\mathbf{f}_k$ are interchanged, and the sign of $f_6$ is reversed. Thus, the submatrix $\mathbf{f}_j$ becomes

$$\mathbf{f}_j = \begin{bmatrix} f_1 & 0 & 0 \\ 0 & -f_3 & f_2 \end{bmatrix}$$

$$= \begin{bmatrix} 1-\xi & 0 & 0 \\ 0 & -(\xi^3 - 2\xi^2 + \xi)L & 2\xi^3 - 3\xi^2 + 1 \end{bmatrix} \qquad (4)$$

and the submatrix $\mathbf{f}_k$ is

$$\mathbf{f}_k = \begin{bmatrix} f_4 & 0 & 0 \\ 0 & -f_6 & f_5 \end{bmatrix}$$
$$= \begin{bmatrix} \xi & 0 & 0 \\ 0 & -(\xi^3 - \xi^2)L & -2\xi^3 + 3\xi^2 \end{bmatrix} \quad (5)$$

For a grid member, we may also have concentrated actions $M_{x'}$, $M_{y'}$, and $P_z$ applied at any point, as shown in Fig. 6.3(a). The moment $M_{x'}$ and the force $P_z$ may be treated in the same manner as $m_{x'}$ and $b_z$, but without integration of the functions in Eqs. (4) and (5). However, for the moment $M_{y'}$ we need the negatives of first derivatives with respect to $x'$ of the functions in the second rows of Eqs. (4) and (5), because positive rotations in the $y'$ sense are equal to negative slopes. Thus, we have

$$p'_{M2} = -M_{y'} \frac{d(-f_3)}{dx'} = \frac{M_{y'}}{L} \frac{df_3}{d\xi} = M_{y'}(3\xi^2 - 4\xi + 1) \quad (6a)$$

$$p'_{M3} = -M_{y'} \frac{df_2}{dx'} = -\frac{M_{y'}}{L} \frac{df_2}{d\xi} = -\frac{6M_{y'}}{L}(\xi^2 - \xi) \quad (6b)$$

$$p'_{M5} = -M_{y'} \frac{d(-f_6)}{dx'} = \frac{M_{y'}}{L} \frac{df_6}{d\xi} = M_{y'}(3\xi^2 - 2\xi) \quad (6c)$$

$$p'_{M6} = -M_{y'} \frac{df_5}{dx'} = -\frac{M_{y'}}{L} \frac{df_5}{d\xi} = \frac{6M_{y'}}{L}(\xi^2 - \xi) = -p'_{M3} \quad (6d)$$

Again, these terms do not require integrations.

Because the grid lies in the $x$-$y$ plane and the nodal displacements are in the sequence $x$-$y$-$z$, rotation-of-axes transformations are the same as those for a plane frame. The resulting terms in stiffness submatrices for structural directions become

$$\mathbf{K}_{jj} = \frac{EI_y}{L^3} \begin{bmatrix} (r_2 c_x^2 + 4c_y^2)L^2 & & \text{Sym.} \\ (r_2 - 4)L^2 c_x c_y & (r_2 c_y^2 + 4c_x^2)L^2 & \\ 6Lc_y & -6Lc_x & 12 \end{bmatrix} \quad (7a)$$

$$\mathbf{K}_{kj} = \frac{EI_y}{L^3} \begin{bmatrix} (-r_2 c_x^2 + 2c_y^2)L^2 & -(r_2 + 2)L^2 c_x c_y & 6Lc_y \\ -(r_2 + 2)L^2 c_x c_y & (-r_2 c_y^2 + 2c_x^2)L^2 & -6Lc_x \\ -6Lc_y & 6Lc_x & -12 \end{bmatrix} \quad (7b)$$

$$\mathbf{K}_{kk} = \frac{EI_y}{L^3} \begin{bmatrix} (r_2 c_x^2 + 4c_y^2)L^2 & & \text{Sym.} \\ (r_2 - 4)L^2 c_x c_y & (r_2 c_y^2 + 4c_x^2)L^2 & \\ -6Lc_y & 6Lc_x & 12 \end{bmatrix} \quad (7c)$$

Similarly, the consistent-mass submatrices for a grid member in structural directions are

### Sec. 6.4   Space Trusses

$$\mathbf{M}_{jj} = \frac{\rho A L}{420} \begin{bmatrix} 140 r_g^2 c_x^2 + 4L^2 c_y^2 & & \text{Sym.} \\ (140 r_g^2 - 4L^2) c_x c_y & 140 r_g^2 c_y^2 + 4L^2 c_x^2 & \\ 22 L c_y & -22 L c_x & 156 \end{bmatrix} \quad (8a)$$

$$\mathbf{M}_{kj} = \frac{\rho A L}{420} \begin{bmatrix} 70 r_g^2 c_x^2 - 3L^2 c_y^2 & (70 r_g^2 + 3L^2) c_x c_y & -13 L c_y \\ (70 r_g^2 + 3L^2) c_x c_y & 70 r_g^2 c_y^2 - 3L^2 c_x^2 & 13 L c_x \\ 13 L c_y & -13 L c_x & 54 \end{bmatrix} \quad (8b)$$

$$\mathbf{M}_{kk} = \frac{\rho A L}{420} \begin{bmatrix} 140 r_g^2 c_x^2 + 4L^2 c_y^2 & & \text{Sym.} \\ (140 r_g^2 - 4L^2) c_x c_y & 140 r_g^2 c_y^2 + 4L^2 c_x^2 & \\ -22 L c_y & 22 L c_x & 156 \end{bmatrix} \quad (8c)$$

Subvectors of equivalent nodal loads in structural directions for a grid member are the same as those in Eqs. (6.2-16) for a plane frame member. However, the actions in each subvector become a moment in the $x$ direction, a moment in the $y$ direction, and a force in the $z$ direction.

Assemblage of stiffnesses, masses, and equivalent nodal loads follows the pattern described in Sec. 6.2 for plane frames. Global indexes $j1$ through $k3$ used as subscripts for this purpose appear in Fig. 6.3(b), and we may calculate their values with Eqs. (6.2-17).

**Example 6.2**

A uniformly distributed body force $b_z \cos \Omega t$ acts in the $z$ direction on the grid member in Fig. 6.3(b). Find the equivalent nodal loads in structural directions at the ends of the member.

For this example we can integrate in accordance with Eq. (6.2-3), as follows:

$$\mathbf{p}_b'(t) = L \int_0^1 \mathbf{f}^T \begin{bmatrix} 0 \\ 0 \\ b_z \end{bmatrix} d\xi \cos \Omega t$$

$$= \{0, -L, 6, 0, L, 6\} \frac{L b_z}{12} \cos \Omega t$$

Then Eqs. (6.2-16) produce

$$\mathbf{p}_b(t) = \{L c_y, -L c_x, 6, -L c_y, L c_x, 6\} \frac{L b_z}{12} \cos \Omega t \quad (b)$$

It is easy to see that these values are correct.

## 6.4 SPACE TRUSSES

A typical member of a space truss appears in Fig. 6.4(a), having universal hinges at joints $j$ and $k$. Due to this idealized type of connection, rotations at the ends of the member are considered to be immaterial for the analysis of the truss. Local

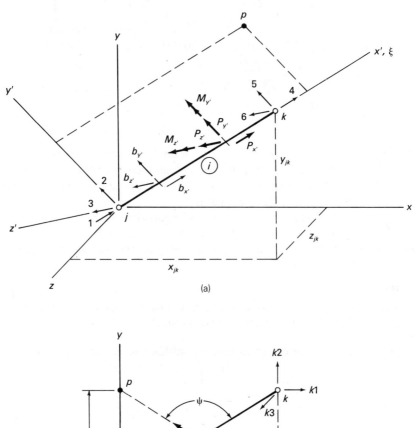

**Figure 6.4** Space truss member: (a) local directions; (b) global directions.

## Sec. 6.4  Space Trusses

axes $y'$ and $z'$, in conjunction with axis $x'$ of the member, define two principal planes of bending. At each end of the member we see three numbered arrows in local directions, representing translations in the $x'$, $y'$, and $z'$ directions. If the member is prismatic, its $6 \times 6$ stiffness matrix for local directions is

$$\mathbf{K}' = \begin{bmatrix} \mathbf{K}'_{jj} & \mathbf{K}'_{jk} \\ \mathbf{K}'_{kj} & \mathbf{K}'_{kk} \end{bmatrix} = \frac{EA}{L} \begin{bmatrix} 1 & & & & & \text{Sym.} \\ 0 & 0 & & & & \\ 0 & 0 & 0 & & & \\ -1 & 0 & 0 & 1 & & \\ 0 & 0 & 0 & 0 & 0 & \\ 0 & 0 & 0 & 0 & 0 & 0 \end{bmatrix} \quad (1)$$

Nonzero terms in this matrix, pertaining to translations 1 and 4 in the $x'$ direction, are drawn from Eq. (3.4-4) for an axial element. However, most of the terms in matrix $\mathbf{K}'$ are zero, because a truss member has no joint stiffnesses in directions perpendicular to its axis.

In a similar manner, we can write the $6 \times 6$ consistent-mass matrix in local directions for the member, as follows:

$$\mathbf{M}' = \begin{bmatrix} \mathbf{M}'_{jj} & \mathbf{M}'_{jk} \\ \mathbf{M}'_{kj} & \mathbf{M}'_{kk} \end{bmatrix} = \frac{\rho A L}{6} \begin{bmatrix} 2 & & & & & \text{Sym.} \\ 0 & 2 & & & & \\ 0 & 0 & 2 & & & \\ 1 & 0 & 0 & 2 & & \\ 0 & 1 & 0 & 0 & 2 & \\ 0 & 0 & 1 & 0 & 0 & 2 \end{bmatrix} \quad (2)$$

Here we use terms for the $x'$ direction that were derived previously for the axial element in Eq. (3.4-5). Also, due to accelerations in the $y'$ and $z'$ directions, the other consistent-mass terms in Eq. (2) are the same as for the $x'$ direction.

Equivalent nodal loads in local directions caused by distributed body forces on a space truss member are calculated as

$$\mathbf{p}'_b(t) = L \int_0^1 \mathbf{f}^T \mathbf{b}'(t) \, d\xi \quad (6.2\text{-}3)$$

For this type of member, we can accommodate three time-varying components of line loads,

$$\mathbf{b}'(t) = \{b_{x'}, b_{y'}, b_{z'}\} \quad (3)$$

Terms in this column vector are forces (per unit length) in the $x'$, $y'$, and $z'$ directions, as shown in Fig. 6.4(a). The matrix of displacement shape functions in Eq. (6.2-3) may again be stated as

$$\mathbf{f} = [\mathbf{f}_j \quad \mathbf{f}_k] \quad (6.2\text{-}5)$$

However, we now have two 3 × 3 submatrices,

$$\mathbf{f}_j = \mathbf{I}_3 f_1 = \begin{bmatrix} 1 & 0 & 0 \\ 0 & 1 & 0 \\ 0 & 0 & 1 \end{bmatrix} f_1 \qquad (4)$$

and

$$\mathbf{f}_k = \mathbf{I}_3 f_4 = \begin{bmatrix} 1 & 0 & 0 \\ 0 & 1 & 0 \\ 0 & 0 & 1 \end{bmatrix} f_4 \qquad (5)$$

where $f_1 = 1 - \xi$ and $f_4 = \xi$, as before.

A space truss member may also be subjected to time-varying concentrated forces and moments at any point. They consist of the forces $P_{x'}$, $P_{y'}$, and $P_{z'}$ and the moments $M_{y'}$ and $M_{z'}$ included in Fig. 6.4(a). (Note that $M_{x'}$ is omitted because a truss member is incapable of resisting an axial moment.) The concentrated forces may be handled in the manner described for distributed forces, but no integrations are required. For the moment $M_{z'}$ we need first derivatives of $f_1$ and $f_4$ with respect to $x'$, but for $M_{y'}$ we must use their negatives, as explained in Sec. 6.3. Thus,

$$p'_{M2} = M_{z'} \frac{df_1}{dx'} = \frac{M_{z'}}{L} \frac{df_1}{d\xi} = -\frac{M_{z'}}{L} \qquad (6a)$$

$$p'_{M3} = -M_{y'} \frac{df_1}{dx'} = -\frac{M_{y'}}{L} \frac{df_1}{d\xi} = \frac{M_{y'}}{L} \qquad (6b)$$

$$p'_{M5} = M_{z'} \frac{df_4}{dx'} = \frac{M_{z'}}{L} \frac{df_4}{d\xi} = \frac{M_{z'}}{L} = -p'_{M2} \qquad (6c)$$

$$p'_{M6} = -M_{y'} \frac{df_4}{dx'} = -\frac{M_{y'}}{L} \frac{df_4}{d\xi} = -\frac{M_{y'}}{L} = -p'_{M3} \qquad (6d)$$

As with concentrated forces, these terms need no integrations.

To form a rotation matrix for a space truss member, we can use a third point $p$ (in addition to $j$ and $k$) to define a principal plane of bending. Figure 6.4(a) shows such a point lying in the $x'$-$y'$ plane and not on the $x'$ axis. Whenever possible, this point would be taken as another joint in the structure, for which the coordinates in space are known. Then the terms in the rotation matrix may be found using properties of the vector (or cross) product, as follows:

$$\mathbf{e}_{z'} = \frac{\mathbf{e}_{x'} \times \mathbf{e}_{jp}}{|\mathbf{e}_{x'} \times \mathbf{e}_{jp}|} \qquad (7a)$$

and

$$\mathbf{e}_{y'} = \mathbf{e}_{z'} \times \mathbf{e}_{x'} \qquad (7b)$$

## Sec. 6.4  Space Trusses

In these expressions the symbol **e** represents a unit vector in the direction indicated by its subscripts. In particular, the unit vector $\mathbf{e}_{x'}$ is

$$\mathbf{e}_{x'} = [c_x \quad c_y \quad c_z] \tag{8a}$$

where

$$c_x = \frac{x_{jk}}{L} \quad c_y = \frac{y_{jk}}{L} \quad c_z = \frac{z_{jk}}{L} \tag{8b}$$

and

$$L = \sqrt{x_{jk}^2 + y_{jk}^2 + z_{jk}^2} \tag{8c}$$

A similar description may be given for the unit vector $\mathbf{e}_{jp}$, using the coordinates of points $j$ and $p$. Collecting the three required unit vectors into a rotation matrix **R**, we have

$$\mathbf{R} = \begin{bmatrix} \mathbf{e}_{x'} \\ \mathbf{e}_{y'} \\ \mathbf{e}_{z'} \end{bmatrix} = \begin{bmatrix} c_x & c_y & c_z \\ \lambda_{21} & \lambda_{22} & \lambda_{23} \\ \lambda_{31} & \lambda_{32} & \lambda_{33} \end{bmatrix} \tag{9}$$

in which the direction cosines of axes $x'$, $y'$, and $z'$ are listed in rows 1, 2, and 3.

For the purpose of transforming stiffnesses and equivalent nodal loads to structural directions, we form a 6 × 6 rotation-of-axes operator $\hat{\mathbf{R}}$ as in Eq. (6.2-9). Using matrix $\hat{\mathbf{R}}$ as indicated in Eq. (6.2-11), we find the member stiffness matrix in structural coordinates to be

$$\mathbf{K} = \hat{\mathbf{R}}^T \mathbf{K}' \hat{\mathbf{R}} = \frac{EA}{L} \begin{bmatrix} c_x^2 & & & & & \\ c_x c_y & c_y^2 & & & \text{Sym.} & \\ c_x c_z & c_y c_z & c_z^2 & & & \\ -c_x^2 & -c_x c_y & -c_x c_z & c_x^2 & & \\ -c_x c_y & -c_y^2 & -c_y c_z & c_x c_y & c_y^2 & \\ -c_x c_z & -c_y c_z & -c_z^2 & c_x c_z & c_y c_z & c_z^2 \end{bmatrix} \tag{10}$$

Notice that only the direction cosines of axis $x'$ (the axis of the member) influence these terms. Applying matrix $\hat{\mathbf{R}}$ in accordance with Eq. (6.2-13) produces the equivalent nodal loads

$$\mathbf{p}_b(t) = \hat{\mathbf{R}}^T \mathbf{p}'_b(t) = \begin{bmatrix} \mathbf{R}^T \mathbf{p}'_{bj}(t) \\ \mathbf{R}^T \mathbf{p}'_{bk}(t) \end{bmatrix} \tag{11}$$

which are now in structural directions. Of course, the detailed forms of the terms in $\mathbf{p}_b(t)$ depend on the nature of the loads on the member and the location of the third point $p$. Finally, the consistent-mass matrix for a space truss member is invariant with rotation of axes, just as it is for a plane truss member (see Sec. 3.5).

As with plane frames in Sec. 6.2 and grids in Sec. 6.3, we can assemble stiffnesses, masses, and equivalent nodal loads by assessing these properties from individual members. Figure 6.4(b) shows the global indexes $j1$ through $k3$ used as subscripts in the assembly process, and their values are again calculated using Eqs. (6.2-17).

At the end of Sec. 3.5, we mentioned that the analytical model for a truss, as described here, is not really suitable for a structure having only a few members. For better accuracy in the dynamic analysis of such trusses, we recommend the component-mode method discussed in Sec. 10.6.

**Example 6.3**

In Fig. 6.4(b) a time-varying force vector $\mathbf{P}(t)$ acts at the midpoint $m$ of member $i$. This force is directed toward point $p$, which is located on the $y$ axis. As in Fig. 6.4(a), the third point $p$, along with points $j$ and $k$, defines the $x'$-$y'$ principal plane of bending. Dimensions appearing in the figure are

$$x_{jk} = x_k - x_j = 4 \qquad y_{jk} = y_k - y_j = 3$$
$$z_{jk} = z_k - z_j = 2 \qquad y_{jp} = y_p - y_j = 2 \tag{a}$$

Find the equivalent nodal loads for both local and global directions due to the concentrated force.

From the given dimensions, the length of the member is

$$L = \sqrt{x_{jk}^2 + y_{jk}^2 + z_{jk}^2} = \sqrt{29} = 5.385 \tag{b}$$

and the direction cosines for its axis are

$$c_x = \frac{x_{jk}}{L} = 0.7428 \qquad c_y = \frac{y_{jk}}{L} = 0.5571 \qquad c_z = \frac{z_{jk}}{L} = 0.3714 \tag{c}$$

For this example the unit vector $\mathbf{e}_{jp}$ is

$$\mathbf{e}_{jp} = [0 \quad 1 \quad 0] \tag{d}$$

Applying Eq. (7a), we obtain

$$\mathbf{e}_{z'} = \frac{[-0.3714 \quad 0 \quad 0.7428]}{\sqrt{(0.3714)^2 + (0.7428)^2}} = [-0.4472 \quad 0 \quad 0.8944] \tag{e}$$

and Eq. (7b) gives

$$\mathbf{e}_{y'} = [-0.4983 \quad 0.8305 \quad -0.2491] \tag{f}$$

Altogether, the rotation matrix becomes

$$\mathbf{R} = \begin{bmatrix} 0.7428 & 0.5571 & 0.3714 \\ -0.4983 & 0.8305 & -0.2491 \\ -0.4472 & 0 & 0.8944 \end{bmatrix} \tag{g}$$

which contains the unit vectors $\mathbf{e}_{x'}$, $\mathbf{e}_{y'}$, and $\mathbf{e}_{z'}$ listed row-wise.

Using the geometry of triangle $jmp$, we can find the angle $\psi$ to be 2.331 rad. Therefore, the components of the force vector $\mathbf{P}(t)$ in local directions are

Sec. 6.5  Space Frames

$$\mathbf{P}'(t) = \begin{bmatrix} P_{x'} \\ P_{y'} \\ P_{z'} \end{bmatrix} = \begin{bmatrix} \cos \psi \\ \sin \psi \\ 0 \end{bmatrix} P(t) = \begin{bmatrix} -0.6889 \\ 0.7249 \\ 0 \end{bmatrix} P(t) \quad \text{(h)}$$

Then the equivalent nodal loads in local directions become

$$\mathbf{p}'_{Pj}(t) = \mathbf{f}_j^T \mathbf{P}'(t) = \mathbf{p}'_{Pk}(t) = \mathbf{f}_k^T \mathbf{P}'(t)$$
$$= \{-0.3445, 0.3625, 0\} P(t) \quad \text{(i)}$$

for which the subscript $P$ replaces the subscript $b$. Also, their counterparts in global directions are

$$\mathbf{p}_{Pj}(t) = \mathbf{R}^T \mathbf{p}'_{Pj}(t) = \mathbf{p}_{Pk}(t) = \mathbf{R}^T \mathbf{p}'_{Pk}(t)$$
$$= \{-0.4365, 0.1091, -0.2182\} P(t) \quad \text{(j)}$$

as given by Eq. (11). Because of the central location of $\mathbf{P}(t)$, the equivalent nodal loads are the same at both ends of the member.

## 6.5 SPACE FRAMES

Figure 6.5(a) depicts a typical member $i$ of a space frame, which is similar to a space truss member; but now rotations are included at joints $j$ and $k$. As before, the principal planes of bending are the $x'$-$y'$ plane and the $x'$-$z'$ plane. Six numbered displacements, indicated at each end of the member, consist of translations and rotations in the $x'$, $y'$, and $z'$ directions. With a prismatic member, the $12 \times 12$ stiffness matrix for local axes is composed of the following $6 \times 6$ submatrices:

$$\mathbf{K}'_{jj} = \frac{E}{L^3} \begin{bmatrix} r_1 I_z & & & & & \text{Sym.} \\ 0 & 12 I_z & & & & \\ 0 & 0 & 12 I_y & & & \\ 0 & 0 & 0 & r_2 L^2 I_y & & \\ 0 & 0 & -6LI_y & 0 & 4L^2 I_y & \\ 0 & 6LI_z & 0 & 0 & 0 & 4L^2 I_z \end{bmatrix} \quad \text{(1a)}$$

and

$$\mathbf{K}'_{kj} = \frac{E}{L^3} \begin{bmatrix} -r_1 I_z & 0 & 0 & 0 & 0 & 0 \\ 0 & -12 I_z & 0 & 0 & 0 & -6LI_z \\ 0 & 0 & -12 I_y & 0 & 6LI_y & 0 \\ 0 & 0 & 0 & -r_2 L^2 I_y & 0 & 0 \\ 0 & 0 & -6LI_y & 0 & 2L^2 I_y & 0 \\ 0 & 6LI_z & 0 & 0 & 0 & 2L^2 I_z \end{bmatrix} \quad \text{(1b)}$$

and

$$\mathbf{K}'_{kk} = \frac{E}{L^3} \begin{bmatrix} r_1 I_z & & & & & \\ 0 & 12 I_z & & & \text{Sym.} & \\ 0 & 0 & 12 I_y & & & \\ 0 & 0 & 0 & r_2 L^2 I_y & & \\ 0 & 0 & 6L I_y & 0 & 4L^2 I_y & \\ 0 & -6L I_z & 0 & 0 & 0 & 4L^2 I_z \end{bmatrix} \quad (1c)$$

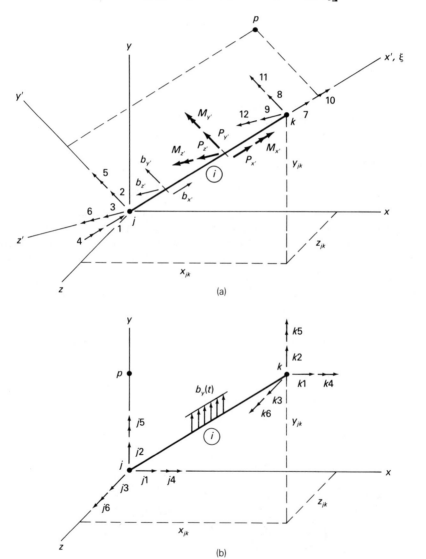

**Figure 6.5** Space frame member: (a) local directions; (b) global directions.

Sec. 6.5  Space Frames

Nonzero terms in these submatrices come from Eqs. (3.4-4), (3.4-12), and (3.4-24) for axial, torsional, and (two) flexural elements. All of the structural parameters in Eqs. (1) have been defined previously.

Similarly, the 12 × 12 consistent-mass matrix $\mathbf{M}'$ for local directions contains the 6 × 6 submatrices

$$\mathbf{M}'_{jj} = \frac{\rho A L}{420} \begin{bmatrix} 140 & & & & & \\ 0 & 156 & & & \text{Sym.} & \\ 0 & 0 & 156 & & & \\ 0 & 0 & 0 & 140 r_g^2 & & \\ 0 & 0 & -22L & 0 & 4L^2 & \\ 0 & 22L & 0 & 0 & 0 & 4L^2 \end{bmatrix} \qquad (2a)$$

and

$$\mathbf{M}'_{kj} = \frac{\rho A L}{420} \begin{bmatrix} 70 & 0 & 0 & 0 & 0 & 0 \\ 0 & 54 & 0 & 0 & 0 & 13L \\ 0 & 0 & 54 & 0 & -13L & 0 \\ 0 & 0 & 0 & 70 r_g^2 & 0 & 0 \\ 0 & 0 & 13L & 0 & -3L^2 & 0 \\ 0 & -13L & 0 & 0 & 0 & -3L^2 \end{bmatrix} \qquad (2b)$$

and

$$\mathbf{M}'_{kk} = \frac{\rho A L}{420} \begin{bmatrix} 140 & & & & & \\ 0 & 156 & & & \text{Sym.} & \\ 0 & 0 & 156 & & & \\ 0 & 0 & 0 & 140 r_g^2 & & \\ 0 & 0 & 22L & 0 & 4L^2 & \\ 0 & -22L & 0 & 0 & 0 & 4L^2 \end{bmatrix} \qquad (2c)$$

Nonzero entries in these submatrices are taken from Eqs. (3.4-5), (3.4-15), and (3.4-26), and all of the structural parameters were given earlier.

Distributed body forces applied in local directions to a space frame member cause the equivalent nodal loads

$$\mathbf{p}'_b(t) = L \int_0^1 \mathbf{f}^T \mathbf{b}'(t) \, d\xi \qquad (6.2\text{-}3)$$

that are also in local directions. As for the space truss member, the three components of force (per unit length) are

$$\mathbf{b}'(t) = \{b_{x'}, b_{y'}, b_{z'}\} \qquad (6.4\text{-}3)$$

which appear at a generic point on the member in Fig. 6.5(a). In Eq. (6.2-3) the matrix of displacement shape functions is once more

$$\mathbf{f} = [\mathbf{f}_j \ \mathbf{f}_k] \quad (6.2\text{-}5)$$

But now the two 3 × 6 submatrices have the forms

$$\mathbf{f}_j = \begin{bmatrix} f_1 & 0 & 0 & 0 & 0 & 0 \\ 0 & f_2 & 0 & 0 & 0 & f_3 \\ 0 & 0 & f_2 & 0 & -f_3 & 0 \end{bmatrix} \quad (3)$$

and

$$\mathbf{f}_k = \begin{bmatrix} f_4 & 0 & 0 & 0 & 0 & 0 \\ 0 & f_5 & 0 & 0 & 0 & f_6 \\ 0 & 0 & f_5 & 0 & -f_6 & 0 \end{bmatrix} \quad (4)$$

Functions $f_1$ through $f_6$ in these arrays were defined previously in Eqs. (6.2-6b) and (6.2-7b).

Figure 6.5(a) also shows the possibility of six time-varying concentrated actions applied at any point on the space frame member. Three force components in local directions are given in the column vector

$$\mathbf{P}'(t) = \{P_{x'}, P_{y'}, P_{z'}\} \quad (5)$$

Due to these forces, the equivalent nodal loads become

$$\mathbf{p}'_P(t) = \mathbf{f}^T \mathbf{P}'(t) \quad (6)$$

which require no integrations. Furthermore, three moment components in local directions are

$$\mathbf{M}'(t) = \{M_{x'}, M_{y'}, M_{z'}\} \quad (7)$$

To calculate equivalent nodal loads caused by these moments, we set up the following matrix-vector multiplication:

$$\mathbf{p}'_M(t) = \mathbf{f}^T_{,x'} \mathbf{M}'(t) \quad (8)$$

which needs no integrations. However, Eq. (8) does require the matrix $\mathbf{f}_{,x'}$, containing appropriate first derivatives of displacement shape functions with respect to $x'$. The first submatrix in $\mathbf{f}_{,x'}$ is

$$\mathbf{f}_{j,x'} = \begin{bmatrix} 0 & 0 & 0 & f_1 & 0 & 0 \\ 0 & 0 & -f_{2,x'} & 0 & f_{3,x'} & 0 \\ 0 & f_{2,x'} & 0 & 0 & 0 & f_{3,x'} \end{bmatrix} \quad (9a)$$

where

$$f_{2,x'} = \frac{df_2}{dx'} = \frac{6}{L}(\xi^2 - \xi) \qquad f_{3,x'} = \frac{df_3}{dx'} = 3\xi^2 - 4\xi + 1 \quad (9b)$$

In addition, the second submatrix is

$$\mathbf{f}_{k,x'} = \begin{bmatrix} 0 & 0 & 0 & f_4 & 0 & 0 \\ 0 & 0 & -f_{5,x'} & 0 & f_{6,x'} & 0 \\ 0 & f_{5,x'} & 0 & 0 & 0 & f_{6,x'} \end{bmatrix} \quad (10a)$$

where

$$f_{5,x'} = \frac{df_5}{dx'} = -\frac{6}{L}(\xi^2 - \xi) \qquad f_{6,x'} = \frac{df_6}{dx'} = 3\xi^2 - 2\xi \quad (10b)$$

Functions $f_1$ and $f_4$ appearing in Eqs. (9a) and (10a) are not differentiated with respect to $x'$, because they are to be simply multiplied by the torque $M_{x'}$. On the other hand, nonzero terms in the second row of each submatrix are the negatives of first derivatives that multiply the moment $M_{y'}$. Finally, nonzero terms in the third row consist of positive values of first derivatives that multiply the moment $M_{z'}$.

To convert stiffnesses, masses, and equivalent nodal loads from local to global directions, we form a 12 × 12 rotation-of-axes transformation matrix $\hat{\mathbf{R}}$, as follows:

$$\hat{\mathbf{R}} = \begin{bmatrix} \mathbf{R} & 0 & 0 & 0 \\ 0 & \mathbf{R} & 0 & 0 \\ 0 & 0 & \mathbf{R} & 0 \\ 0 & 0 & 0 & \mathbf{R} \end{bmatrix} \quad (11)$$

The 3 × 3 rotation matrix $\mathbf{R}$, appearing four times in matrix $\hat{\mathbf{R}}$, is identical to that in Eq. (6.4-9) for the space truss member. Again, the third point $p$ is assumed to lie in the $x'$-$y'$ principal plane of bending, as indicated in Fig. 6.5(a). Using matrix $\hat{\mathbf{R}}$, we apply Eqs. (6.2-11), (6.2-12), and (6.2-13) to obtain the matrices $\mathbf{K}$, $\mathbf{M}$, and $\mathbf{p}_b(t)$ for structural directions.

As for the other types of framed structures in previous sections, the stiffnesses, masses, and equivalent nodal loads are assembled from member contributions. To transfer terms from the member arrays to the structural matrices, we need the global indexes $j1$ through $k6$ shown in Fig. 6.5(b). Their values are calculated by the formulas

$$\begin{aligned} j1 &= 6j - 5 & j2 &= 6j - 4 & j3 &= 6j - 3 \\ j4 &= 6j - 2 & j5 &= 6j - 1 & j6 &= 6j \\ k1 &= 6k - 5 & k2 &= 6k - 4 & k3 &= 6k - 3 \\ k4 &= 6k - 2 & k5 &= 6k - 1 & k6 &= 6k \end{aligned} \quad (12)$$

which are used as subscripts in the assembly process.

### Example 6.4

Figure 6.5(b) shows a time-varying uniformly distributed force of intensity $b_y(t)$, acting in the $y$ direction on the space frame member. Locations of points $j$, $k$, and $p$ are the same as for the space truss member in Example 6.3. Thus, the rotation matrix

$$\mathbf{R} = \begin{bmatrix} 0.7428 & 0.5571 & 0.3714 \\ -0.4983 & 0.8305 & -0.2491 \\ -0.4472 & 0 & 0.8944 \end{bmatrix} \quad (6.4\text{-g})$$

still pertains. We shall determine equivalent nodal loads for both local and global directions caused by the distributed force.

Using the rotation matrix $\mathbf{R}$, we can find the components of $b_y(t)$ in local directions as

$$\mathbf{b}'(t) = \mathbf{R}\,\mathbf{b}(t) = \mathbf{R}\begin{bmatrix} 0 \\ b_y(t) \\ 0 \end{bmatrix} = \begin{bmatrix} 0.5571 \\ 0.8305 \\ 0 \end{bmatrix} b_y(t) \qquad (a)$$

Then apply Eq. (6.2-3) to obtain

$$\mathbf{p}'_{bj}(t) = L \int_0^1 \mathbf{f}_j^T \mathbf{b}'(t)\, d\xi$$

$$= \{0.2786,\ 0.4153,\ 0,\ 0,\ 0,\ 0.06921L\} L b_y(t) \qquad (b)$$

and

$$\mathbf{p}'_{bk}(t) = L \int_0^1 \mathbf{f}_k^T \mathbf{b}'(t)\, d\xi$$

$$= \{0.2786,\ 0.4153,\ 0,\ 0,\ 0,\ -0.06921L\} L b_y(t) \qquad (c)$$

These vectors contain equivalent nodal loads in member-oriented directions at joints $j$ and $k$. Next, we use Eq. (6.2-13) to find

$$\mathbf{p}_{bj}(t) = \begin{bmatrix} \mathbf{R}^T & \mathbf{0} \\ \mathbf{0} & \mathbf{R}^T \end{bmatrix} \mathbf{p}'_{bj}(t)$$

$$= \{0,\ 0.5000,\ 0,\ -0.03095L,\ 0,\ 0.06190L\} L b_y(t) \qquad (d)$$

and

$$\mathbf{p}_{bk}(t) = \begin{bmatrix} \mathbf{R}^T & \mathbf{0} \\ \mathbf{0} & \mathbf{R}^T \end{bmatrix} \mathbf{p}'_{bk}(t)$$

$$= \{0,\ 0.5000,\ 0,\ 0.03095L,\ 0,\ -0.06190L\} L b_y(t) \qquad (e)$$

Terms in these vectors consist of equivalent nodal loads at joints $j$ and $k$ in structure-oriented directions.

## 6.6 PROGRAMS FOR FRAMED STRUCTURES

In a manner analogous to that for Program DYNAPT (see Sec. 5.7), we can develop computer programs for the dynamic analysis of the five other types of framed structures. These programs are named DYNACB for continuous beams,

### Sec. 6.6  Programs for Framed Structures

DYNAPF for plane frames, DYNAGR for grids, DYNAST for space trusses, and DYNASF for space frames. They represent the dynamic counterparts of the static analysis programs described in Ref. 1.

First, let us consider the task of generating Program DYNACB for continuous beams, using Program DYNAPT for plane trusses as a guide. Both types of structures have two displacements per node, but those for a beam consist of a translation in the $y$ direction and a rotation in the $z$ sense, as indicated in Fig. 6.1(a). Modifications of the structural data for plane trusses (see Table 3.2) are minimal. The $y$ coordinates of nodes can be left blank, and we need only add the moment of inertia $ZI(I)$ of the cross section to each of the lines for element information. In addition, the meanings of terms in the nodal restraint list become: $NRL(2J-1)$ = restraint against $y$ translation and $NRL(2J)$ = restraint against $z$ rotation.

Dynamic load data for continuous beams may also be specified as modifications of those for plane trusses (see Table 4.1). However, the two nodal actions for a beam are a force in the $y$ direction and a moment in the $z$ sense, as shown in Fig. 6.6(a). Because forces applied to a flexural element act only in the $y$ direction, the linearly varying line loading depicted in Fig. 6.6(b) requires only two parameters (BL1 and BL2) for its definition. The equivalent

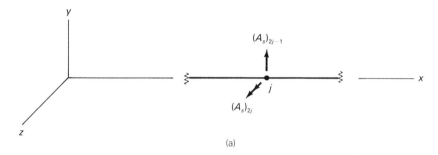

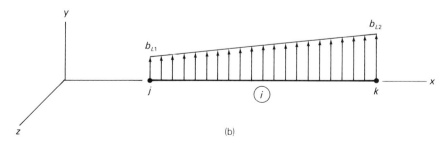

**Figure 6.6**  Loads on continuous beam: (a) nodal loads; (b) line load on element.

nodal loads at points *j* and *k* may be taken as the solution for Prob. 3.4-10. Also, for Program DYNACB the only scale factor needed for ground acceleration is GAY, so the factor GAX must be omitted.

### Example 6.5

Figure 6.7(a) shows a simply supported prismatic beam, divided into four flexural elements of equal lengths. A *moving load P* of constant magnitude traverses the span from left to right, so that its position $x(t)$ is a function of time. We will calculate the translational responses of the beam at node 3 (the midpoint) caused by the load moving at various speeds [3].

Equivalent nodal loads for this problem may be found by multiplying the displacement shape functions for a flexural element [see Eq. (3.4-17)] by $P$, as follows:

$$\mathbf{p}_{bi}(t) = P\mathbf{f}^{\mathrm{T}}[x_i(t)] \tag{a}$$

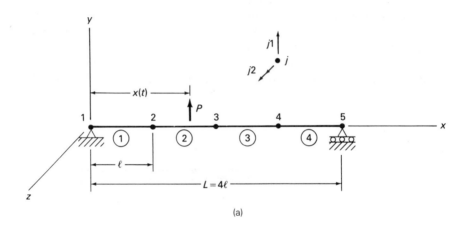

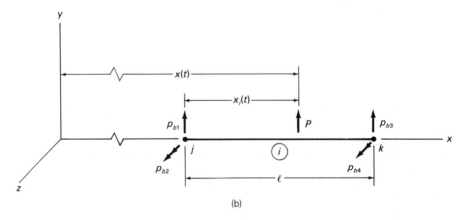

**Figure 6.7**  (a) Simple beam with moving load; (b) equivalent nodal loads.

### Sec. 6.6  Programs for Framed Structures

The term $x_i(t)$ in this formula is the distance that the load has traveled along element $i$ at time $t$, as illustrated in Fig. 6.7(b). Because the functions in matrix **f** are cubic, the equivalent nodal loads given by Eq. (a) will also be at least cubic in time, depending on how $x_i(t)$ varies. For a moving load with constant acceleration, the formula for $x_i(t)$ is

$$x_i(t) = v_{OP}t + a_{OP}\frac{t^2}{2} - (i-1)\ell \tag{b}$$

In this expression $v_{OP}$ is the velocity of the load when it first contacts the beam (at time $t = 0$), and $a_{OP}$ is the value of its constant acceleration. Substitution of Eq. (b) and the functions from Eq. (3.4-17) into Eq. (a) produces the desired equivalent nodal loads $p_{b1}$ through $p_{b4}$ indicated in Fig. 6.7(b).

Of course, we need to extend Program DYNACB to accommodate a moving load. The load parameters NI, N and NEL (see Table 4.1) are augmented with a third parameter, IML, which is an indicator for a moving load. If IML = 0, there is no moving load; but if IML = 1, a moving load is present. We also need the additional load data: (d) Moving load (P, V0P, A0P) to be input for the case when IML = 1. Subprogram DYLOCB reads and writes this data and calculates *nonproportional equivalent nodal loads* in accordance with Eqs. (a), (b), and (3.4-17). We must also modify the loads processed by either Subprogram NORMOD or Subprogram NUMINT by adding to them the equivalent nodal loads due to the moving load. If desired, we could further extend the code to handle more than one moving load simultaneously.

Now let us assume that the beam in Fig. 6.7(a) is reinforced concrete and has the physical properties:

$$E = 3.6 \times 10^3 \text{ k/in.}^2 \qquad \rho = 2.25 \times 10^{-7} \text{ k-s}^2/\text{in.}^4 \qquad L = 4\ell = 240 \text{ in.}$$

$$A = 12 \times 15 = 180 \text{ in.}^2 \qquad I_z = 3375 \text{ in.}^4 \qquad P = 10 \text{ k}$$

where the units are US. Figure 6.8 shows computer plots of $D_{j1}$ at node 3, obtained by running the modified version of Program DYNACB and using Subprogram NUMINT for the response calculations. The case of constant velocity (V0P = $3.585 \times 10^3$ in./s) produces a maximum translation of 0.4036 in. at node 3, and that for constant acceleration (A0P = $1.071 \times 10^5$ in./s$^2$, with zero initial velocity) gives 0.3625 in. Their ratios to the static deflection of $PL^3/48EI_z = 0.2370$ in. (due to the load applied gradually at node 3) are 1.703 and 1.530, respectively. The values of V0P and A0P selected in this example both produce travel times equal to the fundamental period of the beam, which is 66.95 ms.

Second, we shall describe Program DYNAPF for plane frames, using Programs DYNAPT and DYNACB to guide us. As shown in Fig. 6.1(c), a typical node $j$ in a plane frame has three displacements, which are translations in the $x$ and $y$ directions and a rotation in the $z$ sense. Table 6.1 lists the form of structural data required for plane frames. As with continuous beams, each line containing element information must include the moment of inertia ZI(I) of the cross section. Also, a line of nodal restraints has three types, which denote restraints against $x$ translation, $y$ translation, and $z$ rotation, respectively.

Table 6.2 shows dynamic load data for plane frames. In this case, the lines for initial displacements and velocities contain three quantities instead of two.

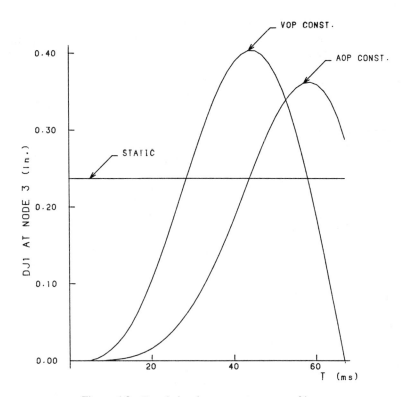

**Figure 6.8** Translational responses at center of beam.

The nodal actions also have three values, consisting of forces in the $x$ and $y$ directions and a moment in the $z$ sense, as indicated in Fig. 6.9(a). Line loads chosen for a plane frame element are the same as those for a member in a plane truss, which can be seen by comparing Fig. 6.9(b) with Fig. 4.10(b). Therefore, the equivalent nodal loads may be obtained by first resolving the line loads into components that are parallel and perpendicular to the member. Using these components and the solutions from Probs. 3.4-1 and 3.4-10 (for axial and

**TABLE 6.1 Structural Data for Plane Frames**

| Type of Data | No. of Lines | Items on Data Lines |
|---|---|---|
| Problem identification | 1 | Descriptive title |
| Structural parameters | 1 | NN, NE, NRN, E, RHO |
| Plane frame data<br>(a) Nodal coordinates<br>(b) Element information<br>(c) Nodal restraints | <br>NN<br>NE<br>NRN | <br>J, X(J), Y(J)<br>I, JN(I), KN(I), AX(I), ZI(I)<br>J, NRL(3J-2), NRL(3J-1), NRL(3J) |

Sec. 6.6    Programs for Framed Structures    269

**TABLE 6.2  Dynamic Load Data for Plane Frames**

| Type of Data | No. of Lines | Items on Data Lines |
|---|---|---|
| Dynamic parameters | 1 | ISOLVE, NTS, DT, DAMPR |
| Initial conditions<br>(a) Condition parameters<br>(b) Displacements<br>(c) Velocities | <br>1<br>NNID<br>NNIV | <br>NNID, NNIV<br>J, D0(3J-2), D0(3J-1), D0(3J)<br>J, V0(3J-2), V0(3J-1), V0(3J) |
| Applied actions<br>(a) Load parameters<br>(b) Nodal loads<br>(c) Line loads | <br>1<br>NLN<br>NEL | <br>NLN, NEL<br>J, AS(3J-2), AS(3J-1), AS(3J)<br>I, BL1, BL2, BL3, BL4 |
| Ground accelerations<br>(a) Acceleration parameter<br>(b) Acceleration factors[a] | <br>1<br>1 | <br>IGA<br>GAX, GAY |
| Forcing function<br>(a) Function parameter<br>(b) Function ordinates | <br>1<br>NFO | <br>NFO<br>K, T(K), FO(K) |

[a] Omit when IGA = 0.

flexural elements), we find equivalent nodal loads in member directions. Then the equivalent nodal loads in structural directions can be computed from Eqs. (6.2-16).

**Example 6.6**

The plane frame in Fig. 6.10 consists of three prismatic members and has an initial load $P_0$ applied by a cable connected to node 1. If the cable suddenly breaks, the frame responds to the initial displacements caused by the load $P_0$ applied statically. Such a response is the sum of free vibrations of the natural modes excited by the initial displacements; and in the presence of damping, they will decay with time.

We shall assume that all elements in the frame are steel W 12 × 85 sections with the following properties:

$$E = 207 \times 10^6 \text{ kN/m}^2 \quad \rho = 7.85 \text{ Mg/m}^3 \quad L = 2 \text{ m}$$
$$A = 1.61 \times 10^{-2} \text{ m}^2 \quad I_z = 3.01 \times 10^{-4} \text{ m}^4$$

for which the units are SI. From static analysis, initial displacements due to the load $P_0 = 10$ kN are: (1) at node 1, $(D_0)_{j1} = 0.5697$ mm, $(D_0)_{j2} = -0.2923$ mm, and $(D_0)_{j3} = -4.739 \times 10^{-5}$; (2) at node 2, $(D_0)_{j1} = 0.3642$ mm, $(D_0)_{j2} = 0.1149$ mm, and $(D_0)_{j3} = 6.598 \times 10^{-5}$. We used these values as input data for Program DYNAPF, with DAMPR = 0.02 and solution by Subprogram NUMINT. Figure 6.11(a) shows computer plots of translational responses $D_{j1}$ and $D_{j2}$ at node 1 due to the load release; and in part (b) we have the bending moments $A_{M3}$ and $A_{M6}$ for the $j$ and $k$ ends (nodes 1 and 2) of element 1. For all of these responses, the maximum (or minimum) values occur at time $t = 0$ and diminish thereafter because of damping.

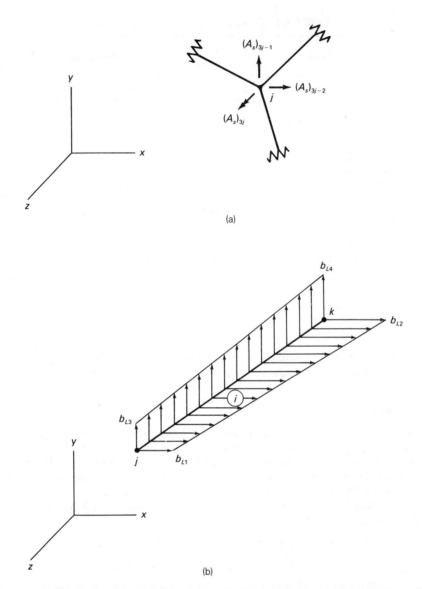

**Figure 6.9** Loads on plane frame: (a) nodal loads; (b) line loads on element.

Third, let us briefly examine Program DYNAGR for the dynamic analysis of grids. Because this type of structure has three displacements per node, the program will be very similar to Program DYNAPF for plane frames. However, the three types of displacements at a typical node $j$ in a grid are rotations in the $x$ and $y$ senses and a translation in the $z$ direction, as indicated in Fig. 6.1(d).

To change the plane-frame structural data in Table 6.1 to that for a grid, we must add the shearing modulus $G$ to the structural parameters. Also, the

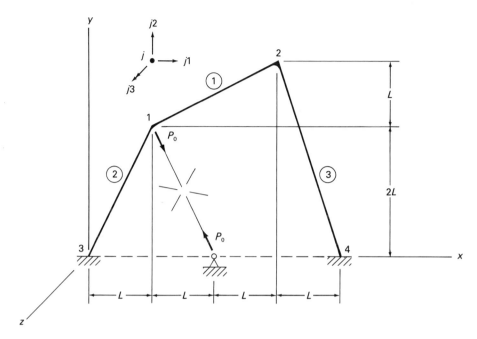

**Figure 6.10** Plane frame with load release.

torsion constant XI(I) and the moment of inertia YI(I) replace the moment of inertia ZI(I) on each line containing element information. Furthermore, the terms in the nodal restraint list become restraints against rotations in the $x$ and $y$ senses and translation in the $z$ direction.

Similarly, the dynamic load data for grids are symbolically the same as for plane frames in Table 6.2, but the meanings are different. When the structure is a grid, the first two initial displacements and velocities at a joint are rotational (in $x$ and $y$ directions); and the third is translational (in the $z$ direction). Nodal loads consist of moments in the $x$ and $y$ senses and a force in the $z$ direction, as shown in Fig. 6.12(a). In addition, the linearly varying line load illustrated in Fig. 6.12(b) acts in the $z$ direction and is defined by only two parameters (BL1 and BL2). Equivalent nodal moments at joints $j$ and $k$ due to the line load are easily converted to structural directions using Eqs. (6.2-16) with $p'_{b1} = p'_{b4} = 0$. Finally, the scale factor GAZ for ground acceleration in the $z$ direction replaces the factors GAX and GAY.

**Example 6.7**

The grid illustrated in Fig. 6.13 is made of brass and has rather small dimensions and loads. The cross section of each prismatic element is a solid square with side 0.25 in., and the applied actions consist of step nodal and line loads. Physical parameters in this example are

$$E = 1.5 \times 10^4 \text{ k/in.}^2 \qquad G = 0.56 \times 10^4 \text{ k/in.}^2 \qquad \rho = 8.10 \times 10^{-7} \text{ k-s}^2/\text{in.}^4$$

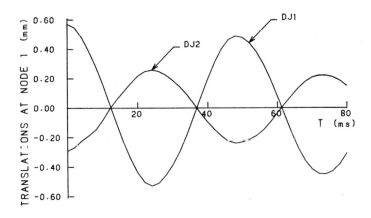

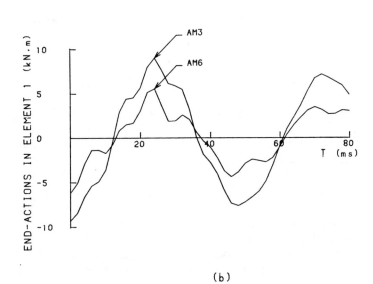

**Figure 6.11** Responses of plane frame: (a) displacements; (b) bending moments.

$L = 4$ in.     $A = 6.25 \times 10^{-2}$ in.$^2$     $I_x = 2I_y = 6.510 \times 10^{-4}$ in.$^4$
$P = 0.02$ k     $b_L = 0.004$ k/in.

where the units are US.

With this data we ran Program DYNAGR, using DAMPR = 0.05 and response calculations by Subprogram NORMOD. Translations of nodes 2 and 3 in the $z$ direction are plotted in Fig. 6.14(a), and the moments $A_{M4}$ and $A_{M5}$ at the $k$ end (node 4) of element 3 appear in part (b) of the figure. The first of these moments is a torque about the $x'$ axis

## Sec. 6.6  Programs for Framed Structures

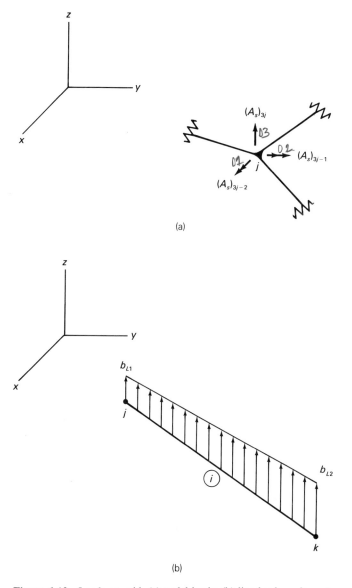

**Figure 6.12** Loads on grid: (a) nodal loads; (b) line load on element.

of the member [see Fig. 6.3(a)], and the second is a bending moment about the $y'$ axis. Maximum (or minimum) values of the nodal translations are 0.1157 in. and 0.08451 in.; and those for the moments are $9.333 \times 10^{-3}$ k-in. and $-8.181 \times 10^{-2}$ k-in.

Fourth, we shall discuss Program DYNAST for space trusses, again using Program DYNAPF for comparisons. In this case, the three types of displace-

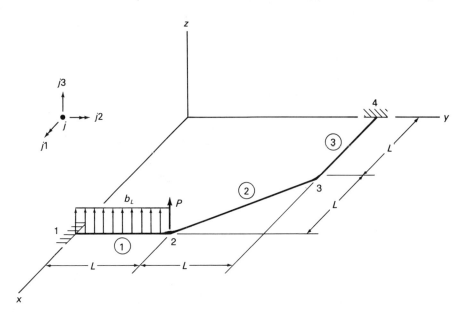

**Figure 6.13** Grid with step loads.

ments at a typical node *j* are translations in the *x*, *y*, and *z* directions, as shown in Fig. 6.1(e).

In order to make the plane-frame structural data in Table 6.1 apply to a space truss, we must add Z(J) to each line of nodal coordinates. On a line of element information, the moment of inertia ZI(I) is replaced by an identifier IP that indicates whether a third point *p* is necessary for locating principal planes of bending. If IP = 0, the $x'$-$y'$ principal plane in Fig. 6.4(a) is taken to be parallel to the *y* axis. However, if IP = 1, the next line of data must contain the coordinates XP, YP, and ZP of the third point *p*. (This type of data is shown in Table 6.3 for space frames.) Last, the terms in a line of nodal restraints denote restraints against *x*, *y*, and *z* translations.

Dynamic load data for space trusses are similar to those for plane frames, but the meanings are different. For a space truss the initial displacements and velocities at a joint refer to translations in the *x*, *y*, and *z* directions. The three nodal loads at a joint are forces in the *x*, *y*, and *z* directions, as depicted in Fig. 6.15(a). Also, the three sets of linearly varying line loads appearing in Fig. 6.15(b) require six parameters (BL1 through BL6) for their definitions. Equivalent nodal loads due to these line loads are easily obtained by extending Eqs. (4.10-4). To accomodate three components of ground acceleration, we must also add the scale factor GAZ to the data table.

### Example 6.8

Figure 6.16 depicts a space truss with nine prismatic elements, having equal cross-sectional areas and composed of high-strength titanium. This structure is subjected to

Sec. 6.6    Programs for Framed Structures                                        275

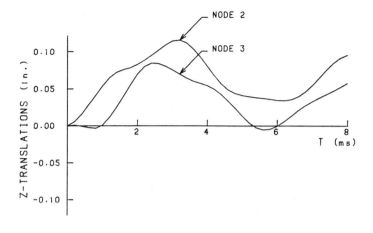

(a)

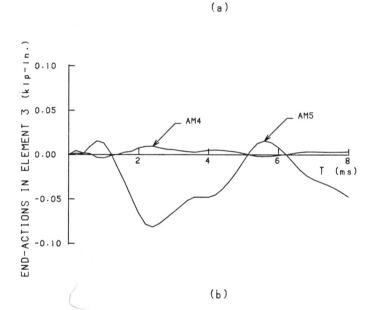

(b)

**Figure 6.14**  Responses of grid: (a) displacements; (b) moments.

three sets of double triangular impulses $P(t)$ at locations indicated in the figure. For this problem the physical parameters are

$$E = 117 \times 10^6 \text{ kN/m}^2 \qquad \rho = 4.49 \text{ Mg/m}^3 \qquad L = 1 \text{ m}$$
$$A = 9 \times 10^{-4} \text{ m}^2 \qquad P_{\max} = 100 \text{ kN}$$

in which SI units are implied.

**TABLE 6.3** Structural Data for Space Frames

| Type of Data | No. of Lines | Items on Data Lines |
|---|---|---|
| Problem identification | 1 | Descriptive title |
| Structural parameters | 1 | NN, NE, NRN, E, G, RHO |
| Space frame data | | |
| (a) Nodal coordinates | NN | J, X(J), Y(J), Z(J) |
| (b) Element information | NE | I, JN(I), KN(I), AX(I), XI(I), YI(I), ZI(I), IP |
| Coordinates[a] of point p | 1 | XP, YP, ZP |
| (c) Nodal restraints | NRN | J, NRL(6J-5), NRL(6J-4), NRL(6J-3) NRL(6J-2), NRL(6J-1), NRL(6J) |

[a] Required when IP = 1.

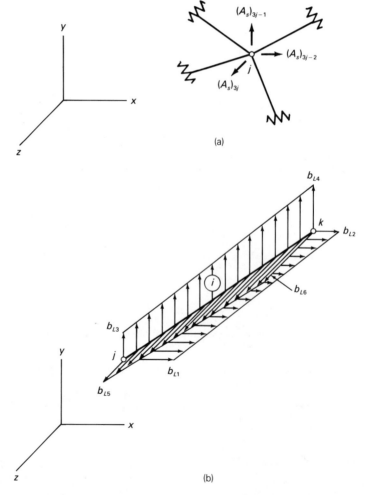

**Figure 6.15** Loads on space truss: (a) nodal loads; (b) line loads on element.

## Sec. 6.6  Programs for Framed Structures

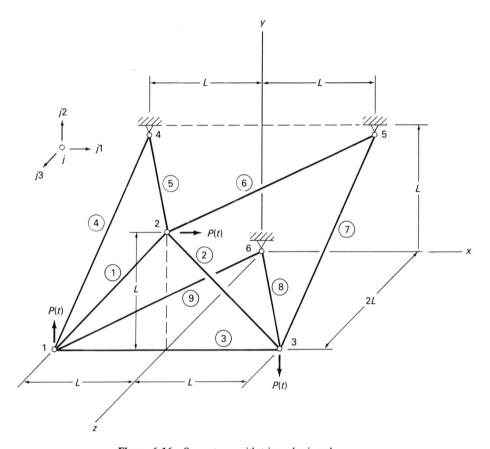

**Figure 6.16** Space truss with triangular impulses.

We used this data in Program DYNAST, with IP = 0 for all elements, DAMPR = 0.02, and called Subprogram NUMINT for numerical evaluation of responses. Figure 6.17(a) shows a computer plot of the applied force $P(t)$; and resulting axial force–time histories for elements 5, 6, and 7 are given in Fig. 6.17(b). Maximum (or minimum) values of the axial forces in these three elements are 325.6 kN, −325.6 kN, and 175.1 kN, respectively.

Fifth, let us build upon all of the previous programs to describe Program DYNASF for space frames. With this type of structure, the six displacements at a typical node $j$ consist of three translations and three rotations in the $x$, $y$, and $z$ directions, as illustrated in Fig. 6.1(f).

Table 6.3 contains the structural data required for space frames. As for grids, the shearing modulus $G$ is added to the line of structural parameters; and each of the lines for nodal coordinates includes $Z(J)$, as for space trusses. Element information includes $AX(I)$, $XI(I)$, $YI(I)$, $ZI(I)$, IP, and (optionally) the coordinates of a third point $p$ for locating principal planes of bending. In the lines

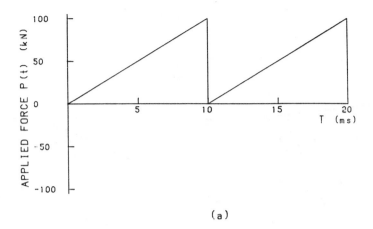

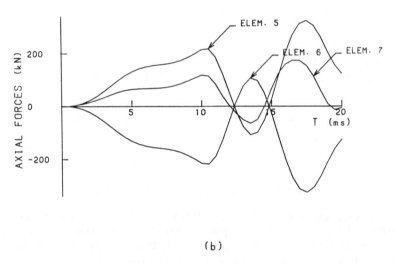

**Figure 6.17** Space truss: (a) triangular impulses; (b) responses.

for nodal restraints, the terms indicate restraints against three translational and three rotational displacements in the $x$, $y$, and $z$ directions.

Dynamic load data for space frames in Table 6.4 are more extensive than for any other type of framed structure. The six types of initial displacements and velocities in the table consist of three translational terms, followed by three rotational terms. For nodal loads we have three force components and three moment components in the $x$, $y$, and $z$ directions, as shown in Fig. 6.18(a). The three sets of linearly varying line loads in Fig. 6.18(b) are chosen to be the same

## Sec. 6.6 Programs for Framed Structures

**TABLE 6.4 Dynamic Load Data for Space Frames**

| Type of Data | No. of Lines | Items on Data Lines |
|---|---|---|
| Dynamic parameters | 1 | ISOLVE, NTS, DT, DAMPR |
| Initial conditions<br>(a) Condition parameters<br>(b) Displacements<br>(c) Velocities | 1<br>NNID<br>NNIV | NNID, NNIV<br>J, D0(6J-5), D0(6J-4), . . . , D0(6J)<br>J, V0(6J-5), V0(6J-4), . . . , V0(6J) |
| Applied actions<br>(a) Load parameters<br>(b) Nodal loads<br>(c) Line loads | 1<br>NLN<br>NEL | NLN, NEL<br>J, AS(6J-5), AS(6J-4), . . . , AS(6J)<br>I, BL1, BL2, . . . , BL6 |
| Ground accelerations<br>(a) Acceleration parameter<br>(b) Acceleration factors[a] | 1<br>1 | IGA<br>GAX, GAY, GAZ |
| Forcing function<br>(a) Function parameter<br>(b) Function ordinates | 1<br>NFO | NFO<br>K, T(K), FO(K) |

[a] Omit when IGA = 0.

as those for a space truss element. However, the equivalent nodal loads for a space frame element must be obtained by first resolving the line loads into components that are parallel to its axis and two principal directions. Using these components and the solutions from Probs. 3.4-1 and 3.4-10 (for axial and flexural elements), we can find equivalent nodal loads in member directions. Then the equivalent nodal loads in structural directions may be computed from Eq. (6.2-13). Finally, all three scale factors GAX, GAY, and GAZ must be given for ground accelerations.

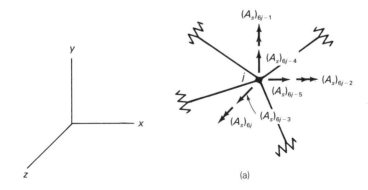

**Figure 6.18** Loads on space frame: (a) nodal loads; (b) line loads on element.

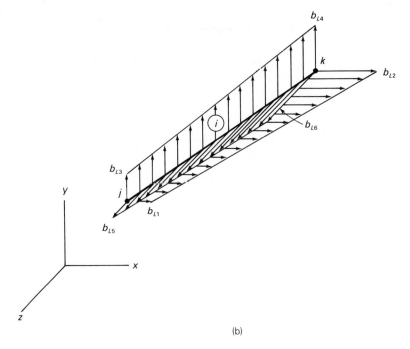

(b)

**Figure 6.18** (*cont.*)

## Example 6.9

The space frame in Fig. 6.19 has three prismatic magnesium elements with equal cross sections that are solid squares with sides 0.1 in. Also shown in the figure is a rigid-body ground acceleration $\ddot{D}_{g1}(t)$ in the $x$ direction. Physical constants for this example are

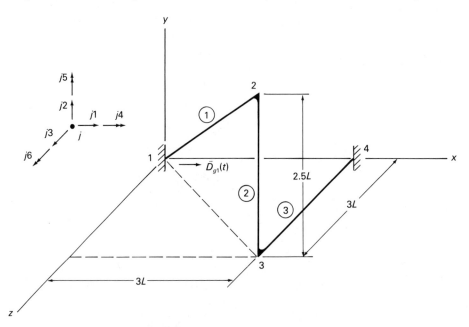

**Figure 6.19** Space frame with ground acceleration.

$E = 6.5 \times 10^3$ k/in.² $\quad G = 2.4 \times 10^3$ k/in.² $\quad \rho = 1.71 \times 10^{-7}$ k-s²/in.⁴
$L = 10$ in. $\quad A = 0.01$ in.² $\quad I_x = 2I_y = 2I_z = 1.667 \times 10^{-5}$ in.⁴
$(\ddot{D}_{g1})_{max} = 10$ in./s²

and the units are US.

Using this data, we ran Program DYNASF, with IP = 0 for all elements, DAMPR = 0.05, and Subprogram NORMOD for response calculations. The plots in Fig. 6.20 represent the forcing function for ground acceleration and time histories of the translational responses at node 2 in the $x$, $y$, and $z$ directions. Maximum (or minimum) values of $D_{j1}$, $D_{j2}$, and $D_{j3}$ are 0.07213 in., 0.04685 in., and $-0.09977$ in., respectively.

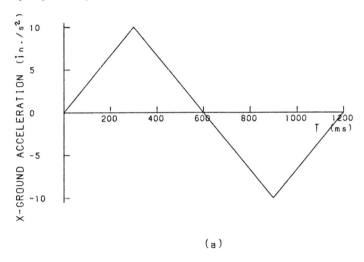

(a)

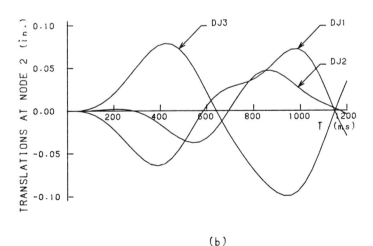

(b)

**Figure 6.20** Space frame: (a) ground acceleration; (b) responses.

## 6.7 GUYAN REDUCTION

The concept of *matrix condensation* [1,2] is a well-known procedure for reducing the number of unknown displacements in a statics problem. With such applications no loss of accuracy results from the reduction, because the method is simply Gaussian elimination of displacements in matrix form. For dynamic analysis, a similar type of condensation was introduced by Guyan [4], which brings in an additional approximation.

Starting with *static reduction*, we write action equations of equilibrium for free displacements in the partitioned form

$$\begin{bmatrix} \mathbf{S}_{AA} & \mathbf{S}_{AB} \\ \mathbf{S}_{BA} & \mathbf{S}_{BB} \end{bmatrix} \begin{bmatrix} \mathbf{D}_A \\ \mathbf{D}_B \end{bmatrix} = \begin{bmatrix} \mathbf{A}_A \\ \mathbf{A}_B \end{bmatrix} \tag{1}$$

Here the subscript $A$ denotes the displacements that are to be eliminated, and the subscript $B$ refers to those that will be retained. Now rewrite Eq. (1) as two sets of equations, as follows:

$$\mathbf{S}_{AA}\mathbf{D}_A + \mathbf{S}_{AB}\mathbf{D}_B = \mathbf{A}_A \tag{2a}$$

$$\mathbf{S}_{BA}\mathbf{D}_A + \mathbf{S}_{BB}\mathbf{D}_B = \mathbf{A}_B \tag{2b}$$

Solving for the vector of *dependent displacements* $\mathbf{D}_A$ in Eq. (2a) yields

$$\mathbf{D}_A = \mathbf{S}_{AA}^{-1}(\mathbf{A}_A - \mathbf{S}_{AB}\mathbf{D}_B) \tag{3}$$

Substitute Eq. (3) into Eq. (2b) and collect terms to obtain

$$\mathbf{S}_{BB}^{*}\mathbf{D}_B = \mathbf{A}_B^{*} \tag{4}$$

in which

$$\mathbf{S}_{BB}^{*} = \mathbf{S}_{BB} - \mathbf{S}_{BA}\mathbf{S}_{AA}^{-1}\mathbf{S}_{AB} \tag{5}$$

and

$$\mathbf{A}_B^{*} = \mathbf{A}_B - \mathbf{S}_{BA}\mathbf{S}_{AA}^{-1}\mathbf{A}_A \tag{6}$$

From Eq. (4) we see that Eqs. (2) have been reduced to a smaller set involving only the *independent displacements* in vector $\mathbf{D}_B$. The *reduced stiffness matrix* $\mathbf{S}_{BB}^{*}$ in Eq. (5) is a modified version of the original submatrix $\mathbf{S}_{BB}$. Also, the *reduced action vector* $\mathbf{A}_B^{*}$ in Eq. (6) contains terms modifying the subvector $\mathbf{A}_B$ that are considered to be equivalent loads of type $B$ due to actions of type $A$. Furthermore, Eq. (3) may now be viewed as the back-substitution formula required to find vector $\mathbf{D}_A$ *exactly* from vector $\mathbf{D}_B$.

Turning next to *dynamic reduction*, we recall from Sec. 4.4 that the undamped equations of motion for free displacements are

$$\mathbf{M}\ddot{\mathbf{D}} + \mathbf{S}\mathbf{D} = \mathbf{A} \tag{4.4-1}$$

Then assume as a new approximation that the displacements of type $A$ are

Sec. 6.7  Guyan Reduction

dependent on those of type $B$, as follows:

$$\mathbf{D}_A = \mathbf{T}_{AB}\mathbf{D}_B \tag{7a}$$

where

$$\mathbf{T}_{AB} = -\mathbf{S}_{AA}^{-1}\mathbf{S}_{AB} \tag{7b}$$

Even for static analysis, this relationship is correct only when actions of type $A$ do not exist [see Eq. (3)]. However, Eq. (7a) follows the finite-element theme of "slave" and "master" displacements. Differentiating Eq. (7a) twice with respect to time produces

$$\ddot{\mathbf{D}}_A = \mathbf{T}_{AB}\ddot{\mathbf{D}}_B \tag{8}$$

For the purpose of reducing the equations of motion to a smaller set, we can form the transformation operator

$$\mathbf{T}_B = \begin{bmatrix} \mathbf{T}_{AB} \\ \mathbf{I}_B \end{bmatrix} \tag{9}$$

in which $\mathbf{I}_B$ is an identity matrix of the same order as $\mathbf{S}_{BB}$. Substituting Eqs. (7a) and (8) into Eq. (4.4-1) and premultiplying the latter by $\mathbf{T}_B^T$ gives

$$\mathbf{M}_{BB}^*\ddot{\mathbf{D}}_B + \mathbf{S}_{BB}^*\mathbf{D}_B = \mathbf{A}_B^* \tag{10}$$

In this equation the matrices $\mathbf{S}_{BB}^*$ and $\mathbf{A}_B^*$ still have the definitions given in Eqs. (5) and (6). However, the *reduced mass matrix* $\mathbf{M}_{BB}^*$ is found to be

$$\mathbf{M}_{BB}^* = \mathbf{T}_B^T \mathbf{M}\, \mathbf{T}_B = \mathbf{M}_{BB} + \mathbf{T}_{AB}^T \mathbf{M}_{AB} + \mathbf{M}_{BA}\mathbf{T}_{AB} + \mathbf{T}_{AB}^T \mathbf{M}_{AA}\mathbf{T}_{AB} \tag{11}$$

As mentioned before, all of the condensed matrices in Eq. (10) are approximate. If damping is to be included in the equations of motion, a reduced damping matrix $\mathbf{C}_{BB}^*$ also can be derived, which has a form analogous to $\mathbf{M}_{BB}^*$ in Eq. (11).

When applying Guyan reduction to framed structures, we usually choose rotations at the joints of beams, plane frames, grids, and space frames as the dependent set of displacements. However, the method can be used in a much more general manner for various discretized continua. That is, any arbitrarily selected set of displacements may be referred to as type $A$, while the remaining displacements become type $B$. The trouble with this generality is that a good choice of "slave" and "master" displacements is not always obvious. Even with framed structures there are cases when joint rotations are more important than translations and should not be eliminated.

### Example 6.10

The fixed beam in Fig. 6.21(a) is divided into three flexural elements, each of which has the same properties $E$, $I$, $\rho$, and $A$. By Guyan reduction, we shall eliminate the rotations at nodes 2 and 3 and retain the translations. The reduction will be followed by a vibrational analysis that is compared against exact results.

For this example, the assembled structural stiffness matrix (without consideration of restraints) is

$$\mathbf{S}_s = \frac{2EI}{\ell^3} \begin{bmatrix} 6 & & & & & & & \\ 3\ell & 2\ell^2 & & & \text{①} & & \text{Sym.} & \\ -6 & -3\ell & 12 & & & & & \\ 3\ell & \ell^2 & 0 & 4\ell^2 & \text{②} & & & \\ 0 & 0 & -6 & -3\ell & 12 & & & \\ 0 & 0 & 3\ell & \ell^2 & 0 & 4\ell^2 & \text{③} & \\ 0 & 0 & 0 & 0 & -6 & -3\ell & 6 & \\ 0 & 0 & 0 & 0 & 3\ell & \ell^2 & -3\ell & 2\ell^2 \end{bmatrix} \begin{matrix} 1 \\ 2 \\ 3 \\ 4 \\ 5 \\ 6 \\ 7 \\ 8 \end{matrix} \quad \text{(a)}$$

$$\phantom{x}\qquad\qquad\quad 1 \quad\; 2 \quad\;\; 3 \quad\;\; 4 \quad\;\; 5 \quad\;\; 6 \quad\;\; 7 \quad\;\; 8$$

Similarly, the assembled structural mass matrix becomes

$$\mathbf{M}_s = \frac{\rho A \ell}{420} \begin{bmatrix} 156 & & & & & & & \\ 22\ell & 4\ell^2 & & & \text{①} & & \text{Sym.} & \\ 54 & 13\ell & 312 & & & & & \\ -13\ell & -3\ell^2 & 0 & 8\ell^2 & & \text{②} & & \\ 0 & 0 & 54 & 13\ell & 312 & & & \\ 0 & 0 & -13\ell & -3\ell^2 & 0 & 8\ell^2 & \text{③} & \\ 0 & 0 & 0 & 0 & 54 & 13\ell & 156 & \\ 0 & 0 & 0 & 0 & -13\ell & -3\ell^2 & -22\ell & 4\ell^2 \end{bmatrix} \begin{matrix} 1 \\ 2 \\ 3 \\ 4 \\ 5 \\ 6 \\ 7 \\ 8 \end{matrix}$$

$$\phantom{x}\qquad\qquad\quad 1 \quad\; 2 \quad\;\; 3 \quad\;\; 4 \quad\;\; 5 \quad\;\; 6 \quad\;\; 7 \quad\;\; 8$$

(b)

Dashed boxes in Eqs. (a) and (b) enclose the contributions of elements 1, 2, and 3, which are drawn from Eqs. (3.4-24) and (3.4-26). Moreover, the joint displacement indexes for the problem [see Fig. 6.21(a)] are listed at the right side and below the matrices.

As the first step, we remove the first, second, seventh, and eighth rows and columns from matrices $\mathbf{S}_s$ and $\mathbf{M}_s$, because displacements 1, 2, 7, and 8 are restrained by supports. Then the remaining 4 × 4 arrays are rearranged to put the rotational terms before the translational terms, as follows:

## Sec. 6.7 Guyan Reduction

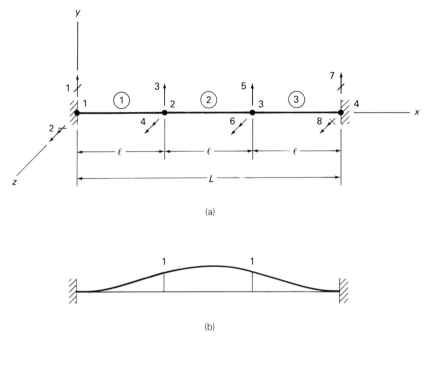

**Figure 6.21** (a) Three-element fixed beam; (b) mode 1; (c) mode 2.

$$\mathbf{S} = \begin{bmatrix} \mathbf{S}_{AA} & \mathbf{S}_{AB} \\ \mathbf{S}_{BA} & \mathbf{S}_{BB} \end{bmatrix} = \frac{2EI}{\ell^3} \begin{bmatrix} 4\ell^2 & & \text{Sym.} & \\ \ell^2 & 4\ell^2 & & \\ \hline 0 & 3\ell & 12 & \\ -3\ell & 0 & -6 & 12 \end{bmatrix} \begin{matrix} 4 \\ 6 \\ 3 \\ 5 \end{matrix} \quad \text{(c)}$$

$$\quad\quad\quad\quad\quad\quad\quad\quad\quad\quad\;\; 4 \quad\; 6 \quad\; 3 \quad\; 5$$

$$\mathbf{M} = \begin{bmatrix} \mathbf{M}_{AA} & \mathbf{M}_{AB} \\ \mathbf{M}_{BA} & \mathbf{M}_{BB} \end{bmatrix} = \frac{\rho A \ell}{420} \begin{bmatrix} 8\ell^2 & & \text{Sym.} & \\ -3\ell^2 & 8\ell^2 & & \\ \hline 0 & -13\ell & 312 & \\ 13\ell & 0 & 54 & 312 \end{bmatrix} \begin{matrix} 4 \\ 6 \\ 3 \\ 5 \end{matrix} \quad \text{(d)}$$
$$\phantom{M}\quad\quad\quad\quad\quad\quad\quad\quad\quad 4 \quad\ 6 \quad\ 3 \quad 5$$

From Eq. (c) the inverse of $\mathbf{S}_{AA}$ is

$$\mathbf{S}_{AA}^{-1} = \frac{\ell}{30EI}\begin{bmatrix} 4 & -1 \\ -1 & 4 \end{bmatrix} \quad\text{(e)}$$

Substituting this array and the other submatrices of $\mathbf{S}$ from Eq. (c) into Eq. (5) produces

$$\mathbf{S}_{BB}^{*} = \frac{12EI}{\ell^3}\begin{bmatrix} 2 & -1 \\ -1 & 2 \end{bmatrix} - \frac{6EI}{5\ell^3}\begin{bmatrix} 4 & 1 \\ 1 & 4 \end{bmatrix} = \frac{6EI}{5\ell^3}\begin{bmatrix} 16 & -11 \\ -11 & 16 \end{bmatrix} \begin{matrix}3\\5\end{matrix} \quad\text{(f)}$$
$$\phantom{xxxxxxxxxxxxxxxxxxxxxxxxxxxxxxxxxxxxxxxxxxxxxxx} 3 \quad\ 5$$

which is the reduced stiffness matrix.

To reduce the mass matrix, we form the transformation matrix $\mathbf{T}_B$ by evaluating $\mathbf{T}_{AB}$ in Eq. (7b) and substituting it into Eq. (9). Thus,

$$\mathbf{T}_{AB} = -\mathbf{S}_{AA}^{-1}\mathbf{S}_{AB} = \frac{1}{5\ell}\begin{bmatrix} 1 & 4 \\ -4 & -1 \end{bmatrix} \quad\text{(g)}$$

and

$$\mathbf{T}_B = \begin{bmatrix} \mathbf{T}_{AB} \\ \mathbf{I}_B \end{bmatrix} = \frac{1}{5\ell}\begin{bmatrix} 1 & 4 \\ -4 & -1 \\ \hline 5\ell & 0 \\ 0 & 5\ell \end{bmatrix} \quad\text{(h)}$$

In this case the submatrix $\mathbf{I}_B$ in the lower partition of matrix $\mathbf{T}_B$ is of order 2 because there are two remaining translational displacements (numbers 3 and 5). Substitution of matrix $\mathbf{M}$ from Eq. (d) and $\mathbf{T}_B$ from Eq. (h) into Eq. (11) yields

$$\mathbf{M}_{BB}^{*} = \frac{\rho A \ell}{2100}\begin{bmatrix} 1696 & 319 \\ 319 & 1696 \end{bmatrix}\begin{matrix}3\\5\end{matrix} \quad\text{(i)}$$
$$\phantom{xxxxxxxxxxxxxxxxxxxxxxxxxxx} 3 \quad\ \ 5$$

which is the reduced mass matrix.

Using the terms in matrices $\mathbf{S}_{BB}^{*}$ and $\mathbf{M}_{BB}^{*}$, we can set up the eigenvalue problem in the form of Eq. (3.6-4). The angular frequencies found by this method are

$$\omega_{1,2} = 22.51,\ 63.26\ \frac{1}{L^2}\sqrt{\frac{EI}{\rho A}} \quad\text{(j)}$$

## Sec. 6.7  Guyan Reduction

where $L = 3\ell$. These values are in error by $+0.63\%$ and $+2.6\%$, respectively; and they constitute upper bounds of the exact angular frequencies [5]. The corresponding mode shapes are

$$\Phi = [\Phi_1 \quad \Phi_2] = \begin{bmatrix} 1 & 1 \\ 1 & -1 \end{bmatrix} \tag{k}$$

which appear in Figs. 6.21(b) and (c).

*Frequency coefficients* $\mu_i$ for prismatic beams with various end conditions are summarized in Table 6.5. In each case the beam is modeled by four flexural elements, and the results for the consistent-mass approach (with and without elimination of rotations) are compared with those for the lumped-mass method (with elimination of rotations). The table shows that the consistent-mass model produces much better accuracy than the lumped-mass model in beam analysis.

### Example 6.11

Figure 6.22(a) shows a rectangular plane frame that has the same properties $E$, $I_z$, $\rho$, and $A$ for each of its three members. Let us find approximations to the fundamental angular frequency for this frame with and without elimination of the rotations at joints 1 and 2. To simplify the analysis, we shall omit axial strains in the members, leaving only the three degrees of freedom indicated in the figure.

The $3 \times 3$ stiffness matrix for unit values of $D_1$, $D_2$, and $D_3$ has the form

$$\mathbf{S} = \begin{bmatrix} \mathbf{S}_{AA} & \mathbf{S}_{AB} \\ \mathbf{S}_{BA} & \mathbf{S}_{BB} \end{bmatrix} = \frac{2EI}{L^3} \begin{bmatrix} 4L^2 & & \text{Sym.} \\ L^2 & 4L^2 & \\ \hline 3L & 3L & 12 \end{bmatrix} \tag{$\ell$}$$

and the accompanying $3 \times 3$ mass matrix is

$$\mathbf{M} = \begin{bmatrix} \mathbf{M}_{AA} & \mathbf{M}_{AB} \\ \mathbf{M}_{BA} & \mathbf{M}_{BB} \end{bmatrix} = \frac{\rho AL}{420} \begin{bmatrix} 8L^2 & & \text{Sym.} \\ -3L^2 & 8L^2 & \\ \hline 22L & 22L & 732 \end{bmatrix} \tag{m}$$

in which $\mathbf{M}_{BB}$ includes $\rho AL$ for the mass of member 1. With these arrays we can set up and solve the eigenvalue problem, which yields the angular frequencies

$$\omega_{1,2,3} = 3.201,\ 15.14,\ 32.68 \frac{1}{L^2} \sqrt{\frac{EI}{\rho A}} \tag{n}$$

Also, the corresponding mode shapes are

$$\Phi = [\Phi_1 \quad \Phi_2 \quad \Phi_3] = \begin{bmatrix} -0.5528 & 1 & 1 \\ -0.5528 & -1 & 1 \\ L & 0 & -0.05436L \end{bmatrix} \tag{o}$$

The shape of mode 1 is illustrated in Fig. 6.22(b).

**TABLE 6.5 Frequency Coefficients $\mu_i$ for Prismatic Beams Modeled by Four Elements**

| Support Conditions | | Mode | Exact (5) | CM–TR | % Error | CM–TO | % Error | LM–TO | % Error |
|---|---|---|---|---|---|---|---|---|---|
| Simple | | 1 | 9.870 | 9.872 | +0.020 | 9.873 | +0.030 | 9.867 | −0.030 |
| | | 2 | 39.48 | 39.63 | +0.38 | 39.76 | +0.71 | 39.19 | −0.73 |
| | | 3 | 88.23 | 90.45 | +2.5 | 94.03 | +6.6 | 83.21 | −5.7 |
| Free | | 1 | 22.37 | 22.41 | +0.18 | 22.46 | +0.40 | 18.91 | −15 |
| | | 2 | 61.67 | 62.06 | +0.63 | 63.12 | −2.4 | 48.00 | −22 |
| | | 3 | 120.9 | 121.9 | +0.83 | 122.4 | −1.2 | 86.84 | −28 |
| Fixed | | 1 | 22.37 | 22.40 | +0.13 | 22.41 | +0.18 | 22.30 | −0.31 |
| | | 2 | 61.67 | 62.24 | +0.92 | 62.77 | +1.8 | 59.25 | −3.9 |
| | | 3 | 120.9 | 123.5 | +2.2 | 124.8 | +3.2 | 97.40 | −19 |
| Cantilever | | 1 | 3.516 | 3.516 | +0.00 | 3.516 | +0.00 | 3.418 | −2.8 |
| | | 2 | 22.03 | 22.06 | +0.14 | 22.09 | +0.27 | 20.09 | −8.8 |
| | | 3 | 61.70 | 62.18 | +0.83 | 62.97 | +2.1 | 53.20 | −14 |
| Propped | | 1 | 15.42 | 15.43 | +0.065 | 15.43 | +0.065 | 15.40 | −0.13 |
| | | 2 | 49.97 | 50.28 | +0.62 | 50.56 | +1.2 | 49.05 | −1.8 |
| | | 3 | 104.2 | 106.6 | +2.3 | 110.5 | +6.0 | 91.53 | −12 |

CM: Consistent Masses
LM: Lumped Masses
TR: Translations and Rotations
TO: Translations Only

$$\omega_i = \frac{\mu_i}{L^2}\sqrt{\frac{EI}{\rho A}}$$

Sec. 6.7  Guyan Reduction

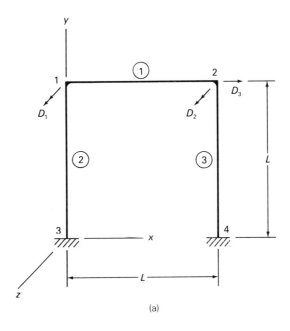

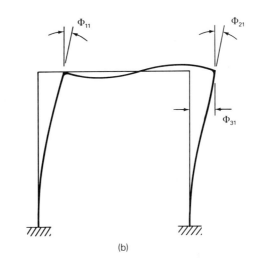

**Figure 6.22**  (a) Plane frame; (b) mode 1.

In preparation for eliminating the rotations, we compute the inverse of submatrix $S_{AA}$ to obtain

$$S_{AA}^{-1} = \frac{L}{30EI} \begin{bmatrix} 4 & -1 \\ -1 & 4 \end{bmatrix} \qquad (p)$$

Then we calculate the reduced stiffness matrix from Eq. (5), as follows:

$$S_{BB}^* = \frac{24EI}{L^3} - \frac{36EI}{5L^3} = \frac{84EI}{5L^3} \qquad (q)$$

which is just a single term. Next, the arrays $\mathbf{T}_{AB}$ and $\mathbf{T}_B$ are found to be

$$\mathbf{T}_{AB} = -\mathbf{S}_{AA}^{-1}\mathbf{S}_{AB} = -\frac{3}{5L}\begin{bmatrix}1\\1\end{bmatrix} \qquad (r)$$

and

$$\mathbf{T}_B = \begin{bmatrix}\mathbf{T}_{AB}\\ \mathbf{I}_B\end{bmatrix} = -\frac{1}{5L}\begin{bmatrix}3\\3\\\hline -5L\end{bmatrix} \qquad (s)$$

Using the latter operator in Eq. (11), we can reduce the mass matrix to the single term

$$\mathbf{M}_{BB}^* = \mathbf{T}_B^T \mathbf{M} \mathbf{T}_B = \frac{569}{350}\rho AL \qquad (t)$$

Now the eigenvalue problem gives

$$\omega_1 = 3.215 \frac{1}{L^2}\sqrt{\frac{EI}{\rho A}} \qquad (u)$$

Comparing this angular frequency against $\omega_1$ for the unreduced 3-DOF problem, we find the relative error to be $+0.16\%$.

## 6.8 CONSTRAINTS AGAINST AXIAL STRAINS

In Example 6.11 we omitted axial strains without difficulty because members of the plane frame were perpendicular to each other. However, for frames with members oriented arbitrarily in a plane or in space, the matter of neglecting axial strains is not so simple. In this section we will introduce axial constraints in plane and space frames, primarily for the purpose of reducing the number of degrees of freedom in dynamic analysis. The number of constraints introduced will be equal to the number of members $m$ for which axial strains are to be neglected. By automatically selecting $m$ of the joint translations to be dependent on the rest of them, we can devise a method for reducing the number of independent equations of motion. This reduction method may be combined with Guyan elimination of joint rotations, which was described in the preceding section. We assume that the frame to be analyzed is *underconstrained*, so that no complications from *redundant constraints* (or superfluous members) need be considered [6].

The axial constraint condition for zero elongation of a typical member $i$ in a plane frame [see Fig. 6.2(b)] may be stated as

$$(D_{j1} - D_{k1})c_x + (D_{j2} - D_{k2})c_y = 0 \qquad (1)$$

## Sec. 6.8  Constraints Against Axial Strains

In this equation the symbols $D_{j1}$ and $D_{j2}$ represent the $x$ and $y$ translations of joint $j$, and $D_{k1}$ and $D_{k2}$ are those at joint $k$. As before, the $x$ and $y$ direction cosines of the axis of the member are denoted by $c_x$ and $c_y$.

Similarly, we write the axial constraint condition for zero elongation of a space frame member [see Fig. 6.5(b)] in the following manner:

$$(D_{j1} - D_{k1})c_x + (D_{j2} - D_{k2})c_y + (D_{j3} - D_{k3})c_z = 0 \tag{2}$$

Here $D_{j3}$ and $D_{k3}$ are translations of joints $j$ and $k$ in the $z$ direction, and $c_z$ is the $z$ direction cosine of the member axis.

Assembling these constraint conditions into a matrix format for either a plane or a space frame gives

$$\mathbf{C} \, \mathbf{D}_t = \mathbf{0} \tag{3}$$

The matrix $\mathbf{C}$ in this expression is called the *constraint matrix*, which contains only positive and negative values of direction cosines for the constrained members. This array is of size $m \times n_t$, where $n_t$ is the number of joint translations. The vector $\mathbf{D}_t$ in Eq. (3) consists of only free joint translations, because no joint rotations are involved at this stage.

Due to the constraint conditions, some of the joint translations in the vector $\mathbf{D}_t$ will be linearly dependent on others. To determine which translations are dependent and which are independent, it is necessary to investigate the rank of matrix $\mathbf{C}$ in a systematic fashion. The rank $r$ and the basis (or vector space) of $\mathbf{C}$ are found using *Gauss–Jordan elimination* with pivoting [7]. Although the rank of a matrix is unique, the basis is not. Therefore, the choice of dependent translations is arbitrary, and pivoting automatically produces the best selection. Thus, it becomes possible (in retrospect) to partition the matrices in Eq. (3) as follows:

$$[\mathbf{C}_{11} \quad \mathbf{C}_{12}] \begin{bmatrix} \mathbf{D}_1 \\ \mathbf{D}_2 \end{bmatrix} = \mathbf{0} \tag{4}$$

In this expanded form the vector $\mathbf{D}_1$ represents $r$ dependent translations, and $\mathbf{D}_2$ contains the remaining $n_i$ independent translations. Because no redundant constraints are included, the rank $r$ of matrix $\mathbf{C}$ will always be equal to the number of members $m$. Therefore, submatrix $\mathbf{C}_{11}$ in Eq. (4) is a square array of size $m \times m$, and submatrix $\mathbf{C}_{12}$ is of size $m \times n_i$.

Multiplying the matrices on the left-hand side of Eq. (4) produces

$$\mathbf{C}_{11} \mathbf{D}_1 + \mathbf{C}_{12} \mathbf{D}_2 = \mathbf{0} \tag{5}$$

Knowing that matrix $\mathbf{C}_{11}$ is square and nonsingular, we can solve for vector $\mathbf{D}_1$ in terms of vector $\mathbf{D}_2$. Thus,

$$\mathbf{D}_1 = \bar{\mathbf{T}}_{12} \mathbf{D}_2 \tag{6}$$

in which the operator $\mathbf{T}_{12}$ is

$$\mathbf{T}_{12} = -\mathbf{C}_{11}^{-1} \mathbf{C}_{12} \tag{7}$$

During the Gauss–Jordan elimination process, the matrix $\mathbf{C}_{11}$ is replaced by an identity matrix; and $\mathbf{C}_{12}$ is replaced by $-\mathbf{T}_{12}$. If the operations are also applied to an identity matrix $\mathbf{I}_m$ of order $m$, it will be replaced by $\mathbf{C}_{11}^{-1}$. We now define this inverse to be the operator

$$\mathbf{T}_{11} = \mathbf{C}_{11}^{-1} \tag{8}$$

and the augmented constraint matrix is

$$\mathbf{C}_I = [\mathbf{C} \quad \mathbf{I}_m] \tag{9}$$

To confirm ideas regarding the constraint matrix, let us consider the plane frame in Fig. 6.23, for which the member information is given in Table 6.6. The augmented constraint matrix [see Eq. (9)] for this case is

$$\mathbf{C}_I = \begin{bmatrix} \overset{1}{0.707} & \overset{2}{0.707} & \overset{4}{-0.707} & \overset{5}{-0.707} & 1.000 & 0 & 0 \\ 0 & -1.000 & 0 & 0 & 0 & 1.000 & 0 \\ 0 & 0 & 0.447 & -0.894 & 0 & 0 & 1.000 \end{bmatrix} \quad \text{(a)}$$

The numbers above the columns of $\mathbf{C}$ indicate the translational displacements

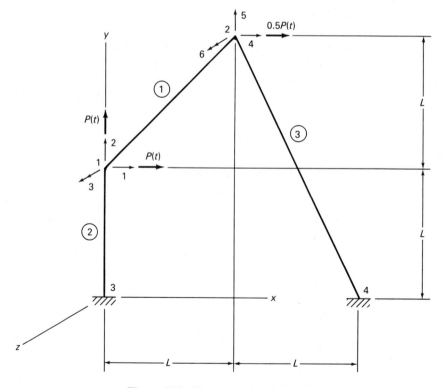

**Figure 6.23** Three-member plane frame.

Sec. 6.8  Constraints Against Axial Strains    293

TABLE 6.6 Member Information for Three-Member Plane Frame

| Member | Joint $j$ | Joint $k$ | $c_x$ | $c_y$ |
|---|---|---|---|---|
| 1 | 1 | 2 | 0.7071 | 0.7071 |
| 2 | 3 | 1 | 0 | 1.000 |
| 3 | 2 | 4 | 0.4472 | −0.8944 |

shown in Fig. 6.18. After the Gauss–Jordan procedure is applied, the matrix in Eq. (a) becomes

$$\mathbf{C}'_I = \begin{bmatrix} \overset{1}{1.000} & \overset{2}{0} & \overset{4}{-1.500} & \overset{5}{0} & 1.414 & 1.000 & -1.118 \\ 0 & 1.000 & 0 & 0 & 0 & -1.000 & 0 \\ 0 & 0 & -0.500 & 1.000 & 0 & 0 & -1.118 \end{bmatrix} \quad (b)$$

Therefore,

$$\mathbf{T}_{11} = \mathbf{C}_{11}^{-1} = \begin{bmatrix} 1.414 & 1.000 & -1.118 \\ 0 & -1.000 & 0 \\ 0 & 0 & -1.118 \end{bmatrix} \quad (c)$$

and

$$\mathbf{T}_{12} = -\mathbf{C}_{11}^{-1}\mathbf{C}_{12} = \overset{4}{\begin{bmatrix} 1.500 \\ 0 \\ 0.500 \end{bmatrix}} \quad (d)$$

Thus, the dependent translations are found to be 1, 2, and 5, whereas the independent translation is automatically chosen to be 4. Note that there is no need to rearrange the augmented constraint matrix during this procedure.

In preparation for a coordinate transformation associated with axial constraints, we shall define a generalized displacement vector $\overline{\mathbf{D}}$, as follows:

$$\overline{\mathbf{D}} = \begin{bmatrix} \mathbf{0} \\ \mathbf{D}_2 \end{bmatrix} \quad (10)$$

The first part of $\overline{\mathbf{D}}$ consists of a null vector, representing member elongations (which are zero); and the second part contains the independent translations $\mathbf{D}_2$. To relate the vector $\overline{\mathbf{D}}$ to $\mathbf{D}_t$, we write

$$\overline{\mathbf{D}} = \mathbf{T}_I \mathbf{D}_t \quad (11a)$$

or

$$\begin{bmatrix} \mathbf{0} \\ \mathbf{D}_2 \end{bmatrix} = \begin{bmatrix} \mathbf{C}_{11} & \mathbf{C}_{12} \\ \mathbf{0} & \mathbf{I}_2 \end{bmatrix} \begin{bmatrix} \mathbf{D}_1 \\ \mathbf{D}_2 \end{bmatrix} \quad (11b)$$

In Eq. (11b) the upper part represents the constraint conditions, and the lower part merely reproduces $\mathbf{D}_2$. Because the generalized displacements are independent and constitute a complete set, there is also an inverse relationship in the form

$$\mathbf{D}_t = \mathbf{T}_C \overline{\mathbf{D}} \tag{12}$$

where

$$\mathbf{T}_C = \mathbf{T}_I^{-1} = \begin{bmatrix} \mathbf{C}_{11}^{-1} & -\mathbf{C}_{11}^{-1}\mathbf{C}_{12} \\ 0 & \mathbf{I}_2 \end{bmatrix} = \begin{bmatrix} \mathbf{T}_{11} & \mathbf{T}_{12} \\ 0 & \mathbf{I}_2 \end{bmatrix} \tag{13}$$

Matrix $\mathbf{T}_C$ is an operator that can be used to transform action equations from the original displacement coordinates to the generalized displacement coordinates for axial constraints. Notice that submatrices $\mathbf{T}_{11}$ and $\mathbf{T}_{12}$ are generated automatically in the Gauss–Jordan procedure described previously.

Let us now restate the action equations of undamped motion for free displacements at the joints of a plane or space frame as

$$\mathbf{M}\ddot{\mathbf{D}} + \mathbf{S}\mathbf{D} = \mathbf{A} \tag{4.4-1}$$

By segregating rotations from translations, we can write this equation in the expanded form

$$\begin{bmatrix} \mathbf{M}_{rr} & \mathbf{M}_{rt} \\ \mathbf{M}_{tr} & \mathbf{M}_{tt} \end{bmatrix} \begin{bmatrix} \ddot{\mathbf{D}}_r \\ \ddot{\mathbf{D}}_t \end{bmatrix} + \begin{bmatrix} \mathbf{S}_{rr} & \mathbf{S}_{rt} \\ \mathbf{S}_{tr} & \mathbf{S}_{tt} \end{bmatrix} \begin{bmatrix} \mathbf{D}_r \\ \mathbf{D}_t \end{bmatrix} = \begin{bmatrix} \mathbf{A}_r \\ \mathbf{A}_t \end{bmatrix} \tag{14}$$

in which the subscripts $r$ and $t$ denote rotations and translations, respectively. Using Guyan reduction from Sec. 6.7, we eliminate the rotations in Eq. (14) and retain the translations, yielding

$$\mathbf{M}_{tt}^* \ddot{\mathbf{D}}_t + \mathbf{S}_{tt}^* \mathbf{D}_t = \mathbf{A}_t^* \tag{15}$$

where

$$\mathbf{S}_{tt}^* = \mathbf{S}_{tt} + \mathbf{T}_{rt}^T \mathbf{S}_{rt} \tag{16}$$

$$\mathbf{A}_t^* = \mathbf{A}_t + \mathbf{T}_{rt}^T \mathbf{A}_r \tag{17}$$

$$\mathbf{M}_{tt}^* = \mathbf{M}_{tt} + \mathbf{T}_{rt}^T \mathbf{M}_{rt} + \mathbf{M}_{tr} \mathbf{T}_{rt} + \mathbf{T}_{rt}^T \mathbf{M}_{rr} \mathbf{T}_{rt} \tag{18}$$

The transformation operator $\mathbf{T}_{rt}$ in these expressions relates the rotations $\mathbf{D}_r$ to the translations $\mathbf{D}_t$, as follows:

$$\mathbf{D}_r = \mathbf{T}_{rt} \mathbf{D}_t \tag{19}$$

where

$$\mathbf{T}_{rt} = -\mathbf{S}_{rr}^{-1} \mathbf{S}_{rt} \tag{20}$$

The operator $\mathbf{T}_C$ [see Eq. (13)] can now be used to transform the reduced equations of motion in Eq. (15) to the generalized displacement coordinates $\overline{\mathbf{D}}$ in Eq. (10). Because axial constraints are to be imposed, we shall omit axial

## Sec. 6.8 Constraints Against Axial Strains

stiffnesses of members from the structural stiffness matrix. Then to satisfy equilibrium at the joints, we must revise Eq. (15) to become

$$\mathbf{M}_{tt}^* \ddot{\mathbf{D}}_t + \mathbf{S}_{tt}^* \mathbf{D}_t = \mathbf{A}_t^* + \mathbf{C}^T \mathbf{Q} \tag{21}$$

in which $\mathbf{Q}$ is a vector of axial forces in the constrained members.

To perform the transformation, we substitute Eq. (12) and its second derivative with respect to time into Eq. (21) and premultiply by $\mathbf{T}_C^T$ to obtain

$$\overline{\mathbf{M}} \ddot{\mathbf{D}} + \overline{\mathbf{S}} \mathbf{D} = \overline{\mathbf{A}} + \overline{\mathbf{C}} \mathbf{Q} \tag{22}$$

Writing this equation in expanded form, we have

$$\begin{bmatrix} \overline{\mathbf{M}}_{11} & \overline{\mathbf{M}}_{12} \\ \overline{\mathbf{M}}_{21} & \overline{\mathbf{M}}_{22} \end{bmatrix} \begin{bmatrix} \mathbf{0} \\ \ddot{\mathbf{D}}_2 \end{bmatrix} + \begin{bmatrix} \overline{\mathbf{S}}_{11} & \overline{\mathbf{S}}_{12} \\ \overline{\mathbf{S}}_{21} & \overline{\mathbf{S}}_{22} \end{bmatrix} \begin{bmatrix} \mathbf{0} \\ \mathbf{D}_2 \end{bmatrix} = \begin{bmatrix} \overline{\mathbf{A}}_1 \\ \overline{\mathbf{A}}_2 \end{bmatrix} + \begin{bmatrix} \mathbf{I}_1 \\ \mathbf{0} \end{bmatrix} \mathbf{Q} \tag{23}$$

Multiplying terms and rearranging the results produces

$$\overline{\mathbf{M}}_{22} \ddot{\mathbf{D}}_2 + \overline{\mathbf{S}}_{22} \mathbf{D}_2 = \overline{\mathbf{A}}_2 \tag{24}$$

and

$$\mathbf{Q} = \overline{\mathbf{M}}_{12} \ddot{\mathbf{D}}_2 + \overline{\mathbf{S}}_{12} \mathbf{D}_2 - \overline{\mathbf{A}}_1 \tag{25}$$

The barred matrices in these equations have the following definitions:

$$\overline{\mathbf{S}}_{12} = \mathbf{T}_{11}^T \mathbf{S}_{11}^* \mathbf{T}_{12} + \mathbf{T}_{11}^T \mathbf{S}_{12}^* \tag{26}$$

$$\overline{\mathbf{S}}_{22} = \mathbf{T}_{12}^T \mathbf{S}_{11}^* \mathbf{T}_{12} + \mathbf{T}_{12}^T \mathbf{S}_{12}^* + \mathbf{S}_{21}^* \mathbf{T}_{12} + \mathbf{S}_{22}^* \tag{27}$$

$$\overline{\mathbf{A}}_1 = \mathbf{T}_{11}^T \mathbf{A}_1^* \tag{28}$$

$$\overline{\mathbf{A}}_2 = \mathbf{T}_{12}^T \mathbf{A}_1^* + \mathbf{A}_2^* \tag{29}$$

$$\overline{\mathbf{M}}_{12} = \mathbf{T}_{11}^T \mathbf{M}_{11}^* \mathbf{T}_{12} + \mathbf{T}_{11}^T \mathbf{M}_{12}^* \tag{30}$$

$$\overline{\mathbf{M}}_{22} = \mathbf{T}_{12}^T \mathbf{M}_{11}^* \mathbf{T}_{12} + \mathbf{T}_{12}^T \mathbf{M}_{12}^* + \mathbf{M}_{21}^* \mathbf{T}_{12} + \mathbf{M}_{22}^* \tag{31}$$

Equation (24) represents a doubly reduced set of equations of motion that can be solved for the dynamic response of the independent translations $\mathbf{D}_2$. Then the vector $\mathbf{D}_2$ and its second derivative $\ddot{\mathbf{D}}_2$ may be substituted into Eq. (25) to determine the vector of axial forces $\mathbf{Q}$ in the members. Next, dependent translations $\mathbf{D}_1$ can be obtained from Eq. (6), and the rotations $\mathbf{D}_r$ are found using Eq. (19). Finally, other internal actions and support reactions may be calculated from known relationships.

If a damping matrix is included in the equations of motion, its reduction is similar to that of the consistent mass matrix. Other topics that could be considered in this section are redundant constraints and nonzero length changes of members. However, these subjects are more complicated and of less interest than the matter of axial constraints, as discussed here.

Whenever axial strains are omitted from analyses of plane or space frames, a loss of accuracy is bound to occur. The significance of such discrepancies will vary from one problem to another. However, for most practical underconstrained frames, the loss of accuracy due to introducing axial constraints is likely to be negligible, except in the columns of tall buildings [8] and similar structures. Moreover, the numerical problem of ill conditioning due to combining large axial stiffnesses with small flexural and torsional stiffnesses is completely avoided. Of course, when the members in a frame are perpendicular to each other (as in Example 6.11), omission of axial strains is easily accomplished without the formal procedure of this section.

**Example 6.12**

For the three-member plane frame in Fig. 6.23, let us first set up the stiffness, mass, and load matrices for the six degrees of freedom shown. Second, we shall use Guyan reduction to eliminate the two joint rotations and retain the four translations. Third, by imposing axial constraints we will eliminate three dependent translations and keep the best single independent translation, which was found to be displacement number 4. Last, we shall calculate the response of the reduced system to a particular set of forcing functions.

From Eq. (6.2-14), we determine member stiffnesses (without axial terms), assemble them, and rearrange the results to produce the following submatrices of the structural stiffness matrix:

$$\mathbf{S}_{rr} = \frac{EI_z}{L} \begin{bmatrix} 6.828 & 1.414 \\ 1.414 & 4.617 \end{bmatrix} \begin{matrix} 3 \\ 6 \end{matrix} \quad \begin{matrix} 3 & 6 \end{matrix} \tag{e}$$

$$\mathbf{S}_{rt} = \frac{EI_z}{L^2} \begin{bmatrix} 3.879 & 2.121 & 2.121 & -2.121 \\ -2.121 & 2.121 & 3.195 & -1.585 \end{bmatrix} \begin{matrix} 3 \\ 6 \end{matrix} \quad \begin{matrix} 1 & 2 & 4 & 5 \end{matrix} \tag{f}$$

$$\mathbf{S}_{tt} = \frac{EI_z}{L^3} \begin{bmatrix} 14.12 & & & \text{Sym.} \\ -2.121 & 2.121 & & \\ -2.121 & 2.121 & 2.980 & \\ 2.121 & -2.121 & -1.692 & 2.336 \end{bmatrix} \begin{matrix} 1 \\ 2 \\ 4 \\ 5 \end{matrix} \quad \begin{matrix} 1 & 2 & 4 & 5 \end{matrix} \tag{g}$$

Similarly, consistent mass matrices for the members are drawn from Eqs. (6.2-15), assembled, and rearranged to give

$$\mathbf{M}_{rr} = \frac{\rho A L^3}{420} \begin{bmatrix} 8 & -3 \\ -3 & 8 \end{bmatrix} \begin{matrix} 3 \\ 6 \end{matrix} \quad \begin{matrix} 3 & 6 \end{matrix} \tag{h}$$

$$\mathbf{M}_{rt} = \frac{\rho A L^2}{420} \begin{bmatrix} 6.444 & 15.56 & -9.192 & 9.192 \\ 9.192 & -9.192 & 35.24 & -5.712 \end{bmatrix} \begin{matrix} 3 \\ 6 \end{matrix} \quad \begin{matrix} 1 & 2 & 4 & 5 \end{matrix} \tag{i}$$

Sec. 6.8   Constraints Against Axial Strains

$$\mathbf{M}_{tt} = \frac{\rho A L}{420} \begin{bmatrix} 304 & & & \text{Sym.} \\ -8 & 288 & & \\ 62 & 8 & 300.8 & \\ 8 & 62 & -1.6 & 291.2 \end{bmatrix} \begin{matrix} 1 \\ 2 \\ 4 \\ 5 \end{matrix} \quad \begin{matrix} 1 & 2 & 4 & 5 \end{matrix} \tag{j}$$

From Fig. 6.23, we see that the parts of the rearranged load vector are

$$\mathbf{A}_r = \{0, 0\} \qquad \mathbf{A}_t = \{1, 1, 0.5, 0\}\, P(t) \tag{k}$$

For the purpose of applying Guyan reduction, we calculate the inverse of submatrix $\mathbf{S}_{rr}$ as

$$\mathbf{S}_{rr}^{-1} = \frac{L}{EI_z} \begin{bmatrix} 0.1564 & -0.04790 \\ -0.04790 & 0.2313 \end{bmatrix} \tag{$\ell$}$$

Then from Eq. (20) the operator $\mathbf{T}_{rt}$ becomes

$$\mathbf{T}_{rt} = -\mathbf{S}_{rr}^{-1} \mathbf{S}_{rt}$$

$$= \frac{1}{L} \begin{bmatrix} -0.7083 & -0.2301 & -0.1787 & 0.2558 \\ 0.6764 & -0.3890 & -0.6374 & 0.2650 \end{bmatrix} \begin{matrix} 3 \\ 6 \end{matrix} \quad \begin{matrix} 1 & 2 & 4 & 5 \end{matrix} \tag{m}$$

Using Eq. (16), we find the reduced stiffness matrix $\mathbf{S}_{tt}^*$ to be

$$\mathbf{S}_{tt}^* = \frac{EI_z}{L^3} \begin{bmatrix} 9.940 & & & \text{Sym.} \\ -2.189 & 0.808 & & \\ -1.463 & 0.391 & 0.565 & \\ 2.552 & -1.017 & -0.303 & 1.374 \end{bmatrix} \begin{matrix} 1 \\ 2 \\ 4 \\ 5 \end{matrix} \quad \begin{matrix} 1 & 2 & 4 & 5 \end{matrix} \tag{n}$$

In addition, Eq. (17) yields

$$\mathbf{A}_t^* = \{1, 1, 0.5, 0\}\, P(t) \tag{o}$$

and Eq. (18) gives

$$\mathbf{M}_{tt}^* = \frac{\rho A L}{420} \begin{bmatrix} 317.9 & & & \text{Sym.} \\ -31.46 & 289.1 & & \\ 81.91 & 1.152 & 262.0 & \\ 1.733 & 62.84 & 6.302 & 293.6 \end{bmatrix} \begin{matrix} 1 \\ 2 \\ 4 \\ 5 \end{matrix} \quad \begin{matrix} 1 & 2 & 4 & 5 \end{matrix} \tag{p}$$

Now we shall further reduce the stiffness, load, and mass matrices to account for axial constraints on the members. By rearranging and partitioning the stiffness matrix in Eq. (n) we find that

$$\mathbf{S}_{11}^* = \frac{EI_z}{L^3} \begin{bmatrix} \overset{1}{9.940} & \overset{2}{} & \overset{5}{\text{Sym.}} \\ -2.189 & 0.808 & \\ 2.552 & -1.017 & 1.374 \end{bmatrix} \begin{matrix} 1 \\ 2 \\ 5 \end{matrix} \qquad (q)$$

$$\mathbf{S}_{21}^* = \frac{EI_z}{L^3}[\overset{1}{-1.463} \quad \overset{2}{0.391} \quad \overset{5}{-0.303}] \; 4 = \mathbf{S}_{12}^{*T} \qquad (r)$$

$$\mathbf{S}_{22}^* = \frac{EI_z}{L^3}[0.565] \; \overset{4}{} 4 \qquad (s)$$

the last of which is just a single term. Similarly, for the load vector in Eq. (o) we have

$$\mathbf{A}_1^* = \{1, 1, 0\}P(t) \qquad \mathbf{A}_2^* = 0.5 P(t) \qquad (t)$$

Furthermore, submatrices of the mass matrix in Eq. (p) are

$$\mathbf{M}_{11}^* = \frac{\rho AL}{420} \begin{bmatrix} \overset{1}{317.9} & \overset{2}{} & \overset{5}{\text{Sym.}} \\ -31.46 & 289.1 & \\ 1.733 & 62.84 & 293.6 \end{bmatrix} \begin{matrix} 1 \\ 2 \\ 5 \end{matrix} \qquad (u)$$

$$\mathbf{M}_{21}^* = \frac{\rho AL}{420}[\overset{1}{81.91} \quad \overset{2}{1.152} \quad \overset{5}{6.302}] \; 4 = \mathbf{M}_{12}^{*T} \qquad (v)$$

$$\mathbf{M}_{22}^* = \frac{\rho AL}{420}[262.0] \; \overset{4}{} 4 \qquad (w)$$

Then the barred matrices in Eqs. (26) through (31) become

$$\overline{\mathbf{S}}_{12} = \frac{EI_z}{L^3} \begin{bmatrix} \overset{4}{20.82} \\ 18.12 \\ -21.17 \end{bmatrix} \begin{matrix} 1 \\ 2 \\ 3 \end{matrix} \qquad \overline{\mathbf{S}}_{22} = \frac{EI_z}{L^3}[22.41] \; \overset{4}{} 4 \qquad (x)$$

$$\overline{\mathbf{A}}_1 = P(t) \begin{bmatrix} 1.414 \\ 0 \\ -1.118 \end{bmatrix} \begin{matrix} 1 \\ 2 \\ 3 \end{matrix} \qquad \overline{\mathbf{A}}_2 = P(t)[2.0] \; 4 \qquad (y)$$

$$\overline{\mathbf{M}}_{12} = \frac{\rho AL}{420} \begin{bmatrix} \overset{4}{791.3} \\ 574.3 \\ -799.7 \end{bmatrix} \begin{matrix} 1 \\ 2 \\ 3 \end{matrix} \qquad \overline{\mathbf{M}}_{22} = \frac{\rho AL}{420}[1305] \; \overset{4}{} 4 \qquad (z)$$

For these arrays the indexes 1, 2, and 3 denote member numbers, whereas index 4 represents the independent joint translation.

We now have all the matrices needed to find the response in Eq. (24) and the axial forces in Eq. (25) for any forcing function. If the function $P(t)$ is a step force $P_1$, Eq. (24) takes the form

$$\frac{1305\rho AL}{420}\ddot{D}_4 + \frac{22.41EI_z}{L^3}D_4 = 2.0P_1 \tag{a'}$$

The eigenvalue problem associated with the homogeneous version of Eq. (a') yields the angular frequency:

$$\omega = \frac{2.686}{L^2}\sqrt{\frac{EI_z}{\rho A}} \tag{b'}$$

and the response to the step force $P_1$ is

$$D_4 = 0.08925\frac{P_1 L^3}{EI_z}(1 - \cos \omega t) \tag{c'}$$

Also,

$$\ddot{D}_4 = 0.08925\frac{P_1 L^3 \omega^2}{EI_z}\cos \omega t$$

$$= 0.6437\frac{P_1}{\rho AL}\cos \omega t \tag{d'}$$

Substitution of $D_4$, $\ddot{D}_4$, and the appropriate barred matrices into Eq. (25) produces

$$\mathbf{Q} = \begin{bmatrix} 0.4442 \\ 1.617 \\ -0.7714 \end{bmatrix} P_1 - \begin{bmatrix} 0.6454 \\ 0.7370 \\ -0.6638 \end{bmatrix} P_1 \cos \omega t \tag{e'}$$

which are the time-varying axial forces in the members. To complete the example, we could also compute the dependent joint translations, the joint rotations, other internal member actions, and support reactions. However, these tasks are straightforward and will be left as exercises for the reader.

## 6.9 PROGRAMS DYPFAC AND DYSFAC

In this section we briefly discuss programs named DYPFAC and DYSFAC for dynamic analyses of plane frames and space frames with axial constraints. These programs use Guyan reduction (see Sec. 6.7) to eliminate the joint rotations and then further reduce the equations of motion by imposing axial constraints (see Sec. 6.8). Response calculations for the reduced system are carried out using the normal-mode method, as described in Chapter 4.

Programs DYPFAC and DYSFAC both have the same outline, as follows:

### Outline of Programs DYPFAC and DYSFAC

1. Read and write structural data
   a. Structural parameters

b. Joint coordinates
c. Joint restraints
d. Member information
2. Form constraint transformation matrices
   a. Assemble constraint matrix
   b. Find dependent translations
   c. Generate transformation matrices
3. Generate, condense, and transform stiffness and mass matrices
   a. Generate member matrices
   b. Assemble structural matrices
   c. Eliminate rotational displacements
   d. Reduce matrices (due to constraints)
4. Determine frequencies and mode shapes
   a. Convert eigenvalue problem to standard, symmetric form
   b. Calculate eigenvalues and eigenvectors
   c. Write natural frequencies (cps)
   d. Transform, normalize, and write modal vectors
   e. Normalize modal vectors with respect to mass matrix
5. Read and write dynamic load data
   a. Dynamic parameters
   b. Initial conditions
   c. Applied actions
   d. Ground accelerations
   e. Forcing function
6. Calculate time histories of displacements
   a. Transform initial conditions and loads to normal coordinates
   b. Determine time histories of normal-mode displacements
   c. Transform independent translations back to physical coordinates
   d. Calculate dependent translations and rotations
7. Write and/or plot results of response calculations
   a. Time histories of independent and dependent joint translations and rotations
   b. Maxima and minima of translations and rotations
   c. Time histories of axial forces and other member end-actions
   d. Maxima and minima of axial forces and end-actions

Both of these programs perform the reductions optionally. Therefore, two new structural parameters, IRO and IAC, must be added to the structural data. Within the logic of the program, if IRO $\neq$ 0 the rotations are eliminated; and if IAC $\neq$ 0 axial constraints are imposed. Further details regarding these programs are given in Ref. 9.

**Example 6.13**

Figure 6.24(a) illustrates an underconstrained plane frame with twelve degrees of free-

## Sec. 6.9 Programs DYPFAC and DYSFAC

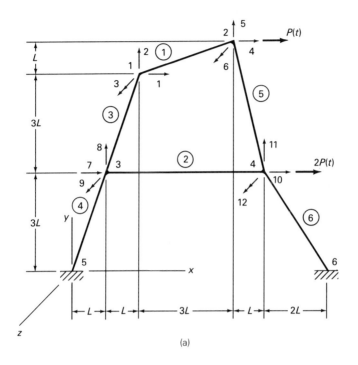

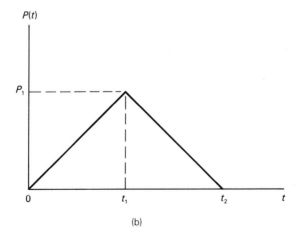

**Figure 6.24** (a) Six-member plane frame; (b) dynamic load.

dom (12 DOF). The frame consists of six rigidly connected prismatic members, all of which have the same values of $E$, $\rho$, $A$, and $I_z$. Dynamic forces $P(t)$ and $2P(t)$ are applied in the $x$ direction at joints 2 and 4, and the time variation of $P(t)$ appears in Fig. 6.24(b). Assuming that the frame is steel, we use the following numerical values for parameters:

$$E = 3.0 \times 10^4 \text{ k/in.}^2 \qquad \rho = 7.35 \times 10^{-7} \text{ k-s}^2/\text{in.}^4$$
$$A = 30 \text{ in.}^2 \qquad I_z = 1.0 \times 10^3 \text{ in.}^4$$
$$L = 50 \text{ in.} \qquad P_1 = 10 \text{ k} \qquad t_2 = 2t_1 = 35 \text{ ms}$$

where US units are implied.

This structure was analyzed by Program DYPFAC three times, as follows: (a) without reduction (12 DOF); (b) with elimination of rotations (8 DOF); and (c) with elimination of rotations, followed by reduction due to axial constraints (2 DOF). Figure 6.25 shows time histories of response for the independent translation $D_4$, as computed

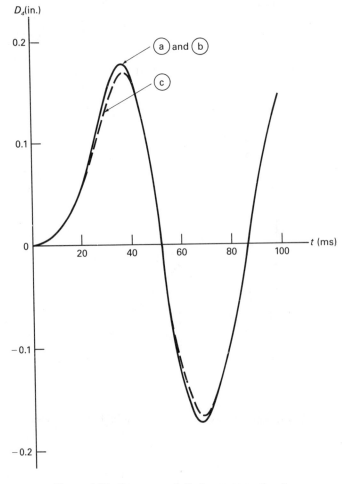

**Figure 6.25** Responses of displacement number 4.

for each of the three cases. We see that the response curve for Case (b) is indistinguishable from that for Case (a). In addition, the response curve for Case (c) differs by only a small amount from the others (error is approximately $-3.3\%$ at the first peak). Thus, the level of accuracy shown by this example seems sufficient for practical purposes.

## REFERENCES

1. Weaver, W., Jr., and Gere, J. M., *Matrix Analysis of Framed Structures*, 2nd ed., Van Nostrand Reinhold, New York, 1980.
2. Weaver, W., Jr., and Johnston, P. R., *Finite Elements for Structural Analysis*, Prentice-Hall, Englewood Cliffs, N. J., 1984.
3. Yoshida, D. M., and Weaver, W., Jr., "Finite-Element Analysis of Beams and Plates with Moving Loads," *Int. Assoc. Bridge Struct. Eng.*, Vol. 31, 1971, pp. 179–195.
4. Guyan, R. J., "Reduction of Stiffness and Mass Matrices," *AIAA J.*, Vol. 3, No. 2, 1965, p. 380.
5. Timoshenko, S. P., Young, D. H., and Weaver, W., Jr, *Vibration Problems in Engineering*, 4th ed., Wiley, New York, 1974.
6. Weaver, W., Jr., and Eisenberger, M., "Dynamics of Frames with Axial Constraints," *ASCE J. Struct. Eng.*, Vol. 109, No. 3, 1983, pp. 773–784.
7. Gere, J. M., and Weaver, W., Jr., *Matrix Algebra for Engineers*, 2nd ed., Brooks/Cole, Monterey, Calif., 1983.
8. Weaver, W., Jr., Nelson, M. F., and Manning, T. A., "Dynamics of Tier Buildings," *ASCE J. Eng. Mech. Div.*, Vol. 94, No. EM6, 1968, pp. 1455–1474.
9. Eisenberger, M., "Static and Dynamic Analysis of Plane and Space Frames with Axial Constraints," *Technical Report 48*, J. A. Blume Earthquake Engineering Center, Stanford, Calif., Oct. 1980.

## PROBLEMS

**6.2-1.** Suppose that a uniformly distributed force $b_{y'}(t)$ acts on half the length (from $x' = 0$ to $x' = L/2$) of the plane frame member in Fig. 6.2(a). Find the equivalent nodal loads at joints $j$ and $k$ for both member and structural directions, assuming that $x_{jk} = 4$ and $y_{jk} = -3$.

**6.2-2.** Let the concentrated force $P_{y'}(t)$ on the plane frame member in Fig. 6.2(a) act at the point where $x' = 3L/4$. For both local and global coordinates, determine the equivalent nodal loads at points $j$ and $k$, with $x_{jk} = y_{jk}$.

**6.2-3.** A moment $M_z(t)$ is applied at the midpoint of the plane frame member in Fig. 6.2(a). If $x_{jk} = 1$ and $y_{jk} = -2$, calculate the equivalent nodal loads at $j$ and $k$ for both member and structural directions.

**6.2-4.** A triangular distribution of force $\xi b_2(t)$ acts in the $y'$ direction on the plane frame member in Fig. 6.2(a). Find the equivalent nodal loads in local and global directions for $x_{jk} = 3$ and $y_{jk} = 4$.

**6.2-5.** Two concentrated forces $P_{y'}(t)$ are applied at the third points of the plane frame member in Fig. 6.2(a). Calculate equivalent nodal loads in member and structural directions, using $x_{jk} = -1$ and $y_{jk} = 2$.

**6.2-6.** Assume that two moments $M_z(t)$ act at the quarter points of the plane frame member in Fig. 6.2(a). Determine equivalent nodal loads in local and global coordinates for $x_{jk} = y_{jk}$.

**6.2-7.** A parabolic distribution of force $\xi^2 b_2(t)$ is applied in the $x'$ direction on the plane frame member in Fig. 6.2(a). For $x_{jk} = 2$ and $y_{jk} = -1$, find the equivalent nodal loads in member and structural directions.

**6.3-1.** Let a uniformly distributed force $b_z(t)$ act on half the length ($L/2 \leq x' \leq L$) of the grid member in Fig. 6.3(a). Calculate equivalent nodal loads for both local and global directions, with $x_{jk} = 3$ and $y_{jk} = 4$.

**6.3-2.** Assume that a concentrated force $P_z(t)$ is applied at the point where $x' = L/3$ on the grid member in Fig. 6.3(a). For $x_{jk} = 3$ and $y_{jk} = -1$, determine the equivalent nodal loads in both member and structural directions.

**6.3-3.** The grid member in Fig. 6.3(a) has a moment $M_{y'}(t)$ applied at the point where $x' = L/4$. Find the equivalent nodal loads for both local and global coordinates, using $x_{jk} = y_{jk}$.

**6.3-4.** Repeat Prob. 6.3-3 for a moment $M_y(t)$ in the $y$ direction.

**6.3-5.** Repeat Prob. 6.3-3 for a moment $M_x(t)$ in the $x$ direction.

**6.3-6.** A triangular distribution of force $(1 - \xi)b_1(t)$ acts in the $z$ direction on the grid member in Fig. 6.3(a). For $x_{jk} = -4$ and $y_{jk} = 3$, obtain the equivalent nodal loads in both member and structural coordinates.

**6.3-7.** Suppose that two concentrated forces $P_z(t)$ are applied at the quarter points of the grid member in Fig. 6.3(a). Calculate equivalent nodal loads in local and global directions for $x_{jk} = 1$ and $y_{jk} = 3$.

**6.3-8.** The grid member in Fig. 6.3(a) is subjected to a moment $M_{y'}(t)$ at $x' = L/3$ and a moment $-M_{y'}(t)$ at $x' = 2L/3$. For $x_{jk} = 1$ and $y_{jk} = -2$, find the equivalent nodal loads in member and structural directions.

**6.4-1.** Suppose that the space truss member in Fig. 6.4(a) has a uniformly distributed force $b_{y'}(t)$ applied over the segment where $L/4 \leq x' \leq 3L/4$. Determine equivalent nodal loads in local and global directions, using the relative coordinates $\mathbf{c}_{jk} = \{3, 2, 1\}$ and $\mathbf{c}_{jp} = \{-1, 3, -2\}$.

**6.4-2.** Repeat Prob. 6.4-1 for a uniformly distributed force $b_{z'}(t)$ applied over the segment where $L/3 \leq x' \leq 2L/3$.

**6.4-3.** Assume that a concentrated force $P_{z'}(t)$ acts at $x' = 2L/5$ on the space truss member in Fig. 6.4(a). For both member and structural directions, find the equivalent nodal loads, with the relative coordinates $\mathbf{c}_{jk} = \{1, 1, 1\}$ and $\mathbf{c}_{jp} = \{-1, 1, -3\}$.

**6.4-4.** Repeat Prob. 6.4-3 if the force acts in the $y'$ direction at the point where $x' = 3L/5$.

**6.4-5.** Let a moment $M_{z'}(t)$ be applied at point $x' = 4L/5$ on the space truss member in Fig. 6.4(a). Calculate the equivalent nodal loads in local and global directions for the relative coordinates $\mathbf{c}_{jk} = \{3, -4, 2\}$ and $\mathbf{c}_{jp} = \{1, 3, 2\}$.

**6.4-6.** Repeat Prob. 6.4-5 with the moment acting in the $y'$ direction at the point $x' = L/5$.

**6.4-7.** Assume that a uniformly distributed force $b(t)$ acts on a space truss member in the direction shown by Fig. 6.4(b). Using the relative coordinates $\mathbf{c}_{jk} = \{2, 2, 3\}$ and $\mathbf{c}_{jp} = \{0, 5, 0\}$, determine equivalent nodal loads in member and structural directions.

**6.4-8.** Repeat Prob. 6.4-7 for a triangular distribution of force $\xi b_2(t)$.

**6.5-1.** Let the space frame member in Fig. 6.5(a) be subjected to a uniformly distributed force $b_{y'}(t)$ over its whole length. Find the equivalent nodal loads for both local and global directions, assuming the relative coordinates $\mathbf{c}_{jk} = \{3, 2, 4\}$ and $\mathbf{c}_{jp} = \{-2, 4, -1\}$.

**6.5-2.** Repeat Prob. 6.5-1 for a uniformly distributed force $b_{z'}(t)$ over half the length $(0 \leq x' \leq L/2)$.

**6.5-3.** Suppose that a concentrated force $P_{y'}(t)$ is applied at $x' = L/4$ on the space frame member in Fig. 6.5(a). Using the relative coordinates $\mathbf{c}_{jk} = \{2, 3, -1\}$ and $\mathbf{c}_{jp} = \{-1, 2, 2\}$, calculate the equivalent nodal loads in member and structural directions.

**6.5-4.** Repeat Prob. 6.5-3 for a force in the $z'$ direction at the point where $x' = 3L/4$.

**6.5-5.** A moment $M_{y'}(t)$ acts on the space frame member in Fig. 6.5(a) at the point where $x' = L/3$. Determine the equivalent nodal loads for both local and global directions, assuming the relative coordinates $\mathbf{c}_{jk} = \{0, 3, 4\}$ and $\mathbf{c}_{jp} = \{0, 0, -2\}$.

**6.5-6.** Repeat Prob. 6.5-5 for a moment in the $z'$ direction applied at the point where $x' = 2L/3$.

**6.5-7.** Assume that a uniformly distributed force $b_x(t)$ acts in the $x$ direction on the space frame member in Fig. 6.5(a). With the relative coordinates $\mathbf{c}_{jk} = \{1, -2, 2\}$ and $\mathbf{c}_{jp} = \{-3, 4, 0\}$, obtain the equivalent nodal loads for member and structural directions.

**6.5-8.** Repeat Prob. 6.5-7 for a triangular distribution of force $(1 - \xi)b_1(t)$ in the $z$ direction.

**6.7-1.** The two-element prismatic beam shown in Fig. P6.7-1 is fixed at point 3 but free to rotate at point 1. Construct the stiffness and consistent mass matrices $\mathbf{S}$ and $\mathbf{M}$ for the three unrestrained displacements. Reduce these matrices by eliminating the rotations and retaining the translation. Then find the angular frequency of vibration for the remaining system, which has only one degree of freedom.

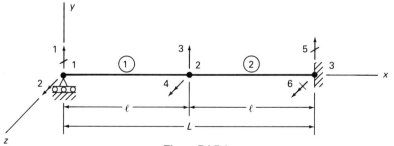

Figure P6.7-1

**6.7-2.** Figure P6.7-2 shows a two-element prismatic beam that is free to translate (but not rotate) at point 1 and is fixed at point 3. Assemble the stiffness and consistent mass matrices **S** and **M** for the three unrestrained displacements, and reduce them by eliminating the rotation and keeping the translations. Solve the eigenvalue problem to obtain angular frequencies and vibrational mode shapes for the reduced system.

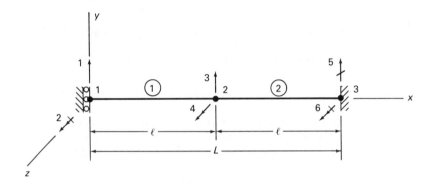

Figure P6.7-2

**6.7-3.** The simply supported beam shown in Fig. P6.7-3 is divided into two flexural elements with equal properties. Set up the stiffness and consistent mass matrices **S** and **M** for the four displacements that are unrestrained. Reduce these matrices by eliminating the rotations and retaining the translation. Then find the angular frequency of the remaining system, which has only one degree of freedom.

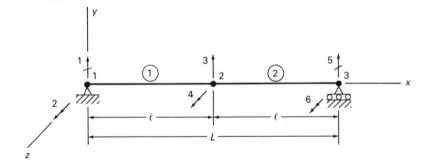

Figure P6.7-3

**6.7-4.** A two-element prismatic beam has no restraints whatsoever, as implied in Fig. P6.7-4. Construct the stiffness and consistent mass matrices **S** and **M** for the six unrestrained displacements, and eliminate the three rotations. For the three remaining translations, calculate the angular frequencies and mode shapes.

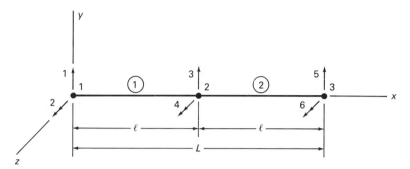

Figure P6.7-4

**6.7-5.** The plane frame shown in Fig. P6.7-5 has the same properties $E$, $\rho$, $A$, and $I_z$ for each of its three prismatic members. Set up the stiffness and consistent mass matrices **S** and **M** for the four displacements shown, neglecting axial strains in the members. Then reduce **S** and **M** by eliminating the rotations $\mathbf{D}_1$ and $\mathbf{D}_2$, while retaining the translations $\mathbf{D}_3$ and $\mathbf{D}_4$. Determine the angular frequencies and mode shapes of the remaining system.

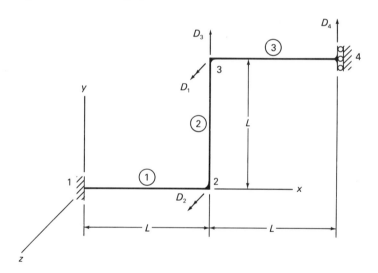

Figure P6.7-5

**6.7-6.** Assume that the plane frame shown in Fig. P6.7-6 has $2EI_z$ and $2\rho A$ for elements 1 and 2; whereas, elements 3 and 4 have $EI_z$ and $\rho A$ for their properties. Construct the stiffness and consistent mass matrices **S** and **M**; and eliminate the rotations $\mathbf{D}_1$, $\mathbf{D}_2$, and $\mathbf{D}_3$, while keeping the translations $\mathbf{D}_4$ and $\mathbf{D}_5$. Calculate the angular frequencies and mode shapes for the remaining system, neglecting axial strains in the members.

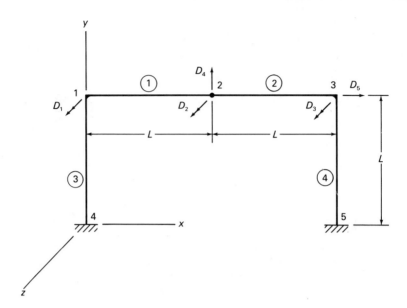

**Figure P6.7-6**

**6.7-7.** Figure P6.7-7 shows a grid consisting of two prismatic members with equal properties. Assemble the stiffness and consistent mass matrices **S** and **M** for the three displacements indicated at joint 2. Reduce these matrices by eliminating the rotations and retaining the translation. Find the angular frequency of the remaining SDOF problem, assuming that $I_x = I_y = J$, $r_g = L/10$, and $E = 5G/2$.

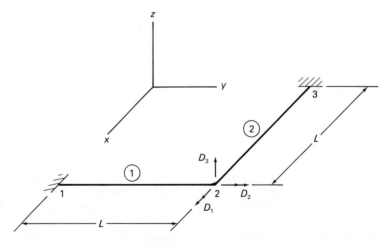

**Figure P6.7-7**

**6.7-8.** A grid with three equal prismatic members appears in Fig. P6.7-8. Determine the stiffness and consistent mass matrices **S** and **M** for the three displacements at joint 2. Then eliminate the rotations, keep the translation, and calculate the angular frequency of the reduced system. For each of the members in this structure, assume that $2I_x = 2I_y = J$, $r_g = L/10$, and $E = 5G/2$.

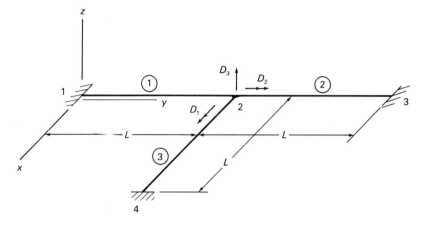

**Figure P6.7-8**

# 7
# Two- and Three-Dimensional Continua

## 7.1 INTRODUCTION

The framed structures discussed in the preceding chapter consist of only slender (or one-dimensional) members. However, we also wish to analyze solids and structures composed of two- and three-dimensional finite elements. In this chapter we shall deal with two-dimensional continua in states of plane stress and plane strain [1] as well as general and axisymmetric solids. Our emphasis will be upon the isoparametric formulations leading to the most commonly used elements [2]. Computer programs cover dynamic analyses of plane-stress and plane-strain problems, general solids, and axisymmetric solids. In all cases the discretized continua are assumed to be linearly elastic with small strains and displacements.

## 7.2 STRESSES AND STRAINS IN CONTINUA

In Sec. 3.2 we introduced the topic of stresses and strains in three dimensions and defined *strain-displacement relationships* with respect to Cartesian coordinates. Now these relationships will be expressed in matrix form as

$$\boldsymbol{\epsilon} = \mathbf{d}\,\mathbf{u} \tag{1}$$

which applies to either static or dynamic analysis. The strain vector in Eq. (1) is

$$\boldsymbol{\epsilon} = \{\epsilon_x,\ \epsilon_y,\ \epsilon_z,\ \gamma_{xy},\ \gamma_{yz},\ \gamma_{zx}\} \tag{2}$$

## Sec. 7.2  Stresses and Strains in Continua

and the displacement vector is

$$\mathbf{u} = \{u, v, w\} \quad (3)$$

Furthermore, the $6 \times 3$ *strain-displacement operator* $\mathbf{d}$ has the form

$$\mathbf{d} = \begin{bmatrix} \dfrac{\partial}{\partial x} & 0 & 0 \\ 0 & \dfrac{\partial}{\partial y} & 0 \\ 0 & 0 & \dfrac{\partial}{\partial z} \\ \dfrac{\partial}{\partial y} & \dfrac{\partial}{\partial x} & 0 \\ 0 & \dfrac{\partial}{\partial z} & \dfrac{\partial}{\partial y} \\ \dfrac{\partial}{\partial z} & 0 & \dfrac{\partial}{\partial x} \end{bmatrix} \quad (4)$$

in which the derivatives are taken from Eqs. (3.2-2) and (3.2-3).

In addition, we developed *stress-strain relationships* for an *isotropic material*.* As before, such relationships can be written:

$$\boldsymbol{\sigma} = \mathbf{E}\,\boldsymbol{\epsilon} \quad (5)$$

where the stress vector is

$$\boldsymbol{\sigma} = \{\sigma_x, \sigma_y, \sigma_z, \tau_{xy}, \tau_{yz}, \tau_{zx}\} \quad (6)$$

and the $6 \times 6$ *stress-strain matrix* $\mathbf{E}$ was given by Eq. (3.2-10).

For the moment, let us specialize our discussion to problems in two dimensions, where the stress vector contains only those terms depicted in Fig. 7.1(a). Thus, in the $x$-$y$ plane we have

$$\boldsymbol{\sigma} = \{\sigma_x, \sigma_y, \tau_{xy}\} \quad (7)$$

Corresponding to these stresses are the strains indicated in Fig. 7.1(b), which shows a displaced and deformed infinitesimal element in the $x$-$y$ plane. Hence, the strain vector is

$$\boldsymbol{\epsilon} = \{\epsilon_x, \epsilon_y, \gamma_{xy}\} \quad (8)$$

in which

$$\epsilon_x = \frac{\partial u}{\partial x} \qquad \epsilon_y = \frac{\partial v}{\partial y} \qquad \gamma_{xy} = \frac{\partial u}{\partial y} + \frac{\partial v}{\partial x} \quad (9)$$

---

*For orthogonally anisotropic (*orthotropic*) materials, see Ref. 2; and for generally *anisotropic* materials, see Ref. 3.

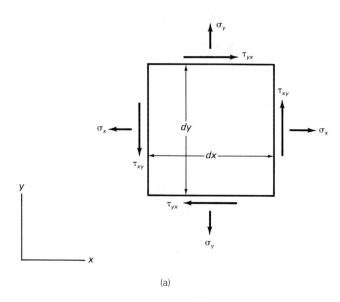

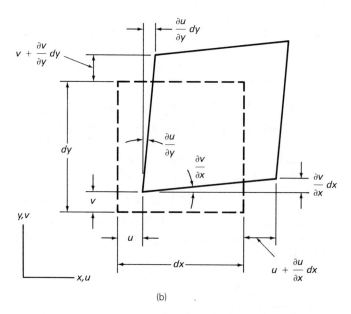

**Figure 7.1** (a) Stresses and (b) strains in two dimensions.

## Sec. 7.2 Stresses and Strains in Continua

These derivatives constitute strain-displacement relationships in two dimensions. As before, they may be written in matrix form as shown in Eq. (1), where the linear differential operator **d** becomes

$$\mathbf{d} = \begin{bmatrix} \dfrac{\partial}{\partial x} & 0 \\ 0 & \dfrac{\partial}{\partial y} \\ \dfrac{\partial}{\partial y} & \dfrac{\partial}{\partial x} \end{bmatrix} \qquad (10)$$

and the vector **u** is

$$\mathbf{u} = \{u, v\} \qquad (11)$$

When a thin plate is loaded with forces in its own plane, the resulting two-dimensional problem is called *plane stress*. Assuming that the plate lies in the x-y plane, we can impose the following conditions of stresses and strains:

$$\sigma_z = \tau_{yz} = \tau_{zx} = 0 \qquad \epsilon_z \neq 0 \qquad \gamma_{yz} = \gamma_{zx} = 0 \qquad (12)$$

In particular, note that the strain normal to the plate is nonzero. Using these conditions in Eqs. (3.2-5), we find that

$$\mathbf{E} = \frac{E}{1 - \nu^2} \begin{bmatrix} 1 & \nu & 0 \\ \nu & 1 & 0 \\ 0 & 0 & \dfrac{1 - \nu}{2} \end{bmatrix} \qquad (13)$$

This is the stress-strain matrix for plane stress in an isotropic material. In addition, the normal strain in the z direction is

$$\epsilon_z = -\frac{\nu}{E}(\sigma_x + \sigma_y) \qquad (14)$$

The case of *plane strain* arises when a long prismatic solid has a constant condition of loading normal to its axis. In this instance, the solid can be analyzed as an infinity of two-dimensional slices of unit thickness, as illustrated in Fig. 3.1(c). However, the conditions given in Eqs. (12) must be changed to

$$\sigma_z \neq 0 \qquad \tau_{yz} = \tau_{zx} = 0 \qquad \epsilon_z = \gamma_{yz} = \gamma_{zx} = 0 \qquad (15)$$

The first expression indicates that the stress $\sigma_z$ in the axial direction is nonzero, even though the corresponding strain $\epsilon_z$ is zero. Substituting these conditions into Eqs. (3.2-5), we obtain

$$\sigma_z = \nu(\sigma_x + \sigma_y) \qquad (16)$$

and

$$\mathbf{E} = \frac{E}{(1+\nu)(1-2\nu)} \begin{bmatrix} 1-\nu & \nu & 0 \\ \nu & 1-\nu & 0 \\ 0 & 0 & \dfrac{1-2\nu}{2} \end{bmatrix} \quad (17)$$

The latter equation gives the stress-strain matrix for the case of plane strain in an isotropic material.

Now we shall consider *rotation of axes* for stresses and strains, starting in three dimensions. For this purpose, the stress vector in Eq. (6) may be recast into the form of a symmetric $3 \times 3$ matrix, as follows:

$$\boldsymbol{\sigma} = \begin{bmatrix} \sigma_x & \tau_{xy} & \tau_{xz} \\ \tau_{yx} & \sigma_y & \tau_{yz} \\ \tau_{zx} & \tau_{zy} & \sigma_z \end{bmatrix} \quad (18)$$

in which complementary shearing stresses are included. Then the rotation-of-axes transformation for stresses can be stated as

$$\boldsymbol{\sigma}' = \mathbf{R}\,\boldsymbol{\sigma}\,\mathbf{R}^T \quad (19)$$

In this equation the $3 \times 3$ matrix $\boldsymbol{\sigma}'$ is similar to $\boldsymbol{\sigma}$ in Eq. (18), but it contains stresses in the directions of inclined (primed) axes shown in Fig. 7.2(a). The $3 \times 3$ rotation matrix $\mathbf{R}$ in Eq. (19) has the form

$$\mathbf{R} = \begin{bmatrix} \lambda_{11} & \lambda_{12} & \lambda_{13} \\ \lambda_{21} & \lambda_{22} & \lambda_{23} \\ \lambda_{31} & \lambda_{32} & \lambda_{33} \end{bmatrix} = \begin{bmatrix} l_1 & m_1 & n_1 \\ l_2 & m_2 & n_2 \\ l_3 & m_3 & n_3 \end{bmatrix} \quad (20)$$

In the latter matrix the terms $l_1$, $m_1$, and so on are slightly more efficient symbols for the direction cosines $\lambda_{11}$, $\lambda_{12}$, and so on, that were explained in Sec. 3.5. Note that the rows of $\mathbf{R}$ contain the $x$, $y$, and $z$ components of the unit vectors $\mathbf{i}'$, $\mathbf{j}'$, and $\mathbf{k}'$ shown in Fig. 7.2(a).

Similarly, the strain vector in Eq. (2) may be restated as the $3 \times 3$ symmetric matrix

$$\boldsymbol{\epsilon} = \begin{bmatrix} \epsilon_x & \gamma_{xy} & \gamma_{xz} \\ \gamma_{yx} & \epsilon_y & \gamma_{yz} \\ \gamma_{zx} & \gamma_{zy} & \epsilon_z \end{bmatrix} \quad (21)$$

which includes dependent shearing strains. Again, the rotation operation is

$$\boldsymbol{\epsilon}' = \mathbf{R}\,\boldsymbol{\epsilon}\,\mathbf{R}^T \quad (22)$$

in which the $3 \times 3$ matrix $\boldsymbol{\epsilon}'$ is similar to $\boldsymbol{\epsilon}$ in Eq. (21), but for inclined axes.

We can obtain *principal normal stresses* and their directions as the solution of an *eigenvalue problem* [4]. Using that method gives

$$\boldsymbol{\sigma}_P = \mathbf{R}_P\,\boldsymbol{\sigma}\,\mathbf{R}_P^T \quad (23)$$

## Sec. 7.2  Stresses and Strains in Continua

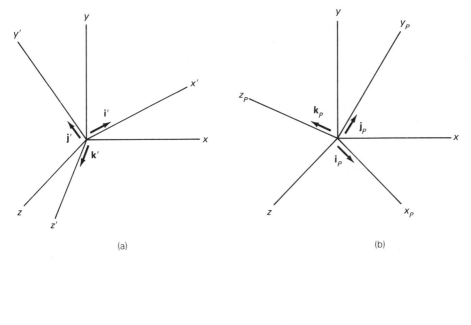

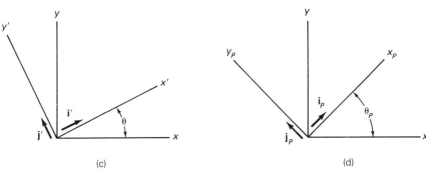

**Figure 7.2** Orthogonal axes—three dimensions: (a) inclined; (b) principal—two dimensions: (c) inclined; (d) principal.

In this equation the symbol $\boldsymbol{\sigma}_P$ represents a diagonal matrix (or *spectral matrix*) of principal normal stresses. Thus,

$$\boldsymbol{\sigma}_P = \begin{bmatrix} \sigma_{P1} & 0 & 0 \\ 0 & \sigma_{P2} & 0 \\ 0 & 0 & \sigma_{P3} \end{bmatrix} \qquad (24)$$

In addition, the symbol $\mathbf{R}_P$ in Eq. (23) denotes the rotation matrix for the *principal axes* shown in Fig. 7.2(b). This matrix is the transpose of the normalized *eigenvector matrix* $\boldsymbol{\Phi}_N$. That is,

$$\mathbf{R}_P = \mathbf{\Phi}_N^T \tag{25}$$

in which the rows of $\mathbf{R}_P$ (or columns of $\mathbf{\Phi}_N$) are normalized to have unit lengths.

Similarly, *principal normal strains* may be calculated as

$$\boldsymbol{\epsilon}_P = \mathbf{R}_P \boldsymbol{\epsilon} \, \mathbf{R}_P^T \tag{26}$$

In this expression the symbol $\boldsymbol{\epsilon}_P$ stands for a diagonal matrix of normal strains corresponding to the stresses in $\boldsymbol{\sigma}_P$.

In subsequent work it will become necessary to transform the stress-strain matrix $\mathbf{E}$ from one set of coordinates to another by rotation of axes. For this purpose, we rewrite the expanded results of Eq. (22) as

$$\boldsymbol{\epsilon}' = \mathbf{T}_\epsilon \boldsymbol{\epsilon} \tag{27}$$

Here the strains $\boldsymbol{\epsilon}$ and $\boldsymbol{\epsilon}'$ are in the form of Eq. (2) instead of Eq. (21). The $6 \times 6$ *strain transformation matrix* $\mathbf{T}_\epsilon$ in Eq. (27) is

$$\mathbf{T}_\epsilon = \begin{bmatrix} l_1^2 & m_1^2 & n_1^2 & l_1 m_1 & m_1 n_1 & n_1 l_1 \\ l_2^2 & m_2^2 & n_2^2 & l_2 m_2 & m_2 n_2 & n_2 l_2 \\ l_3^2 & m_3^2 & n_3^2 & l_3 m_3 & m_3 n_3 & n_3 l_3 \\ 2l_1 l_2 & 2m_1 m_2 & 2n_1 n_2 & l_1 m_2 + l_2 m_1 & m_1 n_2 + m_2 n_1 & n_1 l_2 + n_2 l_1 \\ 2l_2 l_3 & 2m_2 m_3 & 2n_2 n_3 & l_2 m_3 + l_3 m_2 & m_2 n_3 + m_3 n_2 & n_2 l_3 + n_3 l_2 \\ 2l_3 l_1 & 2m_3 m_1 & 2n_3 n_1 & l_3 m_1 + l_1 m_3 & m_3 n_1 + m_1 n_3 & n_3 l_1 + n_1 l_3 \end{bmatrix} \tag{28}$$

The partition lines in matrix $\mathbf{T}_\epsilon$ separate terms pertaining to normal strains from those for shearing strains. Giving symbols to the resulting submatrices, we have

$$\mathbf{T}_\epsilon = \begin{bmatrix} \mathbf{T}_{\epsilon 11} & \mathbf{T}_{\epsilon 12} \\ \mathbf{T}_{\epsilon 21} & \mathbf{T}_{\epsilon 22} \end{bmatrix} \tag{29}$$

in which the subscripts 1 and 2 denote normal and shearing strains, respectively.

In order to determine the form of the *stress transformation matrix* $\mathbf{T}_\sigma$, we equate *virtual strain energy densities* for the primed and unprimed axes in Fig. 7.2(a). Thus,

$$(\delta \boldsymbol{\epsilon}')^T \boldsymbol{\sigma}' = \delta \boldsymbol{\epsilon}^T \boldsymbol{\sigma} \tag{30}$$

Then substitute the transposed incremental form of Eq. (27) into Eq. (30) to obtain

$$\delta \boldsymbol{\epsilon}^T \mathbf{T}_\epsilon^T \boldsymbol{\sigma}' = \delta \boldsymbol{\epsilon}^T \boldsymbol{\sigma} \tag{31}$$

Hence, we conclude that

$$\boldsymbol{\sigma}' = \mathbf{T}_\sigma \boldsymbol{\sigma} \tag{32}$$

where

$$\mathbf{T}_\sigma = \mathbf{T}_\epsilon^{-T} \tag{33}$$

Sec. 7.2  Stresses and Strains in Continua          317

Therefore, the stress transformation matrix $\mathbf{T}_\sigma$ is proven to be the transposed inverse of the strain transformation matrix $\mathbf{T}_\epsilon$. The inversion implied by Eq. (33) is not actually necessary, because expansion of Eq. (19) shows that the parts of $\mathbf{T}_\sigma$ bear the following relationships to the parts of $\mathbf{T}_\epsilon$:

$$\mathbf{T}_\sigma = \begin{bmatrix} \mathbf{T}_{\sigma 11} & \mathbf{T}_{\sigma 12} \\ \mathbf{T}_{\sigma 21} & \mathbf{T}_{\sigma 22} \end{bmatrix} = \begin{bmatrix} \mathbf{T}_{\epsilon 11} & 2\mathbf{T}_{\epsilon 12} \\ \tfrac{1}{2}\mathbf{T}_{\epsilon 21} & \mathbf{T}_{\epsilon 22} \end{bmatrix} \quad (34)$$

As before, the subscripts 1 and 2 on the submatrices of $\mathbf{T}_\sigma$ refer to normal and shearing stresses, respectively.

Now the transformation of $\mathbf{E}'$ to $\mathbf{E}$ can be accomplished by first writing stress-strain relationships in the primed coordinates as

$$\boldsymbol{\sigma}' = \mathbf{E}'\, \boldsymbol{\epsilon}' \quad (35)$$

Next, we substitute Eqs. (27) and (32) into Eq. (35), producing

$$\mathbf{T}_\sigma \boldsymbol{\sigma} = \mathbf{E}'\mathbf{T}_\epsilon \boldsymbol{\epsilon} \quad (36)$$

Then premultiply Eq. (36) with $\mathbf{T}_\sigma^{-1}$, and use Eq. (33) to find that

$$\boldsymbol{\sigma} = \mathbf{T}_\epsilon^T \mathbf{E}' \mathbf{T}_\epsilon \boldsymbol{\epsilon} \quad (37)$$

Thus, we see that

$$\mathbf{E} = \mathbf{T}_\epsilon^T \mathbf{E}' \mathbf{T}_\epsilon \quad (38)$$

which represents the transformation of $\mathbf{E}'$ to $\mathbf{E}$. The reverse transformation is

$$\mathbf{E}' = \mathbf{T}_\sigma \mathbf{E}\, \mathbf{T}_\sigma^T \quad (39)$$

For the purpose of specializing axis rotations to two dimensions, we form a symmetric $2 \times 2$ matrix of stresses, as follows:

$$\boldsymbol{\sigma} = \begin{bmatrix} \sigma_x & \tau_{xy} \\ \tau_{yx} & \sigma_y \end{bmatrix} \quad (40)$$

This matrix can be used in Eq. (19) to determine similar stresses $\boldsymbol{\sigma}'$ for the inclined directions shown in Fig. 7.2(c). For this case the $2 \times 2$ rotation matrix is

$$\mathbf{R} = \begin{bmatrix} l_1 & m_1 \\ l_2 & m_2 \end{bmatrix} = \begin{bmatrix} \cos\theta & \sin\theta \\ -\sin\theta & \cos\theta \end{bmatrix} \quad (41)$$

in which the rows contain the $x$ and $y$ components of vectors $\mathbf{i}'$ and $\mathbf{j}'$. Furthermore, the symmetric $2 \times 2$ matrix of strains becomes

$$\boldsymbol{\epsilon} = \begin{bmatrix} \epsilon_x & \gamma_{xy} \\ \gamma_{yx} & \epsilon_y \end{bmatrix} \quad (42)$$

which can be applied in Eq. (22) to find strains $\boldsymbol{\epsilon}'$ for the inclined axes.

By solving a second-order eigenvalue problem, we can also obtain principal normal stresses and strains in two dimensions. Thus, the eigenvalue problem associated with the 2 × 2 stress matrix in Eq. (40) is

$$(\boldsymbol{\sigma} - \lambda_i \mathbf{I})\boldsymbol{\Phi}_i = \mathbf{0} \qquad (i = 1, 2) \tag{43}$$

In this equation the symbol $\lambda_i$ denotes the $i$th eigenvalue of $\boldsymbol{\sigma}$, and $\boldsymbol{\Phi}_i$ represents the corresponding eigenvector. To find $\lambda_i$ we set the determinant of the coefficient matrix equal to zero, as follows:

$$|\boldsymbol{\sigma} - \lambda_i \mathbf{I}| = 0 \tag{44}$$

or

$$\begin{vmatrix} \sigma_x - \lambda_i & \tau_{xy} \\ \tau_{yx} & \sigma_y - \lambda_i \end{vmatrix} = 0 \tag{45}$$

Expanding this determinant yields

$$\lambda_i^2 - (\sigma_x + \sigma_y)\lambda_i + \sigma_x \sigma_y - \tau_{xy}^2 = 0 \tag{46}$$

From this quadratic equation, we obtain the roots

$$\lambda_{1,2} = \sigma_{P1,2} = \frac{\sigma_x + \sigma_y}{2} \pm \sqrt{\left(\frac{\sigma_x - \sigma_y}{2}\right)^2 + \tau_{xy}^2} \tag{47}$$

which constitute the principal normal stresses $\sigma_{P1}$ and $\sigma_{P2}$. They act in the directions of the axes $x_P$ and $y_P$ shown in Fig. 7.2(d). By substituting them into the homogeneous equations [see Eq. (43)], we can find the eigenvectors, which (when normalized) contain the components of the unit vectors $\mathbf{i}_P$ and $\mathbf{j}_P$.

The strain transformation matrix $\mathbf{T}_\epsilon$ in Eq. (28) can be specialized to two dimensions by removing the third, fifth, and sixth rows and columns to obtain

$$\mathbf{T}_\epsilon = \begin{bmatrix} l_1^2 & m_1^2 & l_1 m_1 \\ l_2^2 & m_2^2 & l_2 m_2 \\ \hline 2l_1 l_2 & 2m_1 m_2 & l_1 m_2 + l_2 m_1 \end{bmatrix} \tag{48}$$

Also, the stress transformation matrix $\mathbf{T}_\sigma$ becomes

$$\mathbf{T}_\sigma = \begin{bmatrix} l_1^2 & m_1^2 & 2l_1 m_1 \\ l_2^2 & m_2^2 & 2l_2 m_2 \\ \hline l_1 l_2 & m_1 m_2 & l_1 m_2 + l_2 m_1 \end{bmatrix} \tag{49}$$

which is related to $\mathbf{T}_\epsilon$ in accordance with Eq. (34).

## 7.3 NATURAL COORDINATES

In this section we define natural coordinates for quadrilaterals and hexahedra in preparation for development of isoparametric elements having these shapes. We begin with a *quadrilateral* that has straight edges, as illustrated in Fig. 7.3(a). The point labeled $g$ is the *geometric center*, for which the coordinates in the $x$-$y$

### Sec. 7.3  Natural Coordinates

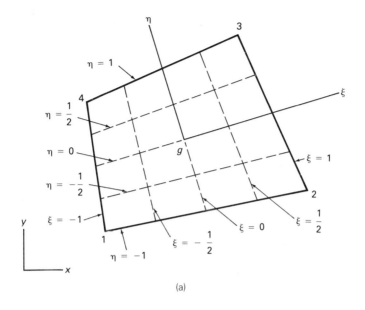

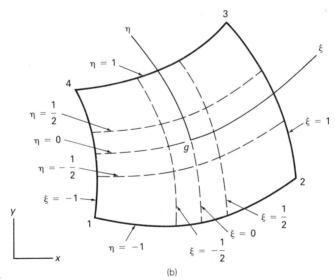

**Figure 7.3** Natural coordinates for quadrilaterals: (a) straight; (b) curved.

plane are

$$x_g = \tfrac{1}{4}(x_1 + x_2 + x_3 + x_4) \qquad y_g = \tfrac{1}{4}(y_1 + y_2 + y_3 + y_4) \qquad (1)$$

In these averaging expressions, $(x_1, y_1)$ are the $x$ and $y$ coordinates of point 1, and so on. Note that the geometric center is not necessarily the same as the centroid of the quadrilateral. Also shown in the figure are the *dimensionless natural coordinates* $\xi$ and $\eta$. Although these skew coordinates have their origin at point $g$, it is important to realize that their directions are different at every point on the element. We see that $\eta = -1$ along edge 1-2, $\xi = 1$ along edge 2-3, $\eta = 1$ along edge 3-4, and $\xi = -1$ along edge 4-1. With linear interpolation in both the $\xi$ and $\eta$ directions, formulas for locating a generic point anywhere on the quadrilateral become

$$x = \sum_{i=1}^{4} f_i x_i \qquad y = \sum_{i=1}^{4} f_i y_i \qquad (2)$$

in which the *geometric interpolation functions* are

$$f_1 = \tfrac{1}{4}(1 - \xi)(1 - \eta) \qquad f_2 = \tfrac{1}{4}(1 + \xi)(1 - \eta)$$
$$f_3 = \tfrac{1}{4}(1 + \xi)(1 + \eta) \qquad f_4 = \tfrac{1}{4}(1 - \xi)(1 + \eta) \qquad (3)$$

These functions give the global coordinates of a point on the quadrilateral in terms of the natural coordinates. However, because Eqs. (2) are bilinear (or hyperbolic), the local coordinates $\xi$ and $\eta$ cannot be expressed directly in terms of the global coordinates $x$ and $y$.

We will need derivatives of the functions $f_1$ through $f_4$ to use in strain-displacement relationships. The chain rule for differentiation of $f(\xi, \eta)$ with respect to $x$ and $y$ gives

$$\frac{\partial f}{\partial x} = \frac{\partial f}{\partial \xi} \frac{\partial \xi}{\partial x} + \frac{\partial f}{\partial \eta} \frac{\partial \eta}{\partial x}$$
$$\frac{\partial f}{\partial y} = \frac{\partial f}{\partial \xi} \frac{\partial \xi}{\partial y} + \frac{\partial f}{\partial \eta} \frac{\partial \eta}{\partial y} \qquad (4)$$

or

$$\begin{bmatrix} f_{,x} \\ f_{,y} \end{bmatrix} = \begin{bmatrix} \xi_{,x} & \eta_{,x} \\ \xi_{,y} & \eta_{,y} \end{bmatrix} \begin{bmatrix} f_{,\xi} \\ f_{,\eta} \end{bmatrix} \qquad (5)$$

where the symbol $f_{,x}$ means differentiation of $f$ with respect to $x$, and so on. Terms in the coefficient matrix of Eq. (5) are not readily available, because we are unable to solve explicitly for $\xi$ and $\eta$ in terms of $x$ and $y$. However, if we take the opposite approach and differentiate $f$ with respect to $\xi$ and $\eta$, the chain rule produces

$$\frac{\partial f}{\partial \xi} = \frac{\partial f}{\partial x} \frac{\partial x}{\partial \xi} + \frac{\partial f}{\partial y} \frac{\partial y}{\partial \xi}$$
$$\frac{\partial f}{\partial \eta} = \frac{\partial f}{\partial x} \frac{\partial x}{\partial \eta} + \frac{\partial f}{\partial y} \frac{\partial y}{\partial \eta} \qquad (6)$$

or

$$\begin{bmatrix} f_{,\xi} \\ f_{,\eta} \end{bmatrix} = \begin{bmatrix} x_{,\xi} & y_{,\xi} \\ x_{,\eta} & y_{,\eta} \end{bmatrix} \begin{bmatrix} f_{,x} \\ f_{,y} \end{bmatrix} \qquad (7)$$

## Sec. 7.3  Natural Coordinates

For this arrangement, terms in the coefficient matrix are easily obtained by differentiating Eqs. (2). This array is called the *Jacobian matrix* **J**, which contains derivatives of the global coordinates with respect to the local coordinates. Thus, we have

$$\mathbf{J} = \begin{bmatrix} J_{11} & J_{12} \\ J_{21} & J_{22} \end{bmatrix} = \begin{bmatrix} x_{,\xi} & y_{,\xi} \\ x_{,\eta} & y_{,\eta} \end{bmatrix} \quad (8)$$

Terms in this 2 × 2 matrix are

$$J_{11} = x_{,\xi} = \sum_{i=1}^{4} f_{i,\xi} x_i \qquad J_{12} = y_{,\xi} = \sum_{i=1}^{4} f_{i,\xi} y_i$$
$$J_{21} = x_{,\eta} = \sum_{i=1}^{4} f_{i,\eta} x_i \qquad J_{22} = y_{,\eta} = \sum_{i=1}^{4} f_{i,\eta} y_i \quad (9)$$

Casting these expressions into matrix form yields

$$\mathbf{J} = \mathbf{D}_L \mathbf{C}_N \quad (10)$$

The array $\mathbf{D}_L$ in this equation contains derivatives with respect to local coordinates, as follows:

$$\mathbf{D}_L = \begin{bmatrix} f_{1,\xi} & f_{2,\xi} & f_{3,\xi} & f_{4,\xi} \\ f_{1,\eta} & f_{2,\eta} & f_{3,\eta} & f_{4,\eta} \end{bmatrix}$$
$$= \frac{1}{4} \begin{bmatrix} -(1-\eta) & (1-\eta) & (1+\eta) & -(1+\eta) \\ -(1-\xi) & -(1+\xi) & (1+\xi) & (1-\xi) \end{bmatrix} \quad (11)$$

and the matrix $\mathbf{C}_N$ consists of nodal coordinates in the arrangement

$$\mathbf{C}_N = \begin{bmatrix} x_1 & y_1 \\ x_2 & y_2 \\ x_3 & y_3 \\ x_4 & y_4 \end{bmatrix} \quad (12)$$

By comparing Eqs. (5) and (7), we can see that the coefficient matrix in the former expression is the *inverse of the Jacobian matrix*. Using the formal definition of the inverse, we can obtain $\mathbf{J}^{-1}$ from $\mathbf{J}$ as

$$\mathbf{J}^{-1} = \frac{\mathbf{J}^a}{|\mathbf{J}|} = \frac{1}{|\mathbf{J}|} \begin{bmatrix} J_{22} & -J_{12} \\ -J_{21} & J_{11} \end{bmatrix} = \frac{1}{|\mathbf{J}|} \begin{bmatrix} y_{,\eta} & -y_{,\xi} \\ -x_{,\eta} & x_{,\xi} \end{bmatrix} \quad (13)$$

where $\mathbf{J}^a$ denotes the adjoint matrix of $\mathbf{J}$ and $|\mathbf{J}|$ is its determinant. The latter quantity is calculated by

$$|\mathbf{J}| = J_{11} J_{22} - J_{21} J_{12} = x_{,\xi} y_{,\eta} - x_{,\eta} y_{,\xi} \quad (14)$$

To determine the derivatives of all of the functions with respect to $x$ and $y$, we can apply Eq. (5) repeatedly. Hence,

$$\begin{bmatrix} f_{i,x} \\ f_{i,y} \end{bmatrix} = \mathbf{J}^{-1} \begin{bmatrix} f_{i,\xi} \\ f_{i,\eta} \end{bmatrix} \quad (i = 1, 2, 3, 4) \tag{15}$$

Altogether, we have

$$\mathbf{D}_G = \mathbf{J}^{-1} \mathbf{D}_L = (\mathbf{D}_L \mathbf{C}_N)^{-1} \mathbf{D}_L \tag{16}$$

The matrix $\mathbf{D}_G$ given by this expression consists of derivatives of $f_i$ with respect to the global coordinates. That is,

$$\mathbf{D}_G = \begin{bmatrix} f_{1,x} & f_{2,x} & f_{3,x} & f_{4,x} \\ f_{1,y} & f_{2,y} & f_{3,y} & f_{4,y} \end{bmatrix} \tag{17}$$

Evaluating terms in $\mathbf{D}_G$, we find

$$D_{G11} = \frac{1}{4|\mathbf{J}|}[-(1-\eta)J_{22} + (1-\xi)J_{12}]$$

$$D_{G12} = \frac{1}{4|\mathbf{J}|}[(1-\eta)J_{22} + (1+\xi)J_{12}]$$

$$D_{G13} = \frac{1}{4|\mathbf{J}|}[(1+\eta)J_{22} - (1+\xi)J_{12}]$$

$$D_{G14} = \frac{1}{4|\mathbf{J}|}[-(1+\eta)J_{22} - (1-\xi)J_{12}]$$

$$D_{G21} = \frac{1}{4|\mathbf{J}|}[(1-\eta)J_{21} - (1-\xi)J_{11}] \tag{18}$$

$$D_{G22} = \frac{1}{4|\mathbf{J}|}[-(1-\eta)J_{21} - (1+\xi)J_{11}]$$

$$D_{G23} = \frac{1}{4|\mathbf{J}|}[-(1+\eta)J_{21} + (1+\xi)J_{11}]$$

$$D_{G24} = \frac{1}{4|\mathbf{J}|}[(1+\eta)J_{21} + (1-\xi)J_{11}]$$

By this approach we are able to solve for $\mathbf{D}_G$ numerically.

Figure 7.3(b) shows a quadrilateral with curved edges that may follow quadratic functions, cubic functions, and so on. Regardless of the complexity of these functions, the natural coordinates $\xi$ and $\eta$ play roles similar to those for a quadrilateral with straight edges. The form of the 2 × 2 Jacobian matrix remains the same, even though the functions to be differentiated are of higher order. Because of the appearance of the determinant of $\mathbf{J}$ in denominator positions, we usually cannot integrate terms explicitly to obtain stiffnesses, consistent masses, and equivalent nodal loads. Instead, it becomes necessary to use

Sec. 7.3  Natural Coordinates   323

numerical integration for both straight-sided and curved quadrilaterals, as explained in Sec. 7.4.

Now let us consider a *hexahedron* with straight edges, as depicted in Fig. 7.4(a). At the geometric center (point $g$) the coordinates are

$$x_g = \frac{1}{8}\sum_{i=1}^{8} x_i \qquad y_g = \frac{1}{8}\sum_{i=1}^{8} y_i \qquad z_g = \frac{1}{8}\sum_{i=1}^{8} z_i \qquad (19)$$

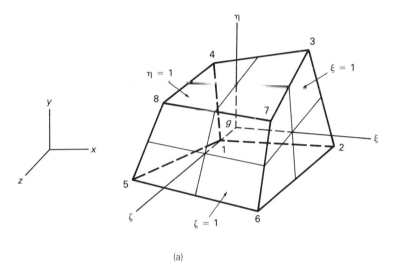

(a)

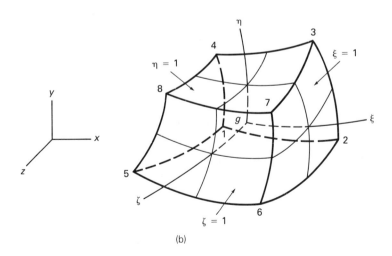

(b)

**Figure 7.4** Natural coordinates for hexahedra: (a) straight; (b) curved.

where $x_i$, $y_i$, and $z_i$ are the Cartesian coordinates of points 1 through 8. The figure also shows *dimensionless natural coordinates* $\xi$, $\eta$, and $\zeta$. Note that $\xi = 1$ on face 2-3-7-6, $\eta = 1$ on face 3-4-8-7, and so on. With linear interpolation in the $\xi$, $\eta$, and $\zeta$ directions, the location of any point in the hexahedron may be written as

$$x = \sum_{i=1}^{8} f_i x_i \qquad y = \sum_{i=1}^{8} f_i y_i \qquad z = \sum_{i=1}^{8} f_i z_i \qquad (20)$$

in which the *geometric interpolation functions* are

$$\begin{aligned}
f_1 &= \tfrac{1}{8}(1 - \xi)(1 - \eta)(1 - \zeta) & f_2 &= \tfrac{1}{8}(1 + \xi)(1 - \eta)(1 - \zeta) \\
f_3 &= \tfrac{1}{8}(1 + \xi)(1 + \eta)(1 - \zeta) & f_4 &= \tfrac{1}{8}(1 - \xi)(1 + \eta)(1 - \zeta) \\
f_5 &= \tfrac{1}{8}(1 - \xi)(1 - \eta)(1 + \zeta) & f_6 &= \tfrac{1}{8}(1 + \xi)(1 - \eta)(1 + \zeta) \\
f_7 &= \tfrac{1}{8}(1 + \xi)(1 + \eta)(1 + \zeta) & f_8 &= \tfrac{1}{8}(1 - \xi)(1 + \eta)(1 + \zeta)
\end{aligned} \qquad (21)$$

Because these interpolation functions are trilinear, the local coordinates $\xi$, $\eta$, and $\zeta$ cannot be expressed in terms of the global coordinates $x$, $y$, and $z$.

In three dimensions the chain rule for differentiation with respect to the natural coordinates for a hexahedron leads to the following $3 \times 3$ *Jacobian matrix*:

$$\mathbf{J} = \begin{bmatrix} J_{11} & J_{12} & J_{13} \\ J_{21} & J_{22} & J_{23} \\ J_{31} & J_{32} & J_{33} \end{bmatrix} = \begin{bmatrix} x_{,\xi} & y_{,\xi} & z_{,\xi} \\ x_{,\eta} & y_{,\eta} & z_{,\eta} \\ x_{,\zeta} & y_{,\zeta} & z_{,\zeta} \end{bmatrix} \qquad (22)$$

Terms in this matrix are found by the differentiations indicated. Thus,

$$\begin{aligned}
J_{11} &= \sum_{i=1}^{8} f_{i,\xi} x_i & J_{12} &= \sum_{i=1}^{8} f_{i,\xi} y_i & J_{13} &= \sum_{i=1}^{8} f_{i,\xi} z_i \\
J_{21} &= \sum_{i=1}^{8} f_{i,\eta} x_i & J_{22} &= \sum_{i=1}^{8} f_{i,\eta} y_i & J_{23} &= \sum_{i=1}^{8} f_{i,\eta} z_i \\
J_{31} &= \sum_{i=1}^{8} f_{i,\zeta} x_i & J_{32} &= \sum_{i=1}^{8} f_{i,\zeta} y_i & J_{33} &= \sum_{i=1}^{8} f_{i,\zeta} z_i
\end{aligned} \qquad (23)$$

As before, these calculations may be arranged in the matrix format given by Eq. (10). In this instance, the matrix $\mathbf{D}_L$ is the following $3 \times 8$ array of derivatives with respect to local coordinates:

$$\mathbf{D}_L = \begin{bmatrix} f_{1,\xi} & f_{2,\xi} & f_{3,\xi} & f_{4,\xi} & f_{5,\xi} & f_{6,\xi} & f_{7,\xi} & f_{8,\xi} \\ f_{1,\eta} & f_{2,\eta} & f_{3,\eta} & f_{4,\eta} & f_{5,\eta} & f_{6,\eta} & f_{7,\eta} & f_{8,\eta} \\ f_{1,\zeta} & f_{2,\zeta} & f_{3,\zeta} & f_{4,\zeta} & f_{5,\zeta} & f_{6,\zeta} & f_{7,\zeta} & f_{8,\zeta} \end{bmatrix}$$

$$= \frac{1}{8} \begin{bmatrix} -(1-\eta)(1-\zeta) & (1-\eta)(1-\zeta) & (1+\eta)(1-\zeta) \\ -(1-\xi)(1-\zeta) & -(1+\xi)(1-\zeta) & (1+\xi)(1-\zeta) \\ -(1-\xi)(1-\eta) & -(1+\xi)(1-\eta) & -(1+\xi)(1+\eta) \end{bmatrix}$$

$$\begin{matrix} -(1+\eta)(1-\zeta) & -(1-\eta)(1+\zeta) & (1-\eta)(1+\zeta) \\ (1-\xi)(1-\zeta) & -(1-\xi)(1+\zeta) & -(1+\xi)(1+\zeta) \\ -(1-\xi)(1+\eta) & (1-\xi)(1-\eta) & (1+\xi)(1-\eta) \end{matrix}$$

$$\left. \begin{matrix} (1+\eta)(1+\zeta) & -(1+\eta)(1+\zeta) \\ (1+\xi)(1+\zeta) & (1-\xi)(1+\zeta) \\ (1+\xi)(1+\eta) & (1-\xi)(1+\eta) \end{matrix} \right] \qquad (24)$$

Also, the matrix $\mathbf{C}_N$ becomes an $8 \times 3$ array of nodal coordinates. Thus,

$$\mathbf{C}_N = \begin{bmatrix} x_1 & y_1 & z_1 \\ x_2 & y_2 & z_2 \\ \cdots & \cdots & \cdots \\ x_8 & y_8 & z_8 \end{bmatrix} \qquad (25)$$

The *inverse of the Jacobian matrix* may be expressed as

$$\mathbf{J}^{-1} = \frac{\mathbf{J}^a}{|\mathbf{J}|} = \frac{1}{|\mathbf{J}|} \begin{bmatrix} J_{11}^a & J_{12}^a & J_{13}^a \\ J_{21}^a & J_{22}^a & J_{23}^a \\ J_{31}^a & J_{32}^a & J_{33}^a \end{bmatrix} \qquad (26)$$

where the symbol $\mathbf{J}^a$ represents the adjoint matrix of $\mathbf{J}$, and $|\mathbf{J}|$ is its determinant. To find the derivatives of all of the functions with respect to global coordinates, we use Eq. (16). In this case, the matrix $\mathbf{D}_G$ consists of the following terms:

$$\mathbf{D}_G = \begin{bmatrix} f_{1,x} & f_{2,x} & f_{3,x} & f_{4,x} & f_{5,x} & f_{6,x} & f_{7,x} & f_{8,x} \\ f_{1,y} & f_{2,y} & f_{3,y} & f_{4,y} & f_{5,y} & f_{6,y} & f_{7,y} & f_{8,y} \\ f_{1,z} & f_{2,z} & f_{3,z} & f_{4,z} & f_{5,z} & f_{6,z} & f_{7,z} & f_{8,z} \end{bmatrix} \qquad (27)$$

A hexahedron with curved edges appears in Fig. 7.4(b). Geometric interpolation functions for this solid may be quadratic, cubic, and so on. The natural coordinates $\xi$, $\eta$, and $\zeta$ will still be used for such a hexahedron in spite of its greater complexity. The $3 \times 3$ Jacobian matrix in Eq. (22) remains the same, although the functions to be differentiated are of higher order. As for a quadrilateral with straight or curved edges, it becomes necessary to use numerical integration to evaluate stiffnesses, consistent masses, and equivalent nodal loads.

## 7.4 NUMERICAL INTEGRATION

The process of computing the value of a definite integral [see Fig. 7.5(a)]

$$I_x = \int_{x_1}^{x_2} f(x) \, dx \tag{1}$$

from a set of numerical values of the integrand is called numerical integration [5]. The problem is solved by representing the integrand by an interpolation formula and then integrating this formula between specified limits. When applied to the integration of a function of a single variable, the method is referred to as *mechanical quadrature*.

If interpolation formulas for numerical integration are polynomials of sufficiently high order relative to those assumed for displacement (or other) functions, the integrations will be exact. Otherwise, the process of numerical integration introduces an additional source of error into finite-element analysis.

The most accurate quadrature formula in common usage is that of Gauss, which involves unequally spaced points that are symmetrically placed. To apply Gauss's method, we usually change the variable from $x$ to the dimensionless coordinate $\xi$ with its origin at the center of the range of integration, as shown in Fig. 7.5(b). The expression for $x$ in terms of $\xi$ is

$$x = \tfrac{1}{2}[(1 - \xi)x_1 + (1 + \xi)x_2] \tag{2}$$

Substitution of Eq. (2) into the function in Eq. (1) gives

$$f(x) = \phi(\xi) \tag{3}$$

Also,

$$dx = \tfrac{1}{2}(x_2 - x_1) \, d\xi \tag{4}$$

Then substituting Eqs. (3) and (4) into Eq. (1) and changing the limits of integration yields

$$I_x = \tfrac{1}{2}(x_2 - x_1) \int_{-1}^{1} \phi(\xi) \, d\xi \tag{5}$$

*Gauss's formula* for determining the integral in Eq. (5) consists of summing the weighted values of $\phi(\xi)$ at $n$ specified points, as follows:

$$I_\xi = \int_{-1}^{1} \phi(\xi) \, d\xi = \sum_{j=1}^{n} R_j \phi(\xi_j)$$
$$= R_1 \phi(\xi_1) + R_2 \phi(\xi_2) + \ldots + R_n \phi(\xi_n) \tag{6}$$

In this expression $\xi_j$ is the location of the *integration point j* relative to the center, $R_j$ is a *weighting factor* for point $j$, and $n$ is the *number of points* at which $\phi(\xi)$ is to be calculated. Values of these parameters are listed in Table 7.1.

## Sec. 7.4 Numerical Integration

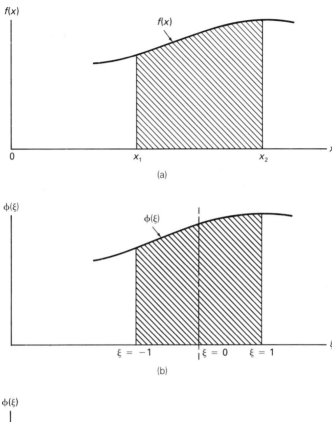

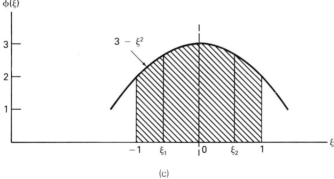

**Figure 7.5** Gaussian quadrature.

Numerical integration by *Gaussian quadrature* is exact for polynomials of degree $2n - 1$. That is, only one integration point is required for the exact integration of a linear function; two points are needed for a cubic polynomial; and so on. For example, consider the quadratic function $\phi(\xi) = 3 - \xi^2$ appearing in Fig. 7.5(c). First, let $n = 1$ and find from Table 7.1 that

$$\xi_1 = 0 \qquad R_1 = 2$$

**TABLE 7.1 Coefficients for Gaussian Quadrature**

| $n$ | $\pm \xi_i$ | $R_i$ |
| --- | --- | --- |
| 1 | 0.0 | 2.0 |
| 2 | 0.5773502692 | 1.0 |
| 3 | 0.7745966692<br>0.0 | 0.5555555556<br>0.8888888889 |
| 4 | 0.8611363116<br>0.3399810436 | 0.3478548451<br>0.6521451549 |
| 5 | 0.9061798459<br>0.5384693101<br>0.0 | 0.2369268851<br>0.4786286705<br>0.5688888889 |
| 6 | 0.9324695142<br>0.6612093865<br>0.2386191861 | 0.1713244924<br>0.3607615730<br>0.4679139346 |
| 7 | 0.9491079123<br>0.7415311856<br>0.4058451514<br>0.0 | 0.1294849662<br>0.2797053915<br>0.3818300505<br>0.4179591837 |
| 8 | 0.9602898565<br>0.7966664774<br>0.5255324099<br>0.1834346425 | 0.1012285363<br>0.2223810345<br>0.3137066459<br>0.3626837834 |

Then from Eq. (6) we have

$$I_\xi = R_1 \phi(\xi_1) = (2)(3) = 6$$

which is approximate. Next, let $n = 2$ and obtain from Table 7.1:

$$\xi_1 = -\xi_2 = -\frac{1}{\sqrt{3}} = -0.5773 \ldots \qquad R_1 = R_2 = 1$$

Hence, we find from Eq. (6)

### Sec. 7.4  Numerical Integration

$$I_\xi = \sum_{j=1}^{2} R_j \phi(\xi_j) = (1)(3 - \xi_1^2) + (1)(3 - \xi_2^2)$$

$$= (2)(2.666\ldots) = 5.333\ldots$$

which is exact.

Now let us apply Gaussian quadrature to quadrilaterals in Cartesian coordinates. The type of integration to be performed is

$$I = \int_{y_1}^{y_2} \int_{x_1}^{x_2} f(x, y)\, dx\, dy \tag{7}$$

However, this integral is more easily evaluated if it is first transformed to the natural coordinates for a quadrilateral. We accomplish this by expressing the function $f$ in terms of $\xi$ and $\eta$ and using the limits $-1$ to $1$ for each of the integrals. In addition, the infinitesimal area $dA = dx\, dy$ must be replaced by an appropriate expression in terms of $d\xi$ and $d\eta$. For this purpose, Fig. 7.6 shows an infinitesimal area $dA$ in the natural coordinates. Vector $\mathbf{r}$ locates a generic point in the Cartesian coordinates $x$ and $y$, as follows:

$$\mathbf{r} = \mathbf{x} + \mathbf{y} = x\mathbf{i} + y\mathbf{j} \tag{8}$$

The rate of change of $\mathbf{r}$ with respect to $\xi$ is

$$\frac{\partial \mathbf{r}}{\partial \xi} = \frac{\partial x}{\partial \xi}\mathbf{i} + \frac{\partial y}{\partial \xi}\mathbf{j} \tag{9}$$

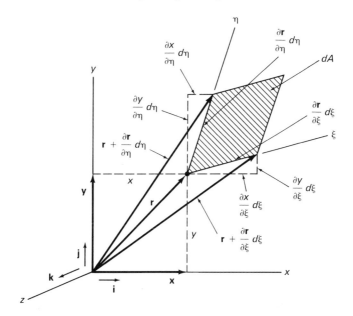

**Figure 7.6**  Infinitesimal area in natural coordinates.

Also, the rate of change of **r** with respect to $\eta$ is

$$\frac{\partial \mathbf{r}}{\partial \eta} = \frac{\partial x}{\partial \eta}\mathbf{i} + \frac{\partial y}{\partial \eta}\mathbf{j} \tag{10}$$

When multiplied by $d\xi$ and $d\eta$, the derivatives in Eqs. (9) and (10) form two adjacent sides of the infinitesimal parallelogram of area $dA$ in the figure. This area may be determined from the following vector triple product

$$dA = \left(\frac{\partial \mathbf{r}}{\partial \xi} d\xi \times \frac{\partial \mathbf{r}}{\partial \eta} d\eta\right) \cdot \mathbf{k} \tag{11}$$

Substitution of Eqs. (9) and (10) into Eq. (11) produces

$$dA = \left(\frac{\partial x}{\partial \xi}\frac{\partial y}{\partial \eta} - \frac{\partial x}{\partial \eta}\frac{\partial y}{\partial \xi}\right) d\xi\, d\eta \tag{12}$$

The expression in the parentheses of Eq. (12) may be written as a $2 \times 2$ determinant. That is,

$$dA = \begin{vmatrix} x_{,\xi} & y_{,\xi} \\ x_{,\eta} & y_{,\eta} \end{vmatrix} d\xi\, d\eta = |\mathbf{J}|\, d\xi\, d\eta \tag{13}$$

in which $\mathbf{J}$ is the Jacobian matrix given in Eq. (7.3-8), and $|\mathbf{J}|$ is its determinant. Thus, the new form of the integral in Eq. (7) becomes

$$I = \int_{-1}^{1}\int_{-1}^{1} f(\xi, \eta)\, |\mathbf{J}|\, d\xi\, d\eta \tag{14}$$

Two successive applications of Gaussian quadrature result in

$$I = \sum_{k=1}^{n}\sum_{j=1}^{n} R_j R_k f(\xi_j, \eta_k)\, |\mathbf{J}(\xi_j, \eta_k)| \tag{15}$$

where $R_j$ and $R_k$ are weighting factors for the function evaluated at the point $(\xi_j, \eta_k)$. Integration points for $n = 1, 2, 3,$ and $4$ each way on a quadrilateral are illustrated in Fig. 7.7.

Next, we turn to hexahedra in Cartesian coordinates, where the type of integral to be evaluated has the form

$$I = \int_{z_1}^{z_2}\int_{y_1}^{y_2}\int_{x_1}^{x_2} f(x, y, z)\, dx\, dy\, dz \tag{16}$$

Before integrating, we rewrite the function in terms of the natural coordinates $\xi$, $\eta$, and $\zeta$ and use the limits $-1$ to $1$ for each of the integrals. Furthermore, we must replace the infinitesimal volume $dV = dx\, dy\, dz$ by an expression involving $d\xi$, $d\eta$, and $d\zeta$. Toward that end, Fig. 7.8 shows an infinitesimal volume $dV$ in the natural coordinates. Also depicted is a vector **r**, which locates a generic point in the space. Thus,

## Sec. 7.4 Numerical Integration

$$\mathbf{r} = x\mathbf{i} + y\mathbf{j} + z\mathbf{k} \tag{17}$$

The rates of change of **r** with respect to $\xi$, $\eta$, and $\zeta$ are

$$\frac{\partial \mathbf{r}}{\partial \xi} = \frac{\partial x}{\partial \xi}\mathbf{i} + \frac{\partial y}{\partial \xi}\mathbf{j} + \frac{\partial z}{\partial \xi}\mathbf{k}$$

$$\frac{\partial \mathbf{r}}{\partial \eta} = \frac{\partial x}{\partial \eta}\mathbf{i} + \frac{\partial y}{\partial \eta}\mathbf{j} + \frac{\partial z}{\partial \eta}\mathbf{k} \tag{18}$$

$$\frac{\partial \mathbf{r}}{\partial \zeta} = \frac{\partial x}{\partial \zeta}\mathbf{i} + \frac{\partial y}{\partial \zeta}\mathbf{j} + \frac{\partial z}{\partial \zeta}\mathbf{k}$$

Let the symbols **a**, **b**, and **c** denote

$$\mathbf{a} = \frac{\partial \mathbf{r}}{\partial \xi} d\xi \qquad \mathbf{b} = \frac{\partial \mathbf{r}}{\partial \eta} d\eta \qquad \mathbf{c} = \frac{\partial \mathbf{r}}{\partial \zeta} d\zeta \tag{19}$$

These vectors are shown in Fig. 7.8 as the edges of the infinitesimal parallelepiped of volume $dV$. This volume may be determined from the following vector triple product:

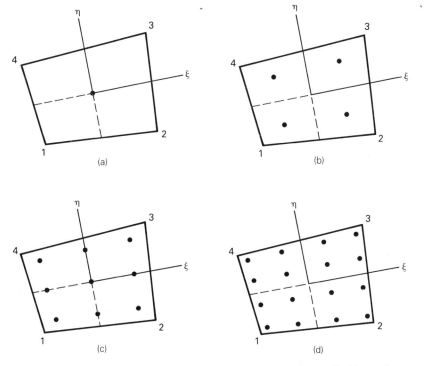

**Figure 7.7** Integration points for quadrilateral: (a) $n = 1$; (b) $n = 2$; (c) $n = 3$; (d) $n = 4$ (each way).

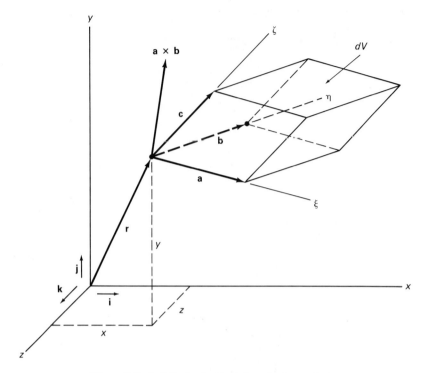

**Figure 7.8** Infinitesimal volume in natural coordinates.

$$dV = (\mathbf{a} \times \mathbf{b}) \cdot \mathbf{c} = |\mathbf{J}|\, d\xi\, d\eta\, d\zeta$$

$$= \begin{vmatrix} \dfrac{\partial x}{\partial \xi} & \dfrac{\partial y}{\partial \xi} & \dfrac{\partial z}{\partial \xi} \\ \dfrac{\partial x}{\partial \eta} & \dfrac{\partial y}{\partial \eta} & \dfrac{\partial z}{\partial \eta} \\ \dfrac{\partial x}{\partial \zeta} & \dfrac{\partial y}{\partial \zeta} & \dfrac{\partial z}{\partial \zeta} \end{vmatrix} d\xi\, d\eta\, d\zeta \quad (20)$$

in which $\mathbf{J}$ is the $3 \times 3$ Jacobian matrix, and $|\mathbf{J}|$ is its determinant. Hence, the revised form of the integral in Eq. (16) becomes

$$I = \int_{-1}^{1}\int_{-1}^{1}\int_{-1}^{1} f(\xi, \eta, \zeta)\, |\mathbf{J}|\, d\xi\, d\eta\, d\zeta \quad (21)$$

Three successive applications of Gaussian quadrature yield

$$I = \sum_{l=1}^{n}\sum_{k=1}^{n}\sum_{j=1}^{n} R_j R_k R_l f(\xi_j, \eta_k, \zeta_l)\, |\mathbf{J}(\xi_j, \eta_k, \zeta_l)| \quad (22)$$

Integration points for $n = 1, 2, 3,$ and $4$ each way number 1, 8, 27, and 64, respectively.

## 7.5 ISOPARAMETRIC QUADRILATERALS FOR PLANE STRESS AND PLANE STRAIN

A finite element is *isoparametric* if the same interpolation formulas define both the geometric and displacement shape functions. Such elements satisfy geometric as well as displacement compatibility conditions. If the geometric interpolation functions are of lower order than the displacement shape functions, the element is called *subparametric*. On the other hand, if the reverse were true, the element would be referred to as *superparametric* [6]. However, most commonly used finite elements are either isoparametric or subparametric. Because isoparametric elements are usually curved, they tend to be more suitable than subparametric elements for modeling curved geometric boundary conditions.

Figure 7.9(a) shows the *rectangular parent R4* of the *isoparametric quadrilateral Q4*, which appears in Fig. 7.9(b). Conversely, the rectangle may be considered as a special case of the quadrilateral, for which the natural coordinates $\xi$ and $\eta$ are orthogonal throughout the element. For either figure the generic displacements at a typical point are

$$\mathbf{u} = \{u, v\} \tag{1}$$

Nodal displacements indicated in Fig. 7.9 consist of $x$ and $y$ translations at each node, as follows:

$$\mathbf{q} = \{q_1, q_2, \ldots, q_8\} = \{u_1, v_1, \ldots, v_4\} \tag{2}$$

For both elements we assume the displacement shape functions

$$u = \sum_{i=1}^{4} f_i u_i \qquad v = \sum_{i=1}^{4} f_i v_i \tag{3}$$

In these expressions the functions $f_1, f_2, f_3$, and $f_4$ are the same as those in the geometric interpolation formulas given by Eqs. (7.3-3). Therefore, the Q4 element is isoparametric, and previous statements regarding differentiation and integration of functions hold true. Equations (3) may also be written in the matrix form

$$\mathbf{u}_i = \mathbf{f}_i \mathbf{q}_i \qquad (i = 1, 2, 3, 4) \tag{4}$$

in which

$$\mathbf{f}_i = \begin{bmatrix} 1 & 0 \\ 0 & 1 \end{bmatrix} f_i \tag{5}$$

The generic displacements $\mathbf{u}_i$ in Eq. (4) represent translations at any point due to the displacements $\mathbf{q}_i$ at node $i$. As a further efficiency of notation, we can write the function $f_i$ as

$$f_i = \tfrac{1}{4}(1 + \xi_0)(1 + \eta_0) \tag{6}$$

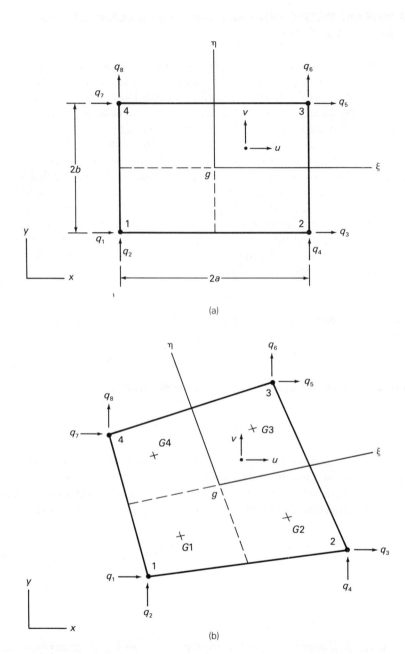

**Figure 7.9** Element Q4: (a) rectangular parent; (b) isoparametric counterpart.

### Sec. 7.5 Isoparametric Quadrilaterals for Plane Stress and Plane Strain

where

$$\xi_0 = \xi_i \xi \qquad \eta_0 = \eta_i \eta \tag{7}$$

Values of $\xi_i$ and $\eta_i$ for this element are listed in Table 7.2

**TABLE 7.2 Nodal Coordinates for Element Q4**

| $i$ | 1 | 2 | 3 | 4 |
|---|---|---|---|---|
| $\xi_i$ | −1 | 1 | 1 | −1 |
| $\eta_i$ | −1 | −1 | 1 | 1 |

Similarly, the strain-displacement relationships for element Q4 may be concisely expressed as

$$\boldsymbol{\xi}_i = \mathbf{B}_i \mathbf{q}_i \qquad (i = 1, 2, 3, 4) \tag{8}$$

where

$$\mathbf{B}_i = \mathbf{d}\, \mathbf{f}_i = \begin{bmatrix} \dfrac{\partial}{\partial x} & 0 \\ 0 & \dfrac{\partial}{\partial y} \\ \dfrac{\partial}{\partial y} & \dfrac{\partial}{\partial x} \end{bmatrix} \mathbf{f}_i = \begin{bmatrix} f_{i,x} & 0 \\ 0 & f_{i,y} \\ f_{i,y} & f_{i,x} \end{bmatrix} \tag{9a}$$

Referring to Eqs. (7.3-17) and (7.3-18), we see that the submatrix $\mathbf{B}_i$ can also be written

$$\mathbf{B}_i = \begin{bmatrix} D_{G1i} & 0 \\ 0 & D_{G2i} \\ D_{G2i} & D_{G1i} \end{bmatrix} \tag{9b}$$

Next, we express the stiffness matrix for the Q4 element (with constant thickness $h$) in Cartesian coordinates as

$$\mathbf{K} = h \int_A \mathbf{B}^T(x, y)\, \mathbf{E}\, \mathbf{B}(x, y)\, dx\, dy \tag{10a}$$

However, in natural coordinates this formula becomes

$$\mathbf{K} = h \int_{-1}^{1} \int_{-1}^{1} \mathbf{B}^T(\xi, \eta)\, \mathbf{E}\, \mathbf{B}(\xi, \eta)\, |\mathbf{J}(\xi, \eta)|\, d\xi\, d\eta \tag{10b}$$

and with two applications of Gaussian quadrature we have

$$\mathbf{K} = h \sum_{k=1}^{n} \sum_{j=1}^{n} R_j R_k \mathbf{B}_{j,k}^T \mathbf{E}\, \mathbf{B}_{j,k} |\mathbf{J}_{j,k}| \tag{10c}$$

In this expression for numerical integration, the matrix $\mathbf{B}_{j,k}$ and the determinant $|\mathbf{J}_{j,k}|$ are evaluated at each integration point, where the coordinates are $(\xi_j, \eta_k)$. Similarly, the consistent mass matrix for element Q4 is

$$\mathbf{M} = \rho h \int_A \mathbf{f}^T(x, y) \mathbf{f}(x, y) \, dx \, dy \tag{11a}$$

or

$$\mathbf{M} = \rho h \int_{-1}^{1} \int_{-1}^{1} \mathbf{f}^T(\xi, \eta) \mathbf{f}(\xi, \eta) \, |\mathbf{J}(\xi, \eta)| \, d\xi \, d\eta \tag{11b}$$

or

$$\mathbf{M} = \rho h \sum_{k=1}^{n} \sum_{j=1}^{n} R_j R_k \mathbf{f}_{j,k}^T \mathbf{f}_{j,k} \, |\mathbf{J}_{j,k}| \tag{11c}$$

in which $\rho$ and $h$ are assumed to be constant.

In addition, equivalent nodal loads due to body forces may be stated as follows:

$$\mathbf{p}_b(t) = h \int_A \mathbf{f}^T(x, y) \mathbf{b}(x, y, t) \, dx \, dy \tag{12a}$$

or

$$\mathbf{p}_b(t) = h \int_{-1}^{1} \int_{-1}^{1} \mathbf{f}^T(\xi, \eta) \mathbf{b}(\xi, \eta, t) \, |\mathbf{J}(\xi, \eta)| \, d\xi \, d\eta \tag{12b}$$

or

$$\mathbf{p}_b(t) = h \sum_{k=1}^{n} \sum_{j=1}^{n} R_j R_k \mathbf{f}_{j,k}^T \mathbf{b}(t)_{j,k} \, |\mathbf{J}_{j,k}| \tag{12c}$$

where both $\mathbf{p}_b$ and $\mathbf{b}$ are functions of time.

Except in special cases, the integrals in Eqs. (10), (11), and (12) must be performed by numerical integration to obtain approximate results. However, if the element is rectangular, direct explicit integration may be used, for either Cartesian or natural coordinates. Also, line loadings with $\xi$ or $\eta$ constant may be handled by explicit line integrations. Of course, if the body forces consist of point loads, no integration is required at all.

By specializing the quadrilateral Q4 to become the rectangle R4, we can derive explicit terms in its stiffness and consistent-mass matrices. For this case the Jacobian matrix and its determinant become

$$\mathbf{J} = \begin{bmatrix} a & 0 \\ 0 & b \end{bmatrix} \qquad |\mathbf{J}| = ab \tag{13}$$

in which $a$ and $b$ are half the width and height of the rectangle in Fig. 7.9(a).

## Sec. 7.5  Isoparametric Quadrilaterals for Plane Stress and Plane Strain

Therefore, Eqs. (10b) and (11b) simplify to

$$\mathbf{K} = abh \int_{-1}^{1}\int_{-1}^{1} \mathbf{B}^T(\xi, \eta)\mathbf{E}\,\mathbf{B}(\xi, \eta)\, d\xi\, d\eta \tag{14}$$

and

$$\mathbf{M} = ab\rho h \int_{-1}^{1}\int_{-1}^{1} \mathbf{f}^T(\xi, \eta)\mathbf{f}(\xi, \eta)\, d\xi\, d\eta \tag{15}$$

The resulting $8 \times 8$ matrices are given in Ref. 2 and need not be repeated here. Finally, the equivalent nodal loads in Eq. (12b) also take the simpler form

$$\mathbf{p}_b(t) = abh \int_{-1}^{1}\int_{-1}^{1} \mathbf{f}^T(\xi, \eta)\mathbf{b}(\xi, \eta, t)\, d\xi\, d\eta \tag{16}$$

### Example 7.1

Derive numerically the consistent mass term $M_{24}$ for the isoparametric Q4 element in Fig. 7.9(b), using Gaussian integration with $n = 2$ each way. Assume that $\rho$ and $h$ are constants and that the coordinates of nodes 1, 2, 3, and 4 are (3, 1), (8, 2), (6, 6), and (2, 5), respectively.

To apply Eq. (11c), we must set up the network of four integration points (or Gauss points) G1, G2, G3, and G4 indicated in Fig. 7.9(b). In particular, for $n = 2$ we have $R_j = R_k = 1$ (from Table 7.1), so that

$$M_{24} = \rho h \sum_{k=1}^{2}\sum_{j=1}^{2} (f_1)_{j,k}(f_2)_{j,k}\, |\mathbf{J}_{j,k}| \tag{a}$$

The functions $f_1$ and $f_2$ are needed for evaluating the term $M_{24}$ because of the arrangement of matrix $\mathbf{f}$ shown in Eq. (5). Substituting $f_1$ and $f_2$ from Eq. (6) into Eq. (a) produces

$$M_{24} = \frac{\rho h}{16} \sum_{k=1}^{2}\sum_{j=1}^{2} [(1 - \xi^2)(1 - \eta)^2]_{j,k}\, |\mathbf{J}_{j,k}| \tag{b}$$

To implement this formula, we first calculate the Jacobian matrix using Eq. (7.3-10), as follows:

$$\mathbf{J} = \mathbf{D}_L \mathbf{C}_N = \frac{1}{4}\begin{bmatrix} -(1-\eta) & (1-\eta) & (1+\eta) & -(1+\eta) \\ -(1-\xi) & -(1+\xi) & (1+\xi) & (1-\xi) \end{bmatrix} \begin{bmatrix} 3 & 1 \\ 8 & 2 \\ 6 & 6 \\ 2 & 5 \end{bmatrix}$$

$$= \frac{1}{4}\begin{bmatrix} 9-\eta & 2 \\ -3-\xi & 8 \end{bmatrix} \tag{c}$$

Then the determinant of $\mathbf{J}$ is

$$|\mathbf{J}| = \tfrac{1}{8}(39 + \xi - 4\eta) \tag{d}$$

From Table 7.1 the values of $\xi$ and $\eta$ at point G1 are $-1/\sqrt{3}$ and $-1/\sqrt{3}$, and so on for the other integration points. Evaluating the terms in Eq. (b) at each of the four integration points and summing the results yields

$$M_{24} = 1.139 \, \rho h \tag{e}$$

which can be finalized using numerical values for $\rho$ and $h$.

Now we shall consider a higher-order quadrilateral element that is based on quadratic geometric and displacement shape functions. Figure 7.10(a) depicts the *rectangular parent R8* of the *isoparametric quadrilateral Q8* illustrated in Fig. 7.10(b). We may consider the rectangle to be a special case of the quadrilateral, for which the natural coordinates $\xi$ and $\eta$ are orthogonal throughout the element. In addition, edges of the rectangle are straight, and nodes 5 through 8 are located at midedges of sides 1-2 through 4-1. For both elements R8 and Q8, the nodal displacement vector is

$$\mathbf{q} = \{q_1, q_2, \ldots, q_{16}\} = \{u_1, v_1, \ldots, v_8\} \tag{17}$$

which contains $x$ and $y$ translations at each node. We also assume the following quadratic displacement shape functions:

$$u = \sum_{i=1}^{8} f_i u_i \qquad v = \sum_{i=1}^{8} f_i v_i \tag{18}$$

where

$$\begin{aligned} f_i &= \tfrac{1}{4}(1 + \xi_0)(1 + \eta_0)(-1 + \xi_0 + \eta_0) & (i = 1, 2, 3, 4) \\ f_i &= \tfrac{1}{2}(1 - \xi^2)(1 + \eta_0) & (i = 5, 7) \\ f_i &= \tfrac{1}{2}(1 + \xi_0)(1 - \eta^2) & (i = 6, 8) \end{aligned} \tag{19}$$

The values of $\xi_i$ and $\eta_i$ required in these formulas [see also Eqs. (7)] are given in Table 7.3.

For element Q8 we take the geometric interpolation functions to be the same as the displacement shape functions in Eqs. (19). Physically, this means that the natural coordinates $\xi$ and $\eta$ are curvilinear, and all edges of the element become quadratic curves [7]. Thus, we locate any point on the quadrilateral (including point $g$) by the formulas

$$x = \sum_{i=1}^{8} f_i x_i \qquad y = \sum_{i=1}^{8} f_i y_i \tag{20}$$

and element Q8 is seen to be isoparametric. Formulations of stiffnesses, consistent masses, and equivalent nodal loads for this element are very similar to those for element Q4 given earlier. Table 7.4 contains the necessary shape functions and their derivatives with respect to $\xi$ and $\eta$. Numerical integration also follows the same pattern as before, even though the local coordinates are curved.

Sec. 7.5  Isoparametric Quadrilaterals for Plane Stress and Plane Strain

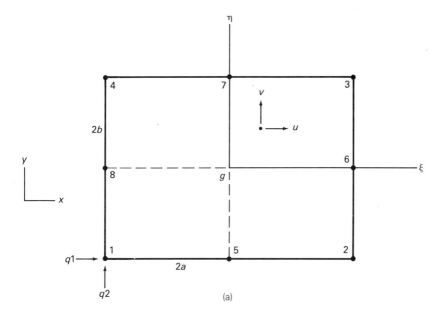

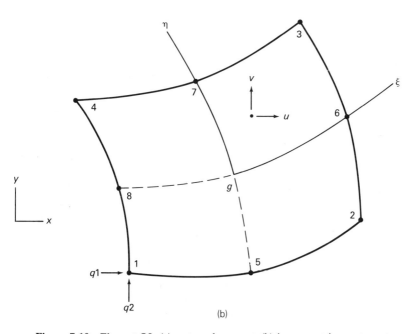

**Figure 7.10**  Element Q8: (a) rectangular parent; (b) isoparametric counterpart.

**TABLE 7.3** Nodal Coordinates for Element Q8

| $i$ | 1 | 2 | 3 | 4 | 5 | 6 | 7 | 8 |
|---|---|---|---|---|---|---|---|---|
| $\xi_i$ | $-1$ | 1 | 1 | $-1$ | 0 | 1 | 0 | $-1$ |
| $\eta_i$ | $-1$ | $-1$ | 1 | 1 | $-1$ | 0 | 1 | 0 |

**TABLE 7.4** Shape Functions and Derivatives for Element Q8

| $i$ | $f_i$ | $f_{i,\xi}$ | $f_{i,\eta}$ |
|---|---|---|---|
| 1 | $(1 - \xi)(1 - \eta)(-\xi - \eta - 1)/4$ | $(2\xi + \eta)(1 - \eta)/4$ | $(1 - \xi)(2\eta + \xi)/4$ |
| 2 | $(1 + \xi)(1 - \eta)(\xi - \eta - 1)/4$ | $(2\xi - \eta)(1 - \eta)/4$ | $(1 + \xi)(2\eta - \xi)/4$ |
| 3 | $(1 + \xi)(1 + \eta)(\xi + \eta - 1)/4$ | $(2\xi + \eta)(1 + \eta)/4$ | $(1 + \xi)(2\eta + \xi)/4$ |
| 4 | $(1 - \xi)(1 + \eta)(-\xi + \eta - 1)/4$ | $(2\xi - \eta)(1 + \eta)/4$ | $(1 - \xi)(2\eta - \xi)/4$ |
| 5 | $(1 - \xi^2)(1 - \eta)/2$ | $-\xi(1 - \eta)$ | $-(1 - \xi^2)/2$ |
| 6 | $(1 + \xi)(1 - \eta^2)/2$ | $(1 - \eta^2)/2$ | $-(1 + \xi)\eta$ |
| 7 | $(1 - \xi^2)(1 + \eta)/2$ | $-\xi(1 + \eta)$ | $(1 - \xi^2)/2$ |
| 8 | $(1 - \xi)(1 - \eta^2)/2$ | $-(1 - \eta^2)/2$ | $-(1 - \xi)\eta$ |

## 7.6 PROGRAM DYNAPS FOR PLANE STRESS AND PLANE STRAIN

Now we shall discuss Program DYNAPS for dynamic analysis of thin plates in plane stress or prismatic solids in plane strain. For this purpose it is assumed that a given continuum has been discretized using either Q4 or Q8 elements, as described in Sec. 7.5. All such quadrilaterals in the analytical model have the same thickness $h$, and the material is taken to be homogeneous and isotropic.

Table 7.5 shows preparation of structural data for Program DYNAPS. This table is similar to Table 3.2 for plane trusses, described previously in Sec. 3.8. However, the second line of Table 7.5 contains four additional structural parameters. The first of these is an indicator for plane stress (IPS = 0) or plane strain (IPS = 1). Next is the number of element nodes NEN for the quadrilateral, where NEN = 4 for Q4 elements and NEN = 8 for Q8 elements. The other two structural parameters included are Poisson's ratio PR and the thickness H. Furthermore, element information now consists of the element number I and node numbers one through four for element Q4 and one through eight for element Q8.

Preparation of dynamic load data appears in Table 7.6, which is similar to Table 4.1 for plane trusses (see Sec. 4.10). However, the load parameters NEL

## Sec. 7.6　Program DYNAPS for Plane Stress and Plane Strain

**TABLE 7.5　Structural Data for Program DYNAPS**

| Type of Data | No. of Lines | Items on Data Lines |
|---|---|---|
| Problem identification | 1 | Descriptive title |
| Structural parameters | 1 | NN, NE, NRN, IPS, NEN, E, PR, RHO, H |
| Plane stress (strain) data (a) Nodal coordinates (b) Element information[a] (c) Nodal restraints | NN NE NRN | J, X(J), Y(J) I, IN(I, 1), IN(I, 2), . . . , IN(I, NEN) J, NRL(2J-1), NRL(2J) |

[a] For sequences of node numbers, see Figs. 7.9(b) and 7.10(b).

**TABLE 7.6　Dynamic Load Data for Program DYNAPS**

| Type of Data | No. of Lines | Items on Data Lines |
|---|---|---|
| Dynamic parameters | 1 | ISOLVE, NTS, DT, DAMPR |
| Initial conditions (a) Condition parameters (b) Displacements (c) Velocities | 1 NNID NNIV | NNID, NNIV J, D0(2J-1), D0(2J) J, V0(2J-1), V0(2J) |
| Applied actions (a) Load parameters (b) Nodal loads (c) Line loads[a] (d) Volume loads | 1 NLN NEL NEV | NLN, NEL, NEV J, AS(2J-1), AS(2J) J, K, BL1, BL2, BL3, BL4 I, BV1, BV2 |
| Ground accelerations (a) Acceleration parameter (b) Acceleration factors | 1 1 | IGA GAX, GAY |
| Forcing function (a) Function parameter (b) Function ordinates | 1 NFO | NFO K, T(K), FO(K) |

[a] Pertains only to element Q4 (NEN = 4). For element Q8 (NEN = 8), we need three node numbers (J, K, L) and six force intensities (BL1, BL2, . . . , BL6).

and NEV in Table 7.6 must be explained. If NEL ≠ 0, at least one line load (force per unit length) exists on the edge *jk* of an element. The types of line loads for quadrilateral elements are illustrated in Fig. 7.11. For the Q4 element a linearly varying line load is specified by the force intensities BL1 through BL4 that are listed in the table and shown in Fig. 7.11(a). On the other hand, the Q8 element in Fig. 7.11(b) has two components of quadratically varying line loads, defined by the force intensities BL1 through BL6 (see the footnote below the table).

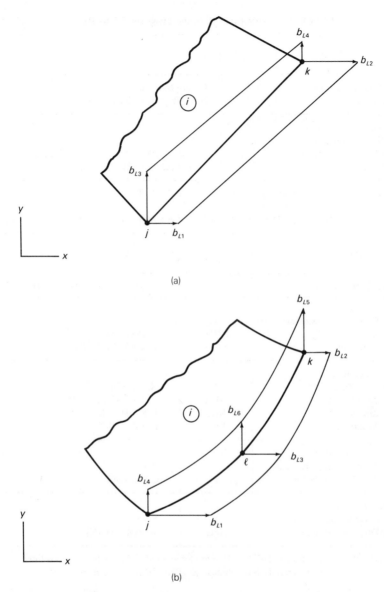

**Figure 7.11** Line loads for quadrilaterals: (a) element Q4; (b) element Q8.

If NEV ≠ 0, at least one element is subjected to volume loads (force per unit volume). Here the notation implies that element I may have a uniformly distributed force BV1 in the $x$ direction and a second uniformly distributed force BV2 in the $y$ direction.

In Program DYNAPS we take $n = 2$ each way to locate points for numerical integration. At each of the four points the computer evaluates the

### Sec. 7.6  Program DYNAPS for Plane Stress and Plane Strain

time-varying stresses $\sigma_x$, $\sigma_y$, $\tau_{xy}$, and $\sigma_z$, the last of which is nonzero for the case of plane strain. However, we found it necessary to use $n = 3$ each way for terms in the consistent-mass matrix of element Q8, which contain products of quadratic functions.

### Example 7.2

Figure 7.12 illustrates the cross section of a machine part that is subjected to a state of plane strain. The part is divided into Q8 elements and has a parabolically distributed step load on edge 1-5, with maximum intensity $b_x$ at node 3. In this example the physical parameters are:

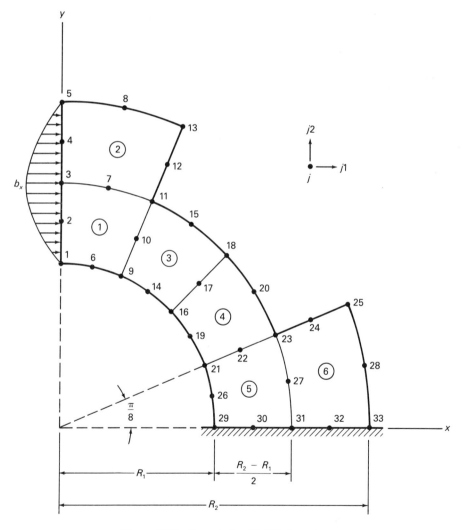

**Figure 7.12** Plane strain with Q8 elements.

$$E = 207 \times 10^6 \text{ kN/m}^2 \qquad \nu = 0.30 \qquad \rho = 7.85 \text{ Mg/m}^3$$
$$R_2 = 2R_1 = 0.02 \text{ m} \qquad h = 1 \text{ m} \qquad b_x = 1 \text{ kN/m}$$

for which the material is steel and the units are SI.

We ran this data with Program DYNAPS, using IPS = 1, NEN = 8, DAMPR = 0.05, and NUMINT for the solution method. Translations of node 5 in the $x$ and $y$ directions appear in the computer plots of Fig. 7.13(a). Also, the normal stresses SX, SY, and SZ near node 29 are plotted in Fig. 7.13(b). Maximum values of the nodal translations are 0.06505 mm and 0.04775 mm; and those for the normal stresses are 56.20 MPa, 116.8 MPa, and 51.89 MPa. (The maximum shearing stress accompanying the normal stresses near node 29 is less than a twentieth of SX.)

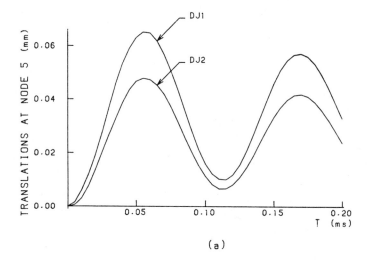

(a)

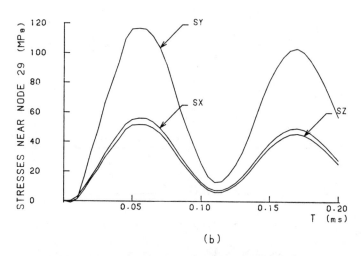

(b)

**Figure 7.13** Responses of plane strain example to step load.

## 7.7 ISOPARAMETRIC HEXAHEDRA FOR GENERAL SOLIDS

Figure 7.14(a) illustrates the *parent rectangular solid RS8* of the *isoparametric hexahedron H8*, shown in Fig. 7.14(b). In the former element, the natural coordinates $\xi$, $\eta$, and $\zeta$ are orthogonal everywhere. For either element the generic displacements at a typical point are

$$\mathbf{u} = \{u, v, w\} \tag{1}$$

Nodal displacements consist of $x$, $y$, and $z$ translations at each corner node, which fill the vector:

$$\mathbf{q} = \{q_1, q_2, q_3, \ldots, q_{24}\} = \{u_1, v_1, w_1, \ldots, w_8\} \tag{2}$$

Trilinear displacement shape functions may be expressed as

$$u = \sum_{i=1}^{8} f_i u_i \quad v = \sum_{i=1}^{8} f_i v_i \quad w = \sum_{i=1}^{8} f_i w_i \tag{3}$$

where

$$f_i = \tfrac{1}{8}(1 + \xi_0)(1 + \eta_0)(1 + \zeta_0) \tag{4}$$

and

$$\xi_0 = \xi_i \xi \quad \eta_0 = \eta_i \eta \quad \zeta_0 = \zeta_i \zeta \tag{5}$$

The formulas represented by Eq. (4) are the same as those given previously in Eqs. (7.3-21). Values of $\xi_i$, $\eta_i$, and $\zeta_i$ required for Eqs. (5) appear in Table 7.7.

For the parent rectangular solid RS8, explicit integrations are feasible, and stiffnesses for an orthotropic material were presented by Melosh [8]. Furthermore, consistent masses for this simplified element are easy to derive and were given in Ref. 2.

Turning now to the more general isoparametric H8 element [9], we take the geometric interpolation functions to be those that were given earlier as Eqs. (7.3-20). Because the geometric and displacement shape functions are the same, the H8 element proves to be isoparametric. Equations (3) can also be stated as the matrix expression:

$$\mathbf{u}_i = \mathbf{f}_i \mathbf{q}_i \quad (i = 1, 2, \ldots, 8) \tag{6}$$

in which

$$\mathbf{f}_i = \begin{bmatrix} 1 & 0 & 0 \\ 0 & 1 & 0 \\ 0 & 0 & 1 \end{bmatrix} f_i \tag{7}$$

As before, the generic displacements $\mathbf{u}_i$ in Eq. (6) denote translations at any point due to the displacements $\mathbf{q}_i$ at node $i$.

In addition, strain-displacement relationships may be written efficiently as:

$$\boldsymbol{\epsilon}_i = \mathbf{B}_i \mathbf{q}_i \quad (i = 1, 2, \ldots, 8) \tag{8}$$

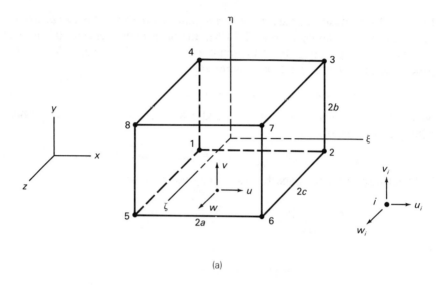

(a)

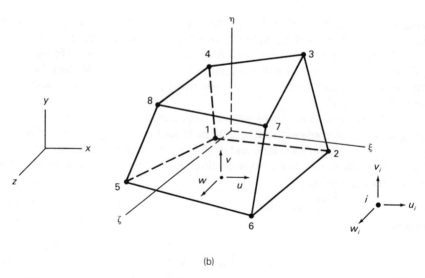

(b)

**Figure 7.14** Element H8: (a) parent rectangular solid; (b) isoparametric counterpart.

## Sec. 7.7  Isoparametric Hexahedra for General Solids

**TABLE 7.7  Nodal Coordinates for Element H8**

| $i$ | $\xi_i$ | $\eta_i$ | $\zeta_i$ |
|---|---|---|---|
| 1 | −1 | −1 | −1 |
| 2 |  1 | −1 | −1 |
| 3 |  1 |  1 | −1 |
| 4 | −1 |  1 | −1 |
| 5 | −1 | −1 |  1 |
| 6 |  1 | −1 |  1 |
| 7 |  1 |  1 |  1 |
| 8 | −1 |  1 |  1 |

where

$$\mathbf{B}_i = \mathbf{d}\,\mathbf{f}_i = \begin{bmatrix} f_{i,x} & 0 & 0 \\ 0 & f_{i,y} & 0 \\ 0 & 0 & f_{i,z} \\ f_{i,y} & f_{i,x} & 0 \\ 0 & f_{i,z} & f_{i,y} \\ f_{i,z} & 0 & f_{i,x} \end{bmatrix} = \begin{bmatrix} D_{G1i} & 0 & 0 \\ 0 & D_{G2i} & 0 \\ 0 & 0 & D_{G3i} \\ D_{G2i} & D_{G1i} & 0 \\ 0 & D_{G3i} & D_{G2i} \\ D_{G3i} & 0 & D_{G1i} \end{bmatrix} \quad (9)$$

Terms appearing in the submatrix $\mathbf{B}_i$ were discussed in Sec. 7.3.

Next, we express the stiffness matrix for element H8 in Cartesian coordinates to be

$$\mathbf{K} = \int_V \mathbf{B}^T(x,y,z)\,\mathbf{E}\,\mathbf{B}(x,y,z)\,dx\,dy\,dz \quad (10a)$$

In natural coordinates, this equation becomes

$$\mathbf{K} = \int_{-1}^{1}\int_{-1}^{1}\int_{-1}^{1} \mathbf{B}^T(\xi,\eta,\zeta)\mathbf{E}\,\mathbf{B}(\xi,\eta,\zeta)\,|\mathbf{J}(\xi,\eta,\zeta)|\,d\xi\,d\eta\,d\zeta \quad (10b)$$

and three applications of Gaussian quadrature give

$$\mathbf{K} = \sum_{l=1}^{n}\sum_{k=1}^{n}\sum_{j=1}^{n} R_j R_k R_l \mathbf{B}_{j,k,l}^T \mathbf{E}\,\mathbf{B}_{j,k,l}\,|\mathbf{J}_{j,k,l}| \quad (10c)$$

This formula for numerical integration implies that the matrix $\mathbf{B}_{j,k,l}$ and the determinant $|\mathbf{J}_{j,k,l}|$ are evaluated at each integration point, where the coordinates are $(\xi_j, \eta_k, \zeta_l)$.

Similarly, the consistent mass matrix for element H8 is

$$\mathbf{M} = \rho \int_V \mathbf{f}^T(x,y,z)\,\mathbf{f}(x,y,z)\,dx\,dy\,dz \quad (11a)$$

or

$$\mathbf{M} = \rho \int_{-1}^{1}\int_{-1}^{1}\int_{-1}^{1} \mathbf{f}^T(\xi,\eta,\zeta)\,\mathbf{f}(\xi,\eta,\zeta)\,|\mathbf{J}(\xi,\eta,\zeta)|\,d\xi\,d\eta\,d\zeta \quad (11b)$$

or

$$\mathbf{M} = \rho \sum_{l=1}^{n} \sum_{k=1}^{n} \sum_{j=1}^{n} R_j R_k R_l \mathbf{f}_{j,k,l}^{\mathrm{T}} \mathbf{f}_{j,k,l} |\mathbf{J}_{j,k,l}| \tag{11c}$$

in which $\rho$ is assumed to be constant.

Furthermore, equivalent nodal loads due to body forces have the form

$$\mathbf{p}_b(t) = \int_V \mathbf{f}^{\mathrm{T}}(x, y, z) \, \mathbf{b}(x, y, z, t) \, dx \, dy \, dz \tag{12a}$$

or

$$\mathbf{p}_b(t) = \int_{-1}^{1} \int_{-1}^{1} \int_{-1}^{1} \mathbf{f}^{\mathrm{T}}(\xi, \eta, \zeta) \mathbf{b}(\xi, \eta, \zeta, t) |\mathbf{J}(\xi, \eta, \zeta)| \, d\xi \, d\eta \, d\zeta \tag{12b}$$

or

$$\mathbf{p}_b(t) = \sum_{l=1}^{n} \sum_{k=1}^{n} \sum_{j=1}^{n} R_j R_k R_l \mathbf{f}_{j,k,l}^{\mathrm{T}} \mathbf{b}(t)_{j,k,l} |\mathbf{J}_{j,k,l}| \tag{12c}$$

where both $\mathbf{p}_b$ and $\mathbf{b}$ are functions of time.

For the rectangular solid RS8, the Jacobian matrix and its determinant specialize to

$$\mathbf{J} = \begin{bmatrix} a & 0 & 0 \\ 0 & b & 0 \\ 0 & 0 & c \end{bmatrix} \qquad |\mathbf{J}| = abc \tag{13}$$

Here the constants $a$, $b$, and $c$ are half the dimensions in the $\xi$, $\eta$, and $\zeta$ directions, as shown in Fig. 7.14(a). Thus, Eqs. (10b), (11b), and (12b) are simplified, as follows:

$$\mathbf{K} = abc \int_{-1}^{1} \int_{-1}^{1} \int_{-1}^{1} \mathbf{B}^{\mathrm{T}}(\xi, \eta, \zeta) \mathbf{E} \, \mathbf{B}(\xi, \eta, \zeta) \, d\xi \, d\eta \, d\zeta \tag{14}$$

and

$$\mathbf{M} = abc\rho \int_{-1}^{1} \int_{-1}^{1} \int_{-1}^{1} \mathbf{f}^{\mathrm{T}}(\xi, \eta, \zeta) \mathbf{f}(\xi, \eta, \zeta) \, d\xi \, d\eta \, d\zeta \tag{15}$$

and

$$\mathbf{p}_b(t) = abc \int_{-1}^{1} \int_{-1}^{1} \int_{-1}^{1} \mathbf{f}^{\mathrm{T}}(\xi, \eta, \zeta) \mathbf{b}(\xi, \eta, \zeta, t) \, d\xi \, d\eta \, d\zeta \tag{16}$$

**Example 7.3**

Assuming that $\rho$ is constant, let us derive the terms in the first column of the consistent mass matrix $\mathbf{M}$ for the rectangular solid element RS8 shown in Fig. 7.14(a). For this purpose, we need functions $f_1$ through $f_8$ given by Eq. (4), with values of $\xi_i$, $\eta_i$, and $\zeta_i$ taken from Table 7.7.

## Sec. 7.7  Isoparametric Hexahedra for General Solids

Considering first the term $M_{11}$, we have from Eq. (15)

$$M_{11} = abc\rho \int_{-1}^{1}\int_{-1}^{1}\int_{-1}^{1} f_1^2\, d\xi\, d\eta\, d\zeta \tag{a}$$

where

$$f_1 = \tfrac{1}{8}(1 - \xi)(1 - \eta)(1 - \zeta) \tag{b}$$

Substitution of Eq. (b) into Eq. (a) produces

$$M_{11} = \frac{abc\rho}{64}\int_{-1}^{1}\int_{-1}^{1}\int_{-1}^{1}(1-\xi)^2(1-\eta)^2(1-\zeta)^2\, d\xi\, d\eta\, d\zeta$$

$$= \frac{8abc\rho}{27} \tag{c}$$

Proceeding in a similar manner, we can find all of the terms in the first column of matrix **M**, as follows:

$$(\mathbf{M})_{\text{col. 1}} = \frac{\rho V}{216}\{8,\, 0,\, 0,\, 4,\, 0,\, 0,\, 2,\, 0,\, 0,\, 4,\, 0,\, 0,\, 4,\, 0,\, 0,\, 2,\, 0,\, 0,\, 1,\, 0,\, 0,\, 2,\, 0,\, 0\} \tag{d}$$

in which the volume $V = 8abc$.

Now let us examine a higher order hexahedral element that is formulated using quadratic geometric and displacement shape functions. The parent rectangular solid (*element RS20*) is illustrated in Fig. 7.15(a), and its isoparametric counterpart (*element H20*) appears in Fig. 7.15(b). For the rectangular solid the natural coordinates $\xi$, $\eta$, and $\zeta$ are orthogonal, and nodes 9 through 20 are located at midlengths of the straight edges. Both elements have the nodal displacement vector

$$\mathbf{q} = \{q_1, q_2, q_3, \ldots, q_{60}\} = \{u_1, v_1, w_1, \ldots, w_{20}\} \tag{17}$$

that contains $x$, $y$, and $z$ translations at each of the 20 nodes. Quadratic displacement shape functions for these elements are

$$u = \sum_{i=1}^{20} f_i u_i \qquad v = \sum_{i=1}^{20} f_i v_i \qquad w = \sum_{i=1}^{20} f_i w_i \tag{18}$$

where

$$f_i = \tfrac{1}{8}(1 + \xi_0)(1 + \eta_0)(1 + \zeta_0)(\xi_0 + \eta_0 + \zeta_0 - 2) \qquad (i = 1, 2, \ldots, 8)$$
$$f_i = \tfrac{1}{4}(1 - \xi^2)(1 + \eta_0)(1 + \zeta_0) \qquad (i = 9, 11, 17, 19)$$
$$f_i = \tfrac{1}{4}(1 - \eta^2)(1 + \zeta_0)(1 + \xi_0) \qquad (i = 10, 12, 18, 20)$$
$$f_i = \tfrac{1}{4}(1 - \zeta^2)(1 + \xi_0)(1 + \eta_0) \qquad (i = 13, 14, 15, 16)$$

$$\tag{19}$$

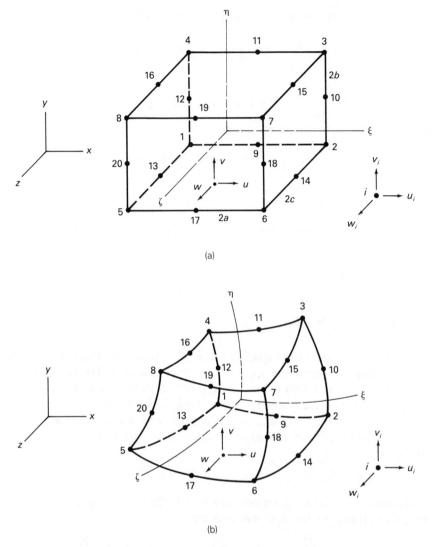

**Figure 7.15** Element H20: (a) parent rectangular solid; (b) isoparametric counterpart.

Values of $\xi_i$, $\eta_i$, and $\zeta_i$ for these formulas are listed in Table 7.8. Explicit integrations are possible for the subparametric parent element.

For the isoparametric H20 element [10] in Fig. 7.15(b), we use geometric interpolation functions that are the same as the displacement shape functions in Eqs. (19). Thus,

$$x = \sum_{i=1}^{20} f_i x_i \qquad y = \sum_{i=1}^{20} f_i y_i \qquad z = \sum_{i=1}^{20} f_i z_i \qquad (20)$$

Sec. 7.8  Program DYNASO for General Solids                                351

**TABLE 7.8  Nodal Coordinates for Element H20**

| $i$ | $\xi_i$ | $\eta_i$ | $\zeta_i$ | $i$ | $\xi_i$ | $\eta_i$ | $\zeta_i$ |
|---|---|---|---|---|---|---|---|
| 1  | −1 | −1 | −1 | 11 |  0 |  1 | −1 |
| 2  |  1 | −1 | −1 | 12 | −1 |  0 | −1 |
| 3  |  1 |  1 | −1 | 13 | −1 | −1 |  0 |
| 4  | −1 |  1 | −1 | 14 |  1 | −1 |  0 |
| 5  | −1 | −1 |  1 | 15 |  1 |  1 |  0 |
| 6  |  1 | −1 |  1 | 16 | −1 |  1 |  0 |
| 7  |  1 |  1 |  1 | 17 |  0 | −1 |  1 |
| 8  | −1 |  1 |  1 | 18 |  1 |  0 |  1 |
| 9  |  0 | −1 | −1 | 19 |  0 |  1 |  1 |
| 10 |  1 |  0 | −1 | 20 | −1 |  0 |  1 |

In this instance, the faces and edges of the element are quadratic surfaces and curves, as indicated in the figure.

Terms in the Jacobian matrix for element H20 are the same as those given in Eqs. (7.3-23), but with the upper index 8 changed to 20. Furthermore, the equations for element H8 will pertain to element H20 if the number 8 is changed to 20 in appropriate locations. Derivatives $f_{i,\xi}$, and so on, required for the development of element H20 are easily obtained and need not be tabulated. For example,

$$f_{1,\xi} = \tfrac{1}{8}(1 + 2\xi + \eta + \zeta)(1 - \eta)(1 - \zeta)$$

Of course, numerical integration is required for this element.

## 7.8 PROGRAM DYNASO FOR GENERAL SOLIDS

In this section we describe Program DYNASO for dynamic analysis of general solids. Before using this program, we must discretize a given solid using either H8 or H20 hexahedral elements (see Sec. 7.7). The material of the solid is assumed to be homogeneous and isotropic.

The manner of preparing structural data for Program DYNASO is shown in Table 7.9. Included among the structural parameters is NEN, which is the number of element nodes for each type of hexahedron. If NEN = 8, the hexahedra are H8 elements; and if NEN = 20, they are H20 elements. Therefore, element information consists of the element number I and node numbers 1 through 8 for element H8 and 1 through 20 for element H20.

Table 7.10 gives preparation of dynamic load data for Program DYNASO. First, we see that there are three possible initial displacements, initial velocities, and applied forces for each node. Also, the element loads implied by the parameters NEL, NEA, and NEV require some explanation. If NEL ≠ 0, at least one line load (force per unit length) exists on the edge *jk* of an element. For element H8 a linearly varying line load is defined by the force intensities BL1

TABLE 7.9 Structural Data for Program DYNASO

| Type of Data | No. of Lines | Items on Data Lines |
|---|---|---|
| Problem identification | 1 | Descriptive title |
| Structural parameters | 1 | NN, NE, NRN, NEN, E, PR, RHO |
| General solid data | | |
| (a) Nodal coordinates | NN | J, X(J), Y(J), Z(J) |
| (b) Element information[a] | NE | I, IN(I, 1), IN(I, 2), ... , IN(I, NEN) |
| (c) Nodal restraints | NRN | J, NRL(3J-2), NRL(3J-1), NRL(3J) |

[a] For sequences of node numbers, see Figs. 7.14(b) and 7.15(b).


TABLE 7.10 Dynamic Load Data for Program DYNASO

| Type of Data | No. of Lines | Items on Data Lines |
|---|---|---|
| Dynamic parameters | 1 | ISOLVE, NTS, DT, DAMPR |
| Initial conditions | | |
| (a) Condition parameters | 1 | NNID, NNIV |
| (b) Displacements | NNID | J, D0(3J-2), D0(3J-1), D0(3J) |
| (c) Velocities | NNIV | J, V0(3J-2), V0(3J-1), V0(3J) |
| Applied actions | | |
| (a) Load parameters | 1 | NLN, NEL, NEA, NEV |
| (b) Nodal loads | NLN | J, AS(3J-2), AS(3J-1), AS(3J) |
| (c) Line loads[a] | NEL | J, K, BL1, BL2, ... , BL6 |
| (d) Area loads[b] | NEA | J, K, L, M, BA1, BA2, ... , BA12 |
| (e) Volume loads | NEV | I, BV1, BV2, BV3 |
| Ground accelerations | | |
| (a) Acceleration parameter | 1 | IGA |
| (b) Acceleration factors | 1 | GAX, GAY, GAZ |
| Forcing function | | |
| (a) Function parameter | 1 | NFO |
| (b) Function ordinates | NFO | K, T(K), FO(K) |

[a] Pertains only to element H8 (NEN = 8). For element H20 (NEN = 20), we need three node numbers and nine force intensities.

[b] For element H20 use eight node numbers and 24 force intensities.

### Sec. 7.8  Program DYNASO for General Solids

through BL6. The first four are as shown in Fig. 7.11(a) for element Q4, and the last two pertain to the $z$ direction. On the other hand, the H20 element may have a quadratically varying line load, specified by BL1 through BL9 (see the first footnote). The first six force intensities have the same meanings as in Fig. 7.11(b) for element Q8, and the last three are for the $z$ direction.

IF NEA $\neq$ 0, at least one element has an area load on one of its surfaces. Figure 7.16 shows the types of area loads for hexahedra. For element H8 such loads on face *jklm* are defined by 12 numbers, of which the first four (BA1 through BA4) denote force (per unit area) in the $x$ direction, as indicated in Fig. 7.16(a). The next four (BA5 through BA8) pertain to the $y$ direction, and the last four (BA9 through BA12) apply to the $z$ direction. It is assumed that each component of area loading has a bilinear variation over the surface *jklm*. On the other hand, we take a biquadratic variation of area loading on a surface of the H20 element. In this case the loads are specified by 24 intensities (see the second footnote). The first eight represent force (per unit area) in the $x$ direction, as depicted in Fig. 7.16(b). The next eight are for the $y$ direction, and the last eight are for the $z$ direction.

Volume loads BV1, BV2, and BV3 on both types of hexahedra simply consist of uniform intensities of force (per unit volume) in the $x$, $y$, and $z$ directions. Also note that GAZ is included in the list of acceleration factors, as for any three-dimensional analysis.

In a manner similar to Program DYNAPS, we use $n = 2$ in each of three ways to locate points for numerical integration. Thus, there are eight such points, at which the time-varying stresses $\sigma_x$, $\sigma_y$, $\sigma_z$, $\tau_{xy}$, $\tau_{yz}$, and $\tau_{zx}$ are determined in Program DYNASO. Again, we had to use $n = 3$ each way for terms in the consistent mass matrix of element H20 to retain sufficient accuracy.

### Example 7.4

The tapered cantilever beam in Fig. 7.17(a) is doubly symmetric, and the parabolas with apexes at the support determine the rate of taper. This beam is made of reinforced concrete and has a rectangular impulse of magnitude $P_1$ and duration $t_1$ applied in the $y$ direction at its free end. Physical parameters are given as follows:

$$E = 3.6 \times 10^3 \text{ k/in.}^2 \qquad \nu = 0.15 \qquad \rho = 2.25 \times 10^{-7} \text{ k-s}^2/\text{in.}^4$$

$$L = 120 \text{ in.} \qquad P_1 = 400 \text{ k} \qquad t_1 = 5 \text{ ms}$$

where the units are seen to be US.

Figure 7.17(b) shows discretization of a quarter of the beam into two H20 elements, with nodal restraints imposed for symmetric and antisymmetric deformations. To define the geometry of this simple network, we need only state that the $x$-coordinates of nodes 9 through 12 are 90 in., those of nodes 13 through 20 are 60 in., and those of nodes 21 through 24 are 30 in.

We used the foregoing data in Program DYNASO with NEN = 20, DAMPR = 0.10, and solution by Subprogram NORMOD (with NMODES = 12). Figure 7.18(a)

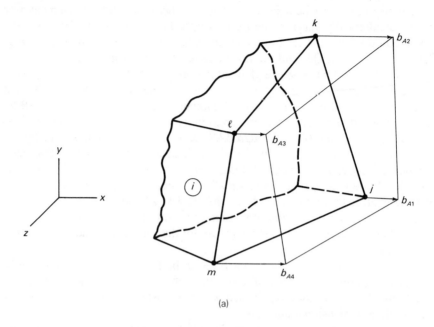

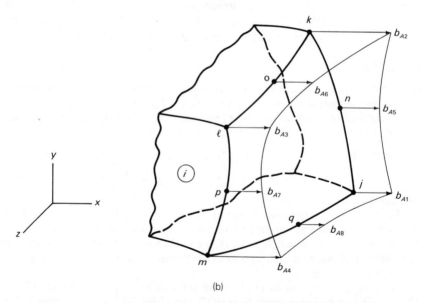

**Figure 7.16** Area loads for hexahedra: (a) element H8; (b) element H20.

## Sec. 7.8 Program DYNASO for General Solids

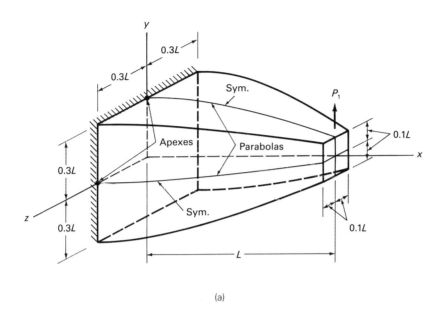

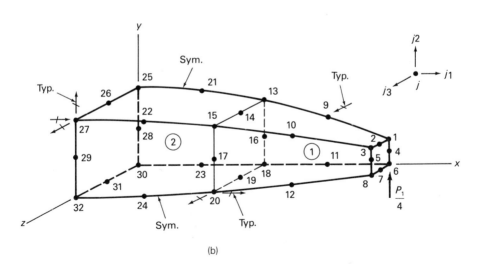

**Figure 7.17** (a) Tapered cantilever beam; (b) H20 elements.

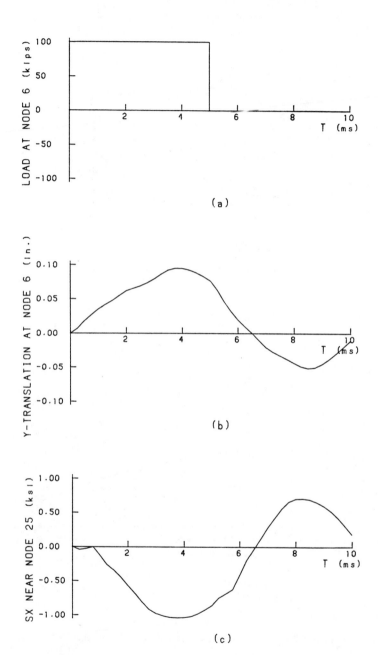

**Figure 7.18** Tapered beam: (a) load; (b) displacement; (c) stress.

### Sec. 7.9 Isoparametric Elements for Axisymmetric Solids

depicts a quarter of the rectangular impulse at node 6; and the $y$-translation at that point is plotted in Fig. 7.18(b), showing a maximum value of 0.09514 in. at $t = 3.75$ ms. Also, the normal stress SX near node 25 is plotted in Fig. 7.18(c), for which the minimum value is $-1.040$ ksi at the same time.

## 7.9 ISOPARAMETRIC ELEMENTS FOR AXISYMMETRIC SOLIDS

An axisymmetric solid is defined as a three-dimensional body that may be developed by rotation of a planar section about an axis. This type of body is sometimes called a *solid of revolution*. Cylindrical coordinates $r$, $z$, and $\theta$ provide a suitable reference frame, as illustrated in Fig. 7.19. We assume that the body is axisymmetric with respect to the $z$ axis and that a typical finite element is a circular ring. This *ring element* may have various cross-sectional shapes, but we will deal only with isoparametric quadrilateral sections. Although nodes are shown as dots on the cross section of a ring element, they are actually *nodal circles*. If the loads on an axisymmetric solid are also axisymmetric, we may analyze a representative cross section as if it were a two-

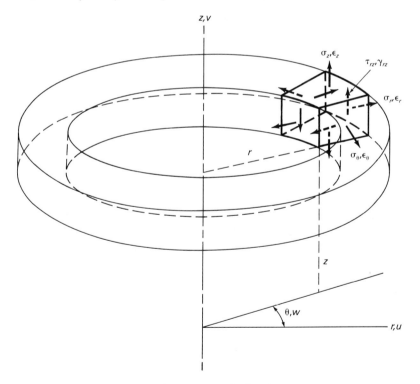

**Figure 7.19** Axisymmetric ring element.

dimensional problem. At first, only axisymmetric patterns of loads will be considered, but nonaxisymmetric loads also will be covered later in the section.

For any point on an axisymmetrically loaded ring element, the generic displacements are

$$\mathbf{u} = \{u, v\} \quad (1)$$

Translations $u$ and $v$ occur in the $r$ and $z$ directions, as indicated in Fig. 7.19. With axisymmetric loads, the translation $w$ in the $\theta$ direction is zero, and the shearing strains $\gamma_{r\theta}$ and $\gamma_{z\theta}$ are also zero. However, the figure shows four types of strains that are nonzero, as follows:

$$\boldsymbol{\epsilon} = \{\epsilon_r, \epsilon_z, \epsilon_\theta, \gamma_{rz}\} \quad (2)$$

Relationships between these strains and the generic displacements in Eq. (1) are seen to be

$$\epsilon_r = \frac{\partial u}{\partial r} \qquad \epsilon_\theta = \frac{2\pi(r+u) - 2\pi r}{2\pi r} = \frac{u}{r}$$
$$\epsilon_z = \frac{\partial v}{\partial z} \qquad \gamma_{rz} = \frac{\partial u}{\partial z} + \frac{\partial v}{\partial r} \quad (3)$$

These relationships are embodied in the differential operator

$$\mathbf{d} = \begin{bmatrix} \dfrac{\partial}{\partial r} & 0 \\ 0 & \dfrac{\partial}{\partial z} \\ \dfrac{1}{r} & 0 \\ \dfrac{\partial}{\partial z} & \dfrac{\partial}{\partial r} \end{bmatrix} \quad (4)$$

In this instance the nonzero term $1/r$ in the third row of matrix $\mathbf{d}$ is a multiplier of $u$, not a derivative.

Corresponding to the strains in Eq. (2), the four types of nonzero stresses depicted in Fig. 7.19 are

$$\boldsymbol{\sigma} = \{\sigma_r, \sigma_z, \sigma_\theta, \tau_{rz}\} \quad (5)$$

For an isotropic material, the stress-strain matrix is

$$\mathbf{E} = \frac{E}{(1+\nu)(1-2\nu)} \begin{bmatrix} 1-\nu & & & \text{Sym.} \\ \nu & 1-\nu & & \\ \nu & \nu & 1-\nu & \\ 0 & 0 & 0 & \dfrac{1-2\nu}{2} \end{bmatrix} \quad (6)$$

This $4 \times 4$ array is similar to the $3 \times 3$ matrix for plane strain in Eq. (7.2-17).

## Sec. 7.9  Isoparametric Elements for Axisymmetric Solids

Figure 7.20(a) shows the cross section of *element AXQ4*, which derives its characteristics from the quadrilateral element Q4 in Sec. 7.5. Bilinear displacement shape functions in matrix **f** are the same as those for element Q4, and the strain-displacement submatrix $\mathbf{B}_i$ becomes

$$\mathbf{B}_i = \begin{bmatrix} f_{i,r} & 0 \\ 0 & f_{i,z} \\ \dfrac{f_i}{r} & 0 \\ f_{i,z} & f_{i,r} \end{bmatrix} \quad (i = 1, 2, 3, 4) \tag{7}$$

which is obtained by using the operator **d** in Eq. (4) on submatrix $\mathbf{f}_i$ from Eq. (7.5-5). The radius $r$ in Eq. (7) is found as

$$r = \sum_{i=1}^{4} f_i r_i \tag{8}$$

In addition, the derivatives $f_{i,r} = D_{G1i}$, and so on, are given by Eqs. (7.3-18), except that $r$ and $z$ replace the coordinates $x$ and $y$.

The stiffness matrix for element AXQ4 may be formulated in natural coordinates as

$$\mathbf{K} = \int_{-1}^{1} \int_{-1}^{1} \int_{0}^{2\pi} \mathbf{B}^T \mathbf{E} \, \mathbf{B} \, |\mathbf{J}| r \, d\theta \, d\xi \, d\eta$$

$$= 2\pi \int_{-1}^{1} \int_{-1}^{1} \mathbf{B}^T \mathbf{E} \, \mathbf{B} \, |\mathbf{J}| r \, d\xi \, d\eta \tag{9}$$

Similarly, the consistent-mass matrix is

$$\mathbf{M} = \rho \int_{-1}^{1} \int_{-1}^{1} \int_{0}^{2\pi} \mathbf{f}^T \mathbf{f} \, |\mathbf{J}| r \, d\theta \, d\xi \, d\eta$$

$$= 2\pi\rho \int_{-1}^{1} \int_{-1}^{1} \mathbf{f}^T \mathbf{f} \, |\mathbf{J}| r \, d\xi \, d\eta \tag{10}$$

Also, equivalent nodal loads due to body forces are

$$\mathbf{p}_b(t) = \int_{-1}^{1} \int_{-1}^{1} \int_{0}^{2\pi} \mathbf{f}^T \mathbf{b}(t) \, |\mathbf{J}| r \, d\theta \, d\xi \, d\eta$$

$$= 2\pi \int_{-1}^{1} \int_{-1}^{1} \mathbf{f}^T \mathbf{b}(t) \, |\mathbf{J}| r \, d\xi \, d\eta \tag{11}$$

Numerical integration is required to evaluate Eqs. (9), (10), and (11).

The cross section of *element AXQ8* appears in Fig. 7.20(b). Its properties are similar to those of the quadrilateral element Q8 in Sec. 7.5. Biquadratic displacement shape functions in matrix **f** are the same as for element Q8.

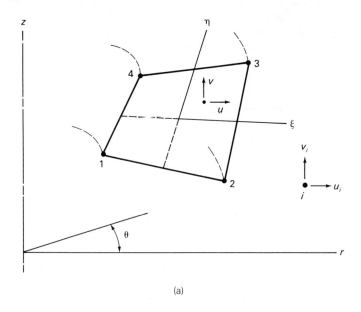

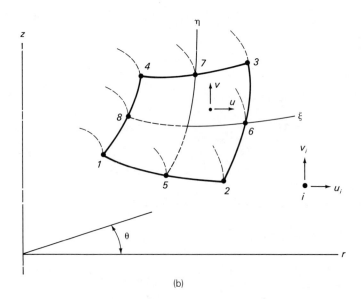

**Figure 7.20** Quadrilateral sections: (a) element AXQ4; (b) element AXQ8.

## Sec. 7.9  Isoparametric Elements for Axisymmetric Solids

Expressions for $\mathbf{B}_i$, $r$, $\mathbf{K}$, $\mathbf{M}$, and $\mathbf{p}_b(t)$ are similar to those for element AXQ4 given by Eqs. (7) through (11), except that $i = 1, 2, \ldots, 8$.

### Example 7.5

Find the consistent mass term $M_{35}$ for the axisymmetric solid element AXQ4 in Fig. 7.20(a), using Gaussian numerical integration with $n = 2$ each way. Let the coordinates of nodes 1, 2, 3, and 4 be (10, 2), (14, 1), (15, 5), and (11, 4), respectively.

From Table 7.1 for $n = 2$, we have $R_j = R_k = 1$, so that the numerical form of Eq. (10) gives

$$M_{35} = 2\pi\rho \sum_{k=1}^{2} \sum_{j=1}^{2} (f_2)_{j,k}(f_3)_{j,k} |\mathbf{J}_{j,k}| r_{j,k} \qquad (a)$$

In this case the bilinear functions $f_2$ and $f_3$ are needed to evaluate the term $M_{35}$ because of the arrangement of matrix $\mathbf{f}$ in Eq. (7.5-5). Substituting $f_2$ and $f_3$ from Eq. (7.5-6) into Eq. (a) yields

$$M_{35} = \frac{\pi\rho}{8} \sum_{k=1}^{2} \sum_{j=1}^{2} [(1 + \xi)^2(1 - \eta^2)]_{j,k} |\mathbf{J}_{j,k}| r_{j,k} \qquad (b)$$

To apply this formula, we first calculate the Jacobian matrix from Eq. (7.3-10), as follows:

$$\mathbf{J} = \mathbf{D}_L \mathbf{C}_N = \frac{1}{4}\begin{bmatrix} -(1-\eta) & (1-\eta) & (1+\eta) & -(1+\eta) \\ -(1-\xi) & -(1+\xi) & (1+\xi) & (1-\xi) \end{bmatrix} \begin{bmatrix} 10 & 2 \\ 14 & 1 \\ 15 & 5 \\ 11 & 4 \end{bmatrix}$$

$$= \frac{1}{2}\begin{bmatrix} 4 & \eta \\ 1 & 3+\xi \end{bmatrix} \qquad (c)$$

Then the determinant of $\mathbf{J}$ is

$$|\mathbf{J}| = \tfrac{1}{4}(12 + 4\xi - \eta) \qquad (d)$$

Evaluating the terms in Eq. (b) at each of the four integration points and summing the results produces

$$M_{35} = 21.06\pi\rho = 66.15\rho \qquad (e)$$

which can be finalized using a numerical value for $\rho$.

Turning now to *nonaxisymmetric loads*, we can divide them into two sets [11]. The first load set is symmetric with respect to a plane containing the axis of revolution, and the second is antisymmetric with respect to that plane. For convenience, the $r$-$z$ plane is taken to be the plane of symmetry. *Fourier decomposition* [12] of the symmetric loads for $m$ harmonic terms produces

$$b_r = \sum_{j=0}^{m} b_{rj} \cos j\theta \qquad b_z = \sum_{j=0}^{m} b_{zj} \cos j\theta$$

$$b_\theta = \sum_{j=0}^{m} b_{\theta j} \sin j\theta \qquad (12)$$

where $b_{rj}$, $b_{zj}$, and $b_{\theta j}$ are functions of $r$ and $z$ only. When $j = 0$, we have $b_\theta = 0$; and Eqs. (12) become the case of axisymmetric loads. Otherwise, $j = 1, 2, \ldots, m$ represent cases of nonaxisymmetric loads that are symmetric with respect to the $r$-$z$ plane. Figure 7.21(a), (b), and (c) show the first harmonic loads for the $r$, $z$, and $\theta$ directions, respectively. If the loads were antisymmetric with respect to the plane of symmetry, the functions $\sin j\theta$ and $\cos j\theta$ would be interchanged.

Generic displacements for nonaxisymmetric loads must include the translation $w$ in the $\theta$ direction. Thus,

$$\mathbf{u} = \{u, v, w\} \tag{13}$$

and we must also have $\gamma_{z\theta}$ and $\gamma_{r\theta}$ in the strain vector, as follows:

$$\boldsymbol{\epsilon} = \{\epsilon_r, \epsilon_z, \epsilon_\theta, \gamma_{rz}, \gamma_{z\theta}, \gamma_{r\theta}\} \tag{14}$$

Strain-displacement relationships developed by Love [13] are

$$\epsilon_r = \frac{\partial u}{\partial r} \qquad \epsilon_z = \frac{\partial v}{\partial z} \qquad \epsilon_\theta = \frac{u}{r} + \frac{1}{r}\frac{\partial w}{\partial \theta}$$

$$\gamma_{rz} = \frac{\partial u}{\partial z} + \frac{\partial v}{\partial r} \qquad \gamma_{z\theta} = \frac{1}{r}\frac{\partial v}{\partial \theta} + \frac{\partial w}{\partial z} \tag{15}$$

$$\gamma_{r\theta} = \frac{1}{r}\frac{\partial u}{\partial \theta} + \frac{\partial w}{\partial r} - \frac{w}{r}$$

Here we see that the radius $r$ appears in the denominators of several expressions. From Eqs. (15) we can form the operator $\mathbf{d}$ as

$$\mathbf{d} = \begin{bmatrix} \dfrac{\partial}{\partial r} & 0 & 0 \\ 0 & \dfrac{\partial}{\partial z} & 0 \\ \dfrac{1}{r} & 0 & \dfrac{1}{r}\dfrac{\partial}{\partial \theta} \\ \dfrac{\partial}{\partial z} & \dfrac{\partial}{\partial r} & 0 \\ 0 & \dfrac{1}{r}\dfrac{\partial}{\partial \theta} & \dfrac{\partial}{\partial z} \\ \dfrac{1}{r}\dfrac{\partial}{\partial \theta} & 0 & \dfrac{\partial}{\partial r} - \dfrac{1}{r} \end{bmatrix} \tag{16}$$

The stress vector for nonaxisymmetric loads must contain $\tau_{z\theta}$ and $\tau_{r\theta}$, as follows:

$$\boldsymbol{\sigma} = \{\sigma_r, \sigma_z, \sigma_\theta, \tau_{rz}, \tau_{z\theta}, \tau_{r\theta}\} \tag{17}$$

## Sec. 7.9 Isoparametric Elements for Axisymmetric Solids

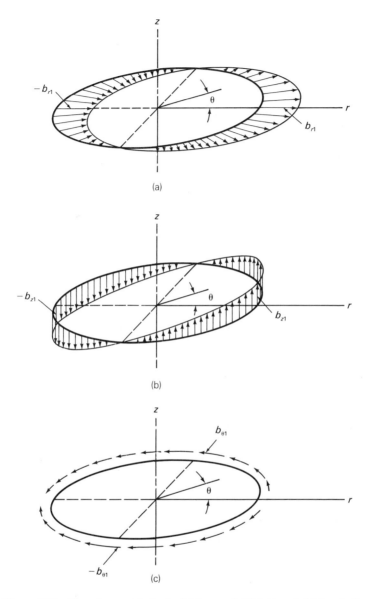

**Figure 7.21** Nonaxisymmetric loads: (a) $b_{r1} \cos \theta$; (b) $b_{z1} \cos \theta$; (c) $b_{\theta 1} \sin \theta$.

Stress-strain relationships are easily extended to cover six types of stresses and the corresponding strains. For example, if the material is isotropic [see Eq. (6)], we add $E_{55} = E_{66} = E/2(1 + \nu)$ to form a 6 × 6 matrix **E**.

The response of an axisymmetric solid to a series of symmetric, harmonic loads consists of a series of symmetric, harmonic, generic displacements that may be expressed as

$$u = \sum_{j=0}^{m} u_j \cos j\theta \qquad v = \sum_{j=0}^{m} v_j \cos j\theta$$

$$w = \sum_{j=0}^{m} w_j \sin j\theta \qquad (18)$$

Again, if the loads were antisymmetric with respect to the plane of symmetry, the functions $\sin j\theta$ and $\cos j\theta$ would be interchanged. Applying the operator $\mathbf{d}$ in Eq. (16) to Eqs. (18) expressed in terms of $\mathbf{f}$, we find a typical partition of the strain-displacement matrix to be

$$(\mathbf{B}_i)_j = \begin{bmatrix} f_{i,r} \cos j\theta & 0 & 0 \\ 0 & f_{i,z} \cos j\theta & 0 \\ \dfrac{f_i}{r} \cos j\theta & 0 & j\dfrac{f_i}{r} \cos j\theta \\ f_{i,z} \cos j\theta & f_{i,r} \cos j\theta & 0 \\ 0 & -j\dfrac{f_i}{r} \sin j\theta & f_{i,z} \sin j\theta \\ -j\dfrac{f_i}{r} \sin j\theta & 0 & \left(f_{i,r} - \dfrac{f_i}{r}\right) \sin j\theta \end{bmatrix} \qquad (19)$$

where $i = 1, 2, \ldots, n_{\text{en}}$ and $j = 0, 1, 2, \ldots, m$.

An element stiffness matrix for each harmonic set of symmetric displacements may be written in natural coordinates as

$$\mathbf{K}_j = \int_{-1}^{1} \int_{-1}^{1} \int_{0}^{2\pi} \mathbf{B}_j^T \mathbf{E} \mathbf{B}_j \, |\mathbf{J}| \, r \, d\theta \, d\xi \, d\eta$$

$$= k\pi \int_{-1}^{1} \int_{-1}^{1} \mathbf{B}_j^T \mathbf{E} \mathbf{B}_j \, |\mathbf{J}| \, r \, d\xi \, d\eta \qquad (j = 0, 1, 2, \ldots, m) \qquad (20)$$

where $k = 2$ for $j = 0$, and $k = 1$ for $j = 1, 2, \ldots, m$. The latter constant ($k = 1$) appears as a consequence of

$$\int_{0}^{2\pi} \cos^2 j\theta \, d\theta = \int_{0}^{2\pi} \sin^2 j\theta \, d\theta = \pi \qquad (21)$$

Similarly, the consistent mass matrix for each harmonic set of symmetric displacements becomes

$$\mathbf{M}_j = \rho \int_{-1}^{1} \int_{-1}^{1} \int_{0}^{2\pi} \mathbf{f}^T \mathbf{c}_j^T \mathbf{c}_j \mathbf{f} \, |\mathbf{J}| r \, d\theta \, d\xi \, d\eta \qquad (22)$$

in which

$$\mathbf{c}_j = \begin{bmatrix} \cos j\theta & 0 & 0 \\ 0 & \cos j\theta & 0 \\ 0 & 0 & \sin j\theta \end{bmatrix} \qquad (j = 0, 1, 2, \ldots, m) \qquad (23)$$

Using Eqs. (21) in Eq. (22), we find that

$$\mathbf{M} = k\pi\rho \int_{-1}^{1} \int_{-1}^{1} \mathbf{f}^T \mathbf{f} |\mathbf{J}| r \, d\xi \, d\eta \qquad (24)$$

This formula is the same for $j = 1, 2, \ldots, m$, and it does not change for antisymmetric displacements. Thus, to determine frequencies and mode shapes for any value of $j$, we use the (variable) stiffness matrix $\mathbf{K}_j$ from Eq. (20) and the (constant) mass matrix $\mathbf{M}$ from Eq. (24). However, in the latter equation note that $k = 2$ for the case of axisymmetric vibrations, where $j = 0$.

Equivalent nodal loads for each harmonic set of symmetric body forces take the form

$$\mathbf{p}_b(t)_j = \int_{-1}^{1} \int_{-1}^{1} \int_{0}^{2\pi} \mathbf{f}^T \mathbf{c}_j^T \mathbf{c}_j \mathbf{b}(t)_j |\mathbf{J}| r \, d\theta \, d\xi \, d\eta$$

$$= k\pi \int_{-1}^{1} \int_{-1}^{1} \mathbf{f}^T \mathbf{b}(t)_j |\mathbf{J}| r \, d\xi \, d\eta \qquad (j = 0, 1, 2, \ldots, m) \qquad (25)$$

where

$$\mathbf{b}(t)_j = \{b_{rj}, b_{zj}, b_{\theta j}\} \qquad (26)$$

Finally, the stresses for each harmonic response are

$$\boldsymbol{\sigma}(t)_j = \mathbf{E}\,\mathbf{B}_j \mathbf{q}(t)_j \qquad (j = 0, 1, 2, \ldots, m) \qquad (27)$$

Of course, such stresses, as well as nodal displacements, must be added at the end of the analysis.

## 7.10 PROGRAM DYAXSO FOR AXISYMMETRIC SOLIDS

Let us now consider Program DYAXSO for dynamic analysis of axisymmetric solids with axisymmetric loads. We assume that such a solid has been discretized into ring elements AXQ4 or AXQ8, which were decribed in the preceding section. As before, the material of the solid is taken to be homogeneous and isotropic.

With very few modifications, Program DYNAPS in Sec. 7.6 can be converted to program DYAXSO. For example, the subprogram in DYAXSO that generates the element stiffness matrix is practically the same as that in Program DYNAPS. However, when calculating stiffness terms there is multiplication by $2\pi r$ instead of $h$. Within the logic of that subprogram, the computer must evaluate not only $\mathbf{B}_{j,k}$ and $|\mathbf{J}_{j,k}|$, but also $r_{j,k}$ at each numerical integration point. Similar comments also apply to the generation of consistent masses and equivalent nodal loads.

Structural data for Program DYNAPS (see Table 7.5) must be altered to account for the fact that the continuum to be analyzed is an axisymmetric solid.

The structural parameters IPS and H must be deleted, and the nodal coordinates $x$ and $y$ are replaced by $r$ and $z$. The only significant changes in the dynamic load data for Program DYNAPS (see Table 7.6) are that line loads become area loads and the acceleration factors GAX and GAY must be replaced by GAZ.

As in Program DYNAPS, we take $n = 2$ each way to locate points for numerical integration on the quadrilateral sections. At each of the four points the computer evaluates the time-varying stresses $\sigma_r$, $\sigma_z$, $\sigma_\theta$, and $\tau_{rz}$. Once more, we needed to use $n = 3$ each way for terms in the consistent mass matrix of element AXQ8.

### Example 7.6

An axisymmetric titanium valve head is discretized using seven AXQ8 elements, as shown in Fig. 7.22. Acting on the lower surfaces of elements 2, 4, and 6 is an explosive internal pressure $p_s$, which is resisted by the valve seat (or restraint) at node 34. Each edge of the finite-element network is divided into equal lengths between the nodes on that edge. For this problem the physical parameters are

$$E = 1.7 \times 10^4 \text{ k/in.}^2 \qquad \nu = 0.33 \qquad \rho = 4.20 \times 10^{-7} \text{ k-s}^2/\text{in.}^4$$
$$L = 0.25 \text{ in.} \qquad (p_s)_{\max} = 1.53 \text{ k/in.}^2$$

and the units are US.

For this example we need to calculate *equivalent nodal loads due to pressure $p_s$* on a surface of an axisymmetric solid element AXQ4 or AXQ8. Considering any of the four surfaces of such an element, we can find the components of $p_s$ in the directions of $r$ and $z$, as follows:

$$\mathbf{b}_s = p_s \mathbf{e}_{\eta'}^{\text{T}} \tag{a}$$

Here the symbol $\mathbf{e}_{\eta'}^{\text{T}}$ denotes the transpose of a unit row vector in the direction of $\eta'$, which is normal to the tangential direction $\xi'$ at the surface. The vector $\mathbf{e}_{\eta'}$ may be found by first calculating a unit row vector in the direction of $\xi'$ as

$$\mathbf{e}_{\xi'} = \frac{1}{c}[r_{,\xi'} \quad z_{,\xi'}] \tag{b}$$

where

$$c = \sqrt{(r_{,\xi'})^2 + (z_{,\xi'})^2} \tag{c}$$

Second, from known orthogonality relationships, we can write the unit normal vector in the form

$$\mathbf{e}_{\eta'} = \frac{1}{c}[-z_{,\xi'} \quad r_{,\xi'}] \tag{d}$$

Then the equivalent nodal loads on surface nodes become

$$\mathbf{p}_b = \int_A \mathbf{f}^{\text{T}} \mathbf{b}_s \, dA = 2\pi \int_{-1}^{1} \mathbf{f}^{\text{T}} \mathbf{b}_s r \, |\mathbf{J}'| \, d\xi' \tag{e}$$

In this formula the determinant of $\mathbf{J}'$ is

## Sec. 7.10 Program DYAXSO for Axisymmetric Solids

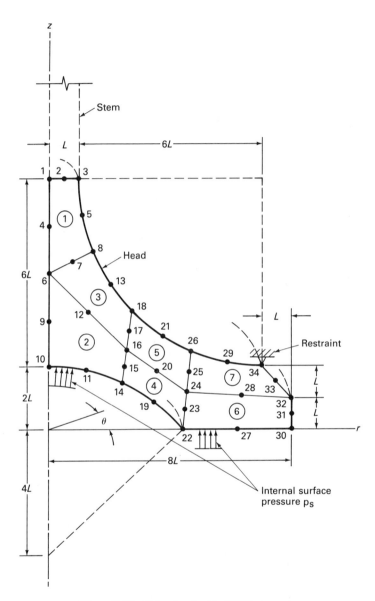

**Figure 7.22** Valve head with AXQ8 elements.

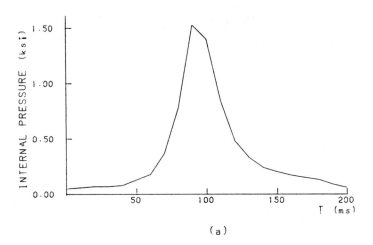

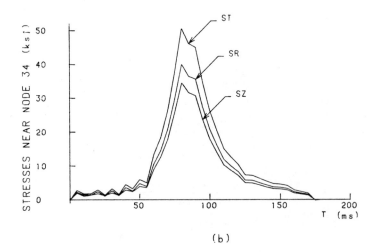

**Figure 7.23** Valve head with internal pressure: (a) load; (b) stresses.

$$|\mathbf{J}'| = \begin{vmatrix} r_{,\xi'} & z_{,\xi'} \\ -\dfrac{z_{,\xi'}}{c} & \dfrac{r_{,\xi'}}{c} \end{vmatrix} \tag{f}$$

This determinant transforms arc length instead of area because its second row is normalized to have unit length. Expanding the determinant produces

$$|\mathbf{J}'| = \frac{1}{c}[(r_{,\xi'})^2 + (z_{,\xi'})^2] = c \tag{g}$$

Substituting Eqs. (a), (d), and (g) into Eq. (e), we find that

$$\mathbf{p}_b = 2\pi p_s \int_{-1}^{1} \mathbf{f}^{\mathrm{T}} \begin{bmatrix} -z_{,\xi'} \\ r_{,\xi'} \end{bmatrix} r\, d\xi' \tag{h}$$

Signs on these equivalent nodal loads are automatically determined by Eq. (h) for pressure in the positive sense of the normal direction $\eta'$.

We processed the foregoing data with Program DYAXSO using NEN = 8, DAMPR = 0.05, and Subprogram NUMINT for responses. The computer plot in Fig. 7.23(a) gives the variation of internal pressure acting on the valve head, and Fig. 7.23(b) shows time histories of the normal stresses SR, SZ, and ST at the integration point near node 34. Maximum values of these stresses are 40.04, 34.54, and 50.67 ksi at time $t = 80$ ms.

## REFERENCES

1. Timoshenko, S. P., and Goodier, J. N., *Theory of Elasticity*, 3rd ed., McGraw-Hill, New York, 1970.
2. Weaver, W., Jr., and Johnston, P. R., *Finite Elements for Structural Analysis*, Prentice-Hall, Englewood Cliffs, N.J., 1984.
3. Lekhnitskii, S. G., *Theory of Elasticity of an Anisotropic Body*, translation from Russian by P. Fern, Holden-Day, San Francisco, 1963.
4. Gere, J. M., and Weaver, W., Jr., *Matrix Algebra for Engineers*, 2nd ed., Brooks-Cole, Monterey, Calif., 1983.
5. Scarborough, J. B., *Numerical Mathematical Analysis*, 6th ed., Johns Hopkins Press, Baltimore, Md., 1966.
6. Zienkiewicz, O. C., *The Finite Element Method*, 4th ed., McGraw-Hill, Maidenhead, Berkshire, England, 1987.
7. Ergatoudis, B., Irons, B. M., and Zienkiewicz, O. C., "Curved Isoparametric 'Quadrilateral' Elements for Finite Element Analysis," *Int. J. Solids Struct.*, Vol. 4, No. 1, 1968, pp. 31–42.
8. Melosh, R. J., "Structural Analysis of Solids," *ASCE J. Struct. Div.*, Vol. 89, No. ST4, 1963, pp. 205–223.
9. Irons, B. M., "Engineering Applications of Numerical Integration in Stiffness Methods," *AIAA J.*, Vol. 4, No. 11, 1966, pp. 2035–2037.
10. Ergatoudis, J., Irons, B. M., and Zienkiewicz, O. C., "Three-Dimensional Stress Analysis of Arch Dams and Their Foundations," *Proc. Symp. Arch Dams* (Inst. Civ. Eng., London), 1968, pp. 37–50.
11. Wilson, E. L., "Structural Analysis of Axisymmetric Solids," *AIAA J.*, Vol. 3, No. 12, 1965, pp. 2269–2274.
12. Sokolnikoff, I. S., and Redheffer, R. M., *Mathematics of Physics and Modern Engineering*, McGraw-Hill, New York, 1966.
13. Love, A. E. H., *The Mathematical Theory of Elasticity*, 4th ed., Cambridge University Press, Cambridge, 1927.

# 8

# Plates and Shells

## 8.1 INTRODUCTION

When a plate is subjected to forces applied in the direction normal to its own plane, it bends and is said to be in a state of flexure. For this type of problem, we deal with flexural and shearing stresses and strains that are somewhat analogous to those in a beam. However, the analysis of a plate is more complicated because it is two-dimensional; whereas a beam is only one-dimensional.

On the other hand, a shell is three-dimensional, and its analysis is even more difficult than that of a plate. In shells we must consider not only flexural and shearing stresses and strains, but also those associated with membrane (or in-plane) deformations.

Finite elements for dynamic analyses of plates and shells will be based upon those for general and axisymmetric solids from the preceding chapter. These specializations will automatically include the effects of shearing deformations and rotary inertias, as in Mindlin's theory of plates [1].

Computer programs in this chapter perform dynamic analyses of plates in bending, general shells, and axisymmetric shells. All structures that we analyze are assumed to be composed of linearly elastic materials with small strains and displacements. Guyan reduction (see Sec. 6.7) is used in the plate and shell programs to eliminate the nodal rotations and retain the translations.

## 8.2 ELEMENT FOR PLATES IN BENDING

It is possible to specialize an isoparametric hexahedron (see Sec. 7.7) to become a plate or a shell element by making one dimension small compared to the other two. This type of modeling was introduced by Ahmad et al. [2] and applies to analyses of both thick and thin plates and shells. For analyses of flat plates, it is also necessary to restrict the other two dimensions of the modified element to lie in a single plane. This section is devoted to the specialization of the isoparametric hexahedron H20 to become a plate-bending quadrilateral called *element PBQ8*. While an H8 hexahedron could also be specialized, the resulting straight-sided quadrilateral would not be suitable for conversion to a shell element later in the chapter.

Figure 8.1(a) shows the original H20 element, which has quadratic interpolation formulas defining its geometry. In order to understand the constraints needed to convert it to a plate-bending element, we first form a flat rectangular solid by making the natural coordinates $\xi$, $\eta$, and $\zeta$ orthogonal and the $\zeta$ dimension small. The resulting element appears in Fig. 8.1(b) as the *rectangular parent PQR8* of element PBQ8 before constraints. Note that groups of three nodes occur at the corners, while pairs of nodes are at midedge locations of element PQR8. By invoking certain constraints, we can convert each group and pair of nodes to a single node on the middle surface, as shown in Fig. 8.1(c). The nodal displacements indicated at point $i$ in that figure are

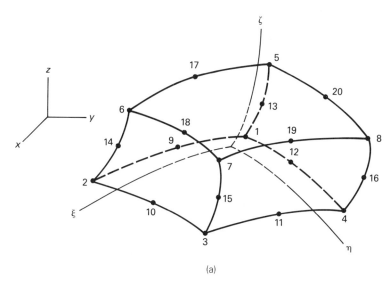

(a)

**Figure 8.1** Specialization of hexahedron: (a) element H20; (b) rectangular parent PQR8 of element PBQ8 before constraints; (c) constrained nodal displacements.

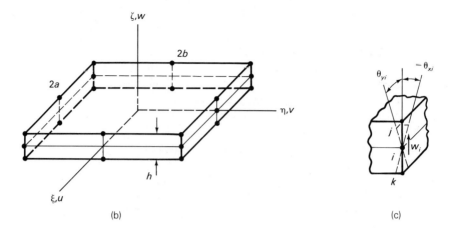

**Figure 8.1** (*cont.*)

$$\mathbf{q}_i = \{q_{i1}, q_{i2}, q_{i3}\} = \{w_i, \theta_{xi}, \theta_{yi}\} \qquad (i = 1, 2, \ldots, 8) \tag{1}$$

where $\theta_{xi}$ and $\theta_{yi}$ are small positive rotations about the $x$ and $y$ axes. Relationships between nodal displacements at a corner of element PQR8, a midedge of PQR8, and a node of element PBQ8 can be seen more clearly in Fig. 8.2. The two types of constraints to be introduced are:

1. Nodes on the same normal to the middle surface have equal translations in the $\zeta$ direction.
2. Normals to the middle surface remain straight (but no longer normal) during deformation.

Using these criteria, we can relate the nine nodal translations in Fig. 8.2(a) to the three nodal displacements in Fig. 8.2(c) by the following $9 \times 3$ *constraint matrix*:

$$\mathbf{C}_{ai} = \begin{bmatrix} 0 & 0 & 0 \\ 0 & 0 & 0 \\ 1 & 0 & 0 \\ 0 & 0 & \dfrac{h_i}{2} \\ 0 & -\dfrac{h_i}{2} & 0 \\ 1 & 0 & 0 \\ 0 & 0 & -\dfrac{h_i}{2} \\ 0 & \dfrac{h_i}{2} & 0 \\ 1 & 0 & 0 \end{bmatrix} \tag{2}$$

## Sec. 8.2  Element for Plates in Bending

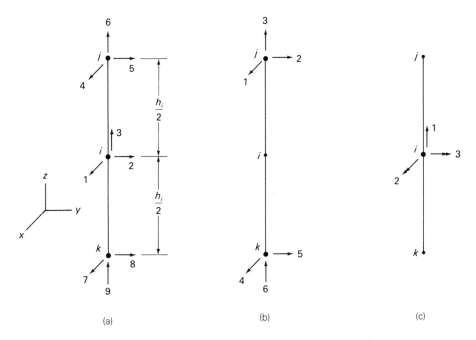

**Figure 8.2** Nodal displacements: (a) corner of PQR8; (b) midedge of PQR8; (c) node of PBQ8.

Similarly, the six nodal translations in Fig. 8.2(b) are related to the three nodal displacements in Fig. 8.2(c) by the constraint matrix

$$\mathbf{C}_{bi} = \begin{bmatrix} 0 & 0 & \frac{h_i}{2} \\ 0 & -\frac{h_i}{2} & 0 \\ 1 & 0 & 0 \\ 0 & 0 & -\frac{h_i}{2} \\ 0 & \frac{h_i}{2} & 0 \\ 1 & 0 & 0 \end{bmatrix} \tag{3}$$

which is of size 6 × 3. If we were to apply each of these constraint matrices in four locations, we would be able to reduce the number of nodal displacements from $(4)(9) + (4)(6) = 60$ to $(8)(3) = 24$. Instead of following this path, however, we will pursue a more direct formulation of element PBQ8 in a manner similar to that in Ref. 3.

Figure 8.3 shows element PBQ8, of constant thickness $h$, with its neutral

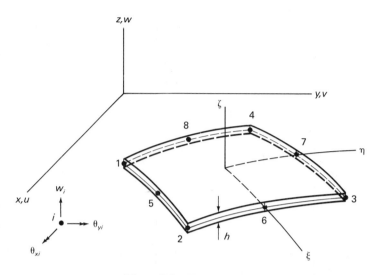

**Figure 8.3** Element PBQ8.

surface lying in the x-y plane. Its geometry is defined to be the same as that for element Q8 in Sec. 7.5. Thus,

$$x = \sum_{i=1}^{8} f_i x_i \qquad y = \sum_{i=1}^{8} f_i y_i \qquad (z = 0) \qquad (7.5\text{–}20)$$

where

$$f_i = \tfrac{1}{4}(1 + \xi_0)(1 + \eta_0)(-1 + \xi_0 + \eta_0) \qquad (i = 1, 2, 3, 4)$$
$$f_i = \tfrac{1}{2}(1 - \xi^2)(1 + \eta_0) \qquad (i = 5, 7) \qquad (7.5\text{–}19)$$
$$f_i = \tfrac{1}{2}(1 + \xi_0)(1 - \eta^2) \qquad (i = 6, 8)$$

Generic displacements at any point off the neutral surface are

$$\mathbf{u} = \{u, v, w\} \qquad (4)$$

We assume that $w$, $\theta_x$, and $\theta_y$ vary quadratically over the element, so that

$$u = z\theta_y = z \sum_{i=1}^{8} f_i \theta_{yi}$$
$$v = -z\theta_x = -z \sum_{i=1}^{8} f_i \theta_{xi} \qquad (5)$$
$$w = \sum_{i=1}^{8} f_i w_i$$

where $z = \zeta h/2$. In these expressions the displacement shape functions $f_i$ are the same as those in Eqs. (7.5–19). Note that the rotations $\theta_x$ and $\theta_y$ are chosen

## Sec. 8.2  Element for Plates in Bending

independently of $w$ and are not related to it by differentiation, as in a beam. In this case the displacement shape functions may be displayed in the matrix form

$$\mathbf{f}_i = \begin{bmatrix} 0 & 0 & \zeta\dfrac{h}{2} \\ 0 & -\zeta\dfrac{h}{2} & 0 \\ 1 & 0 & 0 \end{bmatrix} f_i \qquad (i = 1, 2, \ldots, 8) \tag{6}$$

To isolate terms in submatrix $\mathbf{f}_i$ that multiply $\zeta h/2$, we let

$$\mathbf{f}_{Ai} = \begin{bmatrix} 0 & 0 & 0 \\ 0 & 0 & 0 \\ 1 & 0 & 0 \end{bmatrix} f_i \qquad \mathbf{f}_{Bi} = \begin{bmatrix} 0 & 0 & 1 \\ 0 & -1 & 0 \\ 0 & 0 & 0 \end{bmatrix} f_i \tag{7}$$

Then

$$\mathbf{f}_i = \mathbf{f}_{Ai} + \zeta\dfrac{h}{2}\mathbf{f}_{Bi} \tag{8}$$

and

$$\mathbf{f} = \mathbf{f}_A + \zeta\dfrac{h}{2}\mathbf{f}_B \tag{9}$$

The formula in Eq. (9) will prove useful for obtaining the consistent mass matrix and equivalent nodal loads for element PBQ8.

The 3 × 3 Jacobian matrix required for this element is

$$\mathbf{J} = \begin{bmatrix} x_{,\xi} & y_{,\xi} & 0 \\ x_{,\eta} & y_{,\eta} & 0 \\ 0 & 0 & z_{,\zeta} \end{bmatrix} \tag{10}$$

where $z_{,\zeta} = h/2$ and

$$x_{,\xi} = \sum_{i=1}^{8} f_{i,\xi} x_i \qquad \text{and so on.}$$

The inverse of $\mathbf{J}$ becomes

$$\mathbf{J}^{-1} = \mathbf{J}^* = \begin{bmatrix} \xi_{,x} & \eta_{,x} & 0 \\ \xi_{,y} & \eta_{,y} & 0 \\ 0 & 0 & \zeta_{,z} \end{bmatrix} \tag{11}$$

where $\zeta_{,z} = 2/h$.

We need certain derivatives with respect to local coordinates, which are collected into the following 3 × 3 matrix:

$$\begin{bmatrix} u_{,\xi} & v_{,\xi} & w_{,\xi} \\ u_{,\eta} & v_{,\eta} & w_{,\eta} \\ u_{,\zeta} & v_{,\zeta} & w_{,\zeta} \end{bmatrix} = \sum_{i=1}^{8} \begin{bmatrix} \zeta\frac{h}{2}f_{i,\xi}\theta_{yi} & -\zeta\frac{h}{2}f_{i,\xi}\theta_{xi} & f_{i,\xi}w_i \\ \zeta\frac{h}{2}f_{i,\eta}\theta_{yi} & -\zeta\frac{h}{2}f_{i,\eta}\theta_{xi} & f_{i,\eta}w_i \\ \frac{h}{2}f_i\theta_{yi} & -\frac{h}{2}f_i\theta_{xi} & 0 \end{bmatrix} \quad (12)$$

Transformation of these derivatives to global coordinates is accomplished using the inverse of the Jacobian matrix, as follows:

$$\begin{bmatrix} u_{,x} & v_{,x} & w_{,x} \\ u_{,y} & v_{,y} & w_{,y} \\ u_{,z} & v_{,z} & w_{,z} \end{bmatrix} = \mathbf{J}^{-1} \begin{bmatrix} u_{,\xi} & v_{,\xi} & w_{,\xi} \\ u_{,\eta} & v_{,\eta} & w_{,\eta} \\ u_{,\zeta} & v_{,\zeta} & w_{,\zeta} \end{bmatrix}$$

$$= \sum_{i=1}^{8} \begin{bmatrix} \zeta\frac{h}{2}a_i\theta_{yi} & -\zeta\frac{h}{2}a_i\theta_{xi} & a_i w_i \\ \zeta\frac{h}{2}b_i\theta_{yi} & -\zeta\frac{h}{2}b_i\theta_{xi} & b_i w_i \\ f_i\theta_{yi} & -f_i\theta_{xi} & 0 \end{bmatrix} \quad (13)$$

in which

$$a_i = J^*_{11}f_{i,\xi} + J^*_{12}f_{i,\eta} \qquad b_i = J^*_{21}f_{i,\xi} + J^*_{22}f_{i,\eta} \quad (14)$$

The five types of nonzero strains to be considered for element PBQ8 are

$$\boldsymbol{\epsilon} = \begin{bmatrix} \epsilon_x \\ \epsilon_y \\ \gamma_{xy} \\ \gamma_{yz} \\ \gamma_{zx} \end{bmatrix} = \begin{bmatrix} u_{,x} \\ v_{,y} \\ u_{,y} + v_{,x} \\ v_{,z} + w_{,y} \\ w_{,x} + u_{,z} \end{bmatrix} \quad (15)$$

By inspection of the second version of this strain vector, we can assemble the $i$th part of matrix $\mathbf{B}$ from terms in Eq. (13) as

$$\mathbf{B}_i = \begin{bmatrix} 0 & 0 & \zeta\frac{h}{2}a_i \\ 0 & -\zeta\frac{h}{2}b_i & 0 \\ 0 & -\zeta\frac{h}{2}a_i & \zeta\frac{h}{2}b_i \\ b_i & -f_i & 0 \\ a_i & 0 & f_i \end{bmatrix} \quad (i = 1, 2, \ldots, 8) \quad (16)$$

## Sec. 8.2  Element for Plates in Bending

As with the submatrix $\mathbf{f}_i$, we can isolate terms in $\mathbf{B}_i$ that multiply $\zeta h/2$, as follows:

$$\mathbf{B}_{Ai} = \begin{bmatrix} 0 & 0 & 0 \\ 0 & 0 & 0 \\ 0 & 0 & 0 \\ b_i & -f_i & 0 \\ a_i & 0 & f_i \end{bmatrix} \quad \mathbf{B}_{Bi} = \begin{bmatrix} 0 & 0 & a_i \\ 0 & -b_i & 0 \\ 0 & -a_i & b_i \\ 0 & 0 & 0 \\ 0 & 0 & 0 \end{bmatrix} \quad (17)$$

Then

$$\mathbf{B}_i = \mathbf{B}_{Ai} + \zeta\frac{h}{2}\mathbf{B}_{Bi} \quad (18)$$

and

$$\mathbf{B} = \mathbf{B}_A + \zeta\frac{h}{2}\mathbf{B}_B \quad (19)$$

Equation (19) will be convenient when finding the stiffness matrix for element PBQ8.

Stresses corresponding to the strains in Eq. (15) are

$$\boldsymbol{\sigma} = \{\sigma_x, \sigma_y, \tau_{xy}, \tau_{yz}, \tau_{zx}\} \quad (20)$$

Then the stress-strain matrix for an isotropic material becomes

$$\mathbf{E} = \frac{E}{(1+\nu)(1-2\nu)} \begin{bmatrix} 1-\nu & & & & \text{Sym.} \\ \nu & 1-\nu & & & \\ 0 & 0 & \dfrac{1-2\nu}{2} & & \\ 0 & 0 & 0 & \dfrac{1-2\nu}{2(1.2)} & \\ 0 & 0 & 0 & 0 & \dfrac{1-2\nu}{2(1.2)} \end{bmatrix} \quad (21)$$

This matrix is similar to that in Eq. (3.2–10), but the third row and column (corresponding to $\sigma_z$ and $\epsilon_z$) are omitted. Also, the last two diagonal terms are divided by the *form factor* 1.2 to account for the fact that the transverse shearing stresses produce too little strain energy [4].

We may write the stiffness matrix for element PBQ8 as

$$\mathbf{K} = \int_{-1}^{1}\int_{-1}^{1}\int_{-1}^{1} \mathbf{B}^T \mathbf{E}\, \mathbf{B} \,|\mathbf{J}|\, d\xi\, d\eta\, d\zeta$$

$$= \int_{-1}^{1}\int_{-1}^{1}\int_{-1}^{1} \left(\mathbf{B}_A + \zeta\frac{h}{2}\mathbf{B}_B\right)^T \mathbf{E}\left(\mathbf{B}_A + \zeta\frac{h}{2}\mathbf{B}_B\right)|\mathbf{J}|\, d\xi\, d\eta\, d\zeta \quad (22)$$

In this expression the matrices $\mathbf{B}_A$ and $\mathbf{B}_B$ are both of size $5 \times 24$, but the latter array contains only terms that are multiplied by $\zeta h/2$. Integration of Eq. (22) through the thickness of the element gives

$$\mathbf{K} = \int_{-1}^{1} \int_{-1}^{1} \left(2\mathbf{B}_A^T \mathbf{E} \mathbf{B}_A + \frac{h^2}{6} \mathbf{B}_B^T \mathbf{E} \mathbf{B}_B\right) |\mathbf{J}| \, d\xi \, d\eta \tag{23}$$

which must be evaluated numerically. In this process the factors 2 and $h^2/6$ are multiplied by $h/2$ from the third row of $|\mathbf{J}|$ [see Eq. (10)], producing the factors $h$ and $h^3/12$. Thus, the first part of matrix $\mathbf{K}$ in Eq. (23) is due to transverse *shearing deformations*, whereas the second part is associated with *flexural deformations*.

The consistent mass matrix for element PBQ8 becomes

$$\mathbf{M} = \rho \int_{-1}^{1} \int_{-1}^{1} \int_{-1}^{1} \mathbf{f}^T \mathbf{f} |\mathbf{J}| \, d\xi \, d\eta \, d\zeta$$

$$= \rho \int_{-1}^{1} \int_{-1}^{1} \int_{-1}^{1} \left(\mathbf{f}_A + \zeta \frac{h}{2} \mathbf{f}_B\right)^T \left(\mathbf{f}_A + \zeta \frac{h}{2} \mathbf{f}_B\right) |\mathbf{J}| \, d\xi \, d\eta \, d\zeta \tag{24}$$

In this equation the matrices $\mathbf{f}_A$ and $\mathbf{f}_B$ are both of size $3 \times 24$, but the second has only terms to be multiplied by $\zeta h/2$. Integrating Eq. (24) through the thickness produces

$$\mathbf{M} = \rho \int_{-1}^{1} \int_{-1}^{1} \left(2\mathbf{f}_A^T \mathbf{f}_A + \frac{h^2}{6} \mathbf{f}_B^T \mathbf{f}_B\right) |\mathbf{J}| \, d\xi \, d\eta \tag{25}$$

Again, the factors 2 and $h^2/6$ are multiplied by $h/2$ from $|\mathbf{J}|$. Hence, the first part of matrix $\mathbf{M}$ consists of translational inertias, and the second part gives rotational (or *rotary*) inertias.

Equivalent nodal loads due to body forces on element PBQ8 are calculated using only matrix $\mathbf{f}_A$, as follows:

$$\mathbf{p}_b(t) = \int_{-1}^{1} \int_{-1}^{1} \int_{-1}^{1} \mathbf{f}_A^T \mathbf{b}(t) |\mathbf{J}| \, d\xi \, d\eta \, d\zeta$$

$$= 2 \int_{-1}^{1} \int_{1}^{1} \mathbf{f}_A^T \mathbf{b}(t) |\mathbf{J}| \, d\xi \, d\eta \tag{26}$$

in which

$$\mathbf{b}(t) = \{0, 0, b_z\} \tag{27}$$

and $b_z$ is force per unit volume in the $z$ direction. Alternatively, we may extract the factor $h/2$ from $|\mathbf{J}|$ and rewrite Eq. (26) in the form

$$\mathbf{p}_b(t) = \int_{-1}^{1} \int_{-1}^{1} \mathbf{f}_A^T \mathbf{b}(t) |\bar{\mathbf{J}}| \, d\xi \, d\eta \tag{28}$$

In this expression $\bar{\mathbf{J}}$ has unity in place of $h/2$; and $b_z$ in Eq. (27) now has the meaning of force per unit area. Note that this body force causes no equivalent nodal moments.

After finding the time-varying nodal displacements in the vector $\mathbf{q}(t)$, we can evaluate stresses at any point in each element, as follows:

$$\boldsymbol{\sigma}(t) = \mathbf{E}\,\mathbf{B}\,\mathbf{q}(t) \tag{29}$$

For best accuracy, these stresses should be calculated at the numerical integration points [5].

**Example 8.1**

For the rectangular parent of element PBQ8 (after constraints are imposed), find the consistent mass terms $M_{11}$ and $M_{22}$. In the first case, Eq. (25) specializes to

$$M_{11} = \rho a b h \int_{-1}^{1} \int_{-1}^{1} f_1^2 \, d\xi \, d\eta \tag{a}$$

Substituting the shape function $f_1$ from Eqs. (7.5-19) into Eq. (a) yields

$$M_{11} = \frac{\rho a b h}{16} \int_{-1}^{1} \int_{-1}^{1} (1 - \xi)^2 (1 - \eta)^2 (-1 - \xi - \eta)^2 \, d\xi \, d\eta \tag{b}$$

Performing the integrations indicated in Eq. (b) results in

$$M_{11} = \tfrac{1}{8} \rho a b h \tag{c}$$

which is simply a fraction of the total mass. Similarly, in the second case we have

$$M_{22} = \rho a b \frac{h^3}{12} \int_{-1}^{1} \int_{-1}^{1} f_1^2 \, d\xi \, d\eta$$

$$= \frac{1}{96} \rho a b h^3 \tag{d}$$

which has units of mass moment of inertia.

## 8.3 PROGRAM DYNAPB FOR PLATES IN BENDING

We shall now describe a computer program named DYNAPB for the dynamic analysis of plates in bending, which uses element PBQ8 from the preceding section. This program is constructed by modifying the part of Program DYNAPS pertaining to element Q8 (see Sec. 7.6), because the geometric and displacement shape functions are the same for both. However, in program DYNAPB the matrices $\mathbf{K}$, $\mathbf{M}$, and $\mathbf{p}_b(t)$ must be handled according to the expressions developed in Sec. 8.2.

Table 8.1 shows preparation of structural data for Program DYNAPB. Comparing this data with that in Table 7.5, we see that the structural parameters

**TABLE 8.1 Structural Data for Program DYNAPB**

| Type of Data | No. of Lines | Items on Data Lines |
|---|---|---|
| Problem identification | 1 | Descriptive title |
| Structural parameters | 1 | NN, NE, NRN, E, PR, RHO, H |
| Plate bending data  (a) Nodal coordinates  (b) Element information[a]  (c) Nodal restraints | NN  NE  NRN | J, X(J), Y(J)  I, IN(I, 1), IN(I, 2), ... , IN(I, 8)  J, NRL(3J-2), NRL(3J-1), NRL(3J) |

[a] For sequence of node numbers, see Fig. 8.3.

IPS and NEN are omitted. In addition, there are three possible nodal restraints instead of two per node.

Dynamic load data for Program DYNAPB is given in Table 8.2. Here we have three possible initial displacements, initial velocities, and nodal loads instead of two per node (as in Table 7.6). Line loads act in the $z$ direction along an edge of an element and may vary quadratically. Therefore, we require three node numbers and three force intensities (per unit length) to describe them [see Fig. 7.11(b)]. Also, quadratically varying area loads (force per unit area, in the $z$ direction) need eight force intensities for a complete description [see

**TABLE 8.2 Dynamic Load Data for Program DYNAPB**

| Type of Data | No. of Lines | Items on Data Lines |
|---|---|---|
| Dynamic parameters | 1 | ISOLVE, NTS, DT, DAMPR |
| Initial conditions  (a) Condition parameters  (b) Displacements  (c) Velocities | 1  NNID  NNIV | NNID, NNIV  J, D0(3J-2), D0(3J-1), D0(3J)  J, V0(3J-2), V0(3J-1), V0(3J) |
| Applied actions  (a) Load parameters  (b) Nodal loads  (c) Line loads  (d) Area loads | 1  NLN  NEL  NEA | NLN, NEL, NEA  J, AS(3J-2), AS(3J-1), AS(3J)  J, K, L, BL1, BL2, BL3  I, BA1, BA2, ... , BA8 |
| Ground accelerations  (a) Acceleration parameter  (b) Acceleration factor | 1  1 | IGA  GAZ |
| Forcing function  (a) Function parameter  (b) Function ordinates | 1  NFO | NFO  K, T(K), FO(K) |

### Sec. 8.3 Program DYNAPB for Plates in Bending

Fig. 7.16(b)]. Finally, note that the acceleration factor GAZ is for the $z$ direction only.

As in Program DYNAPS, we take $n = 2$ each way to locate points for numerical integration. With a plate-bending element, the computer evaluates the time-varying stresses $\sigma_x$, $\sigma_y$, $\tau_{xy}$, $\tau_{yz}$, and $\tau_{zx}$ at each of the four integration points. Because the displacement shape functions are quadratic, we found it necessary to use $n = 3$ each way for terms in the consistent-mass matrix of element PBQ8.

**Example 8.2**

Figure 8.4 shows half of a square, symmetric, simply supported plate that is divided into eight PBQ8 elements. Also indicated in the figure is a *moving load P* that travels in the $y$ direction along the centerline, where nodes 5, 8, 13, . . . , 37 are located. We wish to determine the translational responses of the plate at node 21 due to the load moving at constant velocity and constant acceleration [6].

This problem is analogous to Example 6.5 for a moving load on a simply-supported beam divided into four flexural elements. However, the displacement shape functions for the PBQ8 plate element are quadratic instead of cubic. Therefore, the functions given by Eqs. (7.5-19) must be used in Eq. (a) of Example 6.5, and $y_i(t)$ replaces $x_i(t)$ in both Eqs. (a) and (b). Of course, we need to extend Program DYNAPB to handle one or more moving loads, as described previously for Program DYNACB.

Physical parameters in this example are

$$E = 69 \times 10^6 \text{ kN/m}^2 \qquad \nu = 0.33 \qquad \rho = 2.62 \text{ Mg/m}^3$$

$$a = 0.1 \text{ m} \qquad h = 0.025 \text{ m} \qquad P = 20 \text{ kN}$$

for which the material is aluminum and the units are SI. We ran this data on the extended

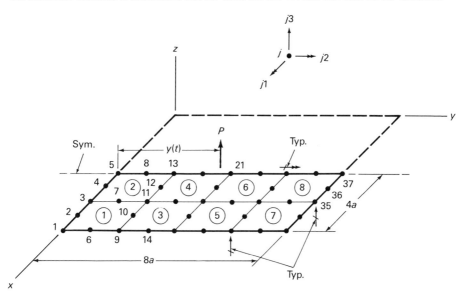

**Figure 8.4** Simply-supported plate with moving load.

version of Program DYNAPB, using Subprogram NORMOD to obtain responses. Computer plots of $D_{j1}$ at node 21 are given in Fig. 8.5. For the case of constant velocity (V0P = 153.7 m/s), the plot shows a maximum translation of 2.320 mm; and for constant acceleration (A0P = 59.05 $\times$ 10$^3$ m/s$^2$, with zero initial velocity), we have a maximum of 2.034 mm. Their ratios to the static deflection of 1.472 mm (due to the load applied gradually at node 21) are 1.576 and 1.382, respectively. As for the beam, the values of V0P and A0P used in this example both give travel times equal to the fundamental period of the plate, which is 5.206 ms.

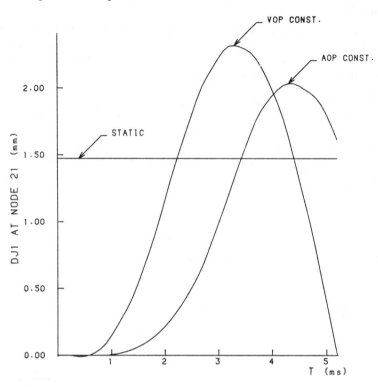

**Figure 8.5** Translational responses at center of plate.

## 8.4 ELEMENT FOR GENERAL SHELLS

In this section we specialize the isoparametric hexahedron H20 to become a curved quadrilateral element for the analysis of general shells. Development of the *shell element SHQ8* is similar to the technique used in obtaining element PBQ8 (see Sec. 8.2) for plate bending. However, the constraint conditions are modified because two additional translations, $u_i$ and $v_i$, occur at each node of the shell element. Thus, the constraint matrix $\mathbf{C}_{ai}$ for a corner node of the rectangular parent element [see Fig. 8.1(b)] has two more columns than before, as follows:

## Sec. 8.4 Element for General Shells

$$\mathbf{C}_{ai} = \begin{bmatrix} 1 & 0 & 0 & 0 & 0 \\ 0 & 1 & 0 & 0 & 0 \\ 0 & 0 & 1 & 0 & 0 \\ 1 & 0 & 0 & 0 & \frac{h_i}{2} \\ 0 & 1 & 0 & -\frac{h_i}{2} & 0 \\ 0 & 0 & 1 & 0 & 0 \\ 1 & 0 & 0 & 0 & -\frac{h_i}{2} \\ 0 & 1 & 0 & \frac{h_i}{2} & 0 \\ 0 & 0 & 1 & 0 & 0 \end{bmatrix} \quad (1)$$

When this 9 × 5 matrix is compared with Eq. (8.2-2), it is seen that columns 1 and 2 have been added. Similarly, the constraint matrix $\mathbf{C}_{bi}$ for a midedge node of the rectangular parent becomes

$$\mathbf{C}_{bi} = \begin{bmatrix} 1 & 0 & 0 & 0 & \frac{h_i}{2} \\ 0 & 1 & 0 & -\frac{h_i}{2} & 0 \\ 0 & 0 & 1 & 0 & 0 \\ 1 & 0 & 0 & 0 & -\frac{h_i}{2} \\ 0 & 1 & 0 & \frac{h_i}{2} & 0 \\ 0 & 0 & 1 & 0 & 0 \end{bmatrix} \quad (2)$$

which is a 6 × 5 array that can be compared with Eq. (8.2-3). With five displacements at each of eight nodes, element SHQ8 has (8)(5) = 40 nodal displacements.

As with the plate element PBQ8, the general shell element SHQ8 will be formulated directly [2, 3]. Figure 8.6(a) shows the geometric layout of element SHQ8, in which the coordinates of any point are

$$\begin{bmatrix} x \\ y \\ z \end{bmatrix} = \sum_{i=1}^{8} f_i \begin{bmatrix} x_i \\ y_i \\ z_i \end{bmatrix} + \sum_{i=1}^{8} f_i \zeta \frac{h_i}{2} \begin{bmatrix} l_{3i} \\ m_{3i} \\ n_{3i} \end{bmatrix} \quad (3)$$

The interpolation functions $f_i$ appearing in Eq. (3) are given by Eqs. (7.5-19).

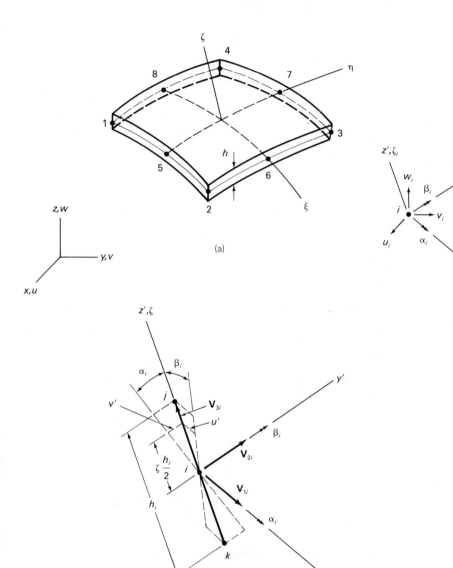

**Figure 8.6** (a) Element SHQ8; (b) nodal vectors.

## Sec. 8.4  Element for General Shells

Thus, we see that the thickness $h$ may vary quadratically over the element. In addition, the terms $l_{3i}$, $m_{3i}$, and $n_{3i}$ are the direction cosines of a vector $\mathbf{V}_{3i}$ that is normal to the middle surface and spans the thickness $h_i$ of the shell at node $i$. Figure 8.6(b) shows this vector, which is obtained as

$$\mathbf{V}_{3i} = \begin{bmatrix} x_j - x_k \\ y_j - y_k \\ z_j - z_k \end{bmatrix} = \begin{bmatrix} l_{3i} \\ m_{3i} \\ n_{3i} \end{bmatrix} h_i \tag{4}$$

Points $j$ and $k$ in the figure are at the surfaces of the shell. In a computer program either the coordinates of points $j$ and $k$ or the direction cosines for $\mathbf{V}_{3i}$ must be given as data.

Generic displacements at any point in the shell element are taken to be in the directions of global axes. Thus,

$$\mathbf{u} = \{u, v, w\} \tag{5}$$

On the other hand, nodal displacements consist of these same translations (in global directions) as well as two small rotations $\alpha_i$ and $\beta_i$ about two local tangential axes $x'$ and $y'$, as indicated in Fig. 8.6(a). Hence,

$$\mathbf{q}_i = \{u_i, v_i, w_i, \alpha_i, \beta_i\} \qquad (i = 1, 2, \ldots, 8) \tag{6}$$

Generic displacements in terms of nodal displacements are

$$\begin{bmatrix} u \\ v \\ w \end{bmatrix} = \sum_{i=1}^{8} f_i \begin{bmatrix} u_i \\ v_i \\ w_i \end{bmatrix} + \sum_{i=1}^{8} f_i \zeta \frac{h_i}{2} \boldsymbol{\mu}_i \begin{bmatrix} \alpha_i \\ \beta_i \end{bmatrix} \tag{7}$$

In this formula the symbol $\boldsymbol{\mu}_i$ denotes the following matrix:

$$\boldsymbol{\mu}_i = \begin{bmatrix} -l_{2i} & l_{1i} \\ -m_{2i} & m_{1i} \\ -n_{2i} & n_{1i} \end{bmatrix} \tag{8}$$

Column 1 in this array contains negative values of the direction cosines of the second tangential vector $\mathbf{V}_{2i}$; and column 2 has the direction cosines for the first tangential vector $\mathbf{V}_{1i}$, as shown in Fig. 8.6(b). These vectors are orthogonal to the vector $\mathbf{V}_{3i}$ and to each other, but the choice for the direction of one of them is arbitrary. To settle the choice, we let

$$\mathbf{V}_{1i} = \mathbf{e}_y \times \mathbf{V}_{3i} \tag{9}$$

Then

$$\mathbf{V}_{2i} = \mathbf{V}_{3i} \times \mathbf{V}_{1i} \tag{10}$$

[If $\mathbf{V}_{3i}$ is parallel to $\mathbf{e}_y$ in Eq. (9), the latter is replaced by $\mathbf{e}_z$.] Figure 8.6(b) shows local generic translations $u'$ and $v'$ (in the directions of $\mathbf{V}_{1i}$ and $\mathbf{V}_{2i}$) due to the nodal rotations $\beta_i$ and $\alpha_i$, respectively. Their values are

$$u' = \zeta \frac{h_i}{2} \beta_i \qquad v' = -\zeta \frac{h_i}{2} \alpha_i \qquad (11)$$

Contributions of these terms to the generic displacements at any point are given by the second summation in Eq. (7).

As for the plate element, the displacement shape functions in Eq. (7) may be cast into the matrix form

$$\mathbf{f}_i = \begin{bmatrix} 1 & 0 & 0 & -\zeta \frac{h_i}{2} l_{2i} & \zeta \frac{h_i}{2} l_{1i} \\ 0 & 1 & 0 & -\zeta \frac{h_i}{2} m_{2i} & \zeta \frac{h_i}{2} m_{1i} \\ 0 & 0 & 1 & -\zeta \frac{h_i}{2} n_{2i} & \zeta \frac{h_i}{2} n_{1i} \end{bmatrix} f_i \qquad (i = 1, 2, \ldots, 8) \qquad (12)$$

In order to isolate terms in submatrix $\mathbf{f}_i$ multiplying $\zeta$, we let

$$\mathbf{f}_{Ai} = \begin{bmatrix} 1 & 0 & 0 & 0 & 0 \\ 0 & 1 & 0 & 0 & 0 \\ 0 & 0 & 1 & 0 & 0 \end{bmatrix} f_i \qquad (13a)$$

and

$$\mathbf{f}_{Bi} = \begin{bmatrix} 0 & 0 & 0 & -l_{2i} & l_{1i} \\ 0 & 0 & 0 & -m_{2i} & m_{1i} \\ 0 & 0 & 0 & -n_{2i} & n_{1i} \end{bmatrix} \frac{h_i}{2} f_i \qquad (13b)$$

Then

$$\mathbf{f}_i = \mathbf{f}_{Ai} + \zeta \mathbf{f}_{Bi} \qquad (14)$$

and

$$\mathbf{f} = \mathbf{f}_A + \zeta \mathbf{f}_B \qquad (15)$$

The last of these formulas will later be used to derive the consistent mass matrix and equivalent nodal loads for element SHQ8.

The 3 × 3 Jacobian matrix required in this formulation is

$$\mathbf{J} = \begin{bmatrix} x_{,\xi} & y_{,\xi} & z_{,\xi} \\ x_{,\eta} & y_{,\eta} & z_{,\eta} \\ x_{,\zeta} & y_{,\zeta} & z_{,\zeta} \end{bmatrix} \qquad (16)$$

We find the derivatives in matrix $\mathbf{J}$ from Eq. (3), as follows:

$$x_{,\xi} = \sum_{i=1}^{8} f_{i,\xi} x_i + \sum_{i=1}^{8} f_{i,\xi} \zeta \frac{h_i}{2} l_{3i}$$

$$x_{,\eta} = \sum_{i=1}^{8} f_{i,\eta} x_i + \sum_{i=1}^{8} f_{i,\eta} \zeta \frac{h_i}{2} l_{3i}$$

## Sec. 8.4  Element for General Shells

$$x_{,\zeta} = \sum_{i=1}^{8} f_i \frac{h_i}{2} l_{3i} \quad \text{and so on.}$$

The inverse of **J** becomes

$$\mathbf{J}^{-1} = \mathbf{J}^* = \begin{bmatrix} \xi_{,x} & \eta_{,x} & \zeta_{,x} \\ \xi_{,y} & \eta_{,y} & \zeta_{,y} \\ \xi_{,z} & \eta_{,z} & \zeta_{,z} \end{bmatrix} \quad (17)$$

We need certain derivatives of the generic displacements [see Eq. (7)] with respect to local coordinates. These derivatives are listed in a column vector of nine terms, as follows:

$$\begin{bmatrix} u_{,\xi} \\ u_{,\eta} \\ u_{,\zeta} \\ v_{,\xi} \\ v_{,\eta} \\ v_{,\zeta} \\ w_{,\xi} \\ w_{,\eta} \\ w_{,\zeta} \end{bmatrix} = \sum_{i=1}^{8} \begin{bmatrix} f_{i,\xi} & 0 & 0 & -\zeta f_{i,\xi} l_{2i} & \zeta f_{i,\xi} l_{1i} \\ f_{i,\eta} & 0 & 0 & -\zeta f_{i,\eta} l_{2i} & \zeta f_{i,\eta} l_{1i} \\ 0 & 0 & 0 & -f_i l_{2i} & f_i l_{1i} \\ 0 & f_{i,\xi} & 0 & -\zeta f_{i,\xi} m_{2i} & \zeta f_{i,\xi} m_{1i} \\ 0 & f_{i,\eta} & 0 & -\zeta f_{i,\eta} m_{2i} & \zeta f_{i,\eta} m_{1i} \\ 0 & 0 & 0 & -f_i m_{2i} & f_i m_{1i} \\ 0 & 0 & f_{i,\xi} & -\zeta f_{i,\xi} n_{2i} & \zeta f_{i,\xi} n_{1i} \\ 0 & 0 & f_{i,\eta} & -\zeta f_{i,\eta} n_{2i} & \zeta f_{i,\eta} n_{1i} \\ 0 & 0 & 0 & -f_i n_{2i} & f_i n_{1i} \end{bmatrix} \begin{bmatrix} u_i \\ v_i \\ w_i \\ \frac{h_i}{2}\alpha_i \\ \frac{h_i}{2}\beta_i \end{bmatrix} \quad (18)$$

Transformation of these derivatives to global coordinates requires that the inverse of the Jacobian matrix be applied. Therefore,

$$\begin{bmatrix} u_{,x} \\ u_{,y} \\ \cdots \\ w_{,z} \end{bmatrix} = \begin{bmatrix} \mathbf{J}^* & 0 & 0 \\ 0 & \mathbf{J}^* & 0 \\ 0 & 0 & \mathbf{J}^* \end{bmatrix} \begin{bmatrix} u_{,\xi} \\ u_{,\eta} \\ \cdots \\ w_{,\zeta} \end{bmatrix} \quad (19)$$

Multiplying the terms in this equation, we obtain

$$\begin{bmatrix} u_{,x} \\ u_{,y} \\ u_{,z} \\ v_{,x} \\ v_{,y} \\ v_{,z} \\ w_{,x} \\ w_{,y} \\ w_{,z} \end{bmatrix} = \sum_{i=1}^{8} \begin{bmatrix} a_i & 0 & 0 & -d_i l_{2i} & d_i l_{1i} \\ b_i & 0 & 0 & -e_i l_{2i} & e_i l_{1i} \\ c_i & 0 & 0 & -g_i l_{2i} & g_i l_{1i} \\ 0 & a_i & 0 & -d_i m_{2i} & d_i m_{1i} \\ 0 & b_i & 0 & -e_i m_{2i} & e_i m_{1i} \\ 0 & c_i & 0 & -g_i m_{2i} & g_i m_{1i} \\ 0 & 0 & a_i & -d_i n_{2i} & d_i n_{1i} \\ 0 & 0 & b_i & -e_i n_{2i} & e_i n_{1i} \\ 0 & 0 & c_i & -g_i n_{2i} & g_i n_{1i} \end{bmatrix} \begin{bmatrix} u_i \\ v_i \\ w_i \\ \alpha_i \\ \beta_i \end{bmatrix} \quad (20)$$

in which

$$a_i = J_{11}^* f_{i,\xi} + J_{12}^* f_{i,\eta} \qquad d_i = \frac{h_i}{2}(a_i \zeta + J_{13}^* f_i)$$

$$b_i = J_{21}^* f_{i,\xi} + J_{22}^* f_{i,\eta} \qquad e_i = \frac{h_i}{2}(b_i \zeta + J_{23}^* f_i) \qquad (21)$$

$$c_i = J_{31}^* f_{i,\xi} + J_{32}^* f_{i,\eta} \qquad g_i = \frac{h_i}{2}(c_i \zeta + J_{33}^* f_i)$$

For element SHQ8 we consider six types of nonzero strains, as follows:

$$\boldsymbol{\epsilon} = \begin{bmatrix} \epsilon_x \\ \epsilon_y \\ \epsilon_z \\ \gamma_{xy} \\ \gamma_{yz} \\ \gamma_{zx} \end{bmatrix} = \begin{bmatrix} u_{,x} \\ v_{,y} \\ w_{,z} \\ u_{,y} + v_{,x} \\ v_{,z} + w_{,y} \\ w_{,x} + u_{,z} \end{bmatrix} \qquad (22)$$

Noting the second version of this strain vector, we may construct the $i$th part of matrix $\mathbf{B}$ from terms in Eq. (20) as

$$\mathbf{B}_i = \begin{bmatrix} a_i & 0 & 0 & -d_i l_{2i} & d_i l_{1i} \\ 0 & b_i & 0 & -e_i m_{2i} & e_i m_{1i} \\ 0 & 0 & c_i & -g_i n_{2i} & g_i n_{1i} \\ b_i & a_i & 0 & -e_i l_{2i} - d_i m_{2i} & e_i l_{1i} + d_i m_{1i} \\ 0 & c_i & b_i & -g_i m_{2i} - e_i n_{2i} & g_i m_{1i} + e_i n_{1i} \\ c_i & 0 & a_i & -d_i n_{2i} - g_i l_{2i} & d_i n_{1i} + g_i l_{1i} \end{bmatrix} \qquad (23)$$

$$(i = 1, 2, \ldots, 8)$$

Similar to the plate element, we can isolate terms in submatrix $\mathbf{B}_i$ that multiply $\zeta$ to find

$$\mathbf{B}_i = \mathbf{B}_{Ai} + \zeta \mathbf{B}_{Bi} \qquad (24)$$

Submatrices $\mathbf{B}_{Ai}$ and $\mathbf{B}_{Bi}$ are composed from Eqs. (21) and (23), but the actual details are omitted. Altogether, we have

$$\mathbf{B} = \mathbf{B}_A + \zeta \mathbf{B}_B \qquad (25)$$

which will be useful when determining the stiffness matrix for the shell element.

The following nonzero stresses in the directions of primed axes will be considered:

$$\boldsymbol{\sigma}' = \{\sigma_{x'},\ \sigma_{y'},\ \tau_{x'y'},\ \tau_{y'z'},\ \tau_{z'x'}\} \qquad (26)$$

and the corresponding strains are

$$\boldsymbol{\epsilon}' = \{\epsilon_{x'},\ \epsilon_{y'},\ \gamma_{x'y'},\ \gamma_{y'z'},\ \gamma_{z'x'}\} \qquad (27)$$

## Sec. 8.4  Element for General Shells

As in a plate element, the normal stress $\sigma_{z'}$ and the strain $\epsilon_{z'}$ have been omitted. Then the stress-strain matrix $\mathbf{E}$ for an isotropic material becomes the same as that for element PBQ8 in Eq. (8.2-21).

To relate local strains in the vector $\boldsymbol{\epsilon}'$ to global strains in the vector $\boldsymbol{\epsilon}$, we can use the $6 \times 6$ strain transformation matrix $\mathbf{T}_\epsilon$ in Eq. (7.2-28), as follows:

$$\boldsymbol{\epsilon}' = \mathbf{T}_\epsilon \boldsymbol{\epsilon} \tag{7.2-27}$$

However, the third row of matrix $\mathbf{T}_\epsilon$ must be deleted, because $\epsilon_{z'}$ is not to be included in vector $\boldsymbol{\epsilon}'$. For the purpose of evaluating $\mathbf{T}_\epsilon$ at an integration point, we need the direction cosines for vectors $\mathbf{V}_1$, $\mathbf{V}_2$, and $\mathbf{V}_3$ at the point. This may be done with the following sequence of calculations:

$$\mathbf{e}_1 = (\mathbf{J}_1)_{\text{norm.}} \qquad \mathbf{e}_3 = (\mathbf{J}_1 \times \mathbf{J}_2)_{\text{norm.}} \qquad \mathbf{e}_2 = \mathbf{e}_3 \times \mathbf{e}_1 \tag{28}$$

In these expressions the vector $(\mathbf{J}_1)_{\text{norm.}}$ denotes the first row of the Jacobian matrix normalized to unit length, and so on.

When calculating stresses in local directions, it is also useful to have

$$\mathbf{B}' = \mathbf{T}_\epsilon \mathbf{B} \tag{29}$$

Matrix $\mathbf{B}'$ will contain only five rows, due to the deletion of the third row of $\mathbf{T}_\epsilon$.

Now we are ready to formulate the stiffness matrix for element SHQ8 using matrix $\mathbf{B}'$, as follows:

$$\mathbf{K} = \int_{-1}^{1} \int_{-1}^{1} \int_{-1}^{1} (\mathbf{B}')^{\mathrm{T}} \mathbf{E} \, \mathbf{B}' \, d\xi \, d\eta \, d\zeta$$

$$= \int_{-1}^{1} \int_{-1}^{1} \int_{-1}^{1} (\mathbf{B}'_A + \zeta \mathbf{B}'_B)^{\mathrm{T}} \mathbf{E} (\mathbf{B}'_A + \zeta \mathbf{B}'_B) \, |\mathbf{J}| \, d\xi \, d\eta \, d\zeta \tag{30}$$

Here the matrices $\mathbf{B}'_A$ and $\mathbf{B}'_B$ are both of size $5 \times 40$, but the latter array contains only terms that are to be multiplied by $\zeta$. Integration of Eq. (30) through the thickness* of the element leads to

$$\mathbf{K} = \int_{-1}^{1} \int_{-1}^{1} [2(\mathbf{B}'_A)^{\mathrm{T}} \mathbf{E} \, \mathbf{B}'_A + \tfrac{2}{3} (\mathbf{B}'_B)^{\mathrm{T}} \mathbf{E} \, \mathbf{B}'_B] \, |\mathbf{J}| \, d\xi \, d\eta \tag{31}$$

The remaining integrals in Eq. (31) must be evaluated numerically, using two integration points in each of the $\xi$ and $\eta$ directions [5]. In this process the factors 2 and $\tfrac{2}{3}$ are multiplied by $h_i/2$ from the third row of $|\mathbf{J}|$, and $\mathbf{B}'_B$ also contains the same constant. Thus, we effectively obtain the factors $h_i$ and $h_i^3/12$ in the two parts of matrix $\mathbf{K}$.

The consistent mass matrix for element SHQ8 is

$$\mathbf{M} = \rho \int_{-1}^{1} \int_{-1}^{1} \int_{-1}^{1} \mathbf{f}^{\mathrm{T}} \mathbf{f} \, |\mathbf{J}| \, d\xi \, d\eta \, d\zeta$$

$$= \rho \int_{-1}^{1} \int_{-1}^{1} \int_{-1}^{1} (\mathbf{f}_A + \zeta \mathbf{f}_B)^{\mathrm{T}} (\mathbf{f}_A + \zeta \mathbf{f}_B) \, |\mathbf{J}| \, d\xi \, d\eta \, d\zeta \tag{32}$$

*To simplify integration through the thickness, terms in matrix $\mathbf{J}$ containing $\zeta$ are neglected.

Recall that the matrices $\mathbf{f}_A$ and $\mathbf{f}_B$ are both of size 3 × 40, but the second has only terms to be multiplied by $\zeta$. Integrating Eq. (32) through the thickness yields

$$\mathbf{M} = \rho \int_{-1}^{1} \int_{-1}^{1} [2\mathbf{f}_A^T \mathbf{f}_A + \tfrac{2}{3}\mathbf{f}_B^T \mathbf{f}_B] \, |\mathbf{J}| \, d\xi \, d\eta \tag{33}$$

Because the factors 2 and $\tfrac{2}{3}$ effectively become $h_i$ and $h_i^3/12$, the first and second parts of matrix $\mathbf{M}$ contain translational and rotational inertias, respectively.

Equivalent nodal loads due to body forces on element SHQ8 may be found using only matrix $\mathbf{f}_A$, as follows:

$$\mathbf{p}_b(t) = \int_{-1}^{1} \int_{-1}^{1} \int_{-1}^{1} \mathbf{f}_A^T \mathbf{b}(t) \, |\mathbf{J}| \, d\xi \, d\eta \, d\zeta$$

$$= 2 \int_{-1}^{1} \int_{-1}^{1} \mathbf{f}_A^T \mathbf{b}(t) \, |\mathbf{J}| \, d\xi \, d\eta \tag{34}$$

In this expression, the load vector $\mathbf{b}(t)$ is assumed to contain components of force (per unit volume) that are uniform through the thickness of the shell. Thus,

$$\mathbf{b}(t) = \{b_x, b_y, b_z\} \tag{35}$$

As for the plate element, these body forces do not cause any equivalent nodal moments.

After the time-varying nodal displacements in the vector $\mathbf{q}(t)$ have been obtained, accompanying stresses in the element may be calculated for local (primed) directions. That is,

$$\boldsymbol{\sigma}'(t) = \mathbf{E} \, \mathbf{T}_\epsilon \mathbf{B} \, \mathbf{q}(t) = \mathbf{E} \, \mathbf{B}' \mathbf{q}(t) \tag{36}$$

Such stresses should be determined at the sampling points for numerical integration.

## 8.5 PROGRAM DYNASH FOR GENERAL SHELLS

In Sec. 8.4 we developed element SHQ8 for the dynamic analysis of general shells. Now we present a computer program called DYNASH, which is based on that element. The easiest way to compose this program is to modify statements in Program DYNASO relating to element H20 (see Sec. 7.8), from which element SHQ8 is derived. But in Program DYNASH, we must use the formulas given in Sec. 8.4 to construct the matrices $\mathbf{K}$, $\mathbf{M}$, and $\mathbf{p}_b(t)$.

Preparation of structural data for Program DYNASH appears in Table 8.3. Comparing this data with that in Table 7.9, we observe that NEN is omitted from the list of structural parameters. However, the symbol H is added to the list for a case where the thickness is constant over the whole shell. We also see that H(J) is included with the nodal coordinates for a case where the thickness varies over an element. The line of normal vectors in the table contains a node number J and

Sec. 8.5  Program DYNASH for General Shells            391

TABLE 8.3  Structural Data for Program DYNASH

| Type of Data | No. of Lines | Items on Data Lines |
|---|---|---|
| Problem identification | 1 | Descriptive title |
| Structural parameters | 1 | NN, NE, NRN, E, PR, RHO, H |
| General shell data <br> (a) Nodal coordinates; $h_j$ <br> (b) Normal vectors <br> (c) Element information[a] <br> (d) Nodal restraints | <br> NN <br> NN <br> NE <br> NRN | <br> J, X(J), Y(J), Z(J), H(J) <br> J, V3X(J), V3Y(J), V3Z(J) <br> I, IN(I, 1), IN(I, 2), . . . , IN(I, 8) <br> J, NRL(5J-4), NRL(5J-3), . . . , NRL(5J) |

[a] For sequence of node numbers, see Fig. 8.6(a).

three orthogonal components of the vector $V_3$ at the node [see Eq. (8.4-4)]. In addition, five possible types of nodal restraints are given in line (d).

Table 8.4 lists the dynamic load data required for Program DYNASH, which is similar to that in Table 7.10 for Program DYNASO. However, now we have five possible initial displacements, initial velocities, and nodal loads instead of three per node. As for element H20, we may have a quadratically varying line load on an edge of the shell element, with components in the $x$, $y$, and $z$ directions. For this purpose, we need three node numbers and nine force intensities (per unit length). Also, the area loads require 24 force intensities (per

TABLE 8.4  Dynamic Load Data for Program DYNASH

| Type of Data | No. of Lines | Items on Data Lines |
|---|---|---|
| Dynamic parameters | 1 | ISOLVE, NTS, DT, DAMPR |
| Initial conditions <br> (a) Condition parameters <br> (b) Displacements <br> (c) Velocities | <br> 1 <br> NNID <br> NNIV | <br> NNID, NNIV <br> J, D0(5J-4), D0(5J-3), . . . , D0(5J) <br> J, V0(5J-4), V0(5J-3), . . . , V0(5J) |
| Applied actions <br> (a) Load parameters <br> (b) Nodal loads <br> (c) Line loads <br> (d) Area loads <br> (e) Volume loads | <br> 1 <br> NLN <br> NEL <br> NEA <br> NEV | <br> NLN, NEL, NEA, NEV <br> J, AS(5J-4), AS(5J-3), . . . , AS(5J) <br> J, K, L, BL1, BL2, . . . , BL9 <br> I, BA1, BA2, . . . , BA24 <br> I, BV1, BV2, BV3 |
| Ground accelerations <br> (a) Acceleration parameter <br> (b) Acceleration factors | <br> 1 <br> 1 | <br> IGA <br> GAX, GAY, GAZ |
| Forcing function <br> (a) Function parameter <br> (b) Function ordinates | <br> 1 <br> NFO | <br> NFO <br> K, T(K), FO(K) |

unit area) to describe quadratically varying components in the $x$, $y$, and $z$ directions.

For element SHQ8 we use $n = 2$ in the $\xi$ and $\eta$ directions [see Fig. 8.6(a)] to locate points for numerical integration. At each of these four points the computer evaluates the time-varying stresses, $\sigma_{x'}$, $\sigma_{y'}$, $\tau_{x'y'}$, $\tau_{y'z'}$, and $\tau_{z'x'}$ in local directions. As for element PBQ8, we needed to use $n = 3$ each way for terms in the consistent-mass matrix of element SHQ8 to retain sufficient accuracy.

**Example 8.3**

A quarter of a doubly symmetric cylindrical roof shell appears in Fig. 8.7. This portion of the shell is divided into four SHQ8 elements of constant thickness. Note that the shell is symmetric with respect to the $x$-$z$ and $y$-$z$ planes. Consequently, nodal restraints must be used to prevent translations across those planes and rotations in the planes, as indicated at nodes 4 and 14. On the other hand, simple supports at the ends of the shell prevent translations in the $x$ and $z$ directions, as at node 20. A rigid-body ground acceleration $\ddot{D}_{g3}(t)$ occurs in the $z$ direction, and we wish to find the response of the structure due to this influence.

For this problem the physical parameters are

$$E = 3.6 \times 10^3 \text{ k/in.}^2 \qquad \nu = 0.15 \qquad \rho = 2.25 \times 10^{-7} \text{ k-s}^2/\text{in.}^4$$

$$L = 100 \text{ in.} \qquad R = 3L \qquad h = 3 \text{ in.} \qquad \phi = 40°$$

$$(\ddot{D}_{g3})_{\max} = 115.9 \text{ in./s}^2$$

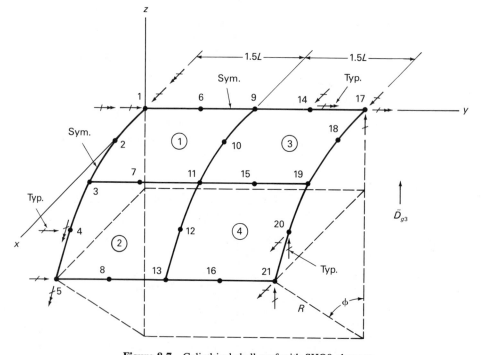

**Figure 8.7** Cylindrical shell roof with SHQ8 elements.

## Sec. 8.5  Program DYNASH for General Shells

where the material is reinforced concrete and the units are US. We used this data in Program DYNASH with DAMPR = 0.05 and solution by Subprogram NUMINT. Figure 8.8(a) shows a computer plot of the ground acceleration, and the resulting translations in the $x$ and $z$ directions at node 5 are given in Fig. 8.8(b). The maximum values of these displacements are DJ1 = 0.8211 in. and DJ3 = 1.430 in. at time $t$ = 720 ms. Also shown in Fig. 8.8(c) are plots of the flexural stresses in the $x'$ and $y'$ directions (at

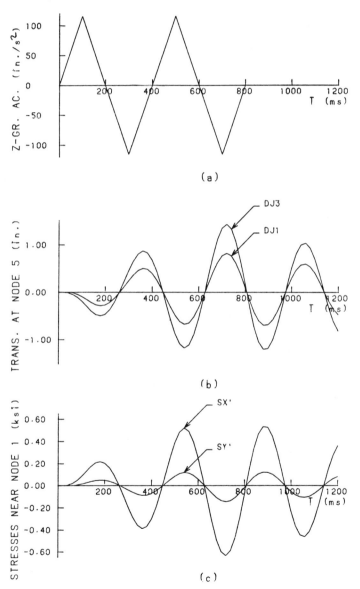

**Figure 8.8** Cylindrical shell roof: (a) ground acceleration; (b) displacements; (c) flexural stresses.

the upper surface) for the integration point near node 1. Their minimum values are SX′ = −0.6350 ksi and SY′ = −0.1456 ksi at time $t$ = 720 ms. Of course, these stresses must be added to the membrane stresses occurring at the same time.

## 8.6 ELEMENT FOR AXISYMMETRIC SHELLS

As shown by Ahmad et al. [7], it is possible to specialize a ring element with an isoparametric cross section to become an axisymmetric shell element by making one dimension small compared to the other. In this section we demonstrate the procedure by specializing element AXQ8 from Sec. 7.9 to form a shell element called AXSH3.

Figure 8.9(a) shows the axisymmetric solid element AXQ8, for which the cross section is an isoparametric quadrilateral with eight nodes. As the first step in the process, we make axes $\xi$ and $\eta$ orthogonal and reduce the $\eta$ dimension to the thickness $h$. Thus, we form the *rectangular parent AXSR3* of *element AXSH3* (before constraints), as shown in Fig. 8.9(b). Next, we may introduce constraints to refer the displacements at each group and pair of nodes to those of a single node on the middle surface, as depicted in Fig. 8.9(c). The nodal displacements indicated at point $i$ in that figure are

$$\mathbf{q}_i = \{q_{i1}, q_{i2}, q_{i3}\} = \{u_i, v_i, \alpha_i\} \qquad (i = 1, 2, 3) \tag{1}$$

where $\alpha_i$ is a small positive rotation about an axis normal to the $\xi$-$\eta$ plane. Figure 8.10(a), (b), and (c) show relationships between nodal displacements at an end of element AXSR3, the middle of AXSR3, and a node of element AXSH3, respectively. The two types of constraints to be invoked are:

1. Nodes on the same normal to the middle surface have equal translations in the $\eta$ direction.
2. Normals to the middle surface remain straight (but no longer normal) during deformation.

With these criteria we can relate the six nodal translations in Fig. 8.10(a) to the three nodal displacements in Fig. 8.10(c) by the following 6 × 3 constraint matrix:

$$\mathbf{C}_{ai} = \begin{bmatrix} 1 & 0 & 0 \\ 0 & 1 & 0 \\ 1 & 0 & -\dfrac{h_i}{2} \\ 0 & 1 & 0 \\ 1 & 0 & \dfrac{h_i}{2} \\ 0 & 1 & 0 \end{bmatrix} \tag{2}$$

Sec. 8.6    Element for Axisymmetric Shells

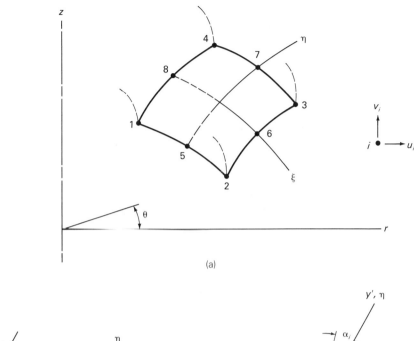

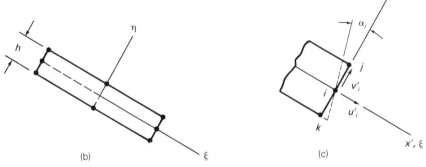

**Figure 8.9** Specialization of axisymmetric solid: (a) element AXQ8; (b) rectangular parent AXSR3 of element AXSH3 before constraints; (c) constrained nodal displacements.

Similarly, the four nodal translations in Fig. 8.10(b) are related to the three nodal displacements in Fig. 8.10(c) by the constraint matrix

$$\mathbf{C}_{bi} = \begin{bmatrix} 1 & 0 & -\dfrac{h_i}{2} \\ 0 & 1 & 0 \\ 1 & 0 & \dfrac{h_i}{2} \\ 0 & 1 & 0 \end{bmatrix} \quad (3)$$

which is of size $4 \times 3$. If we were to apply Eq. (2) at the ends and Eq. (3) at

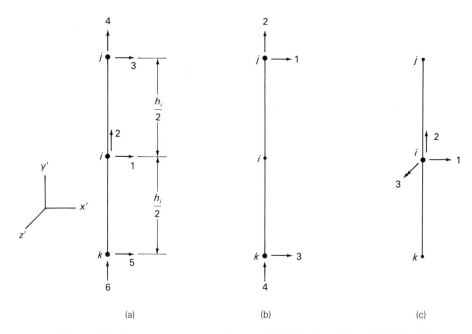

**Figure 8.10** Nodal displacements: (a) end of AXSR3; (b) middle of AXSR3; (c) node of AXSH3.

the middle, we could reduce the number of nodal displacements from 16 to 9. However, we will take a more direct approach, which is similar to those in Secs. 8.2 and 8.4 for plates and general shell elements.

Figure 8.11 shows element AXSH3, for which the coordinates of any point may be stated as

$$\begin{bmatrix} r \\ z \end{bmatrix} = \sum_{i=1}^{3} f_i \begin{bmatrix} r_i \\ z_i \end{bmatrix} + \sum_{i=1}^{3} f_i \eta \frac{h_i}{2} \begin{bmatrix} l_{2i} \\ m_{2i} \end{bmatrix} \tag{4}$$

In this equation the direction cosines for the normal vector $V_2$ are

$$l_{2i} = \cos \gamma_i \qquad m_{2i} = \sin \gamma_i$$

where $\gamma_i$ is the angle between the $r$ axis and the normal at node $i$. The geometric interpolation functions in Eq. (4) have the formulas

$$f_1 = -\frac{\xi}{2}(1 - \xi) \qquad f_2 = \frac{\xi}{2}(1 + \xi) \qquad f_3 = 1 - \xi^2 \tag{5}$$

Therefore, the thickness $h$ may vary quadratically in the $\xi$ direction.

Generic displacements at any point in the element are

$$\mathbf{u} = \{u, v\} \tag{6}$$

assuming that the loads are axisymmetric. These displacements can be expressed in terms of the nodal displacements $u_i$, $v_i$, and $\alpha_i$, as follows:

## Sec. 8.6 Element for Axisymmetric Shells

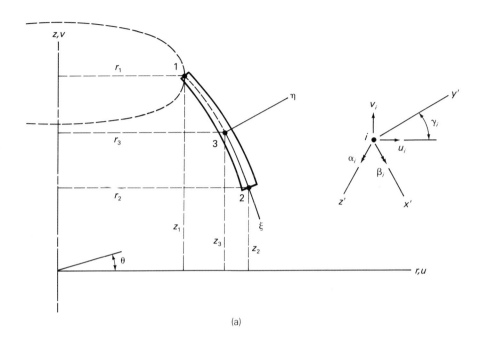

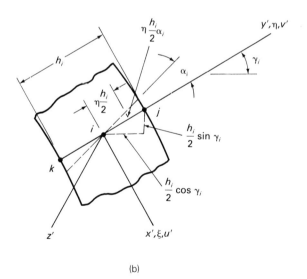

**Figure 8.11** (a) Element AXSH3; (b) nodal rotation.

$$\begin{bmatrix} u \\ v \end{bmatrix} = \sum_{i=1}^{3} f_i \begin{bmatrix} u_i \\ v_i \end{bmatrix} + \sum_{i=1}^{3} f_i \eta \frac{h_i}{2} \begin{bmatrix} -\sin \gamma_i \\ \cos \gamma_i \end{bmatrix} \alpha_i \qquad (7)$$

Geometric details justifying these expressions appear in Fig. 8.11(b). As for the general shell element, we arrange the displacement shape functions from Eq. (7) into the matrix format

$$\mathbf{f}_i = \begin{bmatrix} 1 & 0 & -\eta \frac{h_i}{2} \sin \gamma_i \\ 0 & 1 & \eta \frac{h_i}{2} \cos \gamma_i \end{bmatrix} f_i \qquad (i = 1, 2, 3) \qquad (8)$$

To isolate terms in submatrix $\mathbf{f}_i$ that multiply $\eta$, let us define

$$\mathbf{f}_{Ai} = \begin{bmatrix} 1 & 0 & 0 \\ 0 & 1 & 0 \end{bmatrix} f_i \qquad \mathbf{f}_{Bi} = \begin{bmatrix} 0 & 0 & -\sin \gamma_i \\ 0 & 0 & \cos \gamma_i \end{bmatrix} \frac{h_i}{2} f_i \qquad (9)$$

Then

$$\mathbf{f}_i = \mathbf{f}_{Ai} + \eta \mathbf{f}_{Bi} \qquad (10)$$

and

$$\mathbf{f} = \mathbf{f}_A + \eta \mathbf{f}_B \qquad (11)$$

The formula in Eq. (11) will be used later to determine the consistent mass matrix and equivalent nodal loads for element AXSH3.

The Jacobian matrix required for this element is

$$\mathbf{J} = \begin{bmatrix} r_{,\xi} & z_{,\xi} \\ r_{,\eta} & z_{,\eta} \end{bmatrix} \qquad (12)$$

in which

$$r_{,\xi} = \sum_{i=1}^{3} f_{i,\xi} r_i + \sum_{i=1}^{3} f_{i,\xi} \eta \frac{h_i}{2} \cos \gamma_i \qquad \text{and so on.}$$

The inverse of $\mathbf{J}$ becomes

$$\mathbf{J}^{-1} = \mathbf{J}^* = \begin{bmatrix} \xi_{,r} & \eta_{,r} \\ \xi_{,z} & \eta_{,z} \end{bmatrix} \qquad (13)$$

We will need derivatives of the generic displacements with respect to the local coordinates, as follows:

$$\begin{bmatrix} u_{,\xi} \\ u_{,\eta} \\ v_{,\xi} \\ v_{,\eta} \end{bmatrix} = \sum_{i=1}^{3} \begin{bmatrix} f_{i,\xi} & 0 & -f_{i,\xi} \eta \sin \gamma_i \\ 0 & 0 & -f_i \sin \gamma_i \\ 0 & f_{i,\xi} & f_{i,\xi} \eta \cos \gamma_i \\ 0 & 0 & f_i \cos \gamma_i \end{bmatrix} \begin{bmatrix} u_i \\ v_i \\ \frac{h_i}{2} \alpha_i \end{bmatrix} \qquad (14)$$

## Sec. 8.6 Element for Axisymmetric Shells

These derivatives are transformed to global coordinates by the operation

$$\begin{bmatrix} u_{,r} \\ u_{,z} \\ v_{,r} \\ v_{,z} \end{bmatrix} = \begin{bmatrix} \mathbf{J}^* & \mathbf{0} \\ \mathbf{0} & \mathbf{J}^* \end{bmatrix} \begin{bmatrix} u_{,\xi} \\ u_{,\eta} \\ v_{,\xi} \\ v_{,\eta} \end{bmatrix}$$

$$= \sum_{i=1}^{3} \begin{bmatrix} a_i & 0 & -d_i \sin \gamma_i \\ b_i & 0 & -e_i \sin \gamma_i \\ 0 & a_i & d_i \cos \gamma_i \\ 0 & b_i & e_i \cos \gamma_i \end{bmatrix} \begin{bmatrix} u_i \\ v_i \\ \alpha_i \end{bmatrix} \quad (15)$$

where

$$a_i = J_{11}^* f_{i,\xi} \quad d_i = \frac{h_i}{2}(a_i \eta + J_{12}^* f_i)$$

$$b_i = J_{21}^* f_{i,\xi} \quad e_i = \frac{h_i}{2}(b_i \eta + J_{22}^* f_i) \quad (16)$$

We consider four types of nonzero strains for element AXSH3. They are

$$\boldsymbol{\epsilon} = \begin{bmatrix} \epsilon_r \\ \epsilon_z \\ \epsilon_\theta \\ \gamma_{rz} \end{bmatrix} = \begin{bmatrix} u_{,r} \\ v_{,z} \\ \dfrac{u}{r} \\ u_{,z} + v_{,r} \end{bmatrix} \quad (17)$$

Using the second version of this strain vector, we form the $i$th part of matrix $\mathbf{B}$ from terms in Eq. (15) as

$$\mathbf{B}_i = \begin{bmatrix} a_i & 0 & -d_i \sin \gamma_i \\ 0 & b_i & e_i \cos \gamma_i \\ \dfrac{f_i}{r} & 0 & -\dfrac{1}{2r} f_i \eta h_i \sin \gamma_i \\ b_i & a_i & d_i \cos \gamma_i - e_i \sin \gamma_i \end{bmatrix} \quad (18)$$

$$(i = 1, 2, 3)$$

Similar to the general shell element, we isolate terms in submatrix $\mathbf{B}_i$ multiplying $\eta$, which gives

$$\mathbf{B}_i = \mathbf{B}_{Ai} + \eta \mathbf{B}_{Bi} \quad (19)$$

For the whole matrix, we have

$$\mathbf{B} = \mathbf{B}_A + \eta \mathbf{B}_B \quad (20)$$

The last formula will help us to derive the stiffness matrix for element AXSH3.

Stresses in the directions of primed axes (caused by axisymmetric loads) are

$$\boldsymbol{\sigma}' = \{\sigma_{x'},\ \sigma_{z'},\ \tau_{x'y'}\} \tag{21}$$

and the corresponding strains are

$$\boldsymbol{\epsilon}' = \{\epsilon_{x'},\ \epsilon_{z'},\ \gamma_{x'y'}\} \tag{22}$$

The stress $\sigma_{y'}$ and the strain $\epsilon_{y'}$ normal to the middle surface have been omitted. Thus, the stress-strain matrix for an isotropic material has the form

$$\mathbf{E} = \frac{E}{(1+\nu)(1-2\nu)} \begin{bmatrix} 1-\nu & \nu & 0 \\ \nu & 1-\nu & 0 \\ 0 & 0 & \dfrac{1-2\nu}{2(1.2)} \end{bmatrix} \tag{23}$$

In order to relate local strains in the vector $\boldsymbol{\epsilon}'$ to global strains in the vector $\boldsymbol{\epsilon}$, we specialize the 6 × 6 strain transformation matrix $\mathbf{T}_\epsilon$ in Eq. (7.2-28) to be

$$\mathbf{T}_\epsilon = \begin{bmatrix} l_1^2 & m_1^2 & 0 & l_1 m_1 \\ 0 & 0 & 1 & 0 \\ 2l_1 l_2 & 2m_1 m_2 & 0 & l_1 m_2 + l_2 m_1 \end{bmatrix} \tag{24}$$

which is of size 3 × 4. Then the strain relationship becomes

$$\boldsymbol{\epsilon}' = \mathbf{T}_\epsilon \boldsymbol{\epsilon} \tag{7.2-27}$$

To evaluate matrix $\mathbf{T}_\epsilon$ at an integration point, we must find the direction cosines for unit vectors $\mathbf{e}_1$ and $\mathbf{e}_2$ in the directions of $x'$ and $y'$ at the point. This may be done as follows:

$$\mathbf{e}_1 = (\mathbf{J}_1)_{\text{norm.}} = \{l_1,\ m_1\} \qquad \mathbf{e}_2 = \{l_2,\ m_2\} = \{-m_1,\ l_1\} \tag{25}$$

As for the general shell element, it is also convenient to have

$$\mathbf{B}' = \mathbf{T}_\epsilon \mathbf{B} \tag{8.4-28}$$

for the purpose of calculating stresses in local directions. In this case, matrix $\mathbf{B}'$ will have only three rows, due to the form of $\mathbf{T}_\epsilon$ in Eq. (24).

The element stiffness matrix may now be formulated using matrix $\mathbf{B}'$. Thus,

$$\mathbf{K} = \int_{-1}^{1} \int_{-1}^{1} \int_{0}^{2\pi} (\mathbf{B}')^T \mathbf{E}\, \mathbf{B}' |\mathbf{J}| r\, d\theta\, d\xi\, d\eta$$

$$= 2\pi \int_{-1}^{1} \int_{-1}^{1} (\mathbf{B}'_A + \eta \mathbf{B}'_B)^T \mathbf{E} (\mathbf{B}'_A + \eta \mathbf{B}'_B) |\mathbf{J}| r\, d\xi\, d\eta \tag{26}$$

In this equation the matrices $\mathbf{B}'_A$ and $\mathbf{B}'_B$ are both of size 3 × 9, but the latter has

## Sec. 8.6 Element for Axisymmetric Shells

only terms to be multiplied by $\eta$. Integration of Eq. (26) through the thickness*
of the element gives

$$\mathbf{K} = 2\pi \int_{-1}^{1} [2(\mathbf{B}'_A)^T \mathbf{E} \, \mathbf{B}'_A + \tfrac{2}{3}(\mathbf{B}'_B)^T \mathbf{E} \, \mathbf{B}'_B] |\mathbf{J}| r \, d\xi \tag{27}$$

The remaining integral must be evaluated numerically, using two integration points in the $\xi$ direction [5].

The consistent mass matrix for element AXSH3 has the form

$$\mathbf{M} = \rho \int_{-1}^{1} \int_{-1}^{1} \int_{0}^{2\pi} \mathbf{f}^T \mathbf{f} |\mathbf{J}| r \, d\theta \, d\xi \, d\eta$$

$$= 2\pi\rho \int_{-1}^{1} \int_{-1}^{1} (\mathbf{f}_A + \eta \mathbf{f}_B)^T (\mathbf{f}_A + \eta \mathbf{f}_B) |\mathbf{J}| r \, d\xi \, d\eta \tag{28}$$

In this case the matrices $\mathbf{f}_A$ and $\mathbf{f}_B$ are both of size $2 \times 9$, but the second has only terms that are to be multiplied by $\eta$. Integrating Eq. (28) through the thickness results in

$$\mathbf{M} = 2\pi\rho \int_{-1}^{1} (2\mathbf{f}_A^T \mathbf{f}_A + \tfrac{2}{3} \mathbf{f}_B^T \mathbf{f}_B) |\mathbf{J}| r \, d\xi \tag{29}$$

As before, the first and second parts of $\mathbf{M}$ consist of translational and rotational inertias.

We find equivalent nodal loads caused by body forces using only matrix $\mathbf{f}_A$. Thus,

$$\mathbf{p}_b(t) = \int_{-1}^{1} \int_{-1}^{1} \int_{0}^{2\pi} \mathbf{f}_A^T \mathbf{b}(t) |\mathbf{J}| r \, d\theta \, d\xi \, d\eta$$

$$= 4\pi \int_{-1}^{1} \mathbf{f}_A^T \mathbf{b}(t) |\mathbf{J}| r \, d\xi \tag{30}$$

and the components of force per unit volume (constant through the thickness) are

$$\mathbf{b}(t) = \{b_r, b_z\} \tag{31}$$

These body forces do not produce any equivalent nodal moments.

After solving for time-varying nodal displacements in the vector $\mathbf{q}(t)$, we can determine stresses for local directions in the element using

$$\boldsymbol{\sigma}'(t) = \mathbf{E} \, \mathbf{T}_\epsilon \mathbf{B} \, \mathbf{q}(t) = \mathbf{E} \, \mathbf{B}' \mathbf{q}(t) \tag{8.4-36}$$

These stresses should be calculated at the numerical integration points.

In Sec. 7.9 we expressed *nonaxisymmetric loads* that are symmetric with respect to the $r$-$z$ plane as the Fourier components:

*To simplify integration through the thickness, terms in $r$ and matrix $\mathbf{J}$ containing $\eta$ are neglected.

$$b_r = \sum_{j=0}^{m} b_{rj} \cos j\theta \qquad b_z = \sum_{j=0}^{m} b_{zj} \cos j\theta$$
$$b_\theta = \sum_{j=0}^{m} b_{\theta j} \sin j\theta \qquad (7.9\text{-}12)$$

If the loads were antisymmetric with respect to the $r$-$z$ plane, the functions $\sin j\theta$ and $\cos j\theta$ would be interchanged. The response of an axisymmetric shell to this series of harmonic loads consists of a series of harmonic generic displacements. For element AXSH3 these displacements are expressed as follows:

$$\begin{bmatrix} u \\ v \\ w \end{bmatrix} = \sum_{j=0}^{m} \begin{bmatrix} \cos j\theta & 0 & 0 \\ 0 & \cos j\theta & 0 \\ 0 & 0 & \sin j\theta \end{bmatrix}$$
$$\times \left\{ \sum_{i=1}^{3} f_i \begin{bmatrix} u_i \\ v_i \\ w_i \end{bmatrix}_j + \sum_{i=1}^{3} f_i \eta \frac{h_i}{2} \begin{bmatrix} -\sin \gamma_i & 0 \\ \cos \gamma_i & 0 \\ 0 & 1 \end{bmatrix} \begin{bmatrix} \alpha_i \\ \beta_i \end{bmatrix}_j \right\} \qquad (32)$$

where the angle $\beta_i$ is a small rotation about the $x'$ axis [see Fig. 8.11(a)]. As before, if the loads were antisymmetric with respect to the $r$-$z$ plane, the functions $\sin j\theta$ and $\cos j\theta$ would be interchanged.

Equation (32) may be stated more efficiently as

$$\mathbf{u} = \sum_{j=0}^{m} \mathbf{c}_j \mathbf{f} \, \mathbf{q}_j = \sum_{j=0}^{m} \mathbf{c}_j \sum_{i=1}^{3} \mathbf{f}_i \mathbf{q}_{ij} \qquad (33)$$

in which

$$\mathbf{c}_j = \begin{bmatrix} \cos j\theta & 0 & 0 \\ 0 & \cos j\theta & 0 \\ 0 & 0 & \sin j\theta \end{bmatrix} \qquad (7.9\text{-}33)$$

Also, we put the displacement shape functions into the matrix form

$$\mathbf{f}_i = \begin{bmatrix} 1 & 0 & 0 & -\eta \frac{h_i}{2} \sin \gamma_i & 0 \\ 0 & 1 & 0 & \eta \frac{h_i}{2} \cos \gamma_i & 0 \\ 0 & 0 & 1 & 0 & \eta \frac{h_i}{2} \end{bmatrix} f_i \qquad (i = 1, 2, 3) \qquad (34)$$

As before, let us isolate terms in submatrix $\mathbf{f}_i$ that are multiplied by $\eta$. Hence,

$$\mathbf{f}_{Ai} = \begin{bmatrix} 1 & 0 & 0 & 0 & 0 \\ 0 & 1 & 0 & 0 & 0 \\ 0 & 0 & 1 & 0 & 0 \end{bmatrix} f_i \qquad (35a)$$

## Sec. 8.6  Element for Axisymmetric Shells

and

$$\mathbf{f}_{Bi} = \begin{bmatrix} 0 & 0 & 0 & -\sin\gamma_i & 0 \\ 0 & 0 & 0 & \cos\gamma_i & 0 \\ 0 & 0 & 0 & 0 & 1 \end{bmatrix} \frac{h_i}{2} f_i \quad (35b)$$

Thus, Eqs. (10) and (11) apply equally well to any case of nonaxisymmetric loads, but matrix $\mathbf{f}$ is now of size $3 \times 15$.

For this analysis we must determine the derivatives of the generic displacements in Eq. (32) with respect to local coordinates. They are

$$\begin{bmatrix} u_{,\xi} \\ u_{,\eta} \\ u_{,\theta} \\ v_{,\xi} \\ v_{,\eta} \\ v_{,\theta} \\ w_{,\xi} \\ w_{,\eta} \\ w_{,\theta} \end{bmatrix} = \sum_{i=1}^{3} \begin{bmatrix} f_{i,\xi}c_j & 0 & 0 & -f_{i,\xi}\eta s_i c_j & 0 \\ 0 & 0 & 0 & -f_i s_i c_j & 0 \\ -jf_i s_j & 0 & 0 & jf_i \eta s_i s_j & 0 \\ 0 & f_{i,\xi}c_j & 0 & f_{i,\xi}\eta c_i c_j & 0 \\ 0 & 0 & 0 & f_i c_i c_j & 0 \\ 0 & -jf_i s_j & 0 & -jf_i \eta c_i s_j & 0 \\ 0 & 0 & f_{i,\xi}s_j & 0 & f_{i,\xi}\eta s_j \\ 0 & 0 & 0 & 0 & f_i s_j \\ 0 & 0 & jf_i c_j & 0 & jf_i \eta c_j \end{bmatrix} \begin{bmatrix} u_i \\ v_i \\ w_i \\ \frac{h_i}{2}\alpha_i \\ \frac{h_i}{2}\beta_i \end{bmatrix}_j \quad (36)$$

In the coefficient matrix of this expression, the following abbreviations are used

$$\begin{aligned} s_i &= \sin\gamma_i & s_j &= \sin j\theta \\ c_i &= \cos\gamma_i & c_j &= \cos j\theta \end{aligned} \quad (37)$$

The required Jacobian matrix is

$$\mathbf{J} = \begin{bmatrix} r_{,\xi} & z_{,\xi} & 0 \\ r_{,\eta} & z_{,\eta} & 0 \\ 0 & 0 & 1 \end{bmatrix} \quad (38)$$

In this case the inverse of $\mathbf{J}$ is seen to be

$$\mathbf{J}^{-1} = \mathbf{J}^* = \begin{bmatrix} \xi_{,r} & \eta_{,r} & 0 \\ \xi_{,z} & \eta_{,z} & 0 \\ 0 & 0 & 1 \end{bmatrix} \quad (39)$$

Using this inverse matrix, we can transform the derivatives in Eq. (36) to global coordinates, as follows:

$$\begin{bmatrix} u_{,r} \\ u_{,z} \\ u_{,\theta} \\ \vdots \\ w_{,\theta} \end{bmatrix}_j = \begin{bmatrix} \mathbf{J}^* & 0 & 0 \\ 0 & \mathbf{J}^* & 0 \\ 0 & 0 & \mathbf{J}^* \end{bmatrix} \begin{bmatrix} u_{,\xi} \\ u_{,\eta} \\ u_{,\theta} \\ \vdots \\ w_{,\theta} \end{bmatrix}_j \quad (40)$$

Multiplying the terms in this equation produces

$$
\begin{bmatrix} u_{,r} \\ u_{,z} \\ u_{,\theta} \\ v_{,r} \\ v_{,z} \\ v_{,\theta} \\ w_{,r} \\ w_{,z} \\ w_{,\theta} \end{bmatrix}_j = \sum_{i=1}^{3} \begin{bmatrix} a_i c_j & 0 & 0 & -d_i s_i c_j & 0 \\ b_i c_j & 0 & 0 & -e_i s_i c_j & 0 \\ -j f_i s_j & 0 & 0 & \frac{1}{2}(j f_i \eta s_i s_j h_i) & 0 \\ 0 & a_i c_j & 0 & d_i c_i c_j & 0 \\ 0 & b_i c_j & 0 & e_i c_i c_j & 0 \\ 0 & -j f_i s_j & 0 & -\frac{1}{2}(j f_i \eta c_i s_j h_i) & 0 \\ 0 & 0 & a_i s_j & 0 & d_i s_j \\ 0 & 0 & b_i s_j & 0 & e_i s_j \\ 0 & 0 & j f_i c_j & 0 & \frac{1}{2} j f_i \eta c_j h_i \end{bmatrix} \begin{bmatrix} u_i \\ v_i \\ w_i \\ \alpha_i \\ \beta_i \end{bmatrix}_j \quad (41)
$$

where the constants $a_i$, $b_i$, $d_i$, and $e_i$ are given by Eqs. (16).

For nonaxisymmetric loads on element AXSH3, we consider six types of nonzero strains. Thus,

$$
\boldsymbol{\epsilon}_j = \begin{bmatrix} \epsilon_r \\ \epsilon_z \\ \epsilon_\theta \\ \gamma_{rz} \\ \gamma_{z\theta} \\ \gamma_{r\theta} \end{bmatrix}_j = \begin{bmatrix} u_{,r} \\ v_{,z} \\ \dfrac{1}{r}(u + w_{,\theta}) \\ u_{,z} + v_{,r} \\ w_{,z} + \dfrac{v_{,\theta}}{r} \\ \dfrac{u_{,\theta}}{r} + w_{,r} - \dfrac{w}{r} \end{bmatrix}_j \quad (42)
$$

The strain-displacement relationships shown in the second form of this vector are the same as those in Eqs. (7.9-15). Using these relationships and Eq. (41), we may construct the $i$th part of matrix $\mathbf{B}$ for the $j$th harmonic response, as follows:

$$
(\mathbf{B}_i)_j = \begin{bmatrix} a_i c_j & 0 & 0 & -d_i s_i c_j & 0 \\ 0 & b_i c_j & 0 & e_i c_i c_j & 0 \\ \dfrac{1}{r} f_i c_j & 0 & \dfrac{1}{r} j f_i c_j & -\dfrac{1}{2r} f_i \eta s_i c_j h_i & \dfrac{1}{2r} j f_i \eta c_j h_i \\ b_i c_j & a_i c_j & 0 & (d_i c_i - e_i s_i) c_j & 0 \\ 0 & -\dfrac{1}{r} j f_i s_j & b_i s_j & -\dfrac{1}{2r} j f_i \eta c_i s_j h_i & e_i s_j \\ -\dfrac{1}{r} j f_i s_j & 0 & \left(a_i - \dfrac{f_i}{r}\right) s_j & \dfrac{1}{2r} j f_i \eta s_i s_j h_i & \left(d_i - \dfrac{f_i \eta h_i}{2r}\right) s_j \end{bmatrix}
$$

$$(43)$$

### Sec. 8.6 Element for Axisymmetric Shells

where $i = 1, 2, 3$, and $j = 0, 1, 2, \ldots, m$. As before, we can decompose $(\mathbf{B}_i)_j$ into

$$(\mathbf{B}_i)_j = (\mathbf{B}_{Ai})_j + \eta(\mathbf{B}_{Bi})_j \tag{44}$$

and state the total matrix in the form

$$\mathbf{B}_j = \mathbf{B}_{Aj} + \eta \mathbf{B}_{Bj} \tag{45}$$

which will be used to find stiffnesses.

For the present analysis, nonzero stresses in the directions of primed axes are

$$\boldsymbol{\sigma}' = \{\sigma_{x'}, \sigma_{z'}, \tau_{x'y'}, \tau_{y'z'}, \tau_{z'x'}\} \tag{46}$$

to which the following strains correspond:

$$\boldsymbol{\epsilon}' = \{\epsilon_{x'}, \epsilon_{z'}, \gamma_{x'y'}, \gamma_{y'z'}, \gamma_{z'x'}\} \tag{47}$$

In this case the normal stress $\sigma_{y'}$ and the strain $\epsilon_{y'}$ are omitted. Also, the stress-strain matrix $\mathbf{E}$ for an isotropic material is similar to that for element PBQ8 in Eq. (8.2-21), except that $E_{33}$ instead of $E_{55}$ is divided by 1.2. Additionally, the 6 × 6 strain transformation matrix $\mathbf{T}_\epsilon$ serves to relate $\boldsymbol{\epsilon}'$ to $\boldsymbol{\epsilon}$, as in Eq. (7.2-27), except that the second row is deleted. Another useful matrix is

$$\mathbf{B}'_j = \mathbf{T}_\epsilon \mathbf{B}_j \tag{48}$$

which is of size 5 × 15.

We write the stiffness matrix for each harmonic loading as

$$\mathbf{K}_j = \int_{-1}^{1} \int_{-1}^{1} \int_{0}^{2\pi} (\mathbf{B}'_j)^T \mathbf{E} \, \mathbf{B}'_j \, |\mathbf{J}| r \, d\theta \, d\xi \, d\eta$$

$$= k\pi \int_{-1}^{1} [2(\mathbf{B}'_{Aj})^T \mathbf{E} \, \mathbf{B}'_{Aj} + \tfrac{2}{3}(\mathbf{B}'_{Bj})^T \mathbf{E} \, \mathbf{B}'_{Bj}] \, |\mathbf{J}| r \, d\xi \tag{49}$$

where $k = 2$ for $j = 0$ and $k = 1$ for $j = 1, 2, \ldots, m$. Next, the consistent mass matrix becomes

$$\mathbf{M}_j = \rho \int_{-1}^{1} \int_{-1}^{1} \int_{0}^{2\pi} \mathbf{f}^T \mathbf{c}_j^T \mathbf{c}_j \mathbf{f} \, |\mathbf{J}| r \, d\theta \, d\xi \, d\eta$$

$$= k\pi\rho \int_{-1}^{1} (2\mathbf{f}_A^T \mathbf{f}_A + \tfrac{2}{3}\mathbf{f}_B^T \mathbf{f}_B) \, |\mathbf{J}| r \, d\xi \tag{50}$$

which is the same for $j = 1, 2, \ldots, m$ (see Sec. 7.9). Then equivalent nodal loads for each harmonic set of body forces are

$$\mathbf{p}_b(t)_j = \int_{-1}^{1} \int_{-1}^{1} \int_{0}^{2\pi} \mathbf{f}_A^T \mathbf{c}_j^T \mathbf{c}_j \mathbf{b}(t)_j \, |\mathbf{J}| r \, d\theta \, d\xi \, d\eta$$

$$= k\pi \int_{-1}^{1} 2\mathbf{f}_A^T \mathbf{b}(t)_j \, |\mathbf{J}| r \, d\xi \qquad (j = 0, 1, 2, \ldots, m) \tag{51}$$

in which only matrix $\mathbf{f}_A$ is used. Components of force per unit volume (constant through the thickness) consist of

$$\mathbf{b}(t)_j = \{b_{rj}, b_{zj}, b_{\theta j}\} \tag{52}$$

Finally, the stresses for each harmonic response may be written as

$$\boldsymbol{\sigma}'(t)_j = \mathbf{E}\,\mathbf{B}'_j\,\mathbf{q}(t)_j \qquad (j = 0, 1, 2, \ldots, m) \tag{53}$$

which are in the directions of local axes.

## 8.7 PROGRAM DYAXSH FOR AXISYMMETRIC SHELLS

Let us now consider a computer program named DYAXSH for the dynamic analysis of axisymmetric shells with axisymmetric loads. This program uses element AXSH3 from the preceding section. It is generated by modifying the part of Program DYAXSO dealing with element AXQ8 (see Sec. 7.10), on which element AXSH3 is based. However, in Program DYAXSH we must formulate $\mathbf{K}$, $\mathbf{M}$, and $\mathbf{p}_b(t)$ in accordance with the equations developed in Sec. 8.6.

In Table 8.5 we see how structural data is to be prepared for Program DYAXSH. As for general shell data, the symbol H appears in the list of structural parameters for a case where the thickness is constant over the whole shell. In addition, H(J) is included with the nodal coordinates for a case where thickness varies over an element. That line of data also contains the components V2R(J) and V2Z(J) of the normal vector $\mathbf{V}_2$ for node J [see Eq. (8.6-4)]. Finally, three types of nodal restraints are indicated in line (c).

Dynamic load data for Program DYAXSH is displayed in Table 8.6. Here we have three possible initial displacements, initial velocities, and nodal loads instead of two per node. Area loads for element AXSH3 are similar to those for AXQ8, but only the element number I (not the node numbers J, K, and L) need be given. Also, the acceleration factors GAX and GAY are replaced by GAZ for the axisymmetric shell.

**TABLE 8.5  Structural Data for Program DYAXSH**

| Type of Data | No. of Lines | Items on Data Lines |
|---|---|---|
| Problem identification | 1 | Descriptive title |
| Structural parameters | 1 | NN, NE, NRN, E, PR, RHO, H |
| Axisymmetric shell data<br>(a) Nodal coordinates, etc.<br>(b) Element information[a]<br>(c) Nodal restraints | NN<br>NE<br>NRN | J, R(J), Z(J), V2R(J), V2Z(J), H(J)<br>I, IN(I, 1), IN(I, 2), IN(I, 3)<br>J, NRL(3J-2), NRL(3J-1), NRL(3J) |

[a] For sequence of node numbers, see Fig. 8.11(a).

## Sec. 8.7 Program DYAXSH for Axisymmetric Shells

**TABLE 8.6** Dynamic Load Data for Program DYAXSH

| Type of Data | No. of Lines | Items on Data Lines |
|---|---|---|
| Dynamic parameters | 1 | ISOLVE, NTS, DT, DAMPR |
| Initial conditions<br>(a) Condition parameters<br>(b) Displacements<br>(c) Velocities | 1<br>NNID<br>NNIV | NNID, NNIV<br>J, D0(3J-2), D0(3J-1), D0(3J)<br>J, V0(3J-2), V0(3J-1), V0(3J) |
| Applied actions<br>(a) Load parameters<br>(b) Nodal loads<br>(c) Area loads<br>(d) Volume loads | 1<br>NLN<br>NEA<br>NEV | NLN, NEA, NEV<br>J, AS(3J-2), AS(3J-1), AS(3J)<br>I, BA1, BA2, . . . , BA6<br>I, BV1, BV2 |
| Ground accelerations<br>(a) Acceleration parameter<br>(b) Acceleration factor | 1<br>1 | IGA<br>GAZ |
| Forcing function<br>(a) Function parameter<br>(b) Function ordinates | 1<br>NFO | NFO<br>K, T(K), FO(K) |

With element AXSH3 we use $n = 2$ in the $\xi$ direction [see Fig. 8.11(a)] to locate points for numerical integration. At each of these two points the computer evaluates the time-varying stresses $\sigma_{x'}$, $\sigma_{z'}$, and $\tau_{x'y'}$ in local directions. Again, we had to use $n = 3$ when evaluating terms in the consistent mass matrix for element AXSH3.

**Example 8.4**

The vaporous gas storage tank in Fig. 8.12 has a hemispherical top and a circular

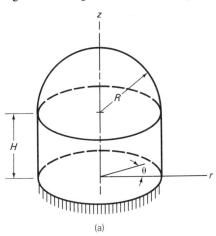

**Figure 8.12** (a) Vaporous gas storage tank; (b) AXSH3 elements.

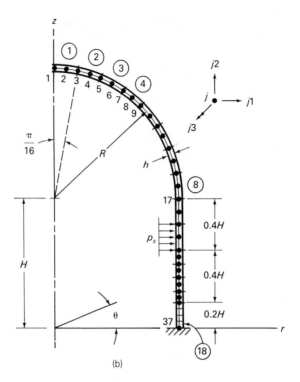

**Figure 8.12** (*cont.*)

cylindrical wall that is fixed at its base. This axisymmetric shell is divided into 18 AXSH3 elements that become progressively shorter toward the base, where high bending stresses are expected. Surface pressure $p_s$ from an interior explosion causes dynamic response of the shell, and we wish to examine stresses in the vicinity of node 37 due to this influence. Physical parameters in this example are

$$E = 207 \times 10^6 \text{ kN/m}^2 \qquad \nu = 0.30 \qquad \rho = 7.85 \text{ Mg/m}^3$$
$$R = H = 2 \text{ m} \qquad h = 0.025 \text{ m} \qquad (p_s)_{\max} = 3 \text{ MPa}$$

where the material is steel and the units are SI.

As in Example 7.6, we need to find *equivalent nodal loads due to pressure* $p_s$, which we assume acts in the direction normal to the middle surface of a shell element (the $\eta$ direction). These equivalent loads are

$$\mathbf{p}_b = 2\pi p_s \int_{-1}^{1} \mathbf{f}^{\mathrm{T}} \begin{bmatrix} -z_{,\xi} \\ r_{,\xi} \end{bmatrix} r \, d\xi \qquad \text{(a)}$$

This formula for $\mathbf{p}_b$ is the same as that in Eq. (7.10-h), but in this case there is no need for the prime on $\xi$. As before, signs for the equivalent nodal loads are automatically determined by Eq. (a) for pressure taken to be positive in the positive sense of the normal direction $\eta$.

Sec. 8.7  Program DYAXSH for Axisymmetric Shells    409

We ran the foregoing data with Program DYAXSH using DAMPR = 0.02 and Subprogram NUMINT for calculating dynamic responses. The computer plot in Fig. 8.13(a) gives the variation of pressure inside the tank, and Fig. 8.13(b) shows time histories of membrane and flexural stresses in the $x'$ direction at the integration point near node 37. The maximum value of the former stress is 205.2 MPa, and the minimum value of the latter stress is $-308.6$ MPa (at the outer surface of the wall). Further refinement of the finite-element layout near the base would produce a somewhat higher flexural stress at that location.

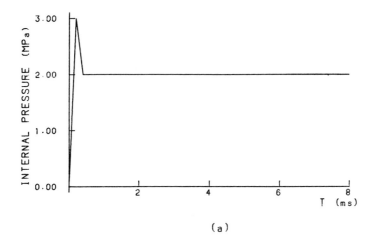

(a)

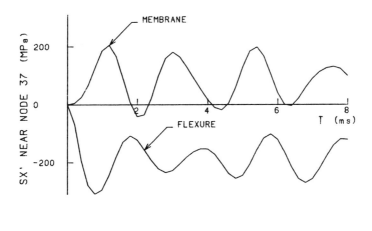

(b)

**Figure 8.13**  Vaporous gas storage tank: (a) load; (b) stresses.

## REFERENCES

1. Mindlin, R. D., "Influence of Rotatory Motion and Shear on Flexural Motions of Isotropic, Elastic Plates," *J. Appl. Mech.*, Vol. 73, 1951, pp. 31–38.
2. Ahmad, S., Irons, B. M., and Zienkiewicz, O. C., "Analysis of Thick and Thin Shell Structures by Curved Finite Elements," *Int. J. Numer. Methods Eng.*, Vol. 3, No. 4, 1971, pp. 575–586.
3. Weaver, W., Jr., and Johnston, P. R., *Finite Elements for Structural Analysis*, Prentice-Hall, Englewood Cliffs, N. J., 1984.
4. Timoshenko, S. P., and Woinowsky-Krieger, S., *Theory of Plates and Shells*, 2nd ed., McGraw-Hill, New York, 1959.
5. Cook, R. D., *Concepts and Applications of Finite Element Analysis*, 2nd ed., Wiley, New York, 1981.
6. Yoshida, D. M., and Weaver, W., Jr., "Finite-Element Analysis of Beams and Plates with Moving Loads," *Int. Assoc. Bridge Struct. Eng.*, Vol. 31, 1971, pp. 179–195.
7. Ahmad, S., Irons, B. M., and Zienkiewicz, O. C., "Curved Thick Shell and Membrane Elements, with Particular Reference to Axisymmetric Problems," *Proc. 2nd Conf. Mat. Methods Struct. Mech.*, Wright-Patterson Air Force Base, Ohio, 1968, pp. 539–572.

# 9

# Rigid Bodies Within Flexible Structures

## 9.1 INTRODUCTION

Occasionally, the analyst encounters a structure containing one or more parts (or bodies) that are very rigid in comparison to the other parts. Such bodies are usually taken to be infinitely rigid [1], and the nodes connecting them to the rest of the structure are constrained to displace in a pattern corresponding to the rigid-body motions. These restrictions on nodal displacements serve to reduce the number of degrees of freedom in a given problem. For example, the joints labeled $A$, $B$, and $C$ in the counterweighted plane truss in Fig. 9.1(a) would ordinarily have a total of six degrees of freedom among them. However, the rigid counterweight (shown hatched in the figure), on which they are located, has only three degrees of freedom. These three displacements are translations in the $x$ and $y$ directions and a rotation in the $z$ sense, as indicated by the arrows labeled 1, 2, and 3 adjacent to the rigid body. Thus, the number of degrees of freedom in the problem is reduced by three due to the presence of the counterweight.

Figure 9.1(b) shows a second example of a building frame, for which the analytical model is taken to be a rectangular space frame containing floor and roof laminae. Each lamina is assumed to have infinite rigidity in its own plane, but zero rigidity normal to the plane. The $x$ and $y$ translations and the $z$ rotation for any joint on a particular body are dictated by the corresponding rigid-body motions of that lamina. These motions are indicated by three numbered arrows at each framing level. Therefore, if $n_j$ is the number of joints at a given level, the number of independent displacements at that level is $3n_j + 3$ instead of $6n_j$.

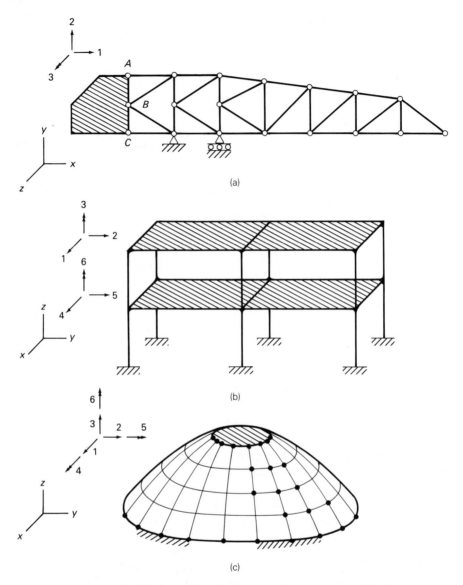

**Figure 9.1** Rigid bodies within flexible continua: (a) counterweighted truss; (b) building with laminae; (c) shell with hub.

Appearing in Fig. 9.1(c) is a third example of a rigid hub at the center of a shell structure. If the geometry of the shell is general, displacements of nodes attached to the hub are constrained to displace according to the rigid-body motions labeled 1 through 6 in the figure. Otherwise, if the shell were axisymmetric, the number of rigid-body motions would be fewer.

## Sec. 9.2    Rigid Bodies in Framed Structures

In this chapter we examine the effects of including rigid bodies within framed structures and discretized continua. It is assumed that the bodies are connected and supported by linearly elastic, flexible materials and that their displacements are small. We also assume that the bodies are not rigidly connected to each other or to the supports, which would require additional constraints.

## 9.2 RIGID BODIES IN FRAMED STRUCTURES

Figure 9.2 shows $x$, $y$, and $z$ axes and six indexes $p1, p2, \ldots, p6$ for components of actions or displacements at a *reference point* $p$ on a three-dimensional rigid body. The figure also depicts a typical joint (or node) $j$ of a structure connected to the body. If the structure is a space frame, there will be

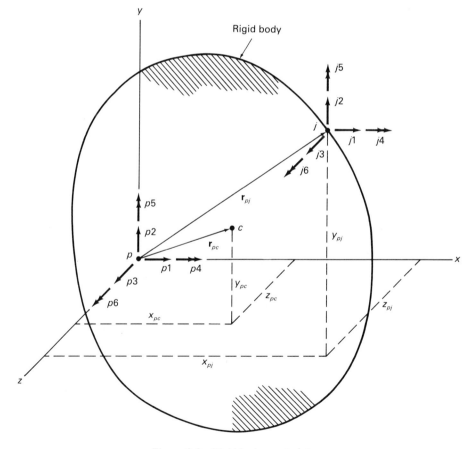

**Figure 9.2**  Rigid-body constraints.

six indexes $j1, j2, \ldots, j6$ for actions or displacements at point $j$, as indicated in the figure. An offset vector $\mathbf{r}_{pj}$ is directed from point $p$ to point $j$ and has scalar components $x_{pj}$, $y_{pj}$, and $z_{pj}$. We may calculate the *statically equivalent actions* at point $p$ due to actions at point $j$ using the concept of *translation of axes* [2, 3]. Thus,

$$\mathbf{A}_p = \mathbf{T}_{pj}\mathbf{A}_j \tag{1}$$

In this equation the symbol $\mathbf{A}_j$ denotes a column vector of six action components at point $j$, as follows:

$$\mathbf{A}_j = \{A_{j1}, A_{j2}, \ldots, A_{j6}\} \tag{2}$$

where the first three are forces, and the last three are moments. Also, the vector $\mathbf{A}_p$ in Eq. (1) contains six statically equivalent action components at point $p$.

$$\mathbf{A}_p = \{A_{p1}, A_{p2}, \ldots, A_{p6}\} \tag{3}$$

Finally, the transformation matrix $\mathbf{T}_{pj}$ in Eq. (1) has the form

$$\mathbf{T}_{pj} = \begin{bmatrix} \mathbf{I}_3 & \mathbf{0} \\ \mathbf{c}_{pj} & \mathbf{I}_3 \end{bmatrix} \tag{4}$$

in which $\mathbf{I}_3$ is an identity matrix of order 3 and

$$\mathbf{c}_{pj} = \begin{bmatrix} 0 & -z_{pj} & y_{pj} \\ z_{pj} & 0 & -x_{pj} \\ -y_{pj} & x_{pj} & 0 \end{bmatrix} \tag{5}$$

This skew-symmetric submatrix contains positive and negative values of the components of the offset vector $\mathbf{r}_{pj}$. These components are arranged in a manner that produces the cross product of $\mathbf{r}_{pj}$ and the force vector $\mathbf{F}_j$ at point $j$.

Similarly, the *kinematically equivalent displacements* at point $j$ may be calculated from those at point $p$ with the relationship

$$\mathbf{D}_j = \mathbf{T}_{pj}^{\mathrm{T}}\mathbf{D}_p \tag{6}$$

Here the symbol $\mathbf{D}_p$ represents a vector of six small displacement components of the rigid body at the reference point $p$. That is,

$$\mathbf{D}_p = \{D_{p1}, D_{p2}, \ldots, D_{p6}\} \tag{7}$$

where the first three are translations, and the last three are rotations. In addition, the vector $\mathbf{D}_j$ consists of six kinematically equivalent displacements at point $j$.

$$\mathbf{D}_j = \{D_{j1}, D_{j2}, \ldots, D_{j6}\} \tag{8}$$

The operator $\mathbf{T}_{pj}^{\mathrm{T}}$ in Eq. (6) is

$$\mathbf{T}_{pj}^{\mathrm{T}} = \begin{bmatrix} \mathbf{I}_3 & \mathbf{c}_{pj}^{\mathrm{T}} \\ \mathbf{0} & \mathbf{I}_3 \end{bmatrix} \tag{9}$$

where

$$\mathbf{c}_{pj}^T = -\mathbf{c}_{pj} = \mathbf{c}_{jp} = \begin{bmatrix} 0 & z_{pj} & -y_{pj} \\ -z_{pj} & 0 & x_{pj} \\ y_{pj} & -x_{pj} & 0 \end{bmatrix} \quad (10)$$

Thus, transposition of $\mathbf{c}_{pj}$ merely changes its sign, which is an inherent property of a skew-symmetric matrix. Note that the transformation in Eq. (6) is analogous to that in Eq. (4.5-1) for small rigid-body ground displacements.

A similar transformation matrix $\mathbf{T}_{pj}$ can be derived for each of the other five types of framed structures. Table 9.1 summarizes these matrices for beams, plane trusses, plane frames, grids, space trusses, and space frames. Each of the first five matrices in the table can be found by deleting appropriate rows and columns from the sixth. For example, point $p$ for a rigid lamina in a plane truss would have three displacements, which are translations in the $x$ and $y$ directions

**TABLE 9.1 Transformation Matrices $\mathbf{T}_{pj}$ for Framed Structures**

1. Beams ($x$-$y$ plane of bending)

$$\mathbf{T}_{pj} = \begin{bmatrix} 1 & 0 \\ x_{pj} & 1 \end{bmatrix}$$

2. Plane trusses (structure in $x$-$y$ plane)

$$\mathbf{T}_{pj} = \begin{bmatrix} 1 & 0 \\ 0 & 1 \\ -y_{pj} & x_{pj} \end{bmatrix}$$

3. Plane frames (structure in $x$-$y$ plane)

$$\mathbf{T}_{pj} = \begin{bmatrix} 1 & 0 & 0 \\ 0 & 1 & 0 \\ -y_{pj} & x_{pj} & 1 \end{bmatrix}$$

4. Grids (structure in $x$-$y$ plane)

$$\mathbf{T}_{pj} = \begin{bmatrix} 1 & 0 & y_{pj} \\ 0 & 1 & -x_{pj} \\ 0 & 0 & 1 \end{bmatrix}$$

5. Space trusses (transposed to save space)

$$\mathbf{T}_{pj}^T = \begin{bmatrix} 1 & 0 & 0 & 0 & z_{pj} & -y_{pj} \\ 0 & 1 & 0 & -z_{pj} & 0 & x_{pj} \\ 0 & 0 & 1 & y_{pj} & -x_{pj} & 0 \end{bmatrix}$$

6. Space frames

$$\mathbf{T}_{pj} = \begin{bmatrix} 1 & 0 & 0 & 0 & 0 & 0 \\ 0 & 1 & 0 & 0 & 0 & 0 \\ 0 & 0 & 1 & 0 & 0 & 0 \\ 0 & -z_{pj} & y_{pj} & 1 & 0 & 0 \\ z_{pj} & 0 & -x_{pj} & 0 & 1 & 0 \\ -y_{pj} & x_{pj} & 0 & 0 & 0 & 1 \end{bmatrix}$$

and a rotation in the $z$ sense. But a joint $j$ in the truss has only two displacements, which are translations in the $x$ and $y$ directions. Therefore, we keep only the first, second, and sixth rows and the first and second columns of the $6 \times 6$ matrix $\mathbf{T}_{pj}$. Deletion of the other rows and columns from the sixth matrix results in the $3 \times 2$ array for plane trusses shown in the table. Note that the $3 \times 3$ transformation matrix pertaining to grids requires not only deletion of rows and columns, but also rearrangement as well. The reason for this is that the $x$- and $y$-rotations at the joint of a grid are taken before the $z$-translation [see Fig. 6.1(d)].

Now let us consider the task of incorporating rigid bodies into our analytical models for framed structures. For this purpose, we use a *member-oriented approach* to transform actions, stiffnesses, and consistent masses at the ends of members to reference points (or *working points*) on the rigid bodies. Also, it is usually necessary to transform mass and mass-moment-of-inertia matrices for the rigid bodies from their mass centers to their working points.

Figure 9.3 shows a space frame member $i$ connected to three-dimensional rigid bodies at both ends. Points $p$ and $q$ on the bodies are taken as working points to which information about joints $j$ and $k$ will be referred. First, any actions in vectors $\mathbf{A}_j$ and $\mathbf{A}_k$ at the ends of the member may be transformed into statically equivalent actions $\mathbf{A}_p$ and $\mathbf{A}_q$ at the working points by the following generalized form of Eq. (1):

$$\begin{bmatrix} \mathbf{A}_p \\ \mathbf{A}_q \end{bmatrix} = \begin{bmatrix} \mathbf{T}_{pj} & 0 \\ 0 & \mathbf{T}_{qk} \end{bmatrix} \begin{bmatrix} \mathbf{A}_j \\ \mathbf{A}_k \end{bmatrix} \tag{11}$$

The action vectors $\mathbf{A}_k$ and $\mathbf{A}_q$ are similar to $\mathbf{A}_j$ and $\mathbf{A}_p$ in Eqs. (2) and (3). Also, the transformation matrix $\mathbf{T}_{qk}$ is the same type as $\mathbf{T}_{pj}$ given in Eq. (4), but for points $q$ and $k$. Equation (11) may be expressed more concisely as

$$\mathbf{A}_{Bi} = \mathbf{T}_i \mathbf{A}_{Mi} \tag{12}$$

in which

$$\mathbf{A}_{Mi} = \{\mathbf{A}_j, \mathbf{A}_k\} \qquad \mathbf{A}_{Bi} = \{\mathbf{A}_p, \mathbf{A}_q\} \tag{13}$$

and

$$\mathbf{T}_i = \begin{bmatrix} \mathbf{T}_{pj} & 0 \\ 0 & \mathbf{T}_{qk} \end{bmatrix} \tag{14}$$

The transformation matrix $\mathbf{T}_i$ is a combined operator that converts the actions in $\mathbf{A}_{Mi}$ (at the ends of the member) to the statically equivalent actions in $\mathbf{A}_{Bi}$ (at the working points of the rigid bodies). These actions are all in the directions of structural axes. Both actual and equivalent nodal loads can be treated in this manner.

Displacements at joints $j$ and $k$ will also be expressed in terms of those at $p$ and $q$ by an extended form of Eq. (6), as follows:

$$\begin{bmatrix} \mathbf{D}_j \\ \mathbf{D}_k \end{bmatrix} = \begin{bmatrix} \mathbf{T}_{pj}^\mathrm{T} & 0 \\ 0 & \mathbf{T}_{qk}^\mathrm{T} \end{bmatrix} \begin{bmatrix} \mathbf{D}_p \\ \mathbf{D}_q \end{bmatrix} \tag{15}$$

## Sec. 9.2  Rigid Bodies in Framed Structures

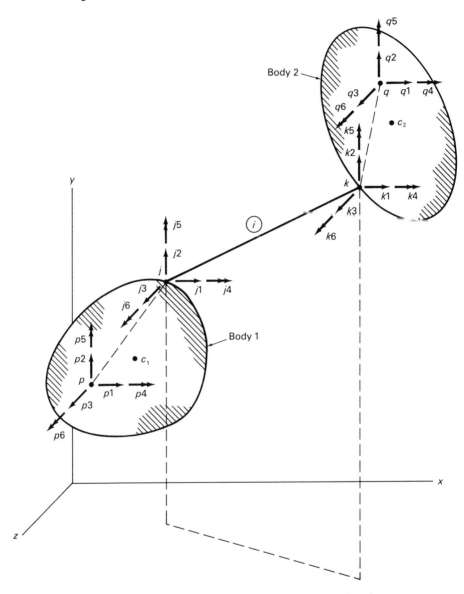

**Figure 9.3** Space frame member with rigid bodies at both ends.

The displacement vectors in this equation correspond to the action vectors in Eqs. (13). Equation (15) may be written more briefly as

$$\mathbf{D}_{Mi} = \mathbf{T}_i^T \mathbf{D}_{Bi} \tag{16}$$

where

$$\mathbf{D}_{Mi} = \{\mathbf{D}_j, \mathbf{D}_k\} \qquad \mathbf{D}_{Bi} = \{\mathbf{D}_p, \mathbf{D}_q\} \tag{17}$$

and

$$\mathbf{T}_i^T = \begin{bmatrix} \mathbf{T}_{pj}^T & \mathbf{0} \\ \mathbf{0} & \mathbf{T}_{qk}^T \end{bmatrix} \quad (18)$$

In addition, we can transform the member stiffness matrix $\mathbf{K}_i$ from joints $j$ and $k$ to the reference points $p$ and $q$. For this purpose, consider the action-displacement relationships

$$\mathbf{A}_{Mi} = \mathbf{K}_i \mathbf{D}_{Mi} \quad (19)$$

Substitution of Eq. (16) for $\mathbf{D}_{Mi}$ gives

$$\mathbf{A}_{Mi} = \mathbf{K}_i \mathbf{T}_i^T \mathbf{D}_{Bi} \quad (20)$$

Use of this expression in Eq. (12) produces

$$\mathbf{A}_{Bi} = \mathbf{T}_i \mathbf{K}_i \mathbf{T}_i^T \mathbf{D}_{Bi} \quad (21)$$

Hence, the matrix relating $\mathbf{A}_{Bi}$ to $\mathbf{D}_{Bi}$ is

$$\mathbf{K}_{Bi} = \mathbf{T}_i \mathbf{K}_i \mathbf{T}_i^T \quad (22)$$

The symbol $\mathbf{K}_{Bi}$ denotes the member stiffness matrix for actions at points $p$ and $q$ due to unit displacements at those points.

Similarly, the consistent-mass matrix $\mathbf{M}_i$ for the member may be transformed from joints $j$ and $k$ to points $p$ and $q$. Reasoning as above, but with accelerations instead of displacements, we can derive the formula

$$\mathbf{M}_{Bi} = \mathbf{T}_i \mathbf{M}_i \mathbf{T}_i^T \quad (23)$$

Here the symbol $\mathbf{M}_i$ represents the consistent-mass matrix for member $i$ at joints $j$ and $k$, while $\mathbf{M}_{Bi}$ is its counterpart for the reference points $p$ and $q$.

Equations of motion for all of the possible rigid bodies in a framed structure may be written as

$$\mathbf{M}_{BB} \ddot{\mathbf{D}}_B + \mathbf{S}_{BB} \mathbf{D}_B = \mathbf{A}_B(t) \quad (24)$$

To form the matrices in this equation, we assemble contributions from individual members by the direct stiffness method, as follows:

$$\mathbf{S}_{BB} = \sum_{i=1}^{n_e} \mathbf{K}_{Bi} \qquad \mathbf{M}_{BB} = \sum_{i=1}^{n_e} \mathbf{M}_{Bi} \qquad \mathbf{A}_B(t) = \sum_{i=1}^{n_e} \mathbf{A}_B(t)_i \quad (25)$$

This assembly process is similar to that described previously in Sec. 3.5, but the motions are at working points instead of joints. If a rigid body does not exist at a particular joint, all of the offset vectors for members framing into that joint are taken to have zero lengths. Also, the rotational displacements are omitted for nonexistent rigid bodies in plane and space trusses.

As yet, the mass matrix $\mathbf{M}_{BB}$ in Eqs. (24) and (25) is devoid of contributions from the rigid bodies themselves, which may be the most important terms. Each rigid body in the analytical model has its *center of mass* located at

## Sec. 9.2  Rigid Bodies in Framed Structures

point $c$, as indicated in Fig. 9.2. If the body is three-dimensional and has six degrees of freedom, its *mass and mass-moment-of-inertia matrix* is

$$\mathbf{M}_c = \begin{bmatrix} m & 0 & 0 & 0 & 0 & 0 \\ 0 & m & 0 & 0 & 0 & 0 \\ 0 & 0 & m & 0 & 0 & 0 \\ 0 & 0 & 0 & I_{xx} & -I_{xy} & -I_{xz} \\ 0 & 0 & 0 & -I_{yx} & I_{yy} & -I_{yz} \\ 0 & 0 & 0 & -I_{zx} & -I_{zy} & I_{zz} \end{bmatrix} \tag{26}$$

In this symmetric $6 \times 6$ array, the symbol $m$ denotes the *mass of the body*, which is computed from

$$m = \int_V \rho \, dV \tag{27}$$

The *mass moment of inertia* $I_{xx}$ is obtained as

$$I_{xx} = \int_V \rho (y_c^2 + z_c^2) \, dV \tag{28}$$

in which $y_c$ and $z_c$ are the $y$- and $z$-distances of a typical point in the body from the center of mass. The *mass product of inertia* $I_{xy}$ is

$$I_{xy} = \int_V \rho x_c y_c \, dV \tag{29}$$

where $x_c$ is the $x$-distance from point $c$. Other moments and products of inertia in matrix $\mathbf{M}_c$ have similar definitions. The negative signs on mass products of inertia in Eq. (26) result from Euler's equations for small motions of rigid bodies [4]. Table 9.2 contains various forms of matrix $\mathbf{M}_c$ required for all types of framed structures.

The matrix $\mathbf{M}_c$ for a rigid body may be transformed to a working point $p$ by an operation similar to that for consistent masses of member $i$ given in Eq. (23). However, only the points $p$ and $c$ are involved, as shown in Fig. 9.2. The required congruence multiplication is

$$\mathbf{M}_p = \mathbf{T}_{pc} \mathbf{M}_c \mathbf{T}_{pc}^T \tag{30}$$

In this equation the transformation matrix $\mathbf{T}_{pc}$ is of the same form as matrix $\mathbf{T}_{pj}$ discussed earlier, except that $c$ replaces $j$. Note that for a rigid body in either a plane truss or a plane frame, the matrix $\mathbf{T}_{pc}$ must be the same (and of size $3 \times 3$). Also, for a three-dimensional body in either a space truss or a space frame, $\mathbf{T}_{pc}$ is again the same (and of size $6 \times 6$).

As the second step in assembling equations of motion, we must add the masses and mass moments of inertia for the rigid bodies to the matrix $\mathbf{M}_{BB}$ in Eq. (24). This gives us an augmented matrix $\mathbf{M}_{BB}^*$, as follows;

**TABLE 9.2  Mass and Mass-Moment-of-Inertia Matrices $\mathbf{M}_c$ for Rigid Bodies**

1. Beams (x-y plane of bending)

$$\mathbf{M}_c = \begin{bmatrix} m & 0 \\ 0 & I_{zz} \end{bmatrix}$$

2. Plane trusses and plane frames (structures in x-y plane)

$$\mathbf{M}_c = \begin{bmatrix} m & 0 & 0 \\ 0 & m & 0 \\ 0 & 0 & I_{zz} \end{bmatrix}$$

3. Grids (structure in x-y plane)

$$\mathbf{M}_c = \begin{bmatrix} I_{xx} & -I_{xy} & 0 \\ -I_{yx} & I_{yy} & 0 \\ 0 & 0 & m \end{bmatrix}$$

4. Space trusses and space frames

$$\mathbf{M}_c = \begin{bmatrix} m & 0 & 0 & 0 & 0 & 0 \\ 0 & m & 0 & 0 & 0 & 0 \\ 0 & 0 & m & 0 & 0 & 0 \\ 0 & 0 & 0 & I_{xx} & -I_{xy} & -I_{xz} \\ 0 & 0 & 0 & -I_{yx} & I_{yy} & -I_{yz} \\ 0 & 0 & 0 & -I_{zx} & -I_{zy} & I_{zz} \end{bmatrix}$$

$$\mathbf{M}_{BB}^* = \mathbf{M}_{BB} + \sum_{k=1}^{n_b} \mathbf{M}_{pk} \tag{31}$$

where $n_b$ is the number of bodies.

After solving the eigenvalue problem for the augmented form of Eq. (24), we can add modal damping to our analytical model. Initial conditions and ground motions may also be included, along with applied actions, if desired. Section 9.3 describes a program named DYRBPF for dynamic analysis of rigid bodies in plane frames, using the member-oriented technique given above.

In any case (especially when the mass of the structural framing is to be neglected), we could use the mass center $c$ of each rigid body as the working point $p$. With grids, space trusses, and space frames, it may also be convenient to use *principal body axes*, for which the mass moment of inertia submatrix within $\mathbf{M}_c$ is diagonal. If *principal mass moments of inertia* and the directions of principal body axes are not known in advance, they may be found by solving an eigenvalue problem of order 2 or 3 for each rigid body. This method is similar to that for principal stresses described previously in Sec. 7.2. If one or more of the implied rigid bodies do not exist, we could also eliminate the displacements at massless nodes by *static reduction*, as discussed in Sec. 6.7.

### Example 9.1

Figure 9.4 shows a portion of a plane truss with two rigid rectangular laminae connected by a prismatic member $i$ at joints $j$ and $k$. Let us develop contributions of the member and the laminae to undamped equations of motion for points $p$ and $q$ on the rigid bodies.

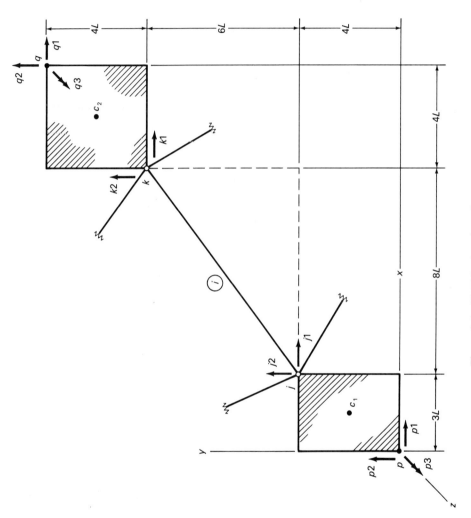

**Figure 9.4** Rigid laminae in a plane truss.

We shall include actions, stiffnesses, and consistent masses from member $i$, as well as masses and mass moments of inertia from the bodies.

First, we consider actions at points $j$ and $k$ in the column vector

$$\mathbf{A}_{Mi} = \{\mathbf{A}_j, \mathbf{A}_k\} = \{A_{j1}, A_{j2}, | A_{k1}, A_{k2}\} \tag{a}$$

Statically equivalent actions at the working points $p$ and $q$ may be obtained using transformation matrices for plane trusses from Table 9.1, as follows:

$$\mathbf{T}_{pj} = \begin{bmatrix} 1 & 0 \\ 0 & 1 \\ -4L & 3L \end{bmatrix} \quad \mathbf{T}_{qk} = \begin{bmatrix} 1 & 0 \\ 0 & 1 \\ 4L & -4L \end{bmatrix} \tag{b}$$

Substituting these arrays into Eq. (14), we form the operator $\mathbf{T}_i$ and use it in Eq. (12). Thus,

$$\mathbf{A}_{Bi} = \mathbf{T}_i \mathbf{A}_{Mi} = \{\mathbf{A}_p, \mathbf{A}_q\}$$
$$= \{A_{j1}, A_{j2}, -L(4A_{j1} - 3A_{j2}), | A_{k1}, A_{k2}, 4L(A_{k1} - A_{k2})\} \tag{c}$$

Second, we set up the 4 × 4 stiffness matrix for member $i$, using Eq. (3.5–26).

$$\mathbf{K}_i = \begin{bmatrix} \mathbf{K}_{jj} & \mathbf{K}_{jk} \\ \mathbf{K}_{kj} & \mathbf{K}_{kk} \end{bmatrix}_i = \frac{EA}{10L} \begin{bmatrix} 0.64 & & & \text{Sym.} \\ 0.48 & 0.36 & & \\ -0.64 & -0.48 & 0.64 & \\ -0.48 & -0.36 & 0.48 & 0.36 \end{bmatrix} \tag{d}$$

Then we transform matrix $\mathbf{K}_i$ to the working points $p$ and $q$ with Eq. (22), producing

$$\mathbf{K}_{Bi} = \mathbf{T}_i \mathbf{K}_i \mathbf{T}_i^T = \begin{bmatrix} \mathbf{K}_{pp} & \mathbf{K}_{pq} \\ \mathbf{K}_{qp} & \mathbf{K}_{qq} \end{bmatrix}_i$$

$$= \frac{EA}{10L} \begin{bmatrix} 0.64 & & & & & \text{Sym.} \\ 0.48 & 0.36 & & & & \\ -1.12L & -0.84L & 1.96L^2 & & & \\ -0.64 & -0.48 & 1.12L & 0.64 & & \\ -0.48 & -0.36 & 0.84L & 0.48 & 0.36 & \\ -0.64L & -0.48L & 1.12L^2 & 0.64L & 0.48L & 0.64L^2 \end{bmatrix} \tag{e}$$

Third, we write the consistent-mass matrix for member $i$ from Eq. (3.5–32). That is,

$$\mathbf{M}_i = \begin{bmatrix} \mathbf{M}_{jj} & \mathbf{M}_{jk} \\ \mathbf{M}_{kj} & \mathbf{M}_{kk} \end{bmatrix}_i = \frac{5\rho AL}{3} \begin{bmatrix} 2 & & & \text{Sym.} \\ 0 & 2 & & \\ 1 & 0 & 2 & \\ 0 & 1 & 0 & 2 \end{bmatrix} \tag{f}$$

Transformation of matrix $\mathbf{M}_i$ to the working points $p$ and $q$ yields

## Sec. 9.2  Rigid Bodies in Framed Structures

$$\mathbf{M}_{Bi} = \mathbf{T}_i \mathbf{M}_i \mathbf{T}_i^T = \begin{bmatrix} \mathbf{M}_{pp} & \mathbf{M}_{pq} \\ \mathbf{M}_{qp} & \mathbf{M}_{qq} \end{bmatrix}_i$$

$$= \frac{5\rho AL}{3} \begin{bmatrix} 2 & & & & & \text{Sym.} \\ 0 & 2 & & & & \\ -8L & 6L & 50L^2 & & & \\ \hline 1 & 0 & -4L & 2 & & \\ 0 & 1 & 3L & 0 & 2 & \\ 4L & -4L & -28L^2 & 8L & -8L & 64L^2 \end{bmatrix} \tag{g}$$

which is found using Eq. (23).

Turning now to the rigid laminae, we take the following mass and mass-moment-of-inertia matrices from Table 9.2 for plane trusses:

$$\mathbf{M}_{c_1} = \frac{m_1}{12} \begin{bmatrix} 12 & 0 & 0 \\ 0 & 12 & 0 \\ 0 & 0 & 25L^2 \end{bmatrix} \qquad \mathbf{M}_{c_2} = \frac{m_2}{3} \begin{bmatrix} 3 & 0 & 0 \\ 0 & 3 & 0 \\ 0 & 0 & 8L^2 \end{bmatrix} \tag{h}$$

Transformation operators required here are the 3 × 3 arrays

$$\mathbf{T}_{pc_1} = \begin{bmatrix} 1 & 0 & 0 \\ 0 & 1 & 0 \\ -y_{pc_1} & x_{pc_1} & 1 \end{bmatrix} = \begin{bmatrix} 1 & 0 & 0 \\ 0 & 1 & 0 \\ -2L & \frac{3L}{2} & 1 \end{bmatrix}$$

$$\mathbf{T}_{qc_2} = \begin{bmatrix} 1 & 0 & 0 \\ 0 & 1 & 0 \\ -y_{qc_2} & x_{qc_2} & 1 \end{bmatrix} = \begin{bmatrix} 1 & 0 & 0 \\ 0 & 1 & 0 \\ 2L & -2L & 1 \end{bmatrix} \tag{i}$$

They are applied to convert matrices $\mathbf{M}_{c_1}$ and $\mathbf{M}_{c_2}$ to the reference points $p$ and $q$, according to Eq. (30). Thus,

$$\mathbf{M}_p = \mathbf{T}_{pc_1} \mathbf{M}_{c_1} \mathbf{T}_{pc_1}^T = \frac{m_1}{6} \begin{bmatrix} 6 & 0 & -12L \\ 0 & 6 & 9L \\ -12L & 9L & 50L^2 \end{bmatrix}$$

$$\mathbf{M}_q = \mathbf{T}_{qc_2} \mathbf{M}_{c_2} \mathbf{T}_{qc_2}^T = \frac{m_2}{3} \begin{bmatrix} 3 & 0 & 6L \\ 0 & 3 & -6L \\ 6L & -6L & 32L^2 \end{bmatrix} \tag{j}$$

The first of these arrays augments submatrix $\mathbf{M}_{ppi}$ in Eq. (g), and the second is added to $\mathbf{M}_{qqi}$. This step fulfills the objectives stated at the beginning of the example.

## 9.3 PROGRAM DYRBPF FOR RIGID BODIES IN PLANE FRAMES

As an example of programming for the member-oriented technique, we shall briefly describe Program DYRBPF for dynamic analysis of rigid bodies in plane frames. To simplify the procedure, we take the mass center of each rigid body as one of the nodes in the structure (as well as the working point for the body). Within the logic of the program, every member is assumed to have a rigid body at each end, unless proven otherwise by the input data. This idea provides the key to easily extending the programs for framed structures to include rigid bodies.

Building upon Program DYNAPF from Sec. 6.6, we can add rigid bodies to plane frames and create Program DYRBPF. For this purpose, the line containing nodal coordinates in the structural data (see Table 6.1) must be augmented by adding BM(J) and BI(J). These terms represent the body mass $m$ and the mass moment of inertia $I_{zz}$ of the body with respect to its center of mass. Also, to the element information we add XCJ(I), YCJ(I), XCK(I), and YCK(I), which denote the $x$ and $y$ components of the offset vectors $\mathbf{r}_{cj}$ and $\mathbf{r}_{ck}$ at ends $j$ and $k$. Of course, if there is no body at one end or the other, such offsets are set equal to zero.

In Program DYRBPF we assemble the stiffness matrix $\mathbf{S}_{BB}$, the mass matrix $\mathbf{M}_{BB}$, and the action vector $\mathbf{A}_B(t)$ by assessing one member at a time [see Eqs. (9.2-25)]. To the second of these matrices we add the mass-inertia terms for the rigid bodies to form the augmented mass matrix $\mathbf{M}_{BB}^*$, as given by Eq. (9.2-31). After the equations for rigid-body motions have been solved, we can find time histories for displacements at the ends of members using Eq. (9.2-16). Then time-varying member end-actions are calculated from these member displacements by premultiplying them with the member stiffness matrix.

**Example 9.2**

Figure 9.5 shows a steel plane frame supporting a single rigid body, which is a concrete cube of size $L$ on each side. We shall analyze the dynamic response of this configuration to ground acceleration $\ddot{D}_{g1}(t)$, caused by an underground blast.

If all members of the frame have W10 × 45 cross sections, we can state their physical properties as

$$E_s = 3.0 \times 10^4 \text{ k/in.}^2 \qquad \rho_s = 7.35 \times 10^{-7} \text{ k-s}^2/\text{in.}^4$$
$$L = 36 \text{ in.} \qquad A = 13.2 \text{ in.}^2 \qquad I_z = 249.0 \text{ in.}^4$$

where the subscript $s$ implies steel. Also, the relevant properties of the rigid body are

$$\rho_c = 2.25 \times 10^{-7} \text{ k-s}^2/\text{in.}^4 \qquad m = \rho_c L^3 = 1.050 \times 10^{-2} \text{ k-s}^2/\text{in.}$$
$$I_{zz} = \tfrac{1}{6}\rho_c L^5 = 2.268 \text{ k-s}^2\text{-in.}$$

for which the subscript $c$ denotes concrete.

We ran the foregoing data on Program DYRBPF, using Subprogram NORMOD with NMODES = 7 and DAMPR = 0.02. This program produces the computer plots in

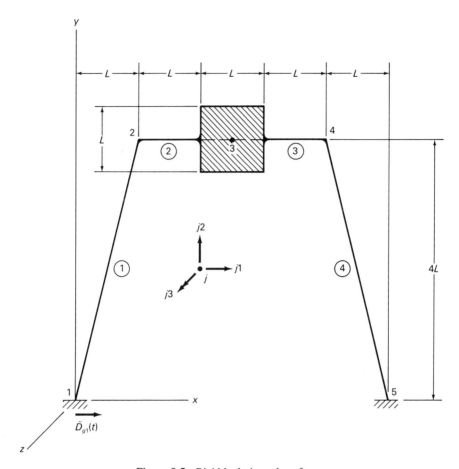

**Figure 9.5** Rigid body in a plane frame.

Fig. 9.6, the first of which depicts the impulsive $x$-ground acceleration due to the underground blast. The plot in Fig. 9.6(b) gives the resulting time history of $x$-translation at node 3 (the center of mass of the rigid body). Also, Fig. 9.6(c) contains plots of time-varying member end-moments AM6 for elements 2 and 4. Maximum (or minimum) values for these three types of responses are 0.03943 in., −157.3 k-in., and 97.06 k-in., all of which occur at time $t = 52$ ms.

## 9.4 RIGID LAMINAE IN MULTISTORY BUILDINGS

Multistory buildings are usually constructed in tiers of one, two, or three stories at a time. Hence, the name *tier buildings* [5, 6] may be used to describe the skyscrapers that abound in large cities. The steel skeletons of these tall buildings consist of space frames that most frequently have their members arranged in an

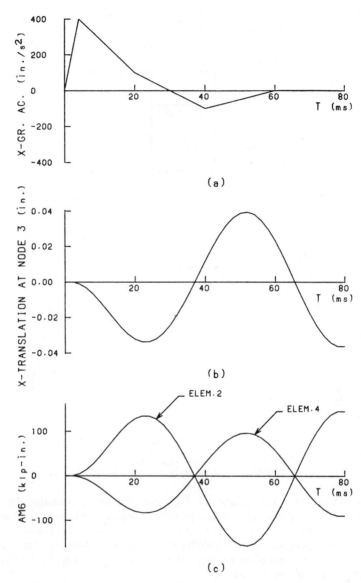

**Figure 9.6** Rigid body in a plane frame: (a) ground acceleration; (b) displacement; (c) end-moments.

orthogonal (or rectangular) pattern. Figure 9.7 shows a typical floor (or roof) plan of a tier building, consisting of x-beams, y-beams, and a slab or deck that we shall consider to be rigid in its own plane. This lamina is flexible in the direction normal to its plane and could be discretized with plate-bending elements. A cruder but more commonly-used approach is to include a tributary strip of the slab in the cross section of each beam. These beams frame into vertical

Sec. 9.4   Rigid Laminae in Multistory Buildings                                   427

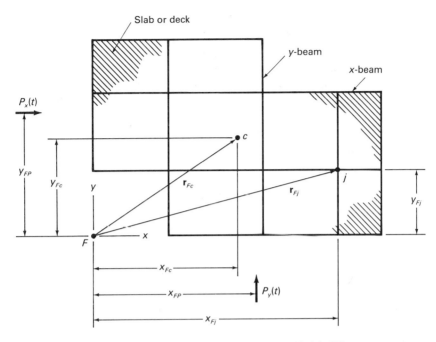

**Figure 9.7**  Typical floor plan of multistory (tier) building.

columns that are parallel to the z axis (not shown in Fig. 9.7). An origin of reference (or structural) coordinates may be located at any convenient point and probably would be taken at the base of the structure. The point labeled $F$ in Fig. 9.7 lies at the intersection of the z axis with the plane of the floor. Also, point $c$ represents the center of mass that is tributary to the floor, and point $j$ is a typical joint at this framing level.

Because each horizontal lamina is taken to be rigid in its own plane, all points at that level are constrained to displace in a rigid-body pattern. Such motions consist of translation in the horizontal plane and rotation about a vertical axis. Thus, the rigid-body motions of the lamina in Fig. 9.7 may be described by the $x$ and $y$ components of the translation of point $F$ (considered to be attached to the body) and its rotation about the $z$ axis. Moreover, the corresponding displacements at any other points on the lamina, such as joint $j$ or the center of mass $c$, are related to the motions of point $F$ by the concepts of Sec. 9.2. On the other hand, the $x$ and $y$ rotations and the $z$ translation at a typical joint $j$ remain as unconstrained displacements.

For simplicity in the analysis, we assume that the geometric layout of the framing is the same at all levels, that there is only one distinct lamina at each level, and that there are no shear walls or lateral bracing in the building. Members of the space frame are taken to be prismatic and of a linearly elastic material. We also assume that joints in the frame are rigid and that displacements relative to ground are small.

Because every member in the analytical model is parallel to one of the structural reference axes, there is no need for rotation-of-axes transformations. Figure 9.8(a) shows a beam with its member axis $x_m$ parallel to the $x$ axis. In addition, its principal bending axes $y_m$ and $z_m$ are assumed to be parallel to axes $y$ and $z$. Action and displacement indexes $j1$, $j2$, and $j3$ at the $j$ end of member $i$ denote rotations in the $x$ and $y$ senses and translation in the $z$ direction. Their counterparts $k1$, $k2$, and $k3$ also appear at the $k$ end of the member. The $6 \times 6$ stiffness matrix $\mathbf{K}_i$ for such a beam is the same as that for the grid member in Sec. 6.3. Thus,

$$\mathbf{K}_i = \begin{bmatrix} \mathbf{K}_{jj} & \mathbf{K}_{jk} \\ \mathbf{K}_{kj} & \mathbf{K}_{kk} \end{bmatrix}_i$$

$$= \frac{1}{L^3} \begin{bmatrix} GI_xL^2 & & & & & \text{Sym.} \\ 0 & 4EI_yL^2 & & & & \\ 0 & -6EI_yL & 12EI_y & & & \\ -GI_xL^2 & 0 & 0 & GI_xL^2 & & \\ 0 & 2EI_yL^2 & -6EI_yL & 0 & 4EI_yL^2 & \\ 0 & 6EI_yL & -12EI_y & 0 & 6EI_yL & 12EI_y \end{bmatrix} \quad (1)$$

As before, the symbol $I_x$ represents the torsion constant, and $I_y$ is the second moment of the cross-sectional area with respect to the $y_m$ axis.

The second type of beam to be considered has its axis parallel to the $y$ axis, as indicated in Fig. 9.8(b). In this case $y_m$ is chosen to be the member axis, and the principal axes $x_m$ and $z_m$ are parallel to axes $x$ and $z$. Action and displacement indexes $j1$ through $k3$ correspond to those for the $x$-beam. Therefore, the $6 \times 6$ member stiffness matrix for the $y$-beam is

$$\mathbf{K}_i = \begin{bmatrix} \mathbf{K}_{jj} & \mathbf{K}_{jk} \\ \mathbf{K}_{kj} & \mathbf{K}_{kk} \end{bmatrix}_i$$

$$= \frac{1}{L^3} \begin{bmatrix} 4EI_xL^2 & & & & & \text{Sym.} \\ 0 & GI_yL^2 & & & & \\ 6EI_xL & 0 & 12EI_x & & & \\ 2EI_xL^2 & 0 & 6EI_xL & 4EI_xL^2 & & \\ 0 & -GI_yL^2 & 0 & 0 & GI_yL^2 & \\ -6EI_xL & 0 & -12EI_x & -6EI_xL & 0 & 12EI_x \end{bmatrix} \quad (2)$$

Note that for this type of member the cross-sectional properties have new symbols because of the orientation of the member axis. That is, $I_y$ is now the torsion constant, while $I_x$ is the second moment of the cross-sectional area with respect to the $x_m$ axis.

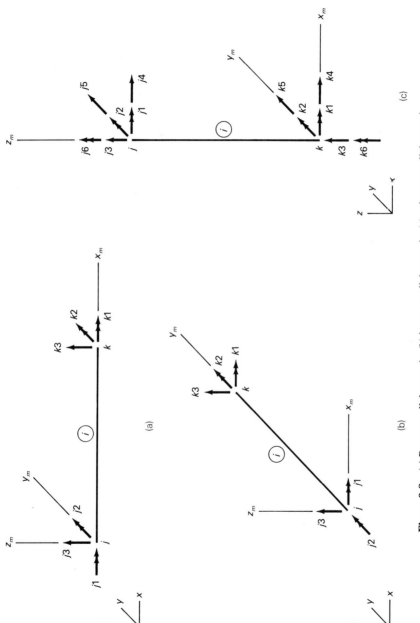

**Figure 9.8** (a) Beam parallel to $x$ axis; (b) beam parallel to $y$ axis; (c) column parallel to $z$ axis.

Figure 9.8(c) depicts a typical column with its member axis $z_m$ parallel to the $z$ reference axis. Its principal bending axes $x_m$ and $y_m$ are assumed to be parallel to axes $x$ and $y$. Action and displacement indexes at joints $j$ and $k$ are numbered in a sequence that expedites the process of transferring contributory terms from the member stiffness matrix to joint and floor stiffness matrices. That is, indexes $j1$, $j2$, and $j3$ at the top of the column correspond to joint displacements that are not directly associated with the rigid-body motions of the floor above. These displacements are the same types as those discussed previously for the beams. Similarly, indexes $k1$, $k2$, and $k3$ at the bottom of the column pertain to joint displacements not directly associated with the rigid-body motions of the floor below. However, the indexes $j4$, $j5$, and $j6$ at the top and $k4$, $k5$, and $k6$ at the bottom correspond to displacements that are directly dependent upon rigid-body motions of the floors above and below. They consist of translations in the $x$ and $y$ directions and a rotation in the $z$ sense at both levels $j$ and $k$.

We can represent the 12 × 12 stiffness matrix for a column as a partitioned array composed of 3 × 3 submatrices, as follows:

$$\mathbf{K}_i = \begin{bmatrix} \mathbf{K}_{11} & & & \text{Sym.} \\ \mathbf{K}_{21} & \mathbf{K}_{22} & & \\ \mathbf{K}_{31} & \mathbf{K}_{32} & \mathbf{K}_{33} & \\ \mathbf{K}_{41} & \mathbf{K}_{42} & \mathbf{K}_{43} & \mathbf{K}_{44} \end{bmatrix} \tag{3}$$

In this matrix subscript 1 denotes action and displacement indexes of types $j1$, $j2$, and $j3$. Subscript 2 represents indexes of types $k1$, $k2$, and $k3$. Subscript 3 stands for indexes $j4$, $j5$, and $j6$. And subscript 4 is for indexes $k4$, $k5$, and $k6$. Note that the stiffness matrix in Eq. (3) is also partitioned to separate the unconstrained displacements at level $j$ from those at level $k$, as well as from the constrained displacements at both levels. The submatrices in Eq. (3) are

$$\mathbf{K}_{11} = \mathbf{K}_{22} = \frac{E}{L} \begin{bmatrix} 4I_x & 0 & 0 \\ 0 & 4I_y & 0 \\ 0 & 0 & A_z \end{bmatrix} \tag{4a}$$

$$\mathbf{K}_{21} = \frac{E}{L} \begin{bmatrix} 2I_x & 0 & 0 \\ 0 & 2I_y & 0 \\ 0 & 0 & -A_z \end{bmatrix} \tag{4b}$$

$$\mathbf{K}_{31} = \mathbf{K}_{32} = -\mathbf{K}_{41} = -\mathbf{K}_{42} = \frac{6E}{L^2} \begin{bmatrix} 0 & -I_y & 0 \\ I_x & 0 & 0 \\ 0 & 0 & 0 \end{bmatrix} \tag{4c}$$

$$\mathbf{K}_{33} = -\mathbf{K}_{43} = \mathbf{K}_{44} = \frac{1}{L^3} \begin{bmatrix} 12EI_y & 0 & 0 \\ 0 & 12EI_x & 0 \\ 0 & 0 & GI_z L^2 \end{bmatrix} \tag{4d}$$

## Sec. 9.4  Rigid Laminae in Multistory Buildings

For this member the cross-sectional properties are defined as follows: $A_z$ = area, $I_z$ = torsion constant, and $I_x$ and $I_y$ = second moments of area with respect to the $x_m$ and $y_m$ axes.

Due to the presence of the rigid laminae, certain *geometric transformations* are required to relate the motions at points $j$ and $c$ to those at point $F$ for each framing level. Figure 9.9(a) shows actions $A_{j4}$, $A_{j5}$, and $A_{j6}$ at point $j$ and their statical equivalents $A_{F1}$, $A_{F2}$, and $A_{F3}$ at point $F$. Values of the latter actions are calculated from the former, as follows:

$$\mathbf{A}_F = \mathbf{T}_{Fj}\mathbf{A}_j \tag{5}$$

This equation is similar to Eq. (9.2-1), but point $F$ replaces point $p$. In this case the vector $\mathbf{A}_j$ contains

$$\mathbf{A}_j = \{A_{j4}, A_{j5}, A_{j6}\} \tag{6}$$

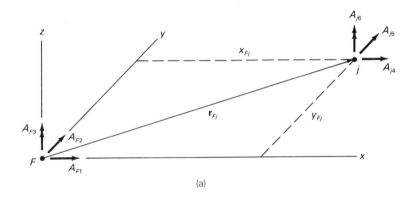

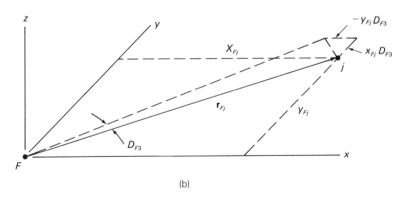

**Figure 9.9** Geometric transformations: (a) action relationships; (b) displacement relationships.

Vector $\mathbf{A}_F$ is

$$\mathbf{A}_F = \{A_{F1}, A_{F2}, A_{F3}\} \tag{7}$$

and the transformation operator $\mathbf{T}_{Fj}$ has the form

$$\mathbf{T}_{Fj} = \begin{bmatrix} 1 & 0 & 0 \\ 0 & 1 & 0 \\ -y_{Fj} & x_{Fj} & 1 \end{bmatrix} \tag{8}$$

Turning next to displacements, we see in Fig. 9.9(b) the effect at point $j$ of a rigid-body rotation $D_{F3}$ at point $F$. This rotation causes a negative $x$-translation and a positive $y$-translation at point $j$. Hence, the kinematically equivalent displacements at $j$ due to rigid-body motions of point $F$ are

$$\mathbf{D}_j = \mathbf{T}_{Fj}^T \mathbf{D}_F \tag{9}$$

Again, this equation is similar to Eq. (9.2-6), but with point $F$ replacing point $p$. Here the vector $\mathbf{D}_F$ contains the terms

$$\mathbf{D}_F = \{D_{F1}, D_{F2}, D_{F3}\} \tag{10}$$

and the vector $\mathbf{D}_j$ is

$$\mathbf{D}_j = \{D_{j4}, D_{j5}, D_{j6}\} \tag{11}$$

Also, the transformation operator becomes

$$\mathbf{T}_{Fj}^T = \begin{bmatrix} 1 & 0 & -y_{Fj} \\ 0 & 1 & x_{Fj} \\ 0 & 0 & 1 \end{bmatrix} \tag{12}$$

which is the transpose of the matrix in Eq. (8).

Considering now the stiffness matrix for the column in Eq. (3), we must transform certain of its submatrices to rigid-body coordinates. This operation requires a combined action-displacement transformation that makes use of the matrix

$$\mathbf{T}_i = \begin{bmatrix} \mathbf{I}_3 & & & \text{Sym.} \\ 0 & \mathbf{I}_3 & & \\ 0 & 0 & \mathbf{T}_{Fj} & \\ 0 & 0 & 0 & \mathbf{T}_{Fk} \end{bmatrix} \tag{13}$$

The transformation matrix $\mathbf{T}_i$ contains four $3 \times 3$ submatrices on the diagonal, of which the first and second are identity matrices. The third and fourth submatrices are of the type given by Eq. (8), and their subscripts denote joint $j$ in the floor above and joint $k$ in the floor below [see Fig. 9.8(c)]. The desired transformation of the column stiffness matrix is:

$$\mathbf{K}_{Fi} = \mathbf{T}_i \mathbf{K}_i \mathbf{T}_i^T \tag{14}$$

## Sec. 9.4  Rigid Laminae in Multistory Buildings

which has the same meaning as Eq. (9.2-22), except that the subscript $F$ replaces $B$. The results of Eq. (14) in expanded form become

$$\mathbf{K}_{Fi} = \begin{bmatrix} \mathbf{K}_{11} & & & \text{Sym.} \\ \mathbf{K}_{21} & \mathbf{K}_{22} & & \\ \mathbf{T}_{Fj}\mathbf{K}_{31} & \mathbf{T}_{Fj}\mathbf{K}_{32} & \mathbf{T}_{Fj}\mathbf{K}_{33}\mathbf{T}_{Fj}^T & \\ \mathbf{T}_{Fk}\mathbf{K}_{41} & \mathbf{T}_{Fk}\mathbf{K}_{42} & \mathbf{T}_{Fk}\mathbf{K}_{43}\mathbf{T}_{Fj}^T & \mathbf{T}_{Fk}\mathbf{K}_{44}\mathbf{T}_{Fk}^T \end{bmatrix} \quad (15)$$

Note that the transformed stiffness matrix $\mathbf{K}_{Fi}$ is still partitioned in the manner indicated for $\mathbf{K}_i$ in Eq. (3). When the submatrices in Eq. (15) are evaluated, we find that the equalities among them are the same as those given in Eqs. (4), provided that reference points at all levels lie on the $z$ axis.

In lieu of more detailed information, we assume that the mass of the building tributary to a given framing level is uniformly distributed over the area of the lamina. In that case the center of mass $c$ will coincide with the centroid of the area, and every floor will have the same *radius of gyration* $r_c$ with respect to point $c$. Thus, the mass and mass-moment-of-inertia matrix with respect to point $c$ for the lamina at level $\ell$ may be written as

$$\mathbf{M}_{c\ell} = m_\ell \begin{bmatrix} 1 & 0 & 0 \\ 0 & 1 & 0 \\ 0 & 0 & r_c^2 \end{bmatrix} \quad (16)$$

where $m_\ell$ is the mass at that level. Furthermore, the mass-inertia matrix with respect to point $F$ is obtained as

$$\mathbf{M}_{F\ell} = \mathbf{T}_{Fc}\mathbf{M}_{c\ell}\mathbf{T}_{Fc}^T = m_\ell \begin{bmatrix} 1 & 0 & -y_{Fc} \\ 0 & 1 & x_{Fc} \\ -y_{Fc} & x_{Fc} & r_F^2 \end{bmatrix} \quad (17)$$

This transformation is of the same type as that in Eq. (9.2-30). The symbol $r_F$ in Eq. (17) denotes the radius of gyration of the floor with respect to point $F$. It is related to that with respect to point $c$ by

$$r_F^2 = r_c^2 + x_{Fc}^2 + y_{Fc}^2 \quad (18)$$

which is a familiar expression for translation of axes from elementary dynamics [4].

Horizontal forces $P_x(t)$ and $P_y(t)$, applied at a typical framing level, appear in Fig. 9.7. In general, these forces are eccentric with respect to the reference point $F$, and the statically equivalent actions at that point are

$$\mathbf{A}_F(t)_\ell = \mathbf{T}_{FP}\mathbf{A}_P(t)_\ell \quad (19)$$

in which

$$\mathbf{A}_P(t)_\ell = \{P_x(t), P_y(t)\}_\ell \quad (20)$$

and

$$\mathbf{T}_{FP} = \begin{bmatrix} 1 & 0 \\ 0 & 1 \\ -y_{FP} & x_{FP} \end{bmatrix} \quad (21)$$

When the forcing influences are horizontal ground accelerations, the resultant inertial forces at level $\ell$ are

$$\mathbf{A}_g(t)_\ell = -m_\ell \{\ddot{D}_{g1}(t), \ddot{D}_{g2}(t)\} \quad (22)$$

where $\ddot{D}_{g1}(t)$ and $\ddot{D}_{g2}(t)$ are accelerations of ground in the $x$ and $y$ directions. The forces in Eq. (22) act through the mass center at each story. Therefore, they are eccentric with respect to point $F$ by the distances $x_{Fc}$ and $y_{Fc}$ (see Fig. 9.7), and these terms replace $x_{FP}$ and $y_{FP}$ in Eq. (21).

The transformation relationships discussed above prove useful for the purpose of writing equations of motion for a tier building. Our method for handling such equations will be described in Chapter 10 when we study the modified tridiagonal method for dynamic analysis by substructures.

## 9.5 RIGID BODIES IN FINITE-ELEMENT NETWORKS

In this section we shall consider rigid bodies that are embedded within finite-element networks of the types described in Chapters 7 and 8. In such a situation, we cannot avoid having more than one connection of an element to a particular rigid body. Therefore, the member-oriented approach for framed structures (see Sec. 9.2) will be abandoned in favor of a *body-oriented method*.

Figure 9.10 illustrates an analytical model consisting of rigid laminae connected by Q8 elements for plane stress or plane strain. By omitting the rigid bodies temporarily and excluding support displacements, we can write equations of undamped motion for the nodal displacements in the finite-element network, as follows:

$$\begin{bmatrix} \mathbf{M}_{AA} & \mathbf{M}_{AF} \\ \mathbf{M}_{FA} & \mathbf{M}_{FF} \end{bmatrix} \begin{bmatrix} \ddot{\mathbf{D}}_A \\ \ddot{\mathbf{D}}_F \end{bmatrix} + \begin{bmatrix} \mathbf{S}_{AA} & \mathbf{S}_{AF} \\ \mathbf{S}_{FA} & \mathbf{S}_{FF} \end{bmatrix} \begin{bmatrix} \mathbf{D}_A \\ \mathbf{D}_F \end{bmatrix} = \begin{bmatrix} \mathbf{A}_A(t) \\ \mathbf{A}_F(t) \end{bmatrix} \quad (1)$$

In this equation the subscript $A$ refers to nodes attached to rigid bodies, and subscript $F$ denotes free nodes (see Fig. 9.10), at which the displacements are independent. For an arbitrary sequence of numbering nodes, the terms contributing to Eq. (1) must be rearranged to put type $A$ first and type $F$ second. Now let us bring in the rigid bodies and recognize that nodes of type $A$ are constrained to move with them. Thus, we need to transform actions, stiffnesses, and consistent masses from nodes $A$ to working points on the rigid bodies. For this purpose, we choose the mass centers as the working points.

Figure 9.10 indicates the mass center $c_k$ of body $k$ and a typical node $j$ where element $i$ is attached. Indexes $k1$, $k2$, and $k3$ at point $c_k$ denote the

## Sec. 9.5  Rigid Bodies in Finite-Element Networks

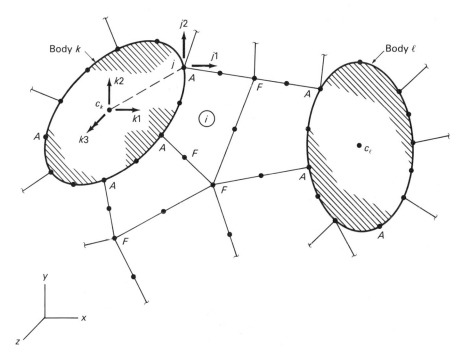

**Figure 9.10** Rigid laminae connected by Q8 elements.

rigid-body motions, which consist of translations in the $x$ and $y$ directions and a rotation in the $z$ sense. At node $j$ the indexes $j1$ and $j2$ represent only translations in the $x$ and $y$ directions (for a $Q8$ element). Corresponding actions at node $j$ are forces in the $x$ and $y$ directions, and their statical equivalents at point $c_k$ become

$$\mathbf{A}_k = \mathbf{T}_{kj}\mathbf{A}_j \tag{2}$$

which is similar to Eq. (9.2–1), but point $c_k$ replaces point $p$. Actions in vectors $\mathbf{A}_j$ and $\mathbf{A}_k$ are

$$\mathbf{A}_j = \{A_{j1}, A_{j2}\} \qquad \mathbf{A}_k = \{A_{k1}, A_{k2}, A_{k3}\} \tag{3}$$

In the latter vector we have forces in the $x$ and $y$ directions and a moment in the $z$ sense. The transformation operator in Eq. (2) is

$$\mathbf{T}_{kj} = \begin{bmatrix} 1 & 0 \\ 0 & 1 \\ -y_{kj} & x_{kj} \end{bmatrix} \tag{4}$$

which has the same form as that for plane trusses in Table 9.1, with $k$ replacing $p$.

Equation (2) refers only to the $j^{\text{th}}$ node on the $k^{\text{th}}$ rigid body, and if all nodes attached to the body are considered, the equation becomes

$$\mathbf{A}_{Bk} = \mathbf{T}_k \mathbf{A}_{Ak} \tag{5}$$

The expanded form of Eq. (5) is

$$\mathbf{A}_{Bk} = \begin{bmatrix} \mathbf{T}_{k1} & \mathbf{T}_{k2} & \cdots & \mathbf{T}_{kn_j} \end{bmatrix} \begin{bmatrix} \mathbf{A}_{A1} \\ \mathbf{A}_{A2} \\ \cdots \\ \mathbf{A}_{An_j} \end{bmatrix}_k \tag{6}$$

where $n_j$ is the number of nodes on the body. Each subvector in $\mathbf{A}_{Ak}$ contains two terms, as in the first of Eqs. (3); and each submatrix in $\mathbf{T}_k$ is a 3 × 2 array, as in Eq. (4).

Considering now all of the rigid bodies in the system, we express the action transformation relationships by

$$\mathbf{A}_B = \mathbf{T}_{BA} \mathbf{A}_A \tag{7}$$

In this equation the vector $\mathbf{A}_A$ contains actions for all of the attached nodes, and $\mathbf{A}_B$ consists of their statical equivalents at mass centers for all of the rigid bodies. The transformation matrix $\mathbf{T}_{BA}$ in Eq. (7) is a large, sparse array containing submatrices of type $\mathbf{T}_k$ in diagonal positions, as follows:

$$\mathbf{T}_{BA} = \begin{bmatrix} \mathbf{T}_1 & & & & & \text{Sym.} \\ 0 & \mathbf{T}_2 & & & & \\ \cdots & \cdots & \cdots & & & \\ 0 & 0 & \cdots & \mathbf{T}_k & & \\ \cdots & \cdots & \cdots & \cdots & \cdots & \\ 0 & 0 & \cdots & 0 & \cdots & \mathbf{T}_{n_b} \end{bmatrix} \tag{8}$$

where $n_b$ is the number of bodies in the system.

Similarly, the kinematically equivalent displacements at node $j$ in Fig. 9.10 may be calculated from those at point $c_k$ by the formula

$$\mathbf{D}_j = \mathbf{T}_{kj}^T \mathbf{D}_k \tag{9}$$

which is of the same form as Eq. (9.2–6). In this case the displacement vectors are

$$\mathbf{D}_j = \{D_{j1}, D_{j2}\} \qquad \mathbf{D}_k = \{D_{k1}, D_{k2}, D_{k3}\} \tag{10}$$

If all nodes on the $k$th rigid body are included, the transformation in Eq. (9) becomes

$$\mathbf{D}_{Ak} = \mathbf{T}_k^T \mathbf{D}_{Bk} \tag{11}$$

for which the expanded form is

### Sec. 9.5  Rigid Bodies in Finite-Element Networks

$$\begin{bmatrix} \mathbf{D}_{A1} \\ \mathbf{D}_{A2} \\ \vdots \\ \mathbf{D}_{An_j} \end{bmatrix}_k = \begin{bmatrix} \mathbf{T}_{k1}^T \\ \mathbf{T}_{k2}^T \\ \vdots \\ \mathbf{T}_{kn_j}^T \end{bmatrix} \mathbf{D}_{Bk} \tag{12}$$

When all rigid bodies are taken into account, we have

$$\mathbf{D}_A = \mathbf{T}_{BA}^T \mathbf{D}_B \tag{13}$$

Here the vector $\mathbf{D}_B$ consists of displacements at mass centers for all of the rigid bodies, and $\mathbf{D}_A$ contains the dependent displacements at all of the attached nodes. Of course, $\mathbf{T}_{BA}^T$ is the transpose of matrix $\mathbf{T}_{BA}$ in Eq. (8).

Now let us return to Eq. (1), consisting of the equations of motion for attached and free nodes, devoid of contributions from rigid bodies. To account for the presence of the rigid bodies, we form a transformation operator $\mathbf{T}$, as follows:

$$\mathbf{T} = \begin{bmatrix} \mathbf{T}_{BA} & \mathbf{0} \\ \mathbf{0} & \mathbf{I}_F \end{bmatrix} \tag{14}$$

in which $\mathbf{I}_F$ is an identity matrix of order equal to the number of free nodal displacements. Then we have

$$\begin{bmatrix} \mathbf{A}_B \\ \mathbf{A}_F \end{bmatrix} = \mathbf{T} \begin{bmatrix} \mathbf{A}_A \\ \mathbf{A}_F \end{bmatrix} \tag{15a}$$

Also,

$$\begin{bmatrix} \mathbf{D}_A \\ \mathbf{D}_F \end{bmatrix} = \mathbf{T}^T \begin{bmatrix} \mathbf{D}_B \\ \mathbf{D}_F \end{bmatrix} \qquad \begin{bmatrix} \ddot{\mathbf{D}}_A \\ \ddot{\mathbf{D}}_F \end{bmatrix} = \mathbf{T}^T \begin{bmatrix} \ddot{\mathbf{D}}_B \\ \ddot{\mathbf{D}}_F \end{bmatrix} \tag{15b}$$

Therefore, premultiplication of Eq. (1) by $\mathbf{T}$ and use of the relationships in Eqs. (15) produces

$$\begin{bmatrix} \mathbf{M}_{BB} & \mathbf{M}_{BF} \\ \mathbf{M}_{FB} & \mathbf{M}_{FF} \end{bmatrix} \begin{bmatrix} \ddot{\mathbf{D}}_B \\ \ddot{\mathbf{D}}_F \end{bmatrix} + \begin{bmatrix} \mathbf{S}_{BB} & \mathbf{S}_{BF} \\ \mathbf{S}_{FB} & \mathbf{S}_{FF} \end{bmatrix} \begin{bmatrix} \mathbf{D}_B \\ \mathbf{D}_F \end{bmatrix} = \begin{bmatrix} \mathbf{A}_B(t) \\ \mathbf{A}_F(t) \end{bmatrix} \tag{16}$$

Terms with the subscript $B$ in this equation refer to motions of the rigid bodies in the system. *Block stiffness and mass submatrices* in Eq. (16) are

$$\mathbf{S}_{BB} = \mathbf{T}_{BA} \mathbf{S}_{AA} \mathbf{T}_{BA}^T \qquad \mathbf{S}_{BF} = \mathbf{S}_{FB}^T = \mathbf{T}_{BA} \mathbf{S}_{AF} \tag{17a}$$

and

$$\mathbf{M}_{BB} = \mathbf{T}_{BA} \mathbf{M}_{AA} \mathbf{T}_{BA}^T \qquad \mathbf{M}_{BF} = \mathbf{M}_{FB}^T = \mathbf{T}_{BA} \mathbf{M}_{AF} \tag{17b}$$

Because the matrix $\mathbf{T}_{BA}$ has the form shown in Eq. (8), we can express Eqs. (17) more explicitly as

$$\mathbf{S}_{BBk,\ell} = \mathbf{T}_k \mathbf{S}_{AAk,\ell} \mathbf{T}_\ell^T \qquad \mathbf{S}_{BFk} = \mathbf{S}_{FBk}^T = \mathbf{T}_k \mathbf{S}_{AFk} \tag{18a}$$

and

$$\mathbf{M}_{BBk,\ell} = \mathbf{T}_k \mathbf{M}_{AAk,\ell} \mathbf{T}_\ell^T \qquad \mathbf{M}_{BFk} = \mathbf{M}_{FBk}^T = \mathbf{T}_k \mathbf{M}_{AFk} \qquad (18b)$$

for $k = 1, 2, \ldots, n_b$ and $\ell = 1, 2, \ldots, n_b$. Matrices $\mathbf{S}_{BBk,\ell}$ and $\mathbf{S}_{AAk,\ell}$ in Eqs. (18a) are submatrices of $\mathbf{S}_{BB}$ and $\mathbf{S}_{AA}$ that contain terms coupling body $k$ with body $\ell$, and so on.

To complete the equations of rigid-body motion, we need only add the mass-inertia matrices for the bodies to matrix $\mathbf{M}_{BB}$, as follows:

$$\mathbf{M}_{BB}^* = \mathbf{M}_{BB} + \sum_{k=1}^{n_b} \mathbf{M}_{c_k} \qquad (19)$$

where $\mathbf{M}_{c_k}$ is drawn from Table 9.2. For a typical two-dimensional body $k$ in Fig. 9.10, the matrix $\mathbf{M}_{c_k}$ is the same as that for plane trusses and plane frames given in the table.

We shall now discuss briefly other types of rigid bodies in other types of discretized continua. If we have three-dimensional rigid bodies embedded in a network of H8 or H20 elements, the transformation matrix $\mathbf{T}_{kj}$ becomes the same as that for space trusses in Table 9.1, except that $p$ is replaced by $k$. Also, the mass-inertia matrix $\mathbf{M}_{c_k}$ for a typical rigid body is the same as that for space trusses and space frames in Table 9.2. Otherwise, the process of setting up equations of undamped motion remains the same as that described above for two-dimensional continua, but numbers of actions and displacements are increased.

Rigid bodies supported by discretized plates in bending are analyzed in a manner analogous to that for grids, except that the theory is body-oriented instead of member-oriented. Also, the sequence of nodal displacements for the plate-bending element is not the same as for the grid element. As shown in Eq. (8.2–1), the $z$-translation $w_i$ is taken before the $x$- and $y$-rotations $\theta_{xi}$ and $\theta_{yi}$.

Three-dimensional rigid bodies connected by networks of general shell elements have characteristics similar to their counterparts in space frames. However, the rotations in the $x$, $y$, and $z$ senses at an attached node are actually dependent components of the independent tangential rotations $\alpha_i$ and $\beta_i$ (see Fig. 8.6).

Finally, rigid rings in axisymmetric solids or shells with axisymmetric loads would be restrained to translate only in the $z$ direction. If loads are nonaxisymmetric, such rings can also translate and rotate in the $r$ and $\theta$ directions and rotate in the $z$ sense.

## 9.6 PROGRAM DYRBPB FOR RIGID BODIES IN PLATE-BENDING CONTINUA

To illustrate the nature of programming for the body-oriented approach, we now discuss Program DYRBPB for dynamic analysis of rigid bodies supported by plates in bending. For this purpose, the structural and dynamic-load data from

### Sec. 9.6 Program DYRBPB for Rigid Bodies in Plate-Bending Continua

Sec. 8.3 (see Tables 8.1 and 8.2) are needed to form action, stiffness, and consistent-mass matrices for all nodes not connected to supports. We also introduce information about the rigid bodies that enables the computer to distinguish attached nodes of type $A$ from free nodes of type $F$. Then the desired matrices may be generated in the partitioned form shown by Eq. (9.5-1).

As mentioned previously, the sequence for numbering displacements at a node of the PBQ8 element is $z$-translation, $x$-rotation, and $y$-rotation. We also use the same sequence of displacements at the mass center $c_k$ of the $k$th rigid body in the analytical model. Therefore, the transformation matrix for corresponding actions at point $c_k$ due to unit actions at node $j$ becomes

$$\mathbf{T}_{kj} = \begin{bmatrix} 1 & 0 & 0 \\ y_{kj} & 1 & 0 \\ -x_{kj} & 0 & 1 \end{bmatrix} \tag{1}$$

This array is taken from the third, fourth, and fifth rows and columns of the last matrix in Table 9.1, with $p$ replaced by $k$.

Similarly, the mass-inertia matrix for rigid body $k$ in a discretized plate is

$$\mathbf{M}_{c_k} = \begin{bmatrix} m & 0 & 0 \\ 0 & I_{xx} & -I_{xy} \\ 0 & -I_{yx} & I_{yy} \end{bmatrix}_k \tag{2}$$

which is drawn from the third, fourth, and fifth rows and columns of the last matrix in Table 9.2.

Supplementary rigid-body data required for Program DYRBPB appear in Table 9.3, which conveys the number of bodies NB in part (a). Then in part (b) we have NB lines of body-node data containing the body number K, the number of nodes NJ(K) attached to the body, and the body-node numbers JB(K, 1) through JB(K, NJ(K)). Properties of the rigid bodies are given in part (c), which indicates NB lines with the body number K, followed by values of the following six terms:

$XC(K) = x$-coordinate of point $c_k$
$YC(K) = y$-coordinate of point $c_k$
$BM(K) = $ mass of body $k$
$XXI(K) = $ mass moment of inertia $I_{xxk}$
$XYI(K) = $ mass product of inertia $I_{xyk}$
$YYI(K) = $ mass moment of inertia $I_{yyk}$

We also need supplementary dynamic-load data for Program DYRBPB, as shown in Table 9.4. Information contained in this table consists of initial conditions and applied actions for the rigid bodies. In part (a) of initial conditions we have the number of bodies with initial displacements NBID and the number of bodies with initial velocities NBIV. In part (b) are NBID lines with the body number K and three possible initial displacements DB0(3K−2), DB0(3K−1), and

**TABLE 9.3  Rigid-Body Data for Program DYRBPB[a]**

| Type of Data | No. of Lines | Items on Data Lines |
|---|---|---|
| Rigid-body data | | |
| (a) Number of bodies | 1 | NB |
| (b) Body nodes | NB | K, NJ(K), JB(K, 1), JB(K, 2), ... , JB(K, NJ(K)) |
| (c) Body properties | NB | K, XC(K), YC(K), BM(K), XXI(K), XYI(K), YYI(K) |

[a] Supplements (and follows) structural data in Table 8.1.

**TABLE 9.4  Rigid-Body Dynamic-Load Data for Program DYRBPB[a]**

| Type of Data | No. of Lines | Items on Data Lines |
|---|---|---|
| Initial conditions | | |
| (a) Condition parameters | 1 | NBID, NBIV |
| (b) Displacements | NBID | K, DB0(3K-2), DB0(3K-1), DB0(3K) |
| (c) Velocities | NBIV | K, VB0(3K-2), VB0(3K-1), VB0(3K) |
| Applied actions | | |
| (a) Load parameter | 1 | NLB |
| (b) Rigid-body loads | NLB | K, AB(3K-2), AB(3K-1), AB(3K) |

[a] Supplements (and precedes) dynamic-load data in Table 8.2.

DB0(3K). Also, in part (c) we see NBIV lines containing K and three possible initial velocities VB0(3K-2), VB0(3K-1), and VB0(3K). Because these initial conditions are given for rigid bodies, their effects on attached nodes must be computed within the program and are not included as dynamic-load data for the structure.

Actions applied directly to the rigid bodies are listed as the second type of data in Table 9.4. The only parameter required in part (a) is the number of loaded bodies NLB. In part (b) the data for rigid-body loads consist of the body number K and applied actions AB(3K-2), AB(3K-1), and AB(3K). This force and the two moment components are assumed to act at the mass center $c_k$, as does the inertial force in the $z$ direction caused by ground acceleration. Of course, actions at any other points on a body could also be handled, but they would require data for locations as well as magnitudes.

As in Program DYNAPB, the rotational displacements at free nodes are eliminated by Guyan reduction. After solving the equations of motion for rigid bodies and free nodes, we can find time histories of displacements for attached nodes with Eq. (9.5-13). Then the displacements in vectors $\mathbf{D}_A$ and $\mathbf{D}_F$ may be used to obtain time-varying stresses in the finite elements.

### Example 9.3

Figure 9.11 depicts half of a symmetric structure, consisting of a quarter of a solid circular disk supported by a plate in bending. The plate is divided into two PBQ8

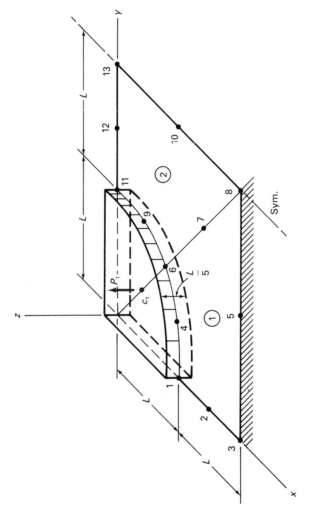

**Figure 9.11** Quarter of solid disk supported by PBQ8 elements.

elements that are fixed along edge 3–5–8 and symmetric with respect to a plane containing edge 8–10–13. A step force of magnitude $P_1 = 0.1$ kN acts in the $z$ direction at the mass center of the first rigid body, which is labeled point $c_1$ in the figure. Physical properties of the plate are

$$E = 103 \times 10^6 \text{ kN/m}^2 \qquad v = 0.34 \qquad \rho = 8.66 \text{ Mg/m}^3$$

$$L = 0.05 \text{ m} \qquad h = 0.002 \text{ m}$$

and both the plate and the solid are made of brass. The rigid body is attached to nodes 1, 4, 6, 9, and 11, and its properties are as follows:

$$R_1 = L = 0.05 \text{ m} \qquad h_1 = \frac{L}{5} = 0.01 \text{ m}$$

$$x_{c_1} = y_{c_1} = \frac{4R_1}{3\pi} = 0.02122 \text{ m} \qquad m_1 = \rho h_1 \frac{\pi R_1^2}{4} = 1.700 \times 10^{-4} \text{ Mg}$$

$$I_{xx} = I_{yy} = m_1 \left[ R_1^2 \left( \frac{1}{4} - \frac{16}{9\pi^2} \right) + \frac{h_1^2}{12} \right] = 3.112 \times 10^{-8} \text{ Mg} \cdot \text{m}^2$$

$$I_{xy} = m_1 \frac{R_1^2}{\pi} \left( \frac{1}{2} - \frac{16}{9\pi} \right) = -8.913 \times 10^{-9} \text{ Mg} \cdot \text{m}^2$$

where the symbols $R_1$, $h_1$, and $m_1$ denote the radius, thickness, and mass of the quarter disk.

We used this data in Program DYRBPB with DAMPR = 0.05 and direct numerical integration by Subprogram NUMINT. For this purpose, half the load was applied in each of two cases to find symmetric and antisymmetric responses. For the symmetric case, restraints against $x$-rotations are required at nodes 10 and 13. However, for the antisymmetric case, we need restraints against $z$-translations and $y$-rotations at those nodes. Figure 9.12(a) shows computer plots of the resulting $z$-translations at point $c_1$ on the rigid body for the symmetric and antisymmetric cases. In part (b) of the figure, we also see the $z$-translations of node 1 for the two cases. Finally, in part (c) are time histories of the flexural stress SX at the integration point near node 3. Of course, the two curves in each of the plots must be added to obtain total values for the responses given.

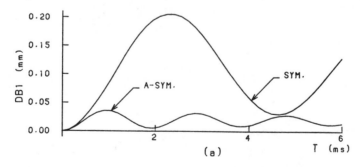

**Figure 9.12** Rigid body in a plate: (a) body translations; (b) nodal translations; (c) flexural stresses.

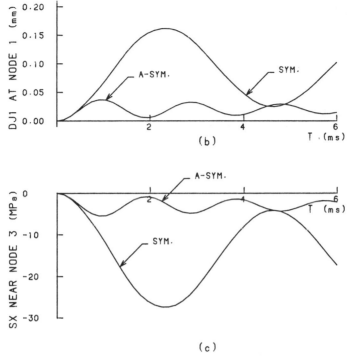

**Figure 9.12** (cont.)

## REFERENCES

1. Weaver, W., Jr., "Dynamics of Elastically-Connected Rigid Bodies," in *Developments in Theoretical and Applied Mechanics,* Vol. 3, ed. W. A. Shaw, Pergamon Press, New York, 1967, pp. 543–562.
2. Weaver, W., Jr., and Gere, J. M., *Matrix Analysis of Framed Structures,* 2nd. ed., Van Nostrand Reinhold, New York, 1980.
3. Weaver, W., Jr., and Johnston, P. R., *Finite Elements for Structural Analysis,* Prentice-Hall, Englewood Cliffs, N.J., 1984.
4. Beer, F. P., and Johnston, E. R., Jr., *Vector Mechanics for Engineers: Dynamics,* 4th ed., McGraw-Hill, New York, 1984.
5. Weaver, W., Jr., and Nelson, M. F., "Three-Dimensional Analysis of Tier Buildings," *ASCE J. Struct. Div.,* Vol. 92, No. ST6, 1966, pp. 385–404.
6. Weaver, W., Jr., Nelson, M. F., and Manning, T. A., "Dynamics of Tier Buildings," *ASCE J. Eng. Mech. Div.,* Vol. 94, No. EM6, 1968, pp. 1455–1474.

# 10

# Substructure Methods

## 10.1 INTRODUCTION

When the number of degrees of freedom in a structure becomes very large, we need to divide the analytical model into substructures. Figure 10.1 illustrates such a case, which is a computer plot of a large radio telescope antenna designed as a space truss. This type of structure consists of a reflector and a support structure. We can take advantage of the facts that the reflector has two planes of symmetry and the support has one such plane. Therefore, we need analyze only a quarter of the former and half of the latter, using appropriate restraints on the planes of symmetry. Also, the more complicated reflector can be divided into substructures and analyzed by one of the methods in this chapter.

For analysis by substructures, we must distinguish techniques that are suitable for statics from those more conducive to dynamics. In static analysis, all nodal displacements for a substructure can be eliminated from equilibrium equations during a frontal reduction procedure [1, 2]. However, in dynamic analysis, we need to retain a certain number of scattered degrees of freedom with low stiffnesses and high inertial actions, for which approximate equations of motion can be written. Three approaches that work well for dynamics are the tridiagonal method (with modifications), the parallel-elimination method, and the component-mode technique. Each of these substructure methods will be described in the ensuing sections of this chapter.

Sec. 10.2  Guyan Reduction Methods

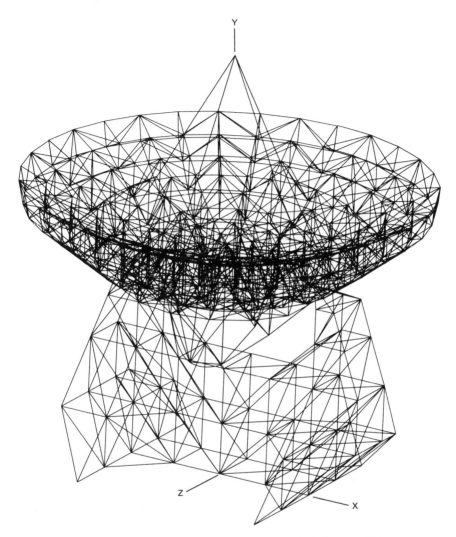

**Figure 10.1**  Radio telescope antenna structure (pointing to zenith).

## 10.2 GUYAN REDUCTION METHODS

### Tridiagonal Method for Substructures in Series

Figure 10.2 shows a two-dimensional discretized continuum divided into substructures that are connected in series. Substructure numbers appear in boxes below the figure. Here the symbol $\ell$ denotes a typical substructure, and $n_s$ is the

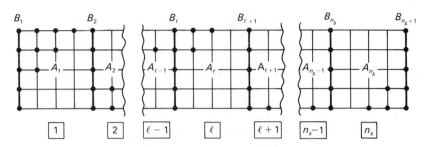

**Figure 10.2** Substructures in series.

number of substructures. For each substructure we have *interior nodes* of type $A$ (or $A_\ell$) and insulating *boundary nodes* of types $B_\ell$ and $B_{\ell+1}$ on each side. The symbol $B_\ell$ indicates boundary nodes common to substructures $\ell - 1$ and $\ell$, while $B_{\ell+1}$ represents those joining substructures $\ell$ and $\ell + 1$. Let us form the column vector of nodal displacements as

$$\mathbf{D} = \left\{ \mathbf{D}_{B_1}, \mathbf{D}_{A_1}, \mathbf{D}_{B_2}, \mathbf{D}_{A_2}, \ldots, \mathbf{D}_{A_{n_s}}, \mathbf{D}_{B_{n_s+1}} \right\} \tag{1}$$

With this sequence, the stiffness matrix for all nodal degrees of freedom becomes

$$\mathbf{S} = \begin{bmatrix} \mathbf{S}_{B_1 B_1} & \mathbf{S}_{B_1 A_1} & & & & & \\ \mathbf{S}_{A_1 B_1} & \mathbf{S}_{A_1 A_1} & \mathbf{S}_{A_1 B_2} & & & \mathbf{0} & \\ & \mathbf{S}_{B_2 A_1} & \mathbf{S}_{B_2 B_2} & \mathbf{S}_{B_2 A_2} & & & \\ & & \mathbf{S}_{A_2 B_2} & \mathbf{S}_{A_2 A_2} & \cdots & & \\ & & & \cdots & \cdots & \cdots & \\ & \mathbf{0} & & & \cdots & \mathbf{S}_{A_{n_s} A_{n_s}} & \mathbf{S}_{A_{n_s} B_{n_s+1}} \\ & & & & & \mathbf{S}_{B_{n_s+1} A_{n_s}} & \mathbf{S}_{B_{n_s+1} B_{n_s+1}} \end{bmatrix} \tag{2}$$

Because matrix $\mathbf{S}$ is a tridiagonal array of submatrices, this approach is referred to as the tridiagonal method. Without writing it, we can also observe that the consistent mass matrix has the same tridiagonal form.

Substructure $\ell$ contributes terms to the equations of motion as follows:

$$\mathbf{M}_\ell \ddot{\mathbf{D}}_\ell + \mathbf{S}_\ell \mathbf{D}_\ell = \mathbf{A}_\ell(t) \tag{3}$$

In expanded form, this equation is

$$\begin{bmatrix} \mathbf{M}_{B_\ell B_\ell} & \mathbf{M}_{B_\ell A_\ell} & \mathbf{0} \\ \mathbf{M}_{A_\ell B_\ell} & \mathbf{M}_{A_\ell A_\ell} & \mathbf{M}_{A_\ell B_{\ell+1}} \\ \mathbf{0} & \mathbf{M}_{B_{\ell+1} A_\ell} & \mathbf{M}_{B_{\ell+1} B_{\ell+1}} \end{bmatrix} \begin{bmatrix} \ddot{\mathbf{D}}_{B_\ell} \\ \ddot{\mathbf{D}}_{A_\ell} \\ \ddot{\mathbf{D}}_{B_{\ell+1}} \end{bmatrix}$$

## Sec. 10.2  Guyan Reduction Methods

$$+ \begin{bmatrix} \mathbf{S}_{B_\ell B_\ell} & \mathbf{S}_{B_\ell A_\ell} & 0 \\ \mathbf{S}_{A_\ell B_\ell} & \mathbf{S}_{A_\ell A_\ell} & \mathbf{S}_{A_\ell B_{\ell+1}} \\ 0 & \mathbf{S}_{B_{\ell+1} A_\ell} & \mathbf{S}_{B_{\ell+1} B_{\ell+1}} \end{bmatrix} \begin{bmatrix} \mathbf{D}_{B_\ell} \\ \mathbf{D}_{A_\ell} \\ \mathbf{D}_{B_{\ell+1}} \end{bmatrix} = \begin{bmatrix} \mathbf{A}_{B_\ell} \\ \mathbf{A}_{A_\ell} \\ \mathbf{A}_{B_{\ell+1}} \end{bmatrix} \quad (4)$$

Now we will apply Guyan reduction (see Sec. 6.7) to decrease the number of degrees of freedom in the $\ell$th substructure. For this purpose, let the displacements of type $A_\ell$ be dependent upon those of types $B_\ell$ and $B_{\ell+1}$. Thus,

$$\mathbf{D}_{A_\ell} = \mathbf{T}_{A_\ell B_\ell} \mathbf{D}_{B_\ell} + \mathbf{T}_{A_\ell B_{\ell+1}} \mathbf{D}_{B_{\ell+1}} \quad (5a)$$

in which

$$\mathbf{T}_{A_\ell B_\ell} = -\mathbf{S}_{A_\ell A_\ell}^{-1} \mathbf{S}_{A_\ell B_\ell} \qquad \mathbf{T}_{A_\ell B_{\ell+1}} = -\mathbf{S}_{A_\ell A_\ell}^{-1} \mathbf{S}_{A_\ell B_{\ell+1}} \quad (5b)$$

Equations (5b) are of the same form as Eq. (6.7-7b). The acceleration relationship similar to Eq. (5a) is

$$\ddot{\mathbf{D}}_{A_\ell} = \mathbf{T}_{A_\ell B_\ell} \ddot{\mathbf{D}}_{B_\ell} + \mathbf{T}_{A_\ell B_{\ell+1}} \ddot{\mathbf{D}}_{B_{\ell+1}} \quad (5c)$$

To reduce the equations of motion in Eq. (3) to a smaller set, we construct the transformation matrix

$$\mathbf{T}_\ell = \begin{bmatrix} \mathbf{I}_{B_\ell} & 0 \\ \mathbf{T}_{A_\ell B_\ell} & \mathbf{T}_{A_\ell B_{\ell+1}} \\ 0 & \mathbf{I}_{B_{\ell+1}} \end{bmatrix} \quad (6)$$

Then we have

$$\mathbf{D}_\ell = \begin{bmatrix} \mathbf{D}_{B_\ell} \\ \mathbf{D}_{A_\ell} \\ \mathbf{D}_{B_{\ell+1}} \end{bmatrix} = \mathbf{T}_\ell \begin{bmatrix} \mathbf{D}_{B_\ell} \\ \mathbf{D}_{B_{\ell+1}} \end{bmatrix} \quad (7a)$$

and

$$\ddot{\mathbf{D}}_\ell = \begin{bmatrix} \ddot{\mathbf{D}}_{B_\ell} \\ \ddot{\mathbf{D}}_{A_\ell} \\ \ddot{\mathbf{D}}_{B_{\ell+1}} \end{bmatrix} = \mathbf{T}_\ell \begin{bmatrix} \ddot{\mathbf{D}}_{B_\ell} \\ \ddot{\mathbf{D}}_{B_{\ell+1}} \end{bmatrix} \quad (7b)$$

Substitution of Eqs. (7) into Eq. (3) and premultiplication of the result by $\mathbf{T}_\ell^T$ produces

$$\mathbf{M}_\ell^* \begin{bmatrix} \ddot{\mathbf{D}}_{B_\ell} \\ \ddot{\mathbf{D}}_{B_{\ell+1}} \end{bmatrix} + \mathbf{S}_\ell^* \begin{bmatrix} \mathbf{D}_{B_\ell} \\ \mathbf{D}_{B_{\ell+1}} \end{bmatrix} = \mathbf{A}_\ell^*(t) \quad (8)$$

In this equation the reduced matrices are

$$\mathbf{S}_\ell^* = \mathbf{T}_\ell^T \mathbf{S}_\ell \mathbf{T}_\ell = \begin{bmatrix} \mathbf{S}_{B_\ell B_\ell}^* & \mathbf{S}_{B_\ell B_{\ell+1}}^* \\ \mathbf{S}_{B_{\ell+1} B_\ell}^* & \mathbf{S}_{B_{\ell+1} B_{\ell+1}}^* \end{bmatrix} \quad (9a)$$

$$\mathbf{M}_\ell^* = \mathbf{T}_\ell^T \mathbf{M}_\ell \mathbf{T}_\ell = \begin{bmatrix} \mathbf{M}_{B_\ell B_\ell}^* & \mathbf{M}_{B_\ell B_{\ell+1}}^* \\ \mathbf{M}_{B_{\ell+1} B_\ell}^* & \mathbf{M}_{B_{\ell+1} B_{\ell+1}}^* \end{bmatrix} \quad (9b)$$

$$\mathbf{A}_\ell^*(t) = \mathbf{T}_\ell^T \mathbf{A}_\ell(t) = \begin{bmatrix} \mathbf{A}_{B_\ell}^* \\ \mathbf{A}_{B_{\ell+1}}^* \end{bmatrix} \quad (9c)$$

Modified submatrices in Eqs. (9) have the definitions

$$\mathbf{S}_{B_\ell B_\ell}^* = \mathbf{S}_{B_\ell B_\ell} + \mathbf{S}_{B_\ell A_\ell} \mathbf{T}_{A_\ell B_\ell} \quad (10a)$$

$$\mathbf{M}_{B_\ell B_\ell}^* = \mathbf{M}_{B_\ell B_\ell} + \mathbf{T}_{A_\ell B_\ell}^T \mathbf{M}_{A_\ell B_\ell} + \mathbf{M}_{B_\ell A_\ell} \mathbf{T}_{A_\ell B_\ell} + \mathbf{T}_{A_\ell B_\ell}^T \mathbf{M}_{A_\ell A_\ell} \mathbf{T}_{A_\ell B_\ell} \quad (10b)$$

$$\mathbf{A}_{B_\ell}^* = \mathbf{A}_{B_\ell} + \mathbf{T}_{A_\ell B_\ell}^T \mathbf{A}_{A_\ell} \quad (10c)$$

and so on.

Finally, the reduced equations of motion for all of the boundary nodes take the form

$$\mathbf{M}_{BB}^* \ddot{\mathbf{D}}_B + \mathbf{S}_{BB}^* \mathbf{D}_B = \mathbf{A}_B^*(t) \quad (11)$$

Substructures contribute to the matrices in Eq. (11) by the usual direct assembly process. The reduced stiffness matrix is still a tridiagonal array of submatrices, as follows:

$$\mathbf{S}_{BB}^* = \begin{bmatrix} \mathbf{S}_{B_1 B_1}^* & \mathbf{S}_{B_1 B_2}^* & & & & \mathbf{0} \\ \mathbf{S}_{B_2 B_1}^* & \mathbf{S}_{B_2 B_2}^{**} & \mathbf{S}_{B_2 B_3}^* & & & \\ & \mathbf{S}_{B_3 B_2}^* & \mathbf{S}_{B_3 B_3}^{**} & \cdots & & \\ & & \cdots & \cdots & \cdots & \\ & & & \cdots & \mathbf{S}_{B_{n_s} B_{n_s}}^{**} & \mathbf{S}_{B_{n_s} B_{n_s+1}}^* \\ \mathbf{0} & & & & \mathbf{S}_{B_{n_s+1} B_{n_s}}^* & \mathbf{S}_{B_{n_s+1} B_{n_s+1}}^* \end{bmatrix} \quad (12)$$

Submatrices of $\mathbf{S}_{BB}^*$ having single asterisks receive contributions from only one substructure, but those with double asterisks have two contributing substructures. The assembled mass matrix $\mathbf{M}_{BB}^*$ in Eq. (11) has the same tridiagonal arrangement of submatrices as the assembled stiffness matrix in Eq. (12). Similarly, the assembled load vector is

$$\mathbf{A}_B^*(t) = \left\{ \mathbf{A}_{B_1}^*, \mathbf{A}_{B_2}^{**}, \mathbf{A}_{B_3}^{**}, \ldots, \mathbf{A}_{B_{n_s}}^{**}, \mathbf{A}_{B_{n_s+1}}^* \right\} \quad (13)$$

which again has single and double contributions. For convenience in computer programming, we assemble the reduced matrices while proceeding from one substructure to the next, thereby increasing the number of type $B$ displacements.

Because the assembled stiffness and mass matrices both have tridiagonal forms, vibrational and dynamic analyses can be more efficient [3] than for other methods where the matrices are filled. The following example demonstrates application of this approach to a beam type of structure.

Sec. 10.2  Guyan Reduction Methods                                                              **449**

**Example 10.1**
The unrestrained beam in Fig. 10.3 is divided into two substructures, each consisting of two equal prismatic flexural elements. Let us determine the coefficient matrices $\mathbf{S}_{BB}^*$ and $\mathbf{M}_{BB}^*$ by the tridiagonal method, as required in Eq. (11).

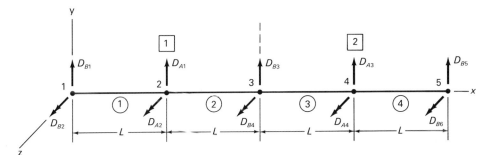

**Figure 10.3** Beam divided into substructures.

For substructure 1, the stiffness matrix in Eqs. (3) and (4) becomes

$$\mathbf{S}_1 = \frac{2EI}{L^3} \begin{bmatrix} 6 & & & & & \text{Sym.} \\ 3L & 2L^2 & & & & \\ \hline -6 & -3L & 12 & & & \\ 3L & L^2 & 0 & 4L^2 & & \\ \hline 0 & 0 & -6 & -3L & 6 & \\ 0 & 0 & 3L & L^2 & -3L & 2L^2 \end{bmatrix} \begin{matrix} B_1 \\ \\ A_1 \\ \\ B_2 \\ \end{matrix} \quad (a)$$

$$\phantom{\mathbf{S}_1 = \frac{2EI}{L^3}} \quad B_1 \qquad\quad A_1 \qquad\quad B_2$$

Terms in this array are drawn from Eq. (3.4-24). Similarly, the consistent-mass matrix for substructure 1 is

$$\mathbf{M}_1 = \frac{\rho A L}{420} \begin{bmatrix} 156 & & & & & \text{Sym.} \\ 22L & 4L^2 & & & & \\ \hline 54 & 13L & 312 & & & \\ -13L & -3L^2 & 0 & 8L^2 & & \\ \hline 0 & 0 & 54 & 13L & 156 & \\ 0 & 0 & -13L & -3L^2 & -22L & 4L^2 \end{bmatrix} \begin{matrix} B_1 \\ \\ A_1 \\ \\ B_2 \\ \end{matrix} \quad (b)$$

$$\phantom{\mathbf{M}_1 = \frac{\rho A L}{420}} \quad B_1 \qquad\quad A_1 \qquad\quad B_2$$

in which the terms are obtained from Eq. (3.4-26).

Next, we use Eqs. (5b) to form $\mathbf{T}_{A_\ell B_\ell}$ and $\mathbf{T}_{A_\ell B_{\ell+1}}$ for substructure 1, as follows:

$$\mathbf{T}_{A_1 B_1} = -\mathbf{S}_{A_1 A_1}^{-1} \mathbf{S}_{A_1 B_1} = \frac{1}{4L} \begin{bmatrix} 2L & L^2 \\ -3 & -L \end{bmatrix} \quad (c)$$

$$\mathbf{T}_{A_1B_2} = -\mathbf{S}_{A_1A_1}^{-1}\mathbf{S}_{A_1B_2} = \frac{1}{4L}\begin{bmatrix} 2L & -L^2 \\ 3 & -L \end{bmatrix} \quad (d)$$

With these submatrices, the transformation operator in Eq. (6) may be written as

$$\mathbf{T}_1 = \begin{bmatrix} \mathbf{I}_{B_1} & \mathbf{0} \\ \mathbf{T}_{A_1B_1} & \mathbf{T}_{A_1B_2} \\ \mathbf{0} & \mathbf{I}_{B_2} \end{bmatrix} = \frac{1}{4L}\begin{bmatrix} 4L & 0 & 0 & 0 \\ 0 & 4L & 0 & 0 \\ 2L & L^2 & 2L & -L^2 \\ -3 & -L & 3 & -L \\ 0 & 0 & 4L & 0 \\ 0 & 0 & 0 & 4L \end{bmatrix}\begin{matrix} B_1 \\ \\ A_1 \\ \\ B_2 \\ \end{matrix} \quad (e)$$

$$\phantom{xxxxxxxxxxxxxxxxxxxxxxxxxx} B_1 \phantom{xxxxxx} B_2$$

Then the reduced stiffness matrix becomes

$$\mathbf{S}_1^* = \mathbf{T}_1^T\mathbf{S}_1\mathbf{T}_1 = \frac{EI}{2L^3}\begin{bmatrix} 3 & & & & \text{Sym.} \\ 3L & 4L^2 & & & \\ -3 & -3L & 3 & & \\ 3L & 2L^2 & -3L & 4L^2 \end{bmatrix}\begin{matrix} B_1 \\ \\ B_2 \end{matrix} \quad (f)$$

$$\phantom{xxxxxxxxxxx} B_1 \phantom{xxxxxxxx} B_2$$

as shown by Eq. (9a). Also, the reduced mass matrix given by Eq. (9b) is

$$\mathbf{M}_1^* = \mathbf{T}_1^T\mathbf{M}_1\mathbf{T}_1 = \frac{\rho AL}{105}\begin{bmatrix} 78 & & & & \text{Sym.} \\ 22L & 8L^2 & & & \\ 27 & 13L & 78 & & \\ -13L & -6L^2 & -22L & 8L^2 \end{bmatrix}\begin{matrix} B_1 \\ \\ B_2 \end{matrix} \quad (g)$$

$$\phantom{xxxxxxxxxxx} B_1 \phantom{xxxxxxxx} B_2$$

For substructure 2, we can proceed in a similar manner and find that

$$\mathbf{S}_2^* = \mathbf{S}_1^* \qquad \mathbf{M}_2^* = \mathbf{M}_1^* \quad (h)$$

Consequently, assembling the reduced stiffness and mass matrices from substructures 1 and 2 yields

$$\mathbf{S}_{BB}^* = \frac{EI}{2L^3}\begin{bmatrix} 3 & & \boxed{1} & & & \text{Sym.} \\ 3L & 4L^2 & & & & \\ -3 & -3L & 6 & & & \\ 3L & 2L^2 & 0 & 8L^2 & \boxed{2} & \\ 0 & 0 & -3 & -3L & 3 & \\ 0 & 0 & 3L & 2L^2 & -3L & 4L^2 \end{bmatrix}\begin{matrix} B_1 \\ \\ B_2 \\ \\ B_3 \end{matrix} \quad (i)$$

$$\phantom{xxxxxxxxx} B_1 \phantom{xxxxxx} B_2 \phantom{xxxxxx} B_3$$

Sec. 10.2  Guyan Reduction Methods    451

and

$$\mathbf{M}_{BB}^* = \frac{\rho A L}{105} \begin{bmatrix} 78 & & & & & \text{Sym.} \\ 22L & 8L^2 & & \boxed{1} & & \\ 27 & 13L & 156 & & & \\ -13L & -6L^2 & 0 & 16L^2 & \boxed{2} & \\ 0 & 0 & 27 & 13L & 78 & \\ 0 & 0 & -13L & -6L^2 & -22L & 8L^2 \end{bmatrix} \begin{matrix} B_1 \\ \\ B_2 \\ \\ B_3 \\ \end{matrix}$$

$$\quad\quad\quad\quad\quad\quad B_1 \quad\quad\quad B_2 \quad\quad\quad B_3$$

(j)

which may now be used in Eq. (11).

### Modified Tridiagonal Method

Now we consider a modification of the tridiagonal method for substructures in series that abandons the idea of insulating boundary nodes. Instead, there are three definitions for types of displacements, as follows: (1) Subscript $A_\ell$ implies dependent displacements in substructure $\ell$; (2) subscript $B_\ell$ represents dependent displacements at boundary nodes joining substructures $\ell$ and $\ell+1$; and (3) subscript $C_\ell$ denotes independent displacements associated with substructures 1 through $\ell$, including boundary nodes common to $\ell$ and $\ell+1$. The dependent displacements of types $A$ and $B$ are to be eliminated, but the independent displacements of type $C$ will be retained. Note that the number of type $C$ displacements increases for each consecutive substructure. Also, they may be scattered within the substructures as well as at boundary nodes.

At each stage in the reduction process, the partially-formed equations of motion are symbolically the same as in Eq. (3). Thus,

$$\mathbf{M}_\ell \ddot{\mathbf{D}}_\ell + \mathbf{S}_\ell \mathbf{D}_\ell = \mathbf{A}_\ell(t) \tag{14}$$

However, the expanded form is now

$$\begin{bmatrix} \mathbf{M}_{AA} & \mathbf{M}_{AB} & \mathbf{M}_{AC} \\ \mathbf{M}_{BA} & \mathbf{M}_{BB} & \mathbf{M}_{BC} \\ \mathbf{M}_{CA} & \mathbf{M}_{CB} & \mathbf{M}_{CC} \end{bmatrix} \begin{bmatrix} \ddot{\mathbf{D}}_A \\ \ddot{\mathbf{D}}_B \\ \ddot{\mathbf{D}}_C \end{bmatrix}_\ell + \begin{bmatrix} \mathbf{S}_{AA} & \mathbf{S}_{AB} & \mathbf{S}_{AC} \\ \mathbf{S}_{BA} & \mathbf{S}_{BB} & \mathbf{S}_{BC} \\ \mathbf{S}_{CA} & \mathbf{S}_{CB} & \mathbf{S}_{CC} \end{bmatrix} \begin{bmatrix} \mathbf{D}_A \\ \mathbf{D}_B \\ \mathbf{D}_C \end{bmatrix}_\ell = \begin{bmatrix} \mathbf{A}_A \\ \mathbf{A}_B \\ \mathbf{A}_C \end{bmatrix}_\ell \tag{15}$$

Let the displacements of types $A_\ell$ be dependent on those of types $B_\ell$ and $C_\ell$, as follows:

$$\mathbf{D}_{A_\ell} = \mathbf{T}_{A_\ell B_\ell} \mathbf{D}_{B_\ell} + \mathbf{T}_{A_\ell C_\ell} \mathbf{D}_{C_\ell} \tag{16a}$$

where

$$\mathbf{T}_{A_\ell B_\ell} = -\mathbf{S}_{A_\ell A_\ell}^{-1} \mathbf{S}_{A_\ell B_\ell} \quad\quad \mathbf{T}_{A_\ell C_\ell} = -\mathbf{S}_{A_\ell A_\ell}^{-1} \mathbf{S}_{A_\ell C_\ell} \tag{16b}$$

Also, we have the acceleration relationship

$$\ddot{\mathbf{D}}_{A_\ell} = \mathbf{T}_{A_\ell B_\ell} \ddot{\mathbf{D}}_{B_\ell} + \mathbf{T}_{A_\ell C_\ell} \ddot{\mathbf{D}}_{C_\ell} \tag{16c}$$

Then the transformation operator becomes

$$\mathbf{T}_\ell = \begin{bmatrix} \mathbf{T}_{AB} & \mathbf{T}_{AC} \\ \mathbf{I}_B & \mathbf{0} \\ \mathbf{0} & \mathbf{I}_C \end{bmatrix}_\ell \quad (17)$$

With this matrix, the displacement and acceleration transformations are

$$\mathbf{D}_\ell = \begin{bmatrix} \mathbf{D}_A \\ \mathbf{D}_B \\ \mathbf{D}_C \end{bmatrix}_\ell = \mathbf{T}_\ell \begin{bmatrix} \mathbf{D}_B \\ \mathbf{D}_C \end{bmatrix}_\ell \quad (18a)$$

and

$$\ddot{\mathbf{D}}_\ell = \begin{bmatrix} \ddot{\mathbf{D}}_A \\ \ddot{\mathbf{D}}_B \\ \ddot{\mathbf{D}}_C \end{bmatrix}_\ell = \mathbf{T}_\ell \begin{bmatrix} \ddot{\mathbf{D}}_B \\ \ddot{\mathbf{D}}_C \end{bmatrix}_\ell \quad (18b)$$

Substituting Eqs. (18) into Eq. (14) and premultiplication of the equation by $\mathbf{T}_\ell^T$ yields

$$\mathbf{M}_\ell^* \begin{bmatrix} \ddot{\mathbf{D}}_B \\ \ddot{\mathbf{D}}_C \end{bmatrix}_\ell + \mathbf{S}_\ell^* \begin{bmatrix} \mathbf{D}_B \\ \mathbf{D}_C \end{bmatrix}_\ell = \mathbf{A}_\ell^*(t) \quad (19)$$

The reduced matrices in this equation have the definitions

$$\mathbf{S}_\ell^* = \mathbf{T}_\ell^T \mathbf{S}_\ell \mathbf{T}_\ell = \begin{bmatrix} \mathbf{S}_{BB}^* & \mathbf{S}_{BC}^* \\ \mathbf{S}_{CB}^* & \mathbf{S}_{CC}^* \end{bmatrix}_\ell \quad (20a)$$

$$\mathbf{M}_\ell^* = \mathbf{T}_\ell^T \mathbf{M}_\ell \mathbf{T}_\ell = \begin{bmatrix} \mathbf{M}_{BB}^* & \mathbf{M}_{BC}^* \\ \mathbf{M}_{CB}^* & \mathbf{M}_{CC}^* \end{bmatrix}_\ell \quad (20b)$$

$$\mathbf{A}_\ell^*(t) = \mathbf{T}_\ell^T \mathbf{A}_\ell(t) = \begin{bmatrix} \mathbf{A}_B^* \\ \mathbf{A}_C^* \end{bmatrix}_\ell \quad (20c)$$

Modified submatrices in Eqs. (20) are the same as those given by Eqs. (10), except that subscript $C_\ell$ replaces $B_{\ell+1}$.

When proceeding from one substructure to the next in a series, displacements of type $B_\ell$ become a subset within those of type $A_{\ell+1}$. Also, for the last substructure, displacements of type $B_{n_s}$ are usually omitted. Thus, the final equations of motion become

$$[\mathbf{M}_{CC}^* \ddot{\mathbf{D}}_C + \mathbf{S}_{CC}^* \mathbf{D}_C = \mathbf{A}_C^*(t)]_{n_s} \quad (21)$$

which involves displacements of type $C$ for the whole structure.

In the modified tridiagonal method, elimination (or release) of displacements of both types $A$ and $B$ causes $\mathbf{S}_{CC}^*$ and $\mathbf{M}_{CC}^*$ in Eq. (21) to be filled. However, if lumped masses are used in conjunction with accelerations of type

Sec. 10.2  Guyan Reduction Methods          453

$C$, the mass matrix is diagonal, as in the next example. For such a case, transformation of the associated eigenvalue problem to standard, symmetric form should involve factoring the mass matrix instead of the stiffness matrix. Also, an efficient overlay technique for stiffness terms is demonstrated in the following example (see also Sec. 10.3).

**Example 10.2**

Figure 10.4 shows a rectangular plane frame divided into two substructures, each of which consists of a beam and the two columns below. For this structure we shall find the reduced stiffness matrix in Eq. (21) by the modified tridiagonal method, using the following assumptions. Tributary masses having values of $m$ and $2m$ are lumped at the two framing levels, as indicated in the figure. Flexural rigidities of beams 1 and 4 are equal to $2EI$ and $4EI$, while those for columns 2, 3, 5, and 6 are equal to $EI$. Members are all prismatic, and axial strains are to be neglected.

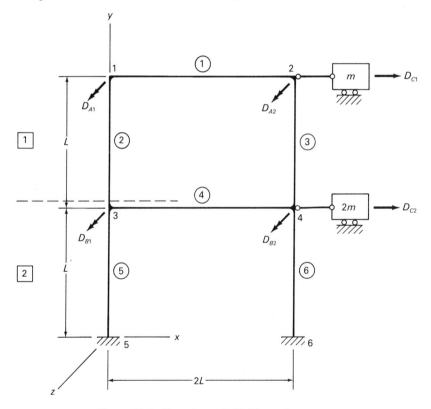

**Figure 10.4** Plane frame divided into substructures.

For substructure 1, we set up the 6 × 6 stiffness matrix in Eqs. (14) and (15), as follows:

$$\mathbf{S}_1 = \frac{2EI}{L^3} \begin{bmatrix} 4L^2 & & & & & & \text{Sym.} \\ L^2 & 4L^2 & & & & & \\ \hline L^2 & 0 & 2L^2 & & & & \\ 0 & L^2 & 0 & 2L^2 & & & \\ \hline 3L & 3L & 3L & 3L & 12 & & \\ -3L & -3L & -3L & -3L & -12 & 12 & \end{bmatrix} \begin{matrix} A_1 \\ \\ B_1 \\ \\ C_1 \\ \end{matrix} \quad \text{(k)}$$

$$\phantom{\mathbf{S}_1 = \frac{2EI}{L^3}} \quad\;\; A_1 \qquad\quad B_1 \qquad\quad C_1$$

which is partitioned according to displacements of types $A_1$, $B_1$, and $C_1$. Then Eqs. (10b) are used to form $\mathbf{T}_{A_\ell B_\ell}$ and $\mathbf{T}_{A_\ell C_\ell}$ for substructure 1. Hence,

$$\mathbf{T}_{A_1 B_1} = -\mathbf{S}_{A_1 A_1}^{-1} \mathbf{S}_{A_1 B_1} = \frac{1}{15} \begin{bmatrix} -4 & 1 \\ 1 & -4 \end{bmatrix} \quad (\ell)$$

$$\mathbf{T}_{A_1 C_1} = -\mathbf{S}_{A_1 A_1}^{-1} \mathbf{S}_{A_1 C_1} = \frac{3}{5L} \begin{bmatrix} -1 & 1 \\ -1 & 1 \end{bmatrix} \quad (m)$$

Using these submatrices, we compose the transformation operator in Eq. (17) to obtain

$$\mathbf{T}_1 = \begin{bmatrix} \mathbf{T}_{A_1 B_1} & \mathbf{T}_{A_1 C_1} \\ \mathbf{I}_{B_1} & \mathbf{0} \\ \mathbf{0} & \mathbf{I}_{C_1} \end{bmatrix} = \frac{1}{15L} \begin{bmatrix} -4L & L & -9 & 9 \\ L & -4L & -9 & 9 \\ \hline 15L & 0 & 0 & 0 \\ 0 & 15L & 0 & 0 \\ \hline 0 & 0 & 15L & 0 \\ 0 & 0 & 0 & 15L \end{bmatrix} \begin{matrix} A_1 \\ \\ B_1 \\ \\ C_1 \end{matrix} \quad (n)$$

$$\phantom{\mathbf{T}_1 = \;\;} B_1 \qquad\;\; C_1$$

Therefore, the reduced stiffness matrix $\mathbf{S}_1^*$ is

$$\mathbf{S}_1^* = \mathbf{T}_1^T \mathbf{S}_1 \mathbf{T}_1 = \frac{2EI}{15L^3} \begin{bmatrix} 26L^2 & & & \text{Sym.} \\ L^2 & 26L^2 & & \\ \hline 36L & 36L & 126 & \\ -36L & -36L & -126 & 126 \end{bmatrix} \begin{matrix} B_1 \\ \\ C_1 \end{matrix} \quad (o)$$

$$\phantom{\mathbf{S}_1^* = \mathbf{T}_1^T \mathbf{S}_1 \mathbf{T}_1 = \frac{2EI}{15L^3}} \; B_1 \qquad\;\; C_1$$

as given by Eq. (20a).

Proceeding from substructure 1 to substructure 2 (see Fig. 10.4), we redefine displacements of type $B$ in the former to become type $A$ in the latter. Additional contributions from substructure 2 to the stiffness matrix are:

$$\mathbf{S}_2 = \frac{2EI}{L^3} \begin{bmatrix} 6L^2 & & & \text{Sym.} \\ 2L^2 & 6L^2 & & \\ \hline 0 & 0 & 0 & \\ 3L & 3L & 0 & 12 \end{bmatrix} \begin{matrix} A_2 \\ \\ C_2 \end{matrix} \quad (p)$$

$$\phantom{\mathbf{S}_2 = \frac{2EI}{L^3}} A_2 \qquad\;\; C_2$$

Sec. 10.2  Guyan Reduction Methods                                                    455

In this instance, there are no displacements of type $B_2$ because the bases of columns are fixed. Also, displacements of type $C_2$ are the same as those of type $C_1$ in the first substructure. Terms in matrix $\mathbf{S}_1^*$ from Eq. (o) may be added to those in matrix $\mathbf{S}_2$ from Eq. (p) to produce

$$\hat{\mathbf{S}}_2 = \mathbf{S}_1^* + \mathbf{S}_2 = \frac{2EI}{15L^3} \begin{bmatrix} 116L^2 & & & \text{Sym.} & A_2 \\ 31L^2 & 116L^2 & & & \\ \hline 36L & 36L & 126 & & \\ 9L & 9L & -126 & 306 & C_2 \end{bmatrix} \quad (q)$$
$$\phantom{xxxxxxxxxxxxxxxxxxxxxxxx} A_2 \phantom{xxxx} C_2$$

This superposition of terms represents an *overlay technique* that will prove useful for computer programming in Sec. 10.3.

In preparation for elimination of type $A_2$ displacements, we determine $\mathbf{T}_{A_2C_2}$ as

$$\mathbf{T}_{A_2C_2} = -\hat{\mathbf{S}}_{A_2A_2}^{-1} \hat{\mathbf{S}}_{A_2C_2} = -\frac{1}{L}\begin{bmatrix} 0.2449 & 0.06122 \\ 0.2449 & 0.06122 \end{bmatrix} \quad (r)$$

Then the abbreviated transformation matrix $\mathbf{T}_2$ becomes

$$\mathbf{T}_2 = \begin{bmatrix} \mathbf{T}_{A_2C_2} \\ \mathbf{I}_{C_2} \end{bmatrix} = -\frac{1}{L}\begin{bmatrix} 0.2449 & 0.06122 \\ 0.2449 & 0.06122 \\ \hline -L & 0 \\ 0 & -L \end{bmatrix} \quad (s)$$

Now we can determine the final reduced stiffness matrix to be

$$\begin{bmatrix} \mathbf{S}_{CC}^* \end{bmatrix}_2 = \mathbf{S}_2^* = \mathbf{T}_2^T \hat{\mathbf{S}}_2 \mathbf{T}_2 = \frac{EI}{L^3}\begin{bmatrix} 14.45 & -17.39 \\ -17.39 & 40.65 \end{bmatrix} \quad (t)$$

This matrix contains actions of type $C$ due to unit displacements of type $C$, with those of types $A$ and $B$ eliminated (or released).

**Parallel Elimination Method**

A more general manner of substructuring does not produce coefficient matrices composed of submatrices in a tridiagonal pattern. For example, the two-dimensional discretized continuum in Fig. 10.5 is divided into four substructures with interior nodes of type $A_\ell$ and insulating boundary nodes of type $B_\ell$ having no special arrangement. In this approach the meaning of subscript $B_\ell$ is that it includes independent nodal displacements for substructures 1 through $\ell$. As in the tridiagonal method, the technique is to eliminate dependent displacements of type $A_\ell$ from each substructure and to retain a growing number of independent displacements of type $B_\ell$.

For each step in the reduction process, substructure $\ell$ contributes the following terms to the equations of motion:

$$\mathbf{M}_\ell \ddot{\mathbf{D}}_\ell + \mathbf{S}_\ell \mathbf{D}_\ell = \mathbf{A}_\ell(t) \quad (22)$$

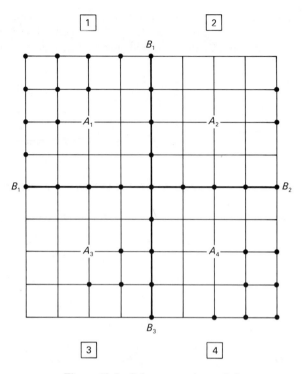

**Figure 10.5** Substructures in parallel.

This expression again contains the same symbols as Eq. (3), but its expanded version is

$$\begin{bmatrix} \mathbf{M}_{AA} & \mathbf{M}_{AB} \\ \mathbf{M}_{BA} & \mathbf{M}_{BB} \end{bmatrix} \begin{bmatrix} \ddot{\mathbf{D}}_A \\ \ddot{\mathbf{D}}_B \end{bmatrix}_\ell + \begin{bmatrix} \mathbf{S}_{AA} & \mathbf{S}_{AB} \\ \mathbf{S}_{BA} & \mathbf{S}_{BB} \end{bmatrix} \begin{bmatrix} \mathbf{D}_A \\ \mathbf{D}_B \end{bmatrix}_\ell = \begin{bmatrix} \mathbf{A}_A \\ \mathbf{A}_B \end{bmatrix}_\ell \quad (23)$$

which is different from Eq. (4). The dependence of type $A_\ell$ displacements upon type $B_\ell$ is written as

$$\mathbf{D}_{A\ell} = \mathbf{T}_{A\ell B\ell} \mathbf{D}_{B\ell} \quad (24)$$

Here the matrix $\mathbf{T}_{A\ell B\ell}$ has the same formula as the first of Eqs. (5b), but the meaning of $B_\ell$ is different. The required transformation operator is

$$\mathbf{T}_\ell = \begin{bmatrix} \mathbf{T}_{AB} \\ \mathbf{I}_B \end{bmatrix}_\ell \quad (25)$$

This has the appearance of Eq. (6.7-9), but it pertains to only one substructure. Now the displacement and acceleration transformations are

$$\mathbf{D}_\ell = \begin{bmatrix} \mathbf{D}_A \\ \mathbf{D}_B \end{bmatrix}_\ell = \mathbf{T}_\ell \mathbf{D}_{B\ell} \quad (26a)$$

and

$$\ddot{\mathbf{D}}_\ell = \begin{bmatrix} \ddot{\mathbf{D}}_A \\ \ddot{\mathbf{D}}_B \end{bmatrix}_\ell = \mathbf{T}_\ell \ddot{\mathbf{D}}_{B\ell} \qquad (26b)$$

Substitution of Eqs. (26) into Eq. (22) and premultiplication of the result by $\mathbf{T}_\ell^T$ gives

$$[\mathbf{M}_{BB}^* \ddot{\mathbf{D}}_B + \mathbf{S}_{BB}^* \mathbf{D}_B = \mathbf{A}_B^*(t)]_\ell \qquad (27)$$

The reduced matrices in this equation are symbolically the same as in Eqs. (10), but displacements of type $B$ are different.

As for the tridiagonal and modified tridiagonal methods, we assemble stiffness, mass, and load matrices in a direct fashion while proceeding from one substructure to the next. We could also devise a modified parallel elimination approach that abandons the notion of insulating boundary nodes and introduces retained displacements of type $C$. In that case, the displacements of types $A$ and $B$ would both become "slaves" to the "master" displacements of type $C$.

## 10.3 MODIFIED TRIDIAGONAL METHOD FOR MULTISTORY BUILDINGS

We now apply the modified tridiagonal method described in Sec. 10.2 to two- and three-dimensional multistory building frames. Most planar building frames can be handled in the high-speed core storage of a large-capacity digital computer without dividing them into substructures. However, we wish to use the same technique for both plane and space frames in order to take advantage of inherent similarities while explaining these two applications.

For a multistory rectangular plane frame, we take the analytical model illustrated in Fig. 10.6. The frame is assumed to have linearly elastic prismatic members, rigid joints, and fixed bases. We also assume that there are no shear walls, diagonal braces, or setbacks in the building. Each substructure $\ell$ in the figure consists of the beams at the framing level $A$ (above) and the columns between levels $A$ and $B$ (below). Displacements of type $A$ appear at the upper level, and those of type $B$ are at the lower level. From the joint displacements labeled in Fig. 10.6, we see that axial strains in the beams are to be omitted; whereas, those in columns are to be retained. The reason for keeping the latter strains is that their influences are known to be significant in analyses of tall buildings [4, 5]. Therefore, at each framing level there is only one translation $D_{F\ell}$ in the $x$ direction and its corresponding lateral force $A_{F\ell}$. Thus, the subscripts in the figure match those for the modified tridiagonal method in Sec. 10.2, except that $F$ replaces $C$. Note that the numerical subscripts on displacements of types $A$ and $B$ (left-to-right) are 1, 2, . . . , $2n_c$, where $n_c$ is the number of columns. Also, the subscripts on those of type $F$ (top-to-bottom) are 1, 2, . . . ,

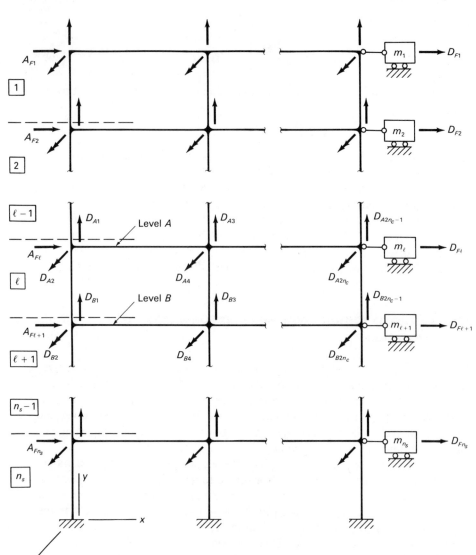

**Figure 10.6** Analytical model for Program DYMSPF.

$n_s$, where $n_s$ is the number of stories. Finally, this analytical model has tributary masses $m_1, m_2, \ldots, m_{n_s}$ lumped at the framing levels, which include contributions from beams and columns.

As for a tier building, every member in Fig. 10.6 is parallel to one of the structural axes; so no rotation-of-axes transformations are required. Figure 10.7(a) depicts a beam with its member axis $x_m$ parallel to the $x$ axis and its principal bending axis $z_m$ parallel to the $z$ axis. Action and displacement indexes

### Sec. 10.3 Modified Tridiagonal Method for Multistory Buildings

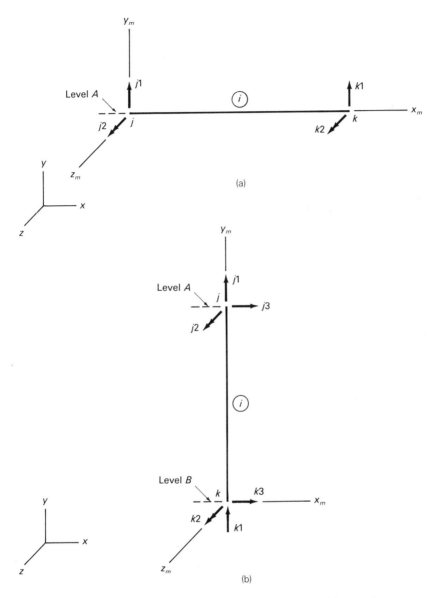

**Figure 10.7** (a) Beam parallel to $x$ axis; (b) column parallel to $y$ axis.

$j1, j2, k1$, and $k2$ imply that the $4 \times 4$ member stiffness matrix is the same as that in Eq. (3.4-24) for a flexural element.

On the other hand, Fig. 10.7(b) shows a column with its member axis $y_m$ parallel to the $y$ axis and its principal bending axis $z_m$ again parallel to the $z$ axis. Action and displacement indexes at ends $j$ and $k$ are numbered in a sequence that is conducive for transferring member stiffnesses to joint and floor stiffness matrices. Thus, indexes $j1$ and $j2$ at the top of the column are the same as those for a beam at level $A$, while $k1$ and $k2$ at the bottom coincide with those for a beam at level $B$. However, the indexes $j3$ at the top and $k3$ at the bottom correspond to the $x$-translations of floors at levels $A$ and $B$. We may write the $6 \times 6$ stiffness matrix for a column as a partitioned array consisting of $2 \times 2$ submatrices, as follows:

$$\mathbf{K}_i = \frac{EI_z}{L^3} \begin{bmatrix} r_1 & & & & & & \text{Sym.} \\ 0 & 4L^2 & & & & & \\ -r_1 & 0 & r_1 & & & & \\ 0 & 2L^2 & 0 & 4L^2 & & & \\ 0 & 6L & 0 & 6L & 12 & & \\ 0 & -6L & 0 & -6L & -12 & 12 \end{bmatrix} \begin{matrix} j1 \\ j2 \\ k1 \\ k2 \\ j3 \\ k3 \end{matrix} \begin{matrix} A \\ \\ B \\ \\ F \end{matrix} \quad (1)$$

$$\begin{matrix} j1 & j2 & k1 & k2 & j3 & k3 \\ A & & B & & F & \end{matrix}$$

which is a rearranged version of Eq. (6.2-1). Note that subscripts in Eq. (1) are taken in the sequence $A$, $B$, and $F$.

When stiffnesses for beams and columns in substructure $\ell$ are assembled, the resulting matrix is

$$\mathbf{S}_\ell = \begin{bmatrix} \mathbf{S}_{AA} & & \text{Sym.} \\ \mathbf{S}_{BA} & \mathbf{S}_{BB} & \\ \mathbf{S}_{FA} & \mathbf{S}_{FB} & \mathbf{S}_{FF} \end{bmatrix}_\ell \quad (2)$$

Beams contribute terms only to submatrix $\mathbf{S}_{AA}$, but columns contribute to all of the submatrices. Matrix $\mathbf{S}_\ell$ augments stiffnesses from previous substructures as a preliminary to reduction. From the modified tridiagonal method, stiffness reduction formulas for each substructure are

$$\mathbf{S}^*_{BB} = \mathbf{S}_{BB} + \mathbf{S}_{BA}\mathbf{T}_{AB} = \mathbf{S}_{BB} - \mathbf{S}_{BA}\mathbf{S}^{-1}_{AA}\mathbf{S}_{AB} \quad (3a)$$

$$\mathbf{S}^*_{BF} = \mathbf{S}_{BF} + \mathbf{S}_{BA}\mathbf{T}_{AF} = \mathbf{S}_{BF} - \mathbf{S}_{BA}\mathbf{S}^{-1}_{AA}\mathbf{S}_{AF} \quad (3b)$$

$$\mathbf{S}^*_{FF} = \mathbf{S}_{FF} + \mathbf{S}_{FA}\mathbf{T}_{AF} = \mathbf{S}_{FF} - \mathbf{S}_{FA}\mathbf{S}^{-1}_{AA}\mathbf{S}_{AF} \quad (3c)$$

where

$$\mathbf{T}_{AB} = -\mathbf{S}^{-1}_{AA}\mathbf{S}_{AB} \qquad \mathbf{T}_{AF} = -\mathbf{S}^{-1}_{AA}\mathbf{S}_{AF} \quad (4)$$

As the series elimination proceeds from one substructure to the next, the number of rows and columns of types $A$ and $B$ remains equal to $2n_c$. Therefore, we redefine displacements of type $B$ in substructure $\ell$ to become type $A$ in substructure $\ell + 1$. This requires placing matrices of type $B$ into positions of type $A$ after Eqs. (3) are applied to each substructure. Then new contributions from

### Sec. 10.3 Modified Tridiagonal Method for Multistory Buildings

the next substructure are added to the residual arrays from the preceding substructure. On the other hand, the number of rows and columns of type $F$ increases from 2 for the first substructure to $n_s$ for the last substructure. For each new story (except the last), we pick up one more lateral displacement, so that matrices $\mathbf{S}_{BF}^*$ and $\mathbf{S}_{FF}^*$ keep expanding in size. This *overlay technique* requires computer core storage for only one substructure and $n_s$ framing levels.

The *forward elimination* procedure is completed at the lowest story, where column bases are assumed to be fixed. At this stage, we have the $n_s$ equations of undamped motion

$$\mathbf{M}_{FF}\ddot{\mathbf{D}}_F + \mathbf{S}_{FF}^*\mathbf{D}_F = \mathbf{A}_F(t) \qquad (5)$$

in which $\mathbf{M}_{FF}$ is a diagonal matrix of lumped masses and $\mathbf{A}_F(t)$ is the vector of lateral forces. Damped or undamped story displacements may then be found using either the normal-mode method or direct numerical integration from Chapter 4 or 5.

After time histories of story displacements $\mathbf{D}_F$ have been determined, other items of interest may be calculated in a *backward-substitution* procedure. Starting at the lowest level and working upward, we compute joint displacements $\mathbf{D}_A$ from

$$\mathbf{D}_A = \mathbf{T}_{AB}\mathbf{D}_B + \mathbf{T}_{AF}\mathbf{D}_F \qquad (6)$$

For this purpose, it is necessary to have saved *back-substitution matrices* $\mathbf{T}_{AB}$ and $\mathbf{T}_{AF}$ for each story during the forward elimination process. When applying Eq. (6) recursively, we redefine displacements $\mathbf{D}_A$ for any particular level $\ell$ to become displacements $\mathbf{D}_B$ for the level $\ell - 1$ above. Of course, the vector $\mathbf{D}_B$ is null for the lowest substructure.

Time histories of member end-actions for beams may be obtained by placing appropriate terms from $\mathbf{D}_A$ into a 4 × 1 member displacement vector $\mathbf{D}_{Mi}$. Premultiplication of this vector by the 4 × 4 beam stiffness matrix $\mathbf{K}_i$ yields:

$$\mathbf{A}_{Mi} = \mathbf{K}_i \mathbf{D}_{Mi} \qquad (7)$$

in which the end-action vector $\mathbf{A}_{Mi}$ contains a shearing force and a bending moment at each end. For columns, time histories of end-actions are found in a similar manner, but vector $\mathbf{D}_{Mi}$ contains two terms each from $\mathbf{D}_A$, $\mathbf{D}_B$, and $\mathbf{D}_F$. Also, the 6 × 6 column stiffness matrix $\mathbf{K}_i$ in Eq. (1) is used in Eq. (7) to produce six member end-actions. Of course, these time-varying end-actions for beams and columns could be added to any static actions existing at time $t = 0$.

Turning now to three-dimensional multistory tier buildings, we must account for the rigid lamina existing at each framing level. Figure 10.8 shows substructure $\ell$, consisting of a rigid lamina, the beams at level $A$, and the columns below. Recall that stiffness matrices for $x$-beams, $y$-beams, and $z$-columns were given in Eqs. (9.4-1), (9.4-2), and (9.4-15) and that the last is transformed to rigid-body coordinates. In addition, formulas for the transformed mass-inertia and story-load matrices appear in Eqs. (9.4-17) and (9.4-19).

**Figure 10.8** Typical story framing in tier building.

For tier buildings, assemblage of stiffnesses and the forward elimination proceed the same as with plane frames. However, the number of rows and columns of types $A$ and $B$ is $3n_c$ because each joint has three unconstrained displacements. Also, the number of rows and columns of type $F$ is $3n_s$ due to the fact that each lamina has three rigid-body displacements. The overlay technique also works the same as for plane frames, and computer core storage need only contain information about one substructure and $3n_s$ rows and columns in matrices of type $F$.

Calculation of story displacements in tier buildings and the backward substitution process follows the sequence given for plane frames. But to obtain column end-actions, we must transform story displacements from the reference point $F$ to joint $j$ at the top and joint $k$ at the bottom. Therefore, we have

$$\mathbf{A}_{Mi} = \mathbf{K}_i \mathbf{D}_{Mi} = \mathbf{K}_i \mathbf{T}_i^T \mathbf{D}_{Fi} \qquad (8)$$

In this expression, $\mathbf{T}_i^T$ is the transpose of the matrix in Eq. (9.4-13), and $\mathbf{D}_{Fi}$ contains story displacements from levels $A$ and $B$ in the last six positions.

In retrospect, it is interesting to note that the axial constraints in the beams of a multistory plane frame are analogous to those due to the rigid laminae in tier buildings. Recall that each framing level of the plane frame in Fig. 10.6 has only one translation in the $x$ direction. This implies infinite axial rigidities for the beams, and we can visualize a one-dimensional rigid body at each framing level. As for the laminae in tier buildings, these unseen constraints in plane frames serve to reduce the number of degrees of freedom in the analytical model.

## 10.4 PROGRAMS DYMSPF AND DYMSTB

In this section we discuss Programs DYMSPF and DYMSTB for dynamic analysis of multistory plane frames and tier buildings. These programs use auxiliary storage as well as core storage to analyze two- and three-dimensional building frames by the modified tridiagonal method.

Starting with the multistory plane frame, we give the outline for Program DYMSPF, as follows:

### Outline of Program DYMSPF

1. Read and write structural data
   a. Structural parameters
   b. Bay widths and story heights
2. Generate story mass and stiffness matrices (by substructures)
   a. Clear stiffness and mass matrices
   b. Read and write story mass and member information
   c. Augment residual stiffness matrices with $\mathbf{S}_\ell$
   d. Calculate reduced stiffness matrices by Eqs. (10.3-3)
   e. Place story information into auxiliary storage
   f. Shift matrices of type $B$ into locations of type $A$
   g. Repeat steps b through f for each story (top-to-bottom)
3. Determine frequencies and mode shapes
   a. Convert eigenvalue problem to standard, symmetric form
   b. Calculate eigenvalues and eigenvectors
   c. Write natural frequencies (cycles per second)
   d. Transform, normalize, and write modal vectors
   e. Normalize modal vectors with respect to mass matrix
4. Read and write dynamic load data
   a. Dynamic parameters
   b. Initial conditions (story displacements and velocities)
   c. Applied actions ($x$-forces at framing levels)

    d. Ground acceleration (in $x$ direction)
    e. Forcing function (piecewise-linear)
5. Calculate story displacements
    a. Set up modal damping matrix
    b. Calculate story displacements for each time step
    c. Write and/or plot story displacement-time histories
    d. Find and write maximum/minimum story displacements
6. Determine member end actions (by substructures)
    a. Retrieve story information from auxiliary storage
    b. Calculate joint displacements $\mathbf{D}_A$
    c. Calculate member end-actions for each time step
    d. Write and/or plot member end-action time histories
    e. Find and write maximum/minimum member end-actions
    f. Shift elements of $\mathbf{D}_A$ into $\mathbf{D}_B$
    g. Repeat steps a through f for each story (bottom-to-top)

Table 10.1 shows preparation of structural data for multistory plane frames. Structural parameters in the table include the number of bays NB and the number of stories NS. For dimensions of bay widths and story heights, Program DYMSPF requires minimal information. Each line of that data gives the number NUM of sequential occurrences, followed by the repeated dimension of the bay width BW (left-to-right) or the story height SH (top-to-bottom). Each of the NS blocks of substructure data contains the superimposed story mass SMASS (top-to-bottom) and member information (left-to-right). For the latter we need give only the number NUM of repetitions, followed by the cross-sectional area of the member AX or AY and its second moment of area ZI with respect to the $z_m$ axis.

TABLE 10.1 Structural Data for Multistory Plane Frames

| Type of Data | No. of Lines | Items on Data Lines |
|---|---|---|
| Problem identification | 1 | Descriptive title |
| Structural parameters | 1 | NB, NS, E, RHO |
| Dimensions | | |
|   (a) Bay widths | [a] | NUM, BW |
|   (b) Story heights | [a] | NUM, SH |
| Substructure data[b] | | |
|   (a) Story mass | 1 | SMASS |
|   (b) Member information | | |
|     1. Beams | [a] | NUM, AX, ZI |
|     2. Columns | [a] | NUM, AY, ZI |

[a] As required.
[b] NS blocks.

### Sec. 10.4  Programs DYMSPF and DYMSTB

Preparation of dynamic load data for multistory plane frames appears in Table 10.2. The dynamic parameters are the same as before, and the initial conditions involve NS story translations and velocities at time $t = 0$. Under applied actions we see the load parameter IAF, indicating whether story loads exist or not. As shown in Fig. 10.6, each story load is a force $A_{F\ell}$ in the $x$ direction at framing level $\ell$. The ground-acceleration and forcing-function data in the table have been discussed before in conjunction with other types of structures.

**TABLE 10.2  Dynamic Load Data for Multistory Plane Frames**

| Type of Data | No. of Lines | Items on Data Lines |
|---|---|---|
| Dynamic parameters | 1 | ISOLVE, NTS, DT, DAMPR |
| Initial conditions<br>(a) Condition parameters<br>(b) Story displacements<br>(c) Story velocities | 1<br>a<br>a | IND, INV<br>D0(1), D0(2), . . . , D0(NS)<br>V0(1), V0(2), . . . , V0(NS) |
| Applied actions<br>(a) Load parameter<br>(b) Story loads | 1<br>a | IAF<br>AF(1), AF(2), . . . , AF(NS) |
| Ground accelerations<br>(a) Acceleration parameter<br>(b) Acceleration factor | 1<br>1 | IGA<br>GAX |
| Forcing function<br>(a) Function parameter<br>(b) Function ordinates | 1<br>NFO | NFO<br>K, T(K), FO(K) |

<sup>a</sup> As required.

### Example 10.3

Figure 10.9 shows a two-bay, ten-story plane building frame having a rectangular layout. Beams in this structure are all steel rolled sections of size W 21 × 55, but the steel columns vary, as follows: (C1) W 8 × 31, (C2) W 10 × 60, (C3) W 12 × 85, and (C4) W 12 × 106. The length $L$ is 144 in., and the mass superimposed at each framing level is 0.06 k-s²/in.

For this frame we have two dynamic loading conditions. The first is an atmospheric blast that causes the triangular force $A_{F\ell}$ in Fig. 10.10(a) at each level, except the top (where it is half as much). The second consists of rigid-body ground acceleration $\ddot{D}_{g1}$ in the $x$ direction that has the sawtooth shape in Fig. 10.10(b). We analyzed the frame for these two loading conditions using Program DYMSPF, with a damping ratio of 0.10. Resulting time histories of the translation $D_{F1}$ at the top level are given in Fig. 10.10(c). Load case (a) causes a maximum response of 4.090 in. at time $t = 0.480$ s, while the maximum excursion for load case (b) is 4.357 in. at time $t = 2.16$ s. After those maxima, the responses diminish because of damping.

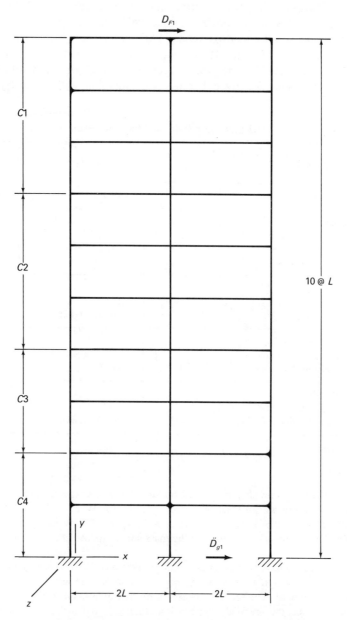

**Figure 10.9** Ten-story example for Program DYMSPF.

### Sec. 10.4 Programs DYMSPF and DYMSTB

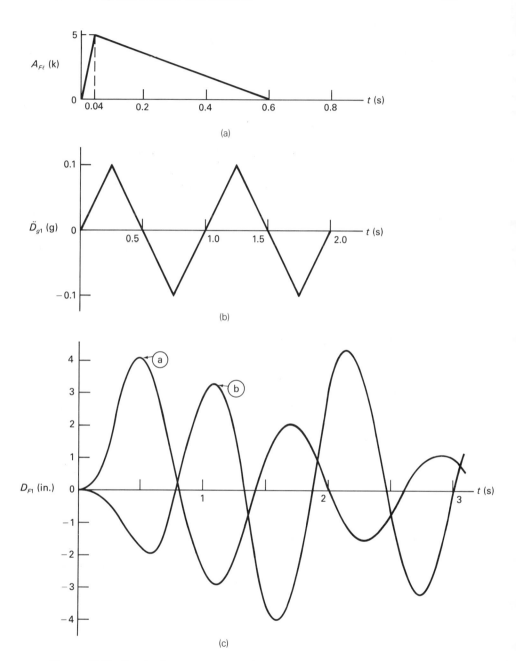

**Figure 10.10** Results for ten-story example: (a) atmospheric blast; (b) ground acceleration; (c) translation $D_{F1}$ due to (a) and (b).

The outline of Program DYMSTB for tier buildings is similar to that of Program DYMSPF for plane frames. However, structural data in part 1 for a tier building is given in the sequence $x$ direction, $y$ direction, and negative $z$ direction. Also, for a typical floor plan, we must provide Boolean data for omitted joints within a rectangular $x$-$y$ pattern. That is, an existing joint is indicated by a one in an integer matrix, while a nonexisting joint is indicated by a zero. In part 2 of the outline, the program calculates the location of the center of mass (point $c$) for each story. Also, the mass-inertia matrix is generated with respect to point $F$, as shown in Eq. (9.4-17). In part 4, the dynamic load data is more extensive than for a plane frame. Three types of initial displacements and velocities are possible for each lamina. In addition, we handle both $x$- and $y$-forces applied at each level, as well as $x$- and $y$-components of ground accelerations. Therefore, the data must also define two independent piecewise-linear forcing functions (for the $x$ and $y$ directions).

If *shear cores*, *bracing*, and *setbacks* are included in the analytical model of a tier building [6], we need more data than that described above. With substructures, it is also possible to analyze soil–structure interaction [7] and to calculate inelastic responses [8] of tier buildings.

## 10.5 COMPONENT-MODE METHOD

The original ideas for the component-mode method are attributed to Hurty [9]. However, Craig and Bampton [10] also made useful improvements. In the latter work, generalized displacements in a substructure consist of a limited number of vibrational mode shapes and a finite number of nodal displacements at insulating boundaries. The component-mode theory produces mass (or dynamic) coupling between the modal and nodal accelerations in a typical substructure.

Figure 10.11 shows a substructure $\ell$ that is arbitrarily located within a discretized continuum. Nodes of type $A$ are indicated at interior positions, while those of type $B$ are at boundary locations. Undamped equations of motion for this substructure may be written as

$$\mathbf{M}_\ell \ddot{\mathbf{D}}_\ell + \mathbf{S}_\ell \mathbf{D}_\ell = \mathbf{A}_\ell(t) \tag{1}$$

which is the same as Eq. (10.2-14). However, the expanded form is now

$$\begin{bmatrix} \mathbf{M}_{AA} & \mathbf{M}_{AB} \\ \mathbf{M}_{BA} & \mathbf{M}_{BB} \end{bmatrix} \begin{bmatrix} \ddot{\mathbf{D}}_A \\ \ddot{\mathbf{D}}_B \end{bmatrix}_\ell + \begin{bmatrix} \mathbf{S}_{AA} & \mathbf{S}_{AB} \\ \mathbf{S}_{BA} & \mathbf{S}_{BB} \end{bmatrix} \begin{bmatrix} \mathbf{D}_A \\ \mathbf{D}_B \end{bmatrix}_\ell = \begin{bmatrix} \mathbf{A}_A \\ \mathbf{A}_B \end{bmatrix}_\ell \tag{2}$$

With boundary nodes restrained and zero loads, we can set up and solve the eigenvalue problem

$$(\mathbf{S}_{AA\ell} - \omega_{i\ell}^2 \mathbf{M}_{AA\ell})\mathbf{\Phi}_{i\ell} = \mathbf{0} \tag{3}$$

in which $\omega_{i\ell}$ is the angular frequency of mode $i$ and $\mathbf{\Phi}_{i\ell}$ is the corresponding mode shape. For this limited vibrational analysis, the modal indexes are $i = 1$,

### Sec. 10.5  Component-Mode Method

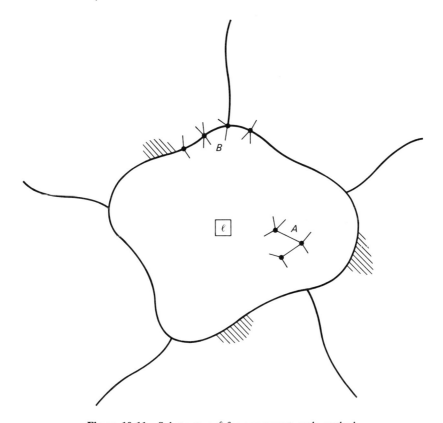

**Figure 10.11** Substructure $\ell$ for component-mode method.

2, ..., $m_r$, where $m_r$ is the number of retained modes. By normalizing the modal vectors with respect to the mass submatrix $\mathbf{M}_{AA\ell}$, we can state the relationship

$$\mathbf{D}_{A\ell} = \mathbf{\Phi}_{N\ell} \mathbf{D}_{N\ell} \tag{4}$$

In this expression, the symbol $\mathbf{D}_{N\ell}$ represents a vector of $m_r$ normal coordinates, and $\mathbf{\Phi}_{N\ell}$ is the normalized modal matrix.

For nonzero boundary displacements (occurring statically), we have

$$\mathbf{D}_{A\ell} = \mathbf{\Phi}_{N\ell} \mathbf{D}_{N\ell} + \mathbf{T}_{AB\ell} \mathbf{D}_{B\ell} \tag{5}$$

where

$$\mathbf{T}_{AB\ell} = -\mathbf{S}_{AA\ell}^{-1} \mathbf{S}_{AB\ell} \tag{6}$$

Now let us define a transformation matrix $\mathbf{T}_\ell$ as

$$\mathbf{T}_\ell = \begin{bmatrix} \mathbf{\Phi}_N & \mathbf{T}_{AB} \\ \mathbf{0} & \mathbf{I}_B \end{bmatrix}_\ell \tag{7}$$

This operator relates the displacements in vector $\mathbf{D}_\ell$ to generalized displacements in a vector $\overline{\mathbf{D}}_\ell$, as follows:

$$\mathbf{D}_\ell = \begin{bmatrix} \mathbf{D}_A \\ \mathbf{D}_B \end{bmatrix}_\ell = \mathbf{T}_\ell \begin{bmatrix} \mathbf{D}_N \\ \mathbf{D}_B \end{bmatrix}_\ell = \mathbf{T}_\ell \overline{\mathbf{D}}_\ell \tag{8a}$$

Note that the new vector $\overline{\mathbf{D}}_\ell$ contains $\mathbf{D}_{N\ell}$ in the first part and $\mathbf{D}_{B\ell}$ in the second part. We also have the relationship

$$\ddot{\mathbf{D}}_\ell = \mathbf{T}_\ell \ddot{\overline{\mathbf{D}}}_\ell \tag{8b}$$

Substitution of Eqs. (8) into Eq. (1) and premultiplication of the result by $\mathbf{T}_\ell^T$ yields

$$\overline{\mathbf{M}}_\ell \ddot{\overline{\mathbf{D}}}_\ell + \overline{\mathbf{S}}_\ell \overline{\mathbf{D}}_\ell = \overline{\mathbf{A}}_\ell(t) \tag{9}$$

In expanded form, this equation is

$$\begin{bmatrix} \mathbf{I}_{m_r} & \overline{\mathbf{M}}_{NB} \\ \overline{\mathbf{M}}_{BN} & \overline{\mathbf{M}}_{BB} \end{bmatrix} \begin{bmatrix} \ddot{\mathbf{D}}_N \\ \ddot{\mathbf{D}}_B \end{bmatrix}_\ell + \begin{bmatrix} \boldsymbol{\omega}_{m_r}^2 & 0 \\ 0 & \overline{\mathbf{S}}_{BB} \end{bmatrix} \begin{bmatrix} \mathbf{D}_N \\ \mathbf{D}_B \end{bmatrix}_\ell = \begin{bmatrix} \overline{\mathbf{A}}_N \\ \overline{\mathbf{A}}_B \end{bmatrix}_\ell \tag{10}$$

The mass submatrices in this equation (without the subscript $\ell$) are

$$\overline{\mathbf{M}}_{MN} = \boldsymbol{\Phi}_N^T \mathbf{M}_{AA} \boldsymbol{\Phi}_N = \mathbf{I}_{m_r} \tag{11a}$$

$$\overline{\mathbf{M}}_{NB} = \boldsymbol{\Phi}_N^T \mathbf{M}_{AA} \mathbf{T}_{AB} + \boldsymbol{\Phi}_N^T \mathbf{M}_{AB} = \overline{\mathbf{M}}_{BN}^T \tag{11b}$$

$$\overline{\mathbf{M}}_{BB} = \mathbf{M}_{BB} + \mathbf{T}_{AB}^T \mathbf{M}_{AB} + \mathbf{M}_{BA} \mathbf{T}_{AB} + \mathbf{T}_{AB}^T \mathbf{M}_{AA} \mathbf{T}_{AB} \tag{11c}$$

Also, the stiffness submatrices become

$$\overline{\mathbf{S}}_{NN} = \boldsymbol{\Phi}_N^T \mathbf{S}_{AA} \boldsymbol{\Phi}_N = \boldsymbol{\omega}_{m_r}^2 \tag{12a}$$

$$\overline{\mathbf{S}}_{NB} = \boldsymbol{\Phi}_N^T \mathbf{S}_{AA} \mathbf{T}_{AB} + \boldsymbol{\Phi}_N^T \mathbf{S}_{AB} = 0 = \overline{\mathbf{S}}_{BN}^T \tag{12b}$$

$$\overline{\mathbf{S}}_{BB} = \mathbf{S}_{BB} + \mathbf{T}_{AB}^T \mathbf{S}_{AB} \tag{12c}$$

and the applied-action subvectors are

$$\overline{\mathbf{A}}_N = \boldsymbol{\Phi}_N^T \mathbf{A}_A \tag{13a}$$

$$\overline{\mathbf{A}}_B = \mathbf{A}_B + \mathbf{T}_{AB}^T \mathbf{A}_A \tag{13b}$$

In Eq. (11a) the symbol $\mathbf{I}_{m_r}$ denotes an identity matrix of order $m_r$. Also, $\boldsymbol{\omega}_{m_r}^2$ in Eq. (12a) is the *spectral matrix* for the substructure, containing squares of angular frequencies $\omega_1^2, \omega_2^2, \ldots, \omega_{m_r}^2$ in diagonal positions. The most important submatrix defined above is $\overline{\mathbf{M}}_{NB}$ in Eq. (11b), which represents *dynamic coupling* terms between accelerations $\ddot{\mathbf{D}}_N$ and $\ddot{\mathbf{D}}_B$.

To assemble equations of motion for all substructures, we simply add the matrices appearing in Eq. (9), using the direct stiffness method. The result is

$$\overline{\mathbf{M}} \ddot{\overline{\mathbf{D}}} + \overline{\mathbf{S}} \overline{\mathbf{D}} = \overline{\mathbf{A}}(t) \tag{14}$$

In this equation the assembled matrices are

Sec. 10.6    Component-Mode Method for Trusses    471

$$\overline{\mathbf{M}} = \sum_{\ell=1}^{n_s} \overline{\mathbf{M}}_\ell \qquad \overline{\mathbf{S}} = \sum_{\ell=1}^{n_s} \overline{\mathbf{S}}_\ell \qquad \overline{\mathbf{A}}(t) = \sum_{\ell=1}^{n_s} \overline{\mathbf{A}}_\ell(t) \qquad (15)$$

where $n_s$ is the number of substructures. Of course, the substructure matrices in Eqs. (15) must be expanded with zeros to become conformable for addition.

Now the equations of motion in Eq. (14) can be solved for the displacements in vector $\overline{\mathbf{D}}$ by either the normal-mode method or direct numerical integration. Then displacements in vector $\mathbf{D}_{A\ell}$ for each substructure are found using Eq. (5) as a back-substitution expression. In the next section we shall apply the component-mode method to trusses, for which some additional complications arise.

## 10.6 COMPONENT-MODE METHOD FOR TRUSSES

In earlier chapters of this book, we considered only axial deformations for dynamic analyses of trusses. However, inertial and body forces also occur along the members in transverse directions, which cause flexural deformations as well. These influences are most significant in trusses composed of only a few members. We have found that the best approach for analyzing such structures is the component-mode method [11, 12]. By this technique, a member with both axial and flexural deformations constitutes a substructure. Treating the member as a simply supported beam, we include a limited number of its exact vibrational mode shapes as flexural displacement shape functions.

Figure 10.12 shows a prismatic plane truss member $i$ with local (primed) and global (unprimed) axes. We shall handle such a member as a specialized finite element acting as a substructure within a plane truss. Flexural displacement shape functions in member $i$ are taken to be the vibrational mode shapes of a simply supported beam [13]. These mode shapes will be superimposed on linear displacement shape functions due to translations at joints (or nodes) $j$ and $k$. Thus, the generic displacements $u'$ and $v'$ in the $x'$ and $y'$ directions may be written in terms of modal and nodal displacements as

$$\mathbf{u}'_i = \mathbf{f}\,\overline{\mathbf{D}}'_i \qquad (1)$$

or

$$\begin{bmatrix} u' \\ v' \end{bmatrix}_i = [\mathbf{f}_A \quad \mathbf{f}_b] \begin{bmatrix} \overline{\mathbf{D}}_A \\ \mathbf{D}'_B \end{bmatrix}_i \qquad (2)$$

In this equation the symbols $\overline{\mathbf{D}}_{Ai}$ and $\mathbf{D}'_{Bi}$ represent the displacement vectors

$$\overline{\mathbf{D}}_{Ai} = \begin{bmatrix} \overline{\mathbf{D}}_{A1} \\ \overline{\mathbf{D}}_{A2} \\ \overline{\mathbf{D}}_{A3} \\ \cdots \end{bmatrix}_i \qquad \mathbf{D}'_{Bi} = \begin{bmatrix} \mathbf{D}'_{B1} \\ \mathbf{D}'_{B2} \\ \mathbf{D}'_{B3} \\ \mathbf{D}'_{B4} \end{bmatrix}_i \qquad (3)$$

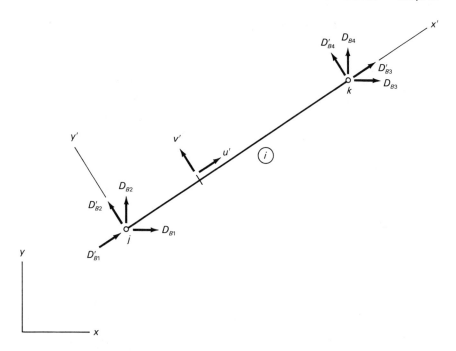

**Figure 10.12** Plane truss member with local and global axes.

Terms in vector $\bar{\mathbf{D}}_{Ai}$ are the amplitudes of a limited number $m_r$ of retained natural mode shapes for a simple beam, as depicted in Figs. 10.13(b)–(d). In addition, the terms in vector $\mathbf{D}'_{Bi}$ are the four translations $D'_{B1}$ through $D'_{B4}$ at joints $j$ and $k$, shown in Figs. 10.12 and 10.13(a).

Matrix $\mathbf{f}$ in Eq. (1) contains displacement shape functions, and its first submatrix is

$$\mathbf{f}_A = \mathbf{\Phi} = \begin{bmatrix} 0 & 0 & 0 & \cdots \\ \sin \pi \xi & \sin 2\pi \xi & \sin 3\pi \xi & \cdots \end{bmatrix} \quad (4)$$

in which the dimensionless coordinate is $\xi = x'/L$. Appearing in the second row of this submatrix are the natural mode shapes for vibrations in a simply supported prismatic beam. To keep them dimensionless, these mode shapes are not normalized with respect to the mass of the member. On the other hand, the second submatrix in $\mathbf{f}$ has the form

$$\mathbf{f}_B = \begin{bmatrix} 1-\xi & 0 & \xi & 0 \\ 0 & 1-\xi & 0 & \xi \end{bmatrix} \quad (5)$$

These linear shape functions result from unit displacements of $D'_{B1}$ through $D'_{B4}$, as indicated in Figs. 10.13(e) and (f).

Strain-displacement relationships for this element may be stated as

$$\mathbf{B}_i = \mathbf{d}_i \mathbf{f} = [\mathbf{B}_A \quad \mathbf{B}_B]_i \quad (6)$$

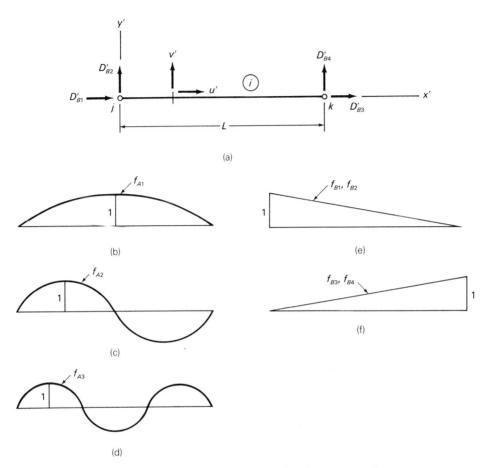

**Figure 10.13** Displacement shape functions for plane truss member.

in which the linear differential operator $\mathbf{d}_i$ is

$$\mathbf{d}_i = \begin{bmatrix} \dfrac{1}{L}\dfrac{\partial}{\partial \xi} & 0 \\ 0 & -\dfrac{y'}{L^2}\dfrac{\partial^2}{\partial \xi^2} \end{bmatrix}_i \tag{7}$$

The first submatrix in $\mathbf{B}_i$ has the resulting form

$$\mathbf{B}_{Ai} = \mathbf{d}_i \mathbf{f}_A = y'\frac{\pi^2}{L^2}\begin{bmatrix} 0 & 0 & 0 & \cdots \\ \sin \pi\xi & 4\sin 2\pi\xi & 9\sin 3\pi\xi & \cdots \end{bmatrix}_i \tag{8}$$

and the second is

$$\mathbf{B}_{Bi} = \mathbf{d}_i \mathbf{f}_B = \frac{1}{L}\begin{bmatrix} -1 & 0 & 1 & 0 \\ 0 & 0 & 0 & 0 \end{bmatrix}_i \tag{9}$$

The stiffness matrix $\bar{\mathbf{S}}'_i$ in local directions for member $i$ has the known form

$$\bar{\mathbf{S}}'_i = \int_V \mathbf{B}_i^T \mathbf{E} \mathbf{B}_i \, dV = \begin{bmatrix} \bar{\mathbf{S}}_{AA} & 0 \\ 0 & \mathbf{S}'_{BB} \end{bmatrix}_i \tag{10}$$

In this equation the stress-strain matrix $\mathbf{E}$ is

$$\mathbf{E} = E\mathbf{I}_2 \tag{11}$$

where $\mathbf{I}_2$ is an identity matrix of order 2. The first submatrix in $\bar{\mathbf{S}}'_i$ becomes

$$\bar{\mathbf{S}}_{AAi} = \int_V \mathbf{B}_{Ai}^T \mathbf{E} \mathbf{B}_{Ai} \, dV = \frac{\pi^4 EI_z}{2L^3} \begin{bmatrix} 1 & & & \text{Sym.} \\ 0 & 16 & & \\ 0 & 0 & 81 & \\ \cdots & \cdots & \cdots & \cdots \end{bmatrix}_i$$

$$= \left(\frac{\rho AL}{2} \omega_{m_r}^2\right)_i \tag{12}$$

and the second submatrix is found to be

$$\mathbf{S}'_{BBi} = \int_V \mathbf{B}_{Bi}^T \mathbf{E} \mathbf{B}_{Bi} \, dV = \frac{EA}{L} \begin{bmatrix} 1 & & & \text{Sym.} \\ 0 & 0 & & \\ -1 & 0 & 1 & \\ 0 & 0 & 0 & 0 \end{bmatrix}_i \tag{13}$$

which is the same as matrix $\mathbf{K}'$ in Eq. (3.5-25).

Next, the consistent mass matrix $\bar{\mathbf{M}}'_i$ (in local directions) for member $i$ is obtained from

$$\bar{\mathbf{M}}'_i = \int_V \rho \mathbf{f}^T \mathbf{f} \, dV = \begin{bmatrix} \bar{\mathbf{M}}_{AA} & \mathbf{M}'_{AB} \\ \mathbf{M}'_{BA} & \mathbf{M}'_{BB} \end{bmatrix}_i \tag{14}$$

The first submatrix in $\bar{\mathbf{M}}'_i$ is

$$\bar{\mathbf{M}}_{AAi} = \int_V \rho \mathbf{\Phi}^T \mathbf{\Phi} \, dV = \left(\frac{\rho AL}{2} \mathbf{I}_{m_r}\right)_i \tag{15}$$

As before, the symbol $\mathbf{I}_{m_r}$ represents an identity matrix of order $m_r$. Other submatrices in $\bar{\mathbf{M}}'_i$ are

$$\mathbf{M}'_{ABi} = \int_V \rho \mathbf{\Phi}^T \mathbf{f}_B \, dV = \frac{\rho AL}{\pi} \begin{bmatrix} 0 & 1 & 0 & 1 \\ 0 & 1/2 & 0 & -1/2 \\ 0 & 1/3 & 0 & 1/3 \\ \cdots & \cdots & \cdots & \cdots \end{bmatrix}_i \tag{16}$$

and

### Sec. 10.6 Component-Mode Method for Trusses

$$\mathbf{M}'_{BBi} = \int_V \rho \mathbf{f}_B^T \mathbf{f}_B \, dV = \frac{\rho A L}{6} \begin{bmatrix} 2 & & & \text{Sym.} \\ 0 & 2 & & \\ 1 & 0 & 2 & \\ 0 & 1 & 0 & 2 \end{bmatrix}_i \tag{17}$$

which is the same as matrix $\mathbf{M}'$ in Eq. (3.5-32).

Also, equivalent modal and nodal loads in local directions are calculated as

$$\mathbf{A}'_i(t) = \int_L \mathbf{f}^T \mathbf{b}'(t) \, dL = \begin{bmatrix} \overline{\mathbf{A}}_A \\ \mathbf{A}'_B \end{bmatrix}_i \tag{18}$$

Here the vector $\mathbf{b}'(t)$ contains body forces (per unit length) in the $x'$ and $y'$ directions, as follows:

$$\mathbf{b}'(t) = \begin{bmatrix} b_{x'} \\ b_{y'} \end{bmatrix} \tag{19}$$

The first subvector in $\overline{\mathbf{A}}'_i(t)$ has the form

$$\overline{\mathbf{A}}_{Ai} = \int_L \mathbf{\Phi}^T \mathbf{b}'(t) \, dL = \int_L \begin{bmatrix} \sin \pi \xi \\ \sin 2\pi \xi \\ \sin 3\pi \xi \\ \cdots \end{bmatrix}_i b_{y'} \, dL \tag{20}$$

and the second is

$$\mathbf{A}'_{Bi} = \int_L \mathbf{f}_B^T \mathbf{b}'(t) \, dL = \int_L \begin{bmatrix} (1 - \xi)b_{x'} \\ (1 - \xi)b_{y'} \\ \xi b_{x'} \\ \xi b_{y'} \end{bmatrix}_i dL \tag{21}$$

For local (or member) axes, the undamped equations of motion for small displacements of the plane truss member are

$$\overline{\mathbf{M}}'_i \ddot{\overline{\mathbf{D}}}'_i + \overline{\mathbf{S}}'_i \overline{\mathbf{D}}'_i = \overline{\mathbf{A}}'_i(t) \tag{22}$$

By rotation of axes for the parts of type $B$, Eq. (22) may be transformed to global (or structural) axes to become

$$\overline{\mathbf{M}}_i \ddot{\overline{\mathbf{D}}}_i + \overline{\mathbf{S}}_i \overline{\mathbf{D}}_i = \overline{\mathbf{A}}_i(t) \tag{23}$$

The displacements $\overline{\mathbf{D}}'_i$ and accelerations $\ddot{\overline{\mathbf{D}}}'_i$ in Eq. (22) are related to the corresponding vectors $\overline{\mathbf{D}}_i$ and $\ddot{\overline{\mathbf{D}}}_i$ in Eq. (23) by the expressions

$$\overline{\mathbf{D}}'_i = \hat{\mathbf{R}}_i \overline{\mathbf{D}}_i \qquad \ddot{\overline{\mathbf{D}}}'_i = \hat{\mathbf{R}}_i \ddot{\overline{\mathbf{D}}}_i \tag{24}$$

In these relationships the rotation-of-axes transformation matrix is

$$\hat{\mathbf{R}}_i = \begin{bmatrix} \mathbf{I}_{m_r} & 0 \\ 0 & \hat{\mathbf{R}}_B \end{bmatrix}_i \quad (25)$$

Note from the form of matrix $\hat{\mathbf{R}}_i$ that only the nodal displacements $\mathbf{D}_{Bi}$ are transformed from structural to member directions. (Displacements $D_{B1}$ through $D_{B4}$ are shown in the $x$ and $y$ directions in Fig. 10.12.) Submatrix $\hat{\mathbf{R}}_{Bi}$ in Eq. (25) is

$$\hat{\mathbf{R}}_{Bi} = \begin{bmatrix} \mathbf{R} & 0 \\ 0 & \mathbf{R} \end{bmatrix}_i \quad (26)$$

where

$$\mathbf{R}_i = \begin{bmatrix} c_x & c_y \\ -c_y & c_x \end{bmatrix}_i \quad (27)$$

Direction cosines in the rotation matrix $\mathbf{R}_i$ were defined in Eqs. (3.5-24). By substituting Eqs. (24) into Eq. (22) and premultiplying the latter by $\hat{\mathbf{R}}_i^T$, we convert it to Eq. (23). In that equation we have the following matrix products:

$$\overline{\mathbf{S}}_i = \hat{\mathbf{R}}_i^T \overline{\mathbf{S}}_i' \hat{\mathbf{R}}_i \quad (28)$$

$$\overline{\mathbf{M}}_i = \hat{\mathbf{R}}_i^T \overline{\mathbf{M}}_i' \hat{\mathbf{R}}_i \quad (29)$$

$$\overline{\mathbf{A}}_i(t) = \hat{\mathbf{R}}_i^T \overline{\mathbf{A}}_i'(t) \quad (30)$$

The results of these operations are as follows:

$$\mathbf{S}_{BBi} = \hat{\mathbf{R}}_{Bi}^T \mathbf{S}_{BBi}' \hat{\mathbf{R}}_{Bi} = \frac{EA}{L} \begin{bmatrix} c_x^2 & & & \text{Sym.} \\ c_x c_y & c_y^2 & & \\ -c_x^2 & -c_x c_y & c_x^2 & \\ -c_x c_y & -c_y^2 & c_x c_y & c_y^2 \end{bmatrix}_i \quad (31)$$

$$\overline{\mathbf{M}}_{ABi} = \overline{\mathbf{M}}_{ABi}' \hat{\mathbf{R}}_{Bi} = \frac{\rho A L}{\pi} \begin{bmatrix} -c_y & c_x & -c_y & c_x \\ -c_y/2 & c_x/2 & c_y/2 & -c_x/2 \\ -c_y/3 & c_x/3 & -c_y/3 & c_x/3 \\ \cdots & \cdots & \cdots & \cdots \end{bmatrix}_i \quad (32)$$

$$\mathbf{A}_{Bi} = \hat{\mathbf{R}}_{Bi}^T \mathbf{A}_{Bi}' = \int_L \begin{bmatrix} (1-\xi)(c_x b_{x'} - c_y b_{y'}) \\ (1-\xi)(c_y b_{x'} + c_x b_{y'}) \\ \xi(c_x b_{x'} - c_y b_{y'}) \\ \xi(c_y b_{x'} + c_x b_{y'}) \end{bmatrix}_i dL \quad (33)$$

Note that submatrix $\mathbf{M}_{BBi} = \mathbf{M}_{BBi}'$ is invariant with rotation of axes.

After the rotation-of-axes transformation for each member is completed, the equations of motion for the whole structure may be assembled to obtain

$$\overline{\mathbf{M}} \ddot{\overline{\mathbf{D}}} + \overline{\mathbf{S}} \overline{\mathbf{D}} = \overline{\mathbf{A}}(t) \quad (34)$$

Sec. 10.7    Programs COMOPT and COMOST

in which

$$\overline{\mathbf{M}} = \sum_{i=1}^{m} \overline{\mathbf{M}}_i \qquad \overline{\mathbf{S}} = \sum_{i=1}^{m} \overline{\mathbf{S}}_i \qquad \overline{\mathbf{A}}(t) = \sum_{i=1}^{m} \overline{\mathbf{A}}_i(t) \tag{35}$$

where $m$ is the number of members. The equations of motion for the structure can now be solved by the normal-mode method or by direct numerical integration. Either approach may be preceded by a vibrational analysis, from which a damping matrix can be established for the whole structure, as before.

Extension of the theory for plane trusses to the analysis of space trusses is straightforward and appears in Ref. 12. The primary change to recognize in three dimensions is that each member has two principal planes of bending, defined as the $x'$-$y'$ plane and the $x'$-$z'$ plane. But the flexural mode shapes in each plane are still the sine functions $\sin \pi\xi$, $\sin 2\pi\xi$, $\sin 3\pi\xi$, and so on. Definition of a principal plane in space may be aided by using a third point $p$ in addition to points $j$ and $k$, as explained in Sec. 6.4.

## 10.7 PROGRAMS COMOPT AND COMOST

Now we briefly discuss Programs COMOPT and COMOST for dynamic analyses of plane and space trusses by the component-mode method of the preceding section. These programs calculate responses to initial conditions and piecewise-linear forcing functions that may be either applied actions or translational ground accelerations. Using the normal-mode method, we determine axial forces and bending moments in the members, as well as time histories of modal and nodal responses.

**Example 10.4**

Figure 10.14 shows a plane truss having only two members, with a step force $P$ applied in the $x$ direction at the quarter point of member 1. Both members are prismatic and have the same values of $\rho$, $E$, $A$, and $I_z$. Although realistic sizes were used, we take the dimensionless parameter $AL^2/I_z$ to be 250 for this problem.

Dimensionless frequencies for the truss solution without flexure are $\omega_1^* = 1.000$ and $\omega_2^* = 3.023$, which have been normalized by dividing them by the first. Table 10.3 contains twelve such dimensionless frequencies for the structure when five vibrational modes are included for each member. Also, Fig. 10.15 shows the corresponding mode shapes of the truss for the first four modes. If members 2 and 1 are taken separately as simply supported beams, their fundamental frequencies become only slightly more than those for modes 1 and 2 in Table 1.

Figure 10.16 gives responses of the dimensionless displacement $D_1^*$, plotted against the dimensionless time $t^*$. Here the dimensionless displacement is obtained by dividing the dynamic value by the static value for the same load. On the other hand, the dimensionless time $t^*$ is time $t$ divided by the period of the first mode for the truss solution without flexure. The curves in Fig. 10.16 are labeled with encircled numbers 0, 1, . . . , 5 to indicate how many flexural modes per member are included in the analytical model. We can see that the solution with no flexure is quite different from those with

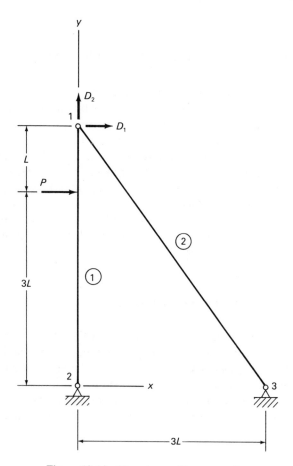

**Figure 10.14** Plane truss with two members.

**TABLE 10.3  Dimensionless Frequencies for Two-Member Truss**

| Mode | $\omega^*$ | Mode | $\omega^*$ |
| --- | --- | --- | --- |
| 1 | 0.204 | 7  | 3.034 |
| 2 | 0.317 | 8  | 3.163 |
| 3 | 0.796 | 9  | 3.628 |
| 4 | 1.189 | 10 | 5.146 |
| 5 | 1.668 | 11 | 5.575 |
| 6 | 2.171 | 12 | 8.267 |

Sec. 10.7  Programs COMOPT and COMOST    **479**

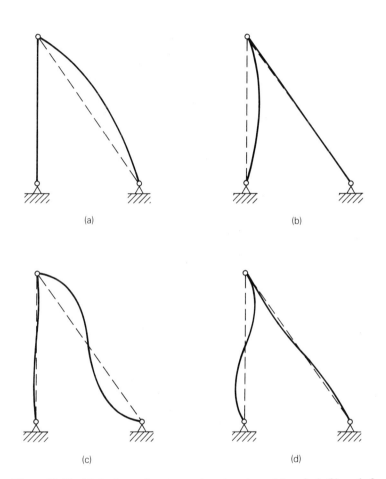

**Figure 10.15**  Mode shapes for two-member plane truss: (a) mode 1; (b) mode 2; (c) mode 3; (d) mode 4.

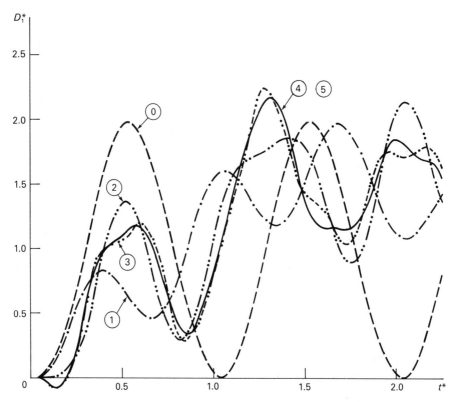

**Figure 10.16** Displacement time histories for $D_1^*$.

flexure included in the members. The first three flexural modes in member 1 tend to dominate the response because the load is applied directly to that member. Adding modes 4 and 5 to each member does not improve the accuracy of the response very much.

**Example 10.5**

Figure 10.17 illustrates a truss with an indefinite number of members, having a step force $P$ applied in the $y$ direction at the center of member 1. By increasing the number of panels from 1 to $n_p$, we can study the effect of this parameter upon the dynamic response $D_2$ of joint 1 in the $y$ direction. All members of this truss have the same values for $\rho$, $E$, $A$, and $I_z$; and we take the dimensionless parameter $AL^2/I_z$ to be 2250.

For this example the truss solutions without flexure will be compared against solutions with one flexural mode included for each member. In Fig. 10.18 we plot the ratio $R$ of the maximum response of $D_2$ for the model with flexure to that without flexure versus the number of panels $n_p$. The ratio $R$ approaches unity as the number of panels increases. Thus, the flexural deformations in the members have little effect on the joint responses when the number of members becomes large.

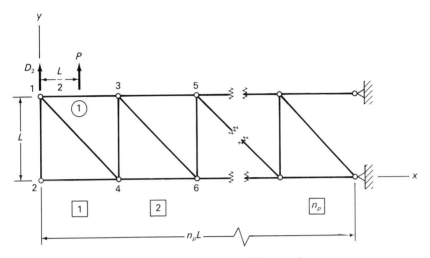

**Figure 10.17** Plane truss with $n_p$ panels.

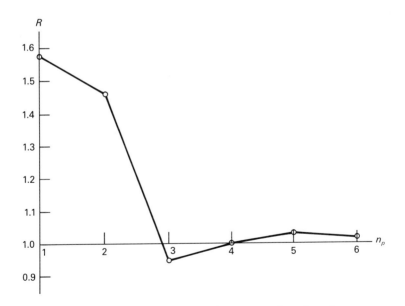

**Figure 10.18** Displacement ratio versus number of panels.

## REFERENCES

1. Irons, B. M., "A Frontal Solution Program," *Int. J. Numer. Methods Eng.*, Vol. 2, No. 1, 1970, pp. 5-32.
2. Hinton, E., and Owen, D. R. J., *Finite Element Programming*, Academic Press, London, 1977.
3. Weaver, W., Jr., and Yoshida, D. M., "The Eigenvalue Problem for Banded Matrices," *J. Comp. Struct.*, Vol. 1, No. 4, 1971, pp. 651–664.
4. Weaver, W., Jr., and Nelson, M. F., "Three-Dimensional Analysis of Tier Buildings," *ASCE J. Struct. Div.*, Vol. 92, No. ST6, 1966, pp. 385–404.
5. Weaver, W., Jr., Nelson, M. F., and Manning, T. A., "Dynamics of Tier Buildings," *ASCE J. Eng. Mech. Div.*, Vol. 94, No. EM6, 1968, pp. 1455–1474.
6. Weaver, W., Jr., Brandow, G. E., and Manning, T. A., "Tier Buildings with Shear Cores, Bracing, and Setbacks," *J. Comp. Struct.*, Vol. 1, Nos. 1/2, 1971, pp. 57–84.
7. Weaver, W., Jr., Brandow, G. E., and Höeg, K., "Three-Dimensional Soil-Structure Response to Earthquakes," *Bull. Seismol. Soc. Am.*, Vol. 63, No. 3, 1973, pp. 1041–1056.
8. Weaver, W., Jr., and Bockholt, J. L., "Inelastic Dynamic Analysis of Tier Buildings," *J. Comp. Struct.*, Vol. 4, No. 3, 1974, pp. 627–645.
9. Hurty, W., C., "Dynamic Analysis of Structural Systems Using Component Modes," *AIAA J.*, Vol. 3, No. 4, 1965, pp. 678–685.
10. Craig, R. R., Jr., and Bampton, M. C. C., "Coupling of Substructures for Dynamic Analysis," *AIAA J.*, Vol. 6, No. 7, 1968, pp. 1313–1319.
11. Weaver, W., Jr., and Loh, C. L., "Dynamics of Trusses by Component-Mode Method," *ASCE J. Struct. Eng.*, Vol. 111, No. 12, 1985, pp. 2526–2575.
12. Loh, C. L., "Dynamics of Trusses by Component-Mode Method," *Ph.D. dissertation*, Department of Civil Engineering, Stanford University, May 1984.
13. Timoshenko, S. P., Young, D. H., and Weaver, W., Jr., *Vibration Problems in Engineering*, 4th ed., Wiley, New York, 1974.

# Notation

## 1. MATRICES AND VECTORS

| Symbol | Definition |
|---|---|
| 0 | Null matrix |
| A | Action vector (*also* coefficient matrix) |
| B | Strain-displacement matrix |
| C | Strain-stress matrix (*also* damping matrix *and* constraint matrix) |
| D | Displacement vector |
| E | Stress-strain matrix |
| F | Flexibility matrix |
| G | Constraint matrix |
| H | Characteristic matrix (*also* Householder matrix) |
| I | Identity matrix |
| J | Jacobian matrix |
| K | Element stiffness matrix |
| L | Lower triangular matrix |
| M | Mass matrix (*also* concentrated moments) |
| P | Concentrated force vector |

| Symbol | Definition |
|---|---|
| Q | Factor in QR algorithm |
| R | Rotation matrix (*also* factor in QR algorithm) |
| S | Stiffness matrix |
| T | Transformation matrix (*also* tridiagonal matrix) |
| U | Upper triangular matrix |
| V | Eigenvector matrix |
| X | Vector of unknowns |
| Y | Vector of unknowns |
| Z | Vector of unknowns |
| $\Lambda$ | Spectral matrix |
| $\Phi$ | Eigenvector matrix |
| b | Body force vector for element |
| d | Linear differential operator for strain-displacement relationships |
| e | Unit vector |
| f | Interpolation function matrix |
| i | Unit vector |
| j | Unit vector |
| k | Unit vector |
| p | Nodal load vector for element |
| q | Nodal displacement vector for element |
| u | Generic displacement vector for element |

## 2. SUBSCRIPTS FOR MATRICES AND VECTORS

| Symbol | Definition |
|---|---|
| A | Nodal displacements eliminated (*also* attached) |
| B | Nodal displacements retained (*also* body *and* boundary) |
| F | Free (*also* floor *or* framing level) |
| L | Lumped |
| M | Member |
| N | Normal coordinates |
| P | Principal coordinates |
| R | Restrained |
| 0 | Initial |

| Symbol | Definition |
|---|---|
| $b$ | Body |
| $d$ | Damped |
| $e$ | Element |
| $f$ | Forced |
| $g$ | Ground |
| $i$ | Index |
| $j$ | Index |
| $k$ | Index |
| $\ell$ | Index |
| $m$ | Number |
| $n$ | Number |
| $p$ | Working (or reference) point |
| $q$ | Working (or reference) point |
| $r$ | Radial direction |
| $s$ | Structure |
| $x$ | $x$ direction |
| $y$ | $y$ direction |
| $z$ | $z$ direction |

## 3. SIMPLE VARIABLES

| Symbol | Definition |
|---|---|
| $A$ | Area |
| $B$ | Constant |
| $C$ | Constant |
| $D$ | Displacement |
| $E$ | Young's modulus of elasticity |
| $G$ | Shearing modulus of elasticity |
| $I$ | Moment of inertia (second moment of area) |
| $J$ | Polar moment of inertia |
| $L$ | Length |
| $M$ | Moment |
| $P$ | Force |
| $R$ | Radius |
| $T$ | Period |
| $U$ | Strain energy |
| $V$ | Potential energy |

| Symbol | Definition |
|---|---|
| $W$ | Work |
| $X$ | Generalized action |
| $a$ | Constant (*also* acceleration) |
| $b$ | Constant |
| $c$ | Constant (*also* damping constant) |
| $d$ | Constant (*also* displacement) |
| $e$ | Base of natural logarithm |
| $f$ | Interpolation function (*also* frequency, cycles/sec) |
| $h$ | Thickness |
| $i$ | Index for . . . (*also* $\sqrt{-1}$) |
| $j$ | Index for . . . |
| $k$ | Index for . . . (*also* spring constant) |
| $\ell$ | Index for . . . (*also* length) |
| $m$ | Number of . . . (*also* mass) |
| $n$ | Number of degrees of freedom (*also* damping parameter) |
| $p$ | Action at element node |
| $q$ | Displacement of element node |
| $r$ | Radius (*also* cylindrical coordinate) |
| $s$ | Segment length |
| $t$ | Time |
| $u$ | Translation in $x$ direction |
| $v$ | Translation in $y$ direction |
| $w$ | Translation in $z$ direction |
| $x$ | Cartesian coordinate |
| $y$ | Cartesian coordinate |
| $z$ | Cartesian coordinate (*also* cylindrical coordinate) |

## 4. GREEK LETTERS

| Symbol | Definition |
|---|---|
| $\Delta$ | Increment |
| $\Sigma$ | Summation |
| $\Phi$ | Function or mode |
| $\Omega$ | Angular frequency |

| Symbol | Definition |
|---|---|
| $\alpha$ | Rotation or angle (*also* Hilber constant) |
| $\beta$ | Magnification factor (*also* Newmark constant) |
| $\gamma$ | Shearing strain (*also* damping ratio *and* Newmark constant) |
| $\delta$ | Increment |
| $\epsilon$ | Normal strain |
| $\zeta$ | Dimensionless coordinate |
| $\eta$ | Dimensionless coordinate |
| $\theta$ | Rotation or angle (*also* Wilson constant) |
| $\lambda$ | Direction cosine (*also* eigenvalue) |
| $\mu$ | Frequency coefficient |
| $\nu$ | Poisson's ratio |
| $\xi$ | Dimensionless coordinate |
| $\pi$ | 3.1416.... |
| $\rho$ | Mass density |
| $\sigma$ | Normal stress |
| $\tau$ | Shearing stress |
| $\phi$ | Curvature |
| $\psi$ | Twist ($d\theta_x/dx$) |
| $\omega$ | Angular frequency (*also* angular velocity) |

## 5. PROGRAM NOTATION

| Symbol | Definition |
|---|---|
| A0( ) | Initial accelerations of nodes |
| A0P | Initial acceleration of moving load |
| AB( ) | Actions applied to bodies |
| AE( ) | Actions at element nodes |
| AF( ) | Actions at free nodes (*also* actions at floors or framing levels) |
| ALPHA | Hilber parameter $\alpha$ |
| AM( ) | Actions at ends of members |
| AN( ) | Actions in normal coordinates |
| AR( ) | Support reactions |
| AS( ) | Actions at structural nodes |
| AX( ) | Cross-sectional areas $A_x$ |

| Symbol | Definition |
|---|---|
| BETA | Newmark parameter $\beta$ |
| BI( ) | Body inertias |
| BL1, BL2, . . . | Intensities of line loads |
| BM( ) | Body masses |
| BS1, BS2, . . . | Intensities of surface loads |
| BV1, BV2, . . . | Intensities of volume loads |
| BW | Bay width |
| CME( , ) | Consistent mass matrix for element |
| CMS( , ) | Consistent mass matrix for structure |
| CV( ) | Characteristic values (eigenvalues) |
| CX, CY, CZ | Direction cosines $c_x$, $c_y$, and $c_z$ |
| D0( ) | Initial displacements of nodes |
| DAMPR | Damping ratio |
| DB0( ) | Initial displacements of bodies |
| DE( ) | Displacements of element nodes |
| DF( ) | Displacements of free nodes (*also* displacements at floors or framing levels) |
| DM( ) | Displacements at ends of members |
| DN( ) | Displacements in normal coordinates |
| DR( ) | Displacements of restraints |
| DS( ) | Displacements of structural nodes |
| DT | Duration of time step $\Delta t$ |
| E | Elasticity modulus |
| E1, E2, . . . | Elasticity constants |
| EL( ) | Element lengths |
| F( ) | Frequencies (cps) |
| FO( ) | Function ordinates |
| G | Shearing modulus |
| GAMMA | Newmark parameter $\gamma$ |
| GAX, GAY, GAZ | Ground acceleration factors for $x$, $y$, and $z$ directions |
| H | Thickness |
| I, J, K, L | Indexes |
| IAC | Indicator for imposing axial constraints |
| IAF | Indicator for actions at floors |

# Notation

| Symbol | Definition |
|---|---|
| ID( ) | Displacement indexes |
| IEO( ) | Element numbers for output of stresses |
| IGA | Indicator for ground accelerations |
| IML | Indicator for moving load |
| IN( ), JN( ) | Indexes for nodes of elements |
| IPL | Indicator for plotting |
| IPS | Indicator for plane stress or plane strain |
| IR, IC | Row and column indexes |
| IRO | Indicator for eliminating rotations |
| ISOLVE | Indicator for method of solution |
| IWR | Indicator for writing |
| JB( , ) | Body-node numbers |
| JNO( ) | Node numbers for output of displacements |
| LN | Loading number |
| NB | Number of bodies (*also* number of bays) |
| NBID | Number of bodies with initial displacements |
| NBIV | Number of bodies with initial velocities |
| NC | Number of columns |
| NDF | Number of degrees of freedom |
| NE | Number of elements |
| NEL | Number of elements with line loads |
| NEN | Number of element nodes |
| NEO | Number of elements for output |
| NES | Number of elements with surface loads |
| NEV | Number of elements with volume loads |
| NFO | Number of function ordinates |
| NJ | Number of joints (or nodes) on a body |
| NLB | Number of loaded bodies |
| NLN | Number of loaded nodes |
| NLS | Number of loading systems |
| NMODES | Number of modes |
| NN | Number of nodes |
| NNA | Number of nodes of type $A$ |

| Symbol | Definition |
|---|---|
| NND | Number of nodal displacements |
| NNF | Number of nodes of type $F$ |
| NNID | Number of nodes with initial displacements |
| NNIV | Number of nodes with initial velocities |
| NNO | Number of nodes for output |
| NNR | Number of nodal restraints |
| NRL( ) | Nodal restraint list |
| NRN | Number of restrained nodes |
| NS | Number of stories |
| NTS | Number of time steps |
| NUM | Number of repetitions |
| OMEGA( ) | Angular frequencies $\omega$ |
| P | Moving load |
| PHI( , ) | Eigenvectors $\Phi$ (mode shapes) |
| PR | Poisson's ratio |
| R( , ) | Rotation matrix |
| RC( ) | Radii of gyration of rigid bodies with respect to centers of mass |
| RHO | Mass density $\rho$ |
| SE( , ) | Element stiffness matrix |
| SH | Story height |
| SS( , ) | Structural stiffness matrix |
| SX, SY, . . . | Stresses |
| T( ) | Times |
| TIME | Time |
| V0( ) | Initial velocities of nodes |
| V0P | Initial velocity of moving load |
| VB0( ) | Initial velocities of bodies |
| VN( ) | Velocities in normal coordinates |
| VS( ) | Velocities of structural nodes |
| X( ), Y( ), Z( ) | Nodal coordinates |
| XC( ), YC( ), ZC( ) | Coordinates of point $c$ |
| XCJ( ), YCJ( ), ZCJ( ) | Components of offset vectors |
| XCK( ), YCK( ), ZCK( ) | Components of offset vectors |
| XI( ) | Torsion constants $I_x$ of cross sections |
| YI( ), ZI( ) | Second moments of area $I_y$ and $I_z$ of cross sections |

# General References

**TEXTBOOKS ON STRUCTURAL DYNAMICS**
(CHRONOLOGICAL ORDER)

1. Rogers, G. L., *Dynamics of Framed Structures*, Wiley, New York, 1959.
2. Norris, C. H., et al., *Structural Design for Dynamic Loads*, McGraw-Hill, New York, 1959.
3. Biggs, J. M., *Introduction to Structural Dynamics*, McGraw-Hill, New York, 1964.
4. Hurty, W. C., and Rubinstein, M. F., *Dynamics of Structures*, Prentice-Hall, Englewood Cliffs, N. J., 1964.
5. Lin, Y. K., *Probabilistic Theory of Structural Dynamics*, McGraw-Hill, New York, 1967.
6. Przemieniecki, J. S., *Theory of Matrix Structural Analysis*, McGraw-Hill, New York, 1968.
7. Rubinstein, M. F., *Structural Systems—Statics, Dynamics, and Stability*, Prentice-Hall, Englewood Cliffs, N. J., 1970.
8. Fryba, L., *Vibration of Solids and Structures under Moving Loads*, Noordhoff, Groningen, The Netherlands, 1972.
9. Fertis, D. G., *Dynamics and Vibrations of Structures*, Wiley, New York, 1973.
10. Clough, R. W., and Penzien, J., *Dynamics of Structures*, McGraw-Hill, New York, 1975.
11. Belytschko, T., Osias, J. R., and Marcal, P. V., *Finite Element Analysis of Transient Nonlinear Structural Behavior*, ASME, AMD, Vol. 14, 1975.
12. Bathe, K. J., and Wilson, E. L., *Numerical Methods in Finite Element Analysis*, Prentice-Hall, Englewood Cliffs, N. J., 1976.

13. Blevins, R. D., *Flow-Induced Vibrations*, Van Nostrand Reinhold, New York, 1977.
14. Simu, E., and Scanlan, R. H., *Wind Effects on Structures*, Wiley, New York, 1978.
15. Meirovitch, L., *Computational Methods in Structural Dynamics*, Sijthoff en Noordhoff, Alphen aan den Rijn, The Netherlands, 1980.
16. Craig, R. R., *Structural Dynamics*, Wiley, New York, 1981.
17. Paz, M., *Structural Dynamics*, 2nd ed., Van Nostrand Reinhold, New York, 1985.


## TEXTBOOKS ON VIBRATIONS
(CHRONOLOGICAL ORDER)

1. Rayleigh, J. W. S., *The Theory of Sound*, Dover, New York, 1945.
2. Den Hartog, J. P., *Mechanical Vibrations*, 4th ed., McGraw-Hill, New York, 1956.
3. Myklestad, N. O., *Fundamentals of Vibration Analysis*, McGraw-Hill, New York, 1956.
4. Jacobsen, L. S., and Ayre, R. S., *Engineering Vibrations*, McGraw-Hill, New York, 1958.
5. Bishop, R. E. D., and Johnson, D. C., *The Mechanics of Vibration*, Cambridge University Press, London, 1960.
6. Tong, K. N., *Theory of Mechanical Vibration*, Wiley, New York, 1960.
7. Church, A. H., *Mechanical Vibrations*, 2nd ed., Wiley, New York, 1963.
8. Crandall, S. H., and Mark, W. D., *Random Vibration in Mechanical Systems*, Academic Press, New York, 1963.
9. Bishop, R. E. D., Gladwell, G. M. L., and Michaelson, S., *The Matrix Analysis of Vibration*, Cambridge University Press, London, 1965.
10. Chen, Y., *Vibrations: Theoretical Methods*, Addison-Wesley, Reading, Mass., 1966.
11. Anderson, R. A., *Fundamentals of Vibrations*, Macmillan, New York, 1967.
12. Vernon, J. B., *Linear Vibration Theory*, Wiley, New York, 1967.
13. Vierck, R. K., *Vibration Analysis*, International Textbook, Scranton, Pa., 1967.
14. Haberman, C. M., *Vibration Analysis*, Charles E. Merrill, Columbus, Ohio, 1968.
15. Thomson, W. T., *Theory of Vibration with Applications*, Prentice-Hall, Englewood Cliffs, N. J., 1972.
16. Timoshenko, S. P., Young, D. H., and Weaver, W., Jr., *Vibration Problems in Engineering*, 4th ed., Wiley, New York, 1974.
17. Meirovitch, L., *Elements of Vibration Analysis*, McGraw-Hill, New York, 1975.
18. Newland, D. E., *An Introduction to Random Vibrations and Spectral Analysis*, Longmans, London, 1975.
19. Tse, F. S., Morse, I. E., and Hinkle, R. T., *Mechanical Vibrations—Theory and Applications*, 2nd ed., Allyn and Bacon, Boston, 1978.

# General References

## TEXTBOOKS ON FINITE ELEMENTS
(CHRONOLOGICAL ORDER)

1. Przemieniecki, J. S., *Theory of Matrix Structural Analysis*, McGraw-Hill, New York, 1968.
2. Desai, C. S., and Abel, J. F., *Introduction to the Finite Element Method*, Van Nostrand Reinhold, New York, 1972.
3. Oden, J. T., *Finite Elements of Nonlinear Continua*, McGraw-Hill, New York, 1972.
4. Martin, H. C., and Carey, G. F., *Introduction to Finite Element Analysis*, McGraw-Hill, New York, 1973.
5. Norrie, D. H., and deVries, G., *The Finite Element Method*, Academic Press, New York, 1973.
6. Strang, G., and Fix, G. J., *An Analysis of the Finite Element Method*, Prentice-Hall, Englewood Cliffs, N. J., 1973.
7. Gallagher, R. H., *Finite Element Analysis Fundamentals*, Prentice-Hall, Englewood Cliffs, N. J., 1975.
8. Bathe, K. J., and Wilson, E. L., *Numerical Methods in Finite Element Analysis*, Prentice-Hall, Englewood Cliffs, N. J., 1976.
9. Hinton, E., and Owen, D. R. J., *Finite Element Programming*, Academic Press, London, 1977.
10. Desai, C. S., *Elementary Finite Element Method*, Prentice-Hall, Englewood Cliffs, N. J., 1979.
11. Cheung, Y. K., and Yeo, M. F., *A Practical Introduction to Finite Element Analysis*, Pitman, London, 1979.
12. Hinton, E., and Owen, D. R. J., *An Introduction to Finite Element Computations*, Pineridge Press, Swansea, Wales (United Kingdom), 1979.
13. Owen, D. R. J., and Hinton, E., *Finite Elements in Plasticity*, Pineridge Press, Swansea, Wales (United Kingdom), 1980.
14. Cook, R. D., *Concepts and Applications of Finite Element Analysis*, 2nd ed., Wiley, New York, 1981.
15. Becker, E. B., et al., *Finite Elements* (five volumes), Prentice-Hall, Englewood Cliffs, N. J., 1981-1984.
16. Bathe, K. J., *Finite Element Procedures in Engineering Analysis*, Prentice-Hall, Englewood Cliffs, N. J., 1982.
17. Huebner, K. H., *The Finite Element Method for Engineers*, 2nd ed., Wiley, New York, 1983.
18. Weaver, W., Jr., and Johnston, P. R., *Finite Elements for Structural Analysis*, Prentice-Hall, Englewood Cliffs, N. J., 1984.
19. Segerlind, L. J., *Applied Finite Element Analysis*, 2nd ed., Wiley, New York, 1985.
20. Zienkiewicz, O. C., *The Finite Element Method*, 4th ed., McGraw-Hill, Maidenhead, Berkshire, England, 1987.

## TEXTBOOKS ON MATRIX ANALYSIS OF STRUCTURES
(CHRONOLOGICAL ORDER)

1. Laursen, H. I., *Matrix Analysis of Structures*, McGraw-Hill, New York, 1966.
2. Martin, H. C., *Introduction to Matrix Methods of Structural Analysis*, McGraw-Hill, New York, 1966.
3. Rubinstein, M. F., *Matrix Computer Analysis of Structures*, Prentice-Hall, Englewood Cliffs, N. J., 1966.
4. Hall, A. S., and Woodhead, R. W., *Frame Analysis*, 2nd ed., Wiley, New York, 1967.
5. Willems, N., and Lucas, W. M., Jr., *Matrix Analysis for Structural Engineers*, Prentice-Hall, Englewood Cliffs, N. J., 1968.
6. Beaufait, F. W., Rowan, W. H., Jr., Hoadley, P. G., and Hackett, R. M., *Computer Methods of Structural Analysis*, Prentice-Hall, Englewood Cliffs, N. J., 1970.
7. Rubinstein, M. F., *Structural Systems—Statics, Dynamics, and Stability*, Prentice-Hall, Englewood Cliffs, N. J., 1970.
8. Wang, C. K., *Matrix Methods of Structural Analysis*, 2nd ed., International Textbook, Scranton, Pa., 1970.
9. Meek, J. L., *Matrix Structural Analysis*, McGraw-Hill, New York, 1971.
10. Kardestuncer, H., *Elementary Matrix Analysis of Structures*, McGraw-Hill, New York, 1974.
11. Vanderbilt, M. D., *Matrix Structural Analysis*, Quantum, New York, 1974.
12. McGuire, W., and Gallagher, R. H., *Matrix Structural Analysis*, Wiley, New York, 1979.
13. Weaver, W., Jr., and Gere, J. M., *Matrix Analysis of Framed Structures*, 2nd ed., Van Nostrand Reinhold, New York, 1980.
14. Meyers, V. J., *Matrix Analysis of Structures*, Harper and Row, New York, 1983.
15. Holzer, S. H., *Computer Analysis of Structures*, Elsevier, New York, 1985.

# Appendix A
# Systems of Units
# and Material Properties

## A.1 SYSTEMS OF UNITS

The two most commonly used systems of units are the International System (*SI units*) and the United States Customary (*US units*). The first of these is called an *absolute system* because the fundamental quantity of mass is independent of where it is measured. On the other hand, the US system has force as a fundamental quantity. It is referred to as a *gravitational system* because the unit of force is defined as the weight of a certain mass, which varies with location on Earth.

In the SI system, the three fundamental units required for structural dynamics are mass (*kilogram*), length (*meter*), and time (*second*). Corresponding to mass is a derived force called a *newton*, which is defined as the force needed to accelerate one kilogram by the amount one meter per second squared. Thus, we have

$$1 \text{ N} = 1 \text{ kg} \cdot \text{m/s}^2$$

which is based on Newton's second law that force = mass × acceleration.

In the US system, we use force (*pound*), length (*foot*), and time (*second*). (Note that the unit of time is the same for both systems.) Corresponding to force is a derived mass, which carries the name *slug*. This quantity is defined as the mass that will be accelerated one foot per second squared when subjected to a force of one pound. Hence,

$$1 \text{ slug} = 1 \text{ lb-s}^2/\text{ft}$$

that comes from the formula mass = force/acceleration.

Table A.1 presents *conversion factors* for calculating quantities in SI units from those in US units. The factors are given to four significant figures, which usually exceeds the accuracy of the numbers to be converted. Note that stress is defined in SI units as the *pascal*. That is,

$$1 \text{ Pa} = 1 \text{ N/m}^2$$

**TABLE A.1  Conversion of US Units to SI Units**

| Quantity | US Units | × Factor | = SI Units |
|---|---|---|---|
| Length | inch (in.) | $2.540 \times 10^{-2}$ | meter (m) |
| Force | kilopound (kip or k) | 4.448 | kilonewton (kN) |
| Moment | kip-inch (k-in.) | $1.130 \times 10^{-1}$ | kilonewton · meter (kN · m) |
| Stress | kip/inch$^2$ (k/in.$^2$ or ksi) | $6.895 \times 10^3$ | kilopascal (kPa) |
| Mass | kip-sec$^2$/inch (k-s$^2$/in.) | $1.751 \times 10^2$ | megagram (Mg) |

For any numerical problem in structural mechanics, we must use a *consistent system of units*. By this we mean that all structural and load parameters must be expressed in the same units within each system. Some examples of consistent units for force, length, and time appear in Table A.2. For instance, in SI(1) we must express an applied force $P$ in newtons (N), a length $L$ in millimeters (mm), the modulus of elasticity $E$ in newtons per square millimeter (N/mm$^2$), an acceleration $\ddot{u}$ in millimeters per second squared (mm/s$^2$), and so on.

**TABLE A.2  Consistent Systems of Units**

| System | | Force | Length | Time |
|---|---|---|---|---|
| SI | (1) | newton | millimeter | second |
| | (2) | **kilonewton** | **meter** | **second** |
| | (3) | meganewton | kilometer | second |
| US | (1) | pound | foot | second |
| | (2) | **kilopound** | **inch** | **second** |
| | (3) | megapound | yard | second |

When programming structural dynamics for a digital computer, it is especially important that the system of units for input data be consistent. Otherwise, units would have to be converted within the logic of the program, thereby restricting its usage. For example, if in US units the length $L$ were given in feet and the modulus $E$ were expressed in pounds per square inch, the program would need to convert either $L$ to inches or $E$ to pounds per square foot.

For all of the numerical examples and problems in this book we use either SI(2) or US(2) in Table A.1. Thus, in SI units we take force $P$ in kilonewtons (kN), length $L$ in meters (m), modulus $E$ in giganewtons per square meter

Sec. A.2  Material Properties

(GN/m² or GPa), acceleration $\ddot{u}$ in meters per second squared (m/s²), and so on. [Note that the force kilonewton corresponds to the mass megagram (Mg).] Also, in US units we give force $P$ in kilopounds (kips or k), length $L$ in inches (in.), modulus $E$ in kips per square inch (k/in.² or ksi), acceleration $\ddot{u}$ in inches per second squared (in./s²), and so on.

## A.2 MATERIAL PROPERTIES

To analyze solids and structures composed of various materials, we need to know certain physical properties. For structural dynamics, the essential material properties are *modulus of elasticity E*, *Poisson's ratio v*, and *mass density ρ*. Table A.3 gives these properties in both US and SI units for some commonly used materials. Note that the shearing modulus $G$ is not listed in the table because it can be derived from $E$ and $\nu$.

**TABLE A.3  Properties of Materials**[a]

| Material | Modulus of Elasticity E | | Poisson's Ratio ν | Mass Density ρ | |
|---|---|---|---|---|---|
| | k/in.² | GPa | | k-s²/in.⁴ | Mg/m³ |
| Aluminum | $1.0 \times 10^4$ | 69 | 0.33 | $2.45 \times 10^{-7}$ | 2.62 |
| Brass | $1.5 \times 10^4$ | 103 | 0.34 | $8.10 \times 10^{-7}$ | 8.66 |
| Bronze | $1.5 \times 10^4$ | 103 | 0.34 | $7.80 \times 10^{-7}$ | 8.34 |
| Cast iron | $1.4 \times 10^4$ | 97 | 0.25 | $6.90 \times 10^{-7}$ | 7.37 |
| Concrete | $3.6 \times 10^3$ | 25 | 0.15 | $2.25 \times 10^{-7}$ | 2.40 |
| Magnesium | $6.5 \times 10^3$ | 45 | 0.35 | $1.71 \times 10^{-7}$ | 1.83 |
| Nickel | $3.0 \times 10^4$ | 207 | 0.31 | $8.25 \times 10^{-7}$ | 8.82 |
| Steel | $3.0 \times 10^4$ | 207 | 0.30 | $7.35 \times 10^{-7}$ | 7.85 |
| Titanium | $1.7 \times 10^4$ | 117 | 0.33 | $4.20 \times 10^{-7}$ | 4.49 |
| Tungsten | $5.5 \times 10^4$ | 379 | 0.20 | $1.80 \times 10^{-6}$ | 19.2 |

[a] Numbers in this table are taken from J. M. Gere and S. P. Timoshenko, *Mechanics of Materials*, 2nd ed., Brooks/Cole, Monterey, Calif., 1984.

# Appendix B
# Eigenvalues
# and Eigenvectors

## B.1 INVERSE ITERATION

All eigenvalue solution routines are iterative because we seek the roots of the *characteristic equation* (see Sec. 3.6), which is a polynomial of order $n$. Homogeneous action equations of motion provide the form of the eigenvalue problem to be solved. Thus, we have

$$\mathbf{S}\,\boldsymbol{\Phi}_i = \omega_i^2 \mathbf{M}\,\boldsymbol{\Phi}_i \qquad (i = 1, 2, \ldots, n) \tag{1}$$

which is a slightly rearranged version of Eq. (3.6-4). *Direct* (or *forward*) *iteration* of Eq. (1) would involve substitution of trial vectors for $\boldsymbol{\Phi}_i$ on the left-hand side and evaluation of $\omega_i^2$ on the right-hand side. This technique [1] converges to the dominant (largest) eigenvalue $\omega_n^2$ and the corresponding eigenvector $\boldsymbol{\Phi}_n$. However, to extract the smallest angular frequency first, we must use *reverse* (or *inverse*) *iteration*. For this second approach, let $\lambda_i = 1/\omega_i^2$ and factor the stiffness matrix $\mathbf{S}$ by the *modified Cholesky method* [2, 3], as follows:

$$\mathbf{S} = \overline{\mathbf{U}}^\mathrm{T} \mathbf{D}\, \overline{\mathbf{U}} \tag{2}$$

In this type of factorization, $\mathbf{D}$ is a diagonal matrix; and $\overline{\mathbf{U}}$ is upper triangular with values of unity in diagonal positions. Substituting Eq. (2) and $\omega_i^2 = 1/\lambda_i$ into Eq. (1) produces

$$\overline{\mathbf{U}}^\mathrm{T} \mathbf{D}\, \overline{\mathbf{U}}\, \lambda_i \boldsymbol{\Phi}_i = \mathbf{M}\, \boldsymbol{\Phi}_i \tag{3}$$

By using trial vectors for $\boldsymbol{\Phi}_i$ on the right-hand side of Eq. (3) and evaluating $\lambda_i$

Sec. B.1   Inverse Iteration                                              **499**

on the left-hand side, we find that the reverse iteration converges to the fundamental mode. To simplify notation, let

$$\mathbf{X}_i = \mathbf{\Phi}_i \qquad \mathbf{Y}_i = \lambda_i \mathbf{\Phi}_i \qquad \mathbf{B}_i = \mathbf{M}\, \mathbf{\Phi}_i \tag{4}$$

Then Eq. (3) becomes

$$\overline{\mathbf{U}}^{\mathrm{T}} \mathbf{D}\, \overline{\mathbf{U}}\, \mathbf{Y}_i = \mathbf{B}_i \tag{5}$$

which will be used in the iteration procedure.

To begin inverse iteration, we first assume an approximate shape of the fundamental mode. The usual arbitrary choice for this starting vector is a column of ones. Thus,

$$(\mathbf{X}_1)_1 = \{1, 1, 1, \ldots, 1\} \tag{6}$$

Substituting this vector into the last of Eqs. (4), we calculate $(\mathbf{B}_1)_1$ as

$$(\mathbf{B}_1)_1 = \mathbf{M}(\mathbf{X}_1)_1 \tag{7}$$

Then solve the simultaneous algebraic equations in Eq. (5) to obtain

$$(\mathbf{Y}_1)_1 \approx (\lambda_1)_1 (\mathbf{X}_1)_1 \tag{8}$$

This expression is only an approximation, unless the estimated mode shape satisfies Eq. (5) exactly. Also, a first approximation to the eigenvalue $\lambda_1$ may be found by dividing any term in vector $(\mathbf{Y}_1)_1$ by the corresponding term in $(\mathbf{X}_1)_1$. That is,

$$(\lambda_1)_1 \approx \frac{(Y_j)_1}{(X_j)_1} \qquad (1 \le j \le n) \tag{9}$$

If $(Y_j)_1$ is chosen to be the largest (positive or negative) term in vector $(\mathbf{Y}_1)_1$, normalization with respect to that value gives us the second trial vector:

$$(\mathbf{X}_1)_2 = \frac{1}{(\lambda_1)_1} (\mathbf{Y}_1)_1 \tag{10}$$

in which the normalization constant is $(\lambda_1)_1$ itself. This procedure is repeated until the eigenvalue $\lambda_1$ and its associated eigenvector $\mathbf{X}_1$ are determined to some specified accuracy.

In the $k$th iteration, the recurrence equations for the steps described above are:

1. Calculate vector $(\mathbf{B}_1)_k$ as

$$(\mathbf{B}_1)_k = \mathbf{M}(\mathbf{X}_1)_k \tag{11}$$

2. Solve for $(\mathbf{Y}_1)_k$ in

$$\overline{\mathbf{U}}^{\mathrm{T}} \mathbf{D}\, \overline{\mathbf{U}}(\mathbf{Y}_1)_k = (\mathbf{B}_1)_k \tag{12}$$

3. Find the new trial vector $(\mathbf{X}_1)_{k+1}$ to be

$$(\mathbf{X}_1)_{k+1} = \frac{1}{(\lambda_1)_k}(\mathbf{Y}_1)_k \tag{13}$$

where $(\lambda_1)_k$ is the largest term in vector $(\mathbf{Y}_1)_k$. To check convergence of the eigenvalue, we use the expression

$$\left|\frac{(\lambda_1)_k - (\lambda_1)_{k-1}}{(\lambda_1)_k}\right| \leq 10^{-n_d} \tag{14}$$

in which $n_d$ is the number of significant digits of accuracy desired.

Bathe [4] recommends calculation of $(\omega_1^2)_k = 1/(\lambda_1)_k$ from the *Rayleigh quotient*

$$(\omega_1^2)_k = \frac{(\mathbf{X}_1^T)_k \mathbf{S}(\mathbf{X}_1)_k}{(\mathbf{X}_1^T)_k \mathbf{M}(\mathbf{X}_1)_k} \tag{15}$$

which produces much faster convergence. Unfortunately, improving the rate of convergence of the eigenvalue has no effect on the rate of convergence of the eigenvector. However, improvement of eigenvector convergence can be attained with spectral shifting, as described later in this appendix.

After the fundamental mode has been determined, it is usually eliminated from the eigenvalue equations by the process of *deflation* [1, 4]. For this technique, we express the orthogonality of the eigenvectors $\mathbf{\Phi}_1$ and $\mathbf{\Phi}_i$ with respect to the mass matrix $\mathbf{M}$, as follows:

$$\mathbf{\Phi}_1^T \mathbf{M} \mathbf{\Phi}_i = 0 \qquad (i = 2, 3, \ldots, n) \tag{16}$$

Adding this *modal constraint* condition to the eigenvalue equations [see Eq. (1)], we can reduce their order from $n$ to $n - 1$. For the reduced equations, the second mode becomes dominant and also may be calculated by inverse iteration. This deflation–iteration sequence is repeated to extract as many modes as desired. However, to retain accuracy in each iteration, we must $\mathbf{M}$-orthogonalize each new vector with those found previously. This may be accomplished using *Gram–Schmidt orthogonalization*, as shown in Ref. 4.

**Example B.1**

Now we shall apply inverse iteration to extract the fundamental mode from the eigenvalue problem for the plane truss in Fig. 3.11(a) (see Sec. 3.5). For this truss the stiffness matrix was found to be

$$\mathbf{S} = s\begin{bmatrix} 1.36 & -0.36 & -0.48 \\ -0.36 & 0.36 & 0.48 \\ -0.48 & 0.48 & 1.64 \end{bmatrix} \tag{a}$$

where $s = EA/L$. Also, the consistent-mass matrix is

Sec. B.1    Inverse Iteration

$$\mathbf{M} = m \begin{bmatrix} 2.72 & 1 & 0 \\ 1 & 3.28 & 0 \\ 0 & 0 & 3.28 \end{bmatrix} \qquad (b)$$

for which $m = \rho A L/6$.

In accordance with Eq. (2), we factor matrix **S** to obtain

$$\mathbf{D} = s \begin{bmatrix} 1.360 & 0 & 0 \\ 0 & 0.2647 & 0 \\ 0 & 0 & 1.000 \end{bmatrix} \qquad (c)$$

and

$$\overline{\mathbf{U}} = \begin{bmatrix} 1 & -0.2647 & -0.3529 \\ 0 & 1 & 1.333 \\ 0 & 0 & 1 \end{bmatrix} \qquad (d)$$

As the starting eigenvector, assume

$$(\mathbf{X}_1)_1 = \{1, 1, 1\} \qquad (e)$$

In the first iteration, we evaluate $(\mathbf{B}_1)_1$ from the last of Eqs. (4), as follows:

$$(\mathbf{B}_1)_1 = \mathbf{M}(\mathbf{X}_1)_1 = m\{3.72, 4.28, 3.28\} \qquad (f)$$

Then solve for $(\mathbf{Y}_1)_1$ from Eq. (5):

$$(\mathbf{Y}_1)_1 = \frac{m}{s}\{8.000, 23.12, -2.426\} \qquad (g)$$

Normalize this vector with respect to its largest term to find the new trial vector

$$(\mathbf{X}_1)_2 = \frac{1}{(\lambda_1)_1}(\mathbf{Y}_1)_1 = \{0.3460, 1.000, -0.1049\} \qquad (h)$$

In this iteration the first approximation to the eigenvalue is

$$(\lambda_1)_1 = 23.12 \frac{m}{s} \qquad (i)$$

which is the largest term in vector $(\mathbf{Y}_1)_1$.

Results of successive iterations are listed in Table B.1. We see that convergence (to four significant digits) for the reciprocal $\lambda_1$ of the eigenvalue $\omega_1^2$ occurs in six cycles of iteration. However, the eigenvector $\mathbf{X}_1$ requires another cycle to attain the same accuracy. Their final values are

$$\lambda_1 = 22.22\frac{m}{s} \qquad \omega_1^2 = \frac{1}{\lambda_1} = 0.04501\frac{s}{m} \qquad (j)$$

and

$$\mathbf{X}_1 = \{0.2314, 1.000, -0.2472\} \qquad (k)$$

which are the same as those found by Program VIBPT in Example 3.4.

TABLE B.1 Iteration of Fundamental Mode

| Cycle | 1 | 2 | 3 | 4 | 5 | 6 | 7 |
|---|---|---|---|---|---|---|---|
| $(X_1)_k$ | $(X_1)_1$ | $(X_1)_2$ | $(X_1)_3$ | $(X_1)_4$ | $(X_1)_5$ | $(X_1)_6$ | $(X_1)_7$ |
| Vector | 1<br>1<br>1 | 0.3460<br>1.000<br>−0.1049 | 0.2469<br>1.000<br>−0.2297 | 0.2334<br>1.000<br>−0.2450 | 0.2316<br>1.000<br>−0.2469 | 0.2314<br>1.000<br>−0.2472 | 0.2314<br>1.000<br>−0.2472 |
| $\frac{s}{m}(\lambda_1)_k$ | 23.12 | 22.54 | 22.27 | 22.23 | 22.22 | 22.22 | 22.22 |

Sec. B.1    Inverse Iteration    503

**Spectral Shifting**

To improve the rate of convergence of eigenvectors, we use the technique known as *spectral shifting*. For this purpose, let the constant $\alpha_i$ be a number close (but not equal) to $\omega_i^2$, which is the eigenvalue to be calculated. Then subtract $\alpha_i \mathbf{M} \, \mathbf{\Phi}_i$ from both sides of Eq. (1), and divide by $\omega_i^2 - \alpha_i$ to obtain

$$\mathbf{S}^* \lambda_i^* \mathbf{\Phi}_i = \mathbf{M} \, \mathbf{\Phi}_i \tag{17}$$

where

$$\mathbf{S}^* = \mathbf{S} - \alpha_i \mathbf{M} \qquad \lambda_i^* = \frac{1}{\omega_i^2 - \alpha_i} \tag{18}$$

By this manipulation, the dominance of $\lambda_i^*$ in Eq. (17) is greater than that of $\lambda_i$ in Eq. (3), because in the second of Eqs. (18) we see that $\lambda_i^* \to \infty$ as $\alpha_i \to \omega_i^2$.

After the spectral shift, inverse iteration proceeds as before, except that the factorization of $\mathbf{S}^*$ is expressed as

$$\mathbf{S}^* = \overline{\mathbf{U}}^{*\mathrm{T}} \mathbf{D}^* \overline{\mathbf{U}}^* \tag{19}$$

Also, the eigenvalue $\omega_i^2$ becomes

$$\omega_i^2 = \frac{1}{\lambda_i^*} + \alpha_i \tag{20}$$

which need only be computed once (after convergence).

Spectral shifting may be used to aid the following tasks:

1. Extraction of modes near an expected resonance ($\alpha_i \to \Omega$), where $\Omega$ is the angular frequency of a periodic forcing function.
2. Extraction of modes corresponding to repeated roots ($\alpha_i \to \omega_m^2$) of multiplicity $m$, including rigid-body modes ($\alpha_i \to 0$)
3. Extraction of sequential modes by using ($\alpha_i \to \omega_i^2$) after deflation

If the stiffness matrix $\mathbf{S}$ in Eq. (1) is semidefinite, one or more rigid-body modes exist. In that case, a small (but finite) spectral shift away from zero makes $\mathbf{S}^*$ in Eq. (19) positive-definite. Then by inverse interation we can extract the first rigid-body mode. Elimination of that mode from the equations by deflation allows iteration for the second rigid-body mode, and so on. A similar procedure is required when calculating nonzero repeated eigenvalues.

**Example B.2**

Let us use a spectral shift to make the second mode dominant for the plane truss in Example B.1. The following shift:

$$\alpha_2 = 0.36 \frac{s}{m} \approx \omega_2^2 \tag{$\ell$}$$

will allow extraction of the second eigenvalue and eigenvector without deflation. From the first of Eqs. (18), the modified stiffness matrix becomes

$$\mathbf{S}^* = \mathbf{S} - \alpha_2 \mathbf{M} = s \begin{bmatrix} 0.3808 & -0.72 & -0.48 \\ -0.72 & -0.8208 & 0.48 \\ -0.48 & 0.48 & 0.4592 \end{bmatrix} \quad \text{(m)}$$

Factorization of $\mathbf{S}^*$ as in Eq. (19) yields

$$\mathbf{D}^* = s \begin{bmatrix} 0.3808 & 0 & 0 \\ 0 & -2.182 & 0 \\ 0 & 0 & -0.06207 \end{bmatrix} \quad \text{(n)}$$

and

$$\overline{\mathbf{U}}^* = \begin{bmatrix} 1 & -1.891 & -1.261 \\ 0 & 1 & 0.1959 \\ 0 & 0 & 1 \end{bmatrix} \quad \text{(o)}$$

As before, we assume the starting eigenvector

$$(\mathbf{X}_2)_1 = \{1, 1, 1\} \quad \text{(p)}$$

In the first iteration, the vector $(\mathbf{B}_2)_1$ is computed to be

$$(\mathbf{B}_2)_1 = \mathbf{M}(\mathbf{X}_2)_1 = m\{3.72, 4.28, 3.28\} \quad \text{(q)}$$

which contains the same values as Eq. (f). Solution of Eq. (17) for vector $(\mathbf{Y}_2)_1$ produces

$$(\mathbf{Y}_2)_1 = \frac{m}{s}\{-82.52, 12.97, -92.68\} \quad \text{(r)}$$

Normalization of this vector with respect to its largest term gives

$$(\mathbf{X}_2)_2 = \frac{1}{(\lambda_2^*)_1}(\mathbf{Y}_2)_1 = \{0.8904, -0.1400, 1.000\} \quad \text{(s)}$$

in which

$$(\lambda_2^*)_1 = -92.68\frac{m}{s} \quad \text{(t)}$$

Table B.2 shows the results of successive iterations. In this case convergence (to four significant digits) for both $\lambda_2^*$ and the second eigenvector $\mathbf{X}_2$ occurs in five cycles. Final values for this mode are

$$\lambda_2^* = -83.25\frac{m}{s} \qquad \omega_2^2 = \frac{1}{\lambda_2^*} + \alpha_2 = 0.3480\frac{s}{m} \quad \text{(u)}$$

and

$$\mathbf{X}_2 = \{0.8673, -0.1715, 1.000\} \quad \text{(v)}$$

which are again equal to those in Example 3.4.

TABLE B.2 Iteration of Second Mode

| Cycle | 1 | 2 | 3 | 4 | 5 |
|---|---|---|---|---|---|
| $(X_2)_k$ | $(X_2)_1$ | $(X_2)_2$ | $(X_2)_3$ | $(X_2)_4$ | $(X_2)_5$ |
| Vector | 1<br>1<br>1 | 0.8904<br>−0.1400<br>1.000 | 0.8677<br>−0.1701<br>1.000 | 0.8673<br>−0.1715<br>1.000 | 0.8673<br>−0.1715<br>1.000 |
| $\frac{s}{m}(\lambda_2^*)_k$ | −92.68 | −84.21 | −83.27 | −83.25 | −83.25 |

## B.2 TRANSFORMATION METHODS

When most or all of the eigenvalues and eigenvectors are desired, *transformation methods* [5] prove to be more efficient than inverse iteration. In this section we describe the Jacobi, Givens, and Householder transformation procedures, as well as QR iteration. To confirm ideas, a numerical example follows the discussion of each approach.

As a preliminary matter, we assume that the expanded eigenvalue problem has been converted to the *standard, symmetric form:*

$$\mathbf{A}\,\mathbf{V} = \mathbf{V}\,\mathbf{\Lambda} \tag{1}$$

as described in Secs. 3.6 and 4.2. In Eq. (1) the symbol $\mathbf{V}$ denotes an orthogonal *modal matrix* of $n$ eigenvectors that are normalized to unit lengths and listed column-wise. Also, the eigenvalues $\lambda_1, \lambda_2, \ldots, \lambda_n$ appear in diagonal positions of the *spectral matrix* $\mathbf{\Lambda}$.

The basic process in all transformation methods is to diagonalize matrix $\mathbf{A}$, as follows:

$$\mathbf{V}^T \mathbf{A}\,\mathbf{V} = \mathbf{\Lambda} \tag{2}$$

When this is accomplished, we have found not only the spectral matrix $\mathbf{\Lambda}$ but also the normalized modal matrix $\mathbf{V}$.

### Jacobi Method

Since its development in the nineteenth century [6], the Jacobi method has enjoyed extensive usage by mathematicians, scientists, and engineers throughout the world. The essential idea in this approach is to zero a selected off-diagonal term $A_{pq} = A_{qp}$ of the coefficient matrix $\mathbf{A}$ in Eq. (1). This is accomplished by operating upon that matrix with a *generalized rotation matrix* $\hat{\mathbf{R}}_k$ in step $k$, as follows:

$$\mathbf{A}_{k+1} = \hat{\mathbf{R}}_k \mathbf{A}_k \hat{\mathbf{R}}_k^T \tag{3}$$

The form of the $n \times n$ rotation matrix is

$$\hat{\mathbf{R}}_k = \begin{bmatrix} 1 & & & & & & \\ & \ddots & & & & & \\ & & \cos\theta_k & & \sin\theta_k & & \\ & & & \ddots & & & \\ & & & & 1 & & \\ & & & & & \ddots & \\ & & -\sin\theta_k & & \cos\theta_k & & \\ & & & & & & \ddots \\ & & & & & & & 1 \end{bmatrix} \begin{array}{l} \\ \\ \text{row } p \\ \\ \\ \\ \text{row } q \\ \\ \end{array} \quad (4)$$

$$\text{col. } p \qquad \text{col. } q$$

In this rather sparse array, we define the terms

$$(\hat{R}_{pp})_k = (\hat{R}_{qq})_k = \cos\theta_k \tag{5a}$$

$$(\hat{R}_{pq})_k = -(\hat{R}_{qp})_k = \sin\theta_k \tag{5b}$$

$$(\hat{R}_{ii})_k = 1 \qquad (i \neq p \text{ or } q) \tag{5c}$$

$$(\hat{R}_{ij})_k = 0 \qquad (i \text{ or } j \neq p \text{ or } q) \tag{5d}$$

The multiplication in Eq. (3) alters only terms in rows and columns $p$ and $q$ of matrix $\mathbf{A}$, which become

$$(A_{pp})_{k+1} = (A_{pp}\cos^2\theta + 2A_{pq}\cos\theta\sin\theta + A_{qq}\sin^2\theta)_k \tag{6a}$$

$$(A_{qq})_{k+1} = (A_{pp}\sin^2\theta - 2A_{pq}\cos\theta\sin\theta + A_{qq}\cos^2\theta)_k \tag{6b}$$

$$(A_{pq})_{k+1} = [(A_{qq} - A_{pp})\cos\theta\sin\theta + A_{pq}(\cos^2\theta - \sin^2\theta)]_k$$
$$= (A_{qp})_{k+1} \tag{6c}$$

$$(A_{ip})_{k+1} = (A_{ip}\cos\theta + A_{iq}\sin\theta)_k = (A_{pi})_{k+1} \tag{6d}$$

$$(A_{jq})_{k+1} = (-A_{jp}\sin\theta + A_{jq}\cos\theta)_k = (A_{qj})_{k+1} \tag{6e}$$

The angle $\theta_k$ is chosen so that the term $(A_{pq})_{k+1}$ becomes zero. Thus, from Eq. (6c) we have

$$\tan 2\theta_k = \left(\frac{2A_{pq}}{A_{pp} - A_{qq}}\right)_k \tag{7}$$

where $|\theta_k| \leq \pi/4$. After $n_r$ rotations, Eq. (3) yields

$$\hat{\mathbf{R}}_{n_r} \cdots \hat{\mathbf{R}}_2 \hat{\mathbf{R}}_1 \mathbf{A}_1 \hat{\mathbf{R}}_1^T \hat{\mathbf{R}}_2^T \cdots \hat{\mathbf{R}}_{n_r}^T \approx \mathbf{\Lambda} \tag{8}$$

Therefore, the matrix of orthonormal eigenvectors is

$$\mathbf{V} = \hat{\mathbf{R}}_1^T \hat{\mathbf{R}}_2^T \cdots \hat{\mathbf{R}}_{n_r}^T \tag{9}$$

and Eq. (8) becomes

$$\mathbf{V}^T \mathbf{A} \mathbf{V} \approx \mathbf{\Lambda} \tag{10}$$

which is an approximation of Eq. (2).

The Jacobi method may be applied selectively to annihilate the largest off-diagonal term in matrix **A**. However, this approach requires searching for that term before it can be put to zero. More commonly, we operate in sweeps, systematically annihilating all of the off-diagonal terms by rows or columns. Each sweep of this kind requires approximately $2n^3$ multiplications. Equations (6) give the formulas to be used for altering either the upper or lower triangular part of matrix **A**.

During a particular sweep, terms that are zeroed do not necessarily remain zeros. However, the square root of the mean of the squares (RMS) for off-diagonal terms is reduced after each annihilation (and even more dramatically after each sweep).

**Example B.3**

With one sweep of the Jacobi method, determine approximately the spectral matrix **Λ** and the modal matrix **V** associated with the 3 × 3 array

$$\mathbf{A} = \begin{bmatrix} 0.64 & -0.48 & 0 \\ -0.48 & 1.44 & -0.48 \\ 0 & -0.48 & 1.92 \end{bmatrix} \tag{a}$$

In the first transformation of matrix **A**, we shall make the term $A_{12} = A_{21}$ equal to zero. Therefore, $p = 1$ and $q = 2$, so Eq. (7) becomes

$$\tan 2\theta_1 = \frac{2A_{12}}{A_{11} - A_{22}} = \frac{(2)(-0.48)}{0.64 - 1.44} = 1.2 \tag{b}$$

From this expression we determine the angle $\theta_1$ and its sine and cosine as

$$\theta_1 = 0.4380 \qquad \sin \theta_1 = 0.4242 \qquad \cos \theta_1 = 0.9056 \tag{c}$$

Then the first generalized rotation matrix in Eq. (4) takes the form

$$\hat{\mathbf{R}}_1 = \begin{bmatrix} 0.9056 & 0.4242 & 0 \\ -0.4242 & 0.9056 & 0 \\ 0 & 0 & 1 \end{bmatrix} \tag{d}$$

and the first transformation in Eq. (3) produces

$$\mathbf{A}_2 = \hat{\mathbf{R}}_1 \mathbf{A}_1 \hat{\mathbf{R}}_1^T = \begin{bmatrix} 0.4152 & 0 & -0.2036 \\ 0 & 1.665 & -0.4347 \\ -0.2036 & -0.4347 & 1.920 \end{bmatrix} \tag{e}$$

As a consequence, the first rotation has reduced the RMS of off-diagonal terms in matrix **A** from 0.48 to 0.3394.

For the second transformation, we make $A_{13} = A_{31}$ equal to zero, so that

$$\tan 2\theta_2 = \frac{2A_{13}}{A_{11} - A_{33}} = \frac{(2)(-0.2036)}{0.4152 - 1.92} = 0.2706 \tag{f}$$

Thus, we have

$$\theta_2 = 0.1321 \qquad \sin \theta_2 = 0.1317 \qquad \cos \theta_2 = 0.9913 \tag{g}$$

and the second rotation matrix is

$$\hat{\mathbf{R}}_2 = \begin{bmatrix} 0.9913 & 0 & 0.1317 \\ 0 & 1 & 0 \\ -0.1317 & 0 & 0.9913 \end{bmatrix} \tag{h}$$

Using $\hat{\mathbf{R}}_2$ in Eq. (3) yields

$$\mathbf{A}_3 = \hat{\mathbf{R}}_2 \mathbf{A}_2 \hat{\mathbf{R}}_2^T = \begin{bmatrix} 0.3881 & -0.05727 & 0 \\ -0.05727 & 1.665 & -0.4309 \\ 0 & -0.4309 & 1.947 \end{bmatrix} \tag{i}$$

for which the RMS of off-diagonal terms is 0.3074.

The third transformation involves zeroing $A_{23} = A_{32}$, which gives

$$\tan 2\theta_3 = \frac{2A_{23}}{A_{22} - A_{33}} = \frac{(2)(-0.4309)}{1.665 - 1.947} = 3.053 \tag{j}$$

Hence,

$$\theta_3 = 0.6272 \qquad \sin \theta_3 = 0.5868 \qquad \cos \theta_3 = 0.8097 \tag{k}$$

So the third rotation matrix becomes

$$\hat{\mathbf{R}}_3 = \begin{bmatrix} 1 & 0 & 0 \\ 0 & 0.8097 & 0.5868 \\ 0 & -0.5868 & 0.8097 \end{bmatrix} \tag{$\ell$}$$

Substituting $\hat{\mathbf{R}}_3$ into Eq. (3) results in

$$\mathbf{A}_4 = \hat{\mathbf{R}}_3 \mathbf{A}_3 \hat{\mathbf{R}}_3^T = \begin{bmatrix} 0.3881 & -0.04637 & 0.03361 \\ -0.04637 & 1.353 & 0 \\ 0.03361 & 0 & 2.259 \end{bmatrix} \tag{m}$$

where the RMS of off-diagonal terms is 0.04051. Note that each off-diagonal term is now at least an order of magnitude smaller than at the beginning of the sweep.

At the end of the sweep, the diagonal terms in matrix $\mathbf{A}_4$ represent good approximations for the eigenvalues of the original matrix. That is,

$$\mathbf{\Lambda} \approx \begin{bmatrix} 0.3881 & 0 & 0 \\ 0 & 1.353 & 0 \\ 0 & 0 & 2.259 \end{bmatrix} \tag{n}$$

In addition, approximate eigenvectors are calculated as

$$\mathbf{V} \approx \hat{\mathbf{R}}_1^T \hat{\mathbf{R}}_2^T \hat{\mathbf{R}}_3^T = \begin{bmatrix} 0.8927 & -0.4135 & 0.1523 \\ 0.4205 & 0.7005 & -0.5767 \\ 0.1317 & 0.5817 & 0.8026 \end{bmatrix} \quad (\text{o})$$

Modal vectors appearing column-wise in this matrix are automatically normalized to have unit lengths.

### Givens Method

Instead of annihilating the $p, q$ term (as in the Jacobi method), Givens [7] proposed annihilating the $p - 1, q$ term instead. Then a zeroed term will remain zero during a forward sweep. For this purpose, we set Eq. (6e) equal to zero to obtain

$$\tan \theta_k = \left( \frac{A_{p-1,q}}{A_{p-1,p}} \right)_k \quad (11)$$

where $|\theta_k| \leq \pi/2$. For the first row of matrix $\mathbf{A}$, rotation in the 2-3 plane gives

$$\tan \theta_1 = \left( \frac{A_{13}}{A_{12}} \right)_1$$

Rotation in the 2-4 plane yields

$$\tan \theta_2 = \left( \frac{A_{14}}{A_{12}} \right)_2$$

and so on. After we clear the first row (and column), matrix $\mathbf{A}$ has the form

$$\mathbf{A}_{n-2} = \begin{bmatrix} x & x & 0 & 0 & 0 & \ldots & 0 \\ x & x & x & x & x & \ldots & x \\ 0 & x & x & x & x & \ldots & x \\ 0 & x & x & x & x & \ldots & x \\ 0 & x & x & x & x & \ldots & x \\ \ldots & \ldots & \ldots & \ldots & \ldots & \ldots & \ldots \\ 0 & x & x & x & x & \ldots & x \end{bmatrix} \quad (12)$$

This technique produces a *tridiagonal matrix* $\mathbf{T}$ in a finite number of steps, consisting of $n_r = (n - 2)(n - 1)/2$ rotations. The entire sweep requires approximately $4n^3/3$ multiplications.

While the Givens method does not lead directly to the spectral matrix, we consider it a useful preliminary to the Householder method, which is described next.

**Example B.4**

By the Givens method, transform the following 4 × 4 matrix to tridiagonal form:

$$\mathbf{A} = \begin{bmatrix} 1.36 & -0.48 & -1.00 & 0 \\ -0.48 & 1.64 & 0 & 0 \\ -1.00 & 0 & 1.36 & 0.48 \\ 0 & 0 & 0.48 & 1.64 \end{bmatrix} \tag{p}$$

We start by annihilating the term $A_{13}$ with a rotation in the 2-3 plane. For this purpose, Eq. (11) results in

$$\theta_1 = \tan^{-1} \frac{A_{13}}{A_{12}} = \tan^{-1} \frac{-1.00}{-0.48} = 1.123 \tag{q}$$

Then we have

$$\sin \theta_1 = 0.9015 \qquad \cos \theta_1 = 0.4327 \tag{r}$$

and the first rotation matrix $\hat{\mathbf{R}}_1$ becomes

$$\hat{\mathbf{R}}_1 = \begin{bmatrix} 1 & 0 & 0 & 0 \\ 0 & 0.4327 & 0.9015 & 0 \\ 0 & -0.9015 & 0.4327 & 0 \\ 0 & 0 & 0 & 1 \end{bmatrix} \tag{s}$$

Using this operator in Eq. (3) gives

$$\mathbf{A}_2 = \hat{\mathbf{R}}_1 \mathbf{A}_1 \hat{\mathbf{R}}_1^T = \begin{bmatrix} 1.360 & -1.109 & 0 & 0 \\ -1.109 & 1.412 & -0.1092 & 0.4327 \\ 0 & -0.1092 & 1.588 & 0.2077 \\ 0 & 0.4327 & 0.2077 & 1.640 \end{bmatrix} \tag{t}$$

In this case only one transformation is required to clear the first row (and column) outside of the triangular part. The term $A_{14} = A_{41}$ was zero initially and remained unaffected by the operation above.

The second step in this example is to annihilate element $A_{24}$ by a rotation in the 3-4 plane. Thus, Eq. (11) produces

$$\theta_2 = \tan^{-1} \frac{A_{24}}{A_{23}} = \tan^{-1} \frac{0.4327}{-0.1092} = -1.324 \tag{u}$$

From this we get

$$\sin \theta_2 = -0.9696 \qquad \cos \theta_2 = 0.2447 \tag{v}$$

Then the second rotation matrix $\hat{\mathbf{R}}_2$ is

$$\hat{\mathbf{R}}_2 = \begin{bmatrix} 1 & 0 & 0 & 0 \\ 0 & 1 & 0 & 0 \\ 0 & 0 & 0.2447 & -0.9696 \\ 0 & 0 & 0.9696 & 0.2447 \end{bmatrix} \tag{w}$$

Sec. B.2  Transformation Methods

Substituting this matrix into Eq. (3) yields

$$\mathbf{A}_3 = \hat{\mathbf{R}}_2 \mathbf{A}_2 \hat{\mathbf{R}}_2^T = \begin{bmatrix} 1.360 & -1.109 & 0 & 0 \\ -1.109 & 1.412 & -0.4463 & 0 \\ 0 & -0.4463 & 1.538 & -0.1953 \\ 0 & 0 & -0.1953 & 1.689 \end{bmatrix} \quad (x)$$

which is now in tridiagonal form.

### Householder Method

A Householder transformation operator [8] causes all of the terms in a vector to become zero except the first, which becomes the length of the vector itself. The operator has the form

$$\mathbf{P} = \mathbf{I} - 2\mathbf{e}\,\mathbf{e}^T \quad (13)$$

in which $\mathbf{e}$ is a column vector of unit length. The matrix $\mathbf{P}$ is both symmetric and orthogonal, so that

$$\mathbf{P} = \mathbf{P}^T = \mathbf{P}^{-1} \quad (14)$$

Hence, $\mathbf{P}$ is equal to its own inverse.

Consider a vector $\mathbf{a}$ to be converted to $\mathbf{b} = \{\pm s, 0, 0, \ldots, 0\}$, where $s$ is the length of $\mathbf{a}$. That is,

$$\mathbf{P}\,\mathbf{a} = \mathbf{b} \quad (15)$$

We form the matrix $\mathbf{P}$ by first creating $\mathbf{e}$ in Eq. (13) from the vector $\mathbf{a}$, as follows:

$$s^2 = \sum_{j=1}^{n} a_j^2 \quad (16a)$$

$$\mathbf{c} = \{a_1 \pm s, a_2, a_3, \ldots, a_n\} \quad (16b)$$

$$\mathbf{e} = \frac{\mathbf{c}}{\sqrt{\mathbf{c}^T \mathbf{c}}} \quad (16c)$$

In Eq. (16b) the sign of $s$ is taken to be the same as that of $a_1$, so that there is no possibility of getting zero for the first term in vector $\mathbf{c}$. The unit vector $\mathbf{e}$ in Eq. (13) is the result of normalizing vector $\mathbf{c}$ to unit length, as indicated in Eq. (16c).

Now let us express the product $\mathbf{P}\,\mathbf{a}$ in Eq. (15) in terms of the vector $\mathbf{c}$. Using Eqs. (13) and (16c), we obtain

$$\mathbf{P}\,\mathbf{a} = \left(\mathbf{I} - 2\frac{\mathbf{c}\,\mathbf{c}^T}{\mathbf{c}^T \mathbf{c}}\right)\mathbf{a} = \mathbf{a} - \frac{2\mathbf{c}(\mathbf{c}^T \mathbf{a})}{(\mathbf{c}^T \mathbf{c})} \quad (17)$$

Results of the inner products (in parentheses) from the second form of Eq. (17)

are

$$\mathbf{c}^T\mathbf{a} = (a_1 \pm s)a_1 + \sum_{j=2}^{n} a_j^2 = s^2 \pm a_1 s$$

$$\mathbf{c}^T\mathbf{c} = (a_1 \pm s)^2 + \sum_{j=2}^{n} a_j^2 = s^2 \pm 2a_1 s + s^2 = 2(s^2 \pm a_1 s)$$

Substitution of these expressions into Eq. (17) produces

$$\mathbf{P}\mathbf{a} = \mathbf{a} - \frac{2\mathbf{c}(s^2 \pm a_1 s)}{2(s^2 \pm a_1 s)} = \mathbf{a} - \mathbf{c}$$

$$= \{\pm s, 0, 0, \ldots, 0\} = \mathbf{b} \tag{18}$$

as desired.

In a similar manner, the last $n - 2$ terms in the first column of the coefficient matrix $\mathbf{A}$ may be zeroed. With this objective, we let

$$s_1^2 = \sum_{j=2}^{n} A_{j,1}^2 \tag{19a}$$

$$\mathbf{c}_1 = \{0, A_{21} \pm s_1, A_{31}, A_{41}, \ldots, A_{n1}\} \tag{19b}$$

$$\mathbf{e}_1 = \frac{\mathbf{c}_1}{\sqrt{\mathbf{c}_1^T \mathbf{c}_1}} \tag{19c}$$

$$\mathbf{P}_1 = \mathbf{I} - 2\mathbf{e}_1 \mathbf{e}_1^T \tag{19d}$$

Then

$$\mathbf{P}_1 \mathbf{A}_1 = \begin{bmatrix} x & x & x & x & \ldots & x \\ s_1 & x & x & x & \ldots & x \\ 0 & x & x & x & \ldots & x \\ 0 & x & x & x & \ldots & x \\ \ldots & \ldots & \ldots & \ldots & \ldots & \ldots \\ 0 & x & x & x & \ldots & x \end{bmatrix} \tag{20}$$

Also, the last $n - 2$ terms in the first row of matrix $\mathbf{A}$ may be zeroed as well by

$$\mathbf{A}_2 = \mathbf{P}_1 \mathbf{A}_1 \mathbf{P}_1^T = \begin{bmatrix} x & s_1 & 0 & 0 & \ldots & 0 \\ s_1 & x & x & x & \ldots & x \\ 0 & x & x & x & \ldots & x \\ 0 & x & x & x & \ldots & x \\ \ldots & \ldots & \ldots & \ldots & \ldots & \ldots \\ 0 & x & x & x & \ldots & x \end{bmatrix} \tag{21}$$

Sec. B.2  Transformation Methods

Recurrence equations for the $k$th step are the same as those in Eqs. (19) and (21), except that $k$ replaced 1 and the lower limit on $j$ is $k + 1$.

After $n - 2$ Householder transformations, we evolve the tridiagonal matrix

$$\mathbf{T} = \mathbf{H}^T \mathbf{A} \mathbf{H} \tag{22}$$

where the *Householder matrix* is

$$\mathbf{H} = \mathbf{P}_1^T \mathbf{P}_2^T \ldots \mathbf{P}_{n-2}^T \tag{23}$$

Equations (22) and (23) require approximately $2n^3/3$ multiplications.

**Example B.5**

From the same matrix $\mathbf{A}$ used in Example B.4, we shall create a tridiagonal matrix $\mathbf{T}$, using Householder transformations. To begin, let us annihilate the first row and column outside the tridiagonal region, as shown by Eqs. (19) and (21).

$$s_1^2 = \sum_{j=2}^{n} A_{j,1}^2 = (-0.48)^2 + (-1.00)^2 = 1.230 \tag{y}$$

$$s_1 = \pm 1.109 \tag{z}$$

$$\mathbf{c}_1 = \{0, A_{21} \pm s_1, A_{31}, A_{41}\} = \{0, -1.589, -1.000, 0\} \tag{a'}$$

$$\mathbf{e}_1 = \frac{\mathbf{c}_1}{\sqrt{\mathbf{c}_1^T \mathbf{c}_1}} = \frac{\mathbf{c}_1}{1.878} = \{0, -0.8464, -0.5326, 0\} \tag{b'}$$

$$\mathbf{P}_1 = \mathbf{I} - 2\mathbf{e}_1 \mathbf{e}_1^T = \begin{bmatrix} 1 & 0 & 0 & 0 \\ 0 & -0.4327 & -0.9015 & 0 \\ 0 & -0.9015 & 0.4327 & 0 \\ 0 & 0 & 0 & 1 \end{bmatrix} \tag{c'}$$

$$\mathbf{A}_2 = \mathbf{P}_1 \mathbf{A}_1 \mathbf{P}_1^T = \begin{bmatrix} 1.360 & 1.109 & 0 & 0 \\ 1.109 & 1.412 & 0.1092 & -0.4327 \\ 0 & 0.1092 & 1.588 & 0.2077 \\ 0 & -0.4327 & 0.2077 & 1.640 \end{bmatrix} \tag{d'}$$

Matrix $\mathbf{A}_2$ is the same as that found before in Eq. (t) by the Givens method, except for the signs on several off-diagonal terms.

Similarly, for the second row and column we have

$$s_2^2 = \sum_{j=3}^{n} A_{j,2}^2 = (0.1092)^2 + (-0.4327)^2 = 0.1992 \tag{e'}$$

$$s_2 = \pm 0.4463 \tag{f'}$$

$$\mathbf{c}_2 = \{0, 0, A_{32} \pm s_2, A_{42}\} = \{0, 0, 0.5555, -0.4327\} \tag{g'}$$

$$\mathbf{e}_2 = \frac{\mathbf{c}_2}{\sqrt{\mathbf{c}_2^T \mathbf{c}_2}} = \frac{\mathbf{c}_2}{0.70414} = \{0, 0, 0.7889, -0.6145\} \tag{h'}$$

$$P_2 = I - 2e_2 e_2^T = \begin{bmatrix} 1 & 0 & 0 & 0 \\ 0 & 1 & 0 & 0 \\ 0 & 0 & -0.2447 & 0.9696 \\ 0 & 0 & 0.9696 & 0.2447 \end{bmatrix} \quad (i')$$

$$A_3 = P_2 A_2 P_2^T = \begin{bmatrix} 1.360 & 1.109 & 0 & 0 \\ 1.109 & 1.412 & -0.4463 & 0 \\ 0 & -0.4463 & 1.538 & 0.1953 \\ 0 & 0 & 0.1953 & 1.689 \end{bmatrix} \quad (j')$$

Again, matrix $A_3$ is the same as that in Eq. (x), except for the signs on off-diagonal terms. Thus, we see that the desired tridiagonal matrix is

$$T = A_3 = H^T A H \quad (k')$$

where

$$H = P_1^T P_2^T = \begin{bmatrix} 1 & 0 & 0 & 0 \\ 0 & -0.4327 & 0.2206 & -0.8741 \\ 0 & -0.9015 & -0.1059 & 0.4196 \\ 0 & 0 & 0.9696 & 0.2447 \end{bmatrix} \quad (\ell')$$

### QR Algorithm

Assume that Householder transformations have converted Eq. (1) to

$$T W = W \Lambda \quad (24)$$

where

$$W = H^T V \quad (25a)$$

or

$$V = H W \quad (25b)$$

Now let us factor $T = T_1$ into the form

$$T_1 = Q_1 R_1 \quad (26)$$

which is known as *Givens factorization*. The symbol $Q_1$ in Eq. (26) denotes an orthogonal matrix obtained by $n - 1$ Jacobi rotations, and $R_1$ is an upper triangular array. The QR algorithm [9] derives its name from the factors in Eq. (26).

Premultiplication of Eq. (26) by $Q_1^T$ and postmultiplication by $Q_1$ itself gives

$$Q_1^T T_1 Q_1 = R_1 Q_1 = T_2 \quad (27)$$

The new matrix $T_2$ is another tridiagonal array having smaller off-diagonal terms

## Sec. B.2  Transformation Methods

than $T_1$. Recursively, we have

$$T_{k+1} = Q_k^T T_k Q_k = R_k Q_k \to \Lambda \tag{28}$$

which iterates to the spectral matrix $\Lambda$. The modal matrix $V$ is calculated from Eq. (25b), in which

$$W = Q_1 Q_2 \ldots Q_{n_f} \tag{29}$$

where $n_f$ is the number of factorizations.

The recurrence algorithm expressed by Eq. (28) indicates that we must generate $R_k$ and $Q_k$ and multiply them in the sequence shown. From that equation we see that the definition for $R_k$ is

$$R_k = Q_k^T T_k \tag{30}$$

This formula gives us the means for determining both of the desired matrices, as will be explained next.

At the operational level for this method, we wish to annihilate a lower-triangular term $(T_{ji})_k$ of the tridiagonal matrix $T_k$, where $j = i + 1$. For this purpose, we set equal to zero the product of row $j$ of a Jacobi rotation matrix and column $i$ of matrix $T_k$, as follows:

$$-\sin \theta_i (T_{ii})_k + \cos \theta_i (T_{ji})_k = 0 \tag{31}$$

From this expression, we have

$$\tan \theta_i = \left(\frac{T_{ji}}{T_{ii}}\right)_k \tag{32}$$

Consequently,

$$\sin \theta_i = \frac{(T_{ji})_k}{C_i} \qquad \cos \theta_i = \frac{(T_{ii})_k}{C_i} \tag{33}$$

where

$$C_i = \sqrt{(T_{ii})_k^2 + (T_{ji})_k^2} \tag{34}$$

Then the Jacobi rotation matrix $(\hat{R}_i)_k$ premultiplies matrix $T_k$ (for $i = 1, 2, \ldots, n - 1$) to produce

$$(\hat{R}_{n-1} \ldots \hat{R}_2 \hat{R}_1)_k T_k = R_k \tag{35}$$

After the first operation in this sequence, matrix $T_k$ is no longer tridiagonal; and after $n - 1$ operations it becomes the upper triangular matrix $R_k$ in Eq. (28). Comparing Eq. (35) with Eq. (30), we see that

$$Q_k^T = (\hat{R}_{n-1} \ldots \hat{R}_2 \hat{R}_1)_k \tag{36a}$$

or

$$Q_k = (\hat{R}_1^T \hat{R}_2^T \ldots \hat{R}_{n-1}^T)_k \tag{36b}$$

Therefore, matrix $\mathbf{Q}_k$ is defined as the product of the transposes of $n - 1$ Jacobi rotation matrices.

**Example B.6**

Now we shall apply one cycle of QR iteration to the tridiagonal matrix $\mathbf{T}$ obtained by the Householder method in Example B.5. For the first rotation (in the 2-1 plane), Eqs. (34) and (33) give

$$C_1 = \sqrt{T_{11}^2 + T_{21}^2} = \sqrt{(1.360)^2 + (1.109)^2} = 1.755 \quad \text{(m')}$$

$$\sin \theta_1 = \frac{T_{21}}{C_1} = \frac{1.109}{1.755} = 0.6320 \quad \text{(n')}$$

$$\cos \theta_1 = \frac{T_{11}}{C_1} = \frac{1.360}{1.755} = 0.7749 \quad \text{(o')}$$

Then the first $4 \times 4$ Jacobi rotation matrix becomes

$$(\hat{\mathbf{R}}_1)_1 = \begin{bmatrix} 0.7749 & 0.6320 & 0 & 0 \\ -0.6320 & 0.7749 & 0 & 0 \\ 0 & 0 & 1 & 0 \\ 0 & 0 & 0 & 1 \end{bmatrix} \quad \text{(p')}$$

Table B.3 shows this matrix in the upper left-hand position. After multiplying it with matrix $\mathbf{T} = \mathbf{T}_1$ (in the upper right-hand position), we list the product $(\hat{\mathbf{R}}_1)_1 \mathbf{T}_1$ in the position below $\mathbf{T}_1$.

For the second rotation (in the 3-2 plane), we have

$$C_2 = \sqrt{T_{22}^2 + T_{32}^2} = \sqrt{(0.3935)^2 + (-0.4463)^2} = 0.5950 \quad \text{(q')}$$

$$\sin \theta_2 = \frac{T_{32}}{C_2} = \frac{-0.4463}{0.5950} = -0.7501 \quad \text{(r')}$$

$$\cos \theta_2 = \frac{T_{22}}{C_2} = \frac{0.3935}{0.5950} = 0.6613 \quad \text{(s')}$$

Table B.3 contains the second Jacobi rotation matrix $(\hat{\mathbf{R}}_2)_1$ below the first, and the result of multiplying it with $(\hat{\mathbf{R}}_1)_1 \mathbf{T}_1$ is shown below the latter matrix.

The third rotation (in the 4-3 plane) involves

$$C_3 = \sqrt{T_{33}^2 + T_{43}^2} = \sqrt{(0.7578)^2 + (0.1953)^2} = 0.7826 \quad \text{(t')}$$

$$\sin \theta_3 = \frac{T_{43}}{C_3} = \frac{0.1953}{0.7826} = 0.2495 \quad \text{(u')}$$

$$\cos \theta_3 = \frac{T_{33}}{C_3} = \frac{0.7590}{0.7826} = 0.9684 \quad \text{(v')}$$

The third rotation matrix $(\hat{\mathbf{R}}_3)_1$ also appears in Table B.3, along with the product

$$\mathbf{R}_1 = (\hat{\mathbf{R}}_3 \hat{\mathbf{R}}_2 \hat{\mathbf{R}}_1)_1 \mathbf{T}_1 \quad \text{(w')}$$

In addition, the matrix

**TABLE B.3  First Cycle of QR Iteration**

$(\hat{\mathbf{R}}_1)_1$

$$\begin{bmatrix} 0.7776 & 0.6341 & 0 & 0 \\ -0.6341 & 0.7776 & 0 & 0 \\ 0 & 0 & 1 & 0 \\ 0 & 0 & 0 & 1 \end{bmatrix}$$

$\mathbf{T}_1$

$$\begin{bmatrix} 1.360 & 1.109 & 0 & 0 \\ 1.109 & 1.412 & -0.4463 & 0 \\ 0 & -0.4463 & 1.538 & 0.1953 \\ 0 & 0 & 0.1953 & 1.689 \end{bmatrix}$$

$(\hat{\mathbf{R}}_2)_1$

$$\begin{bmatrix} 1 & 0 & 0 & 0 \\ 0 & 0.6613 & -0.7501 & 0 \\ 0 & 0.7501 & 0.6613 & 0 \\ 0 & 0 & 0 & 1 \end{bmatrix}$$

$(\hat{\mathbf{R}}_1)_1 \mathbf{T}_1$

$$\begin{bmatrix} 1.755 & 1.752 & -0.2821 & 0 \\ 0 & 0.3935 & -0.3459 & 0 \\ 0 & -0.4463 & 1.538 & 0.1953 \\ 0 & 0 & 0.1953 & 1.689 \end{bmatrix}$$

$(\hat{\mathbf{R}}_3)_1$

$$\begin{bmatrix} 1 & 0 & 0 & 0 \\ 0 & 1 & 0 & 0 \\ 0 & 0 & 0.9684 & 0.2495 \\ 0 & 0 & -0.2495 & 0.9684 \end{bmatrix}$$

$(\hat{\mathbf{R}}_2 \hat{\mathbf{R}}_1)_1 \mathbf{T}_1$

$$\begin{bmatrix} 1.755 & 1.752 & -0.2821 & 0 \\ 1.7550 & 0.5950 & -1.383 & -0.1465 \\ 0 & 0 & 0 & 0.7578 & 0.1291 \\ 0 & 0 & 0.1953 & 1.689 \end{bmatrix}$$

$\mathbf{Q}_1 = (\hat{\mathbf{R}}_1^T \hat{\mathbf{R}}_2^T \hat{\mathbf{R}}_3^T)_1$

$$\begin{bmatrix} 0.7749 & -0.4180 & -0.4591 & 0.1183 \\ 0.6320 & 0.5152 & 0.5629 & -0.1450 \\ 0 & -0.7501 & 0.6404 & -0.1650 \\ 0 & 0 & 0.2495 & 0.9684 \end{bmatrix}$$

$\mathbf{R}_1 = (\hat{\mathbf{R}}_3 \hat{\mathbf{R}}_2 \hat{\mathbf{R}}_1)_1 \mathbf{T}_1$

$$\begin{bmatrix} 1.755 & 1.752 & -0.2821 & 0 \\ 0 & 0.5950 & -1.383 & -0.1465 \\ 0 & 0 & 0.7826 & 0.5466 \\ 0 & 0 & 0 & 1.604 \end{bmatrix}$$

$$\mathbf{Q}_1 = (\hat{\mathbf{R}}_1^T \hat{\mathbf{R}}_2^T \hat{\mathbf{R}}_3^T)_1 \quad (x')$$

is listed in the lower left-hand position.

Using Eq. (27), we now compute a second tridiagonal matrix, as follows:

$$\mathbf{T}_2 = \mathbf{R}_1 \mathbf{Q}_1 = \begin{bmatrix} 2.468 & 0.3760 & 0.0000 & 0.0000 \\ 0.3760 & 1.342 & -0.5870 & 0.0000 \\ 0 & -0.5870 & 0.6375 & 0.4001 \\ 0 & 0 & 0.4001 & 1.553 \end{bmatrix} \quad (y')$$

Note that computed zeros must appear in the upper triangular part of matrix $\mathbf{T}_2$ to satisfy the symmetry guaranteed by the congruence transformation in Eq. (27). We also see that the RMS of off-diagonal terms is reduced from 0.6993 (for matrix $\mathbf{T}_1$) to 0.4645 (for matrix $\mathbf{T}_2$). Spectral shifting can be used in successive cycles to accelerate the rate of convergence, as explained in Sec. B.1.

## REFERENCES

1. Timoshenko, S. P., Young, D. H., and Weaver, W., Jr., *Vibration Problems in Engineering*, 4th ed., Wiley, New York, 1974.
2. Weaver, W., Jr., and Gere, J. M., *Matrix Analysis of Framed Structures*, 2nd ed., Van Nostrand Reinhold, New York, 1980.

3. Weaver, W., Jr., and Johnston, P. R., *Finite Elements for Structural Analysis*, Prentice-Hall, Englewood Cliffs, N.J., 1984.
4. Bathe, K. J., *Finite Element Procedures in Engineering Analysis*, Prentice-Hall, Englewood Cliffs, N.J., 1982.
5. Wilkinson, J. H., *The Algebraic Eigenvalue Problem*, Oxford University Press, London, 1965.
6. Jacobi, C. G. J., "Über ein leichtes Verfahren die in der Theorie der Säculärstörungen vorkommenden Gleichungen numerisch aufzulösen," *Crelle's J.*, Vol. 30, 1846, pp. 51–94.
7. Givens, W., "Numerical Computations of the Characteristic Values of a Real Symmetric Matrix," *Report No. ORNL-1574*, Oak Ridge National Laboratory, Oak Ridge, Tenn., 1954.
8. Martin, R. S., Reinsch, C., and Wilkinson, J. H., "Householder's Tridiagonalization of a Symmetric Matrix," *Numer. Math.*, Vol. 11, 1968, pp. 181–195.
9. Parlett, B. N., and Kahan, W., "On the Convergence of a Practical QR Algorithm," *Proc. IFIP Cong.*, 1968.

# Appendix C
# Flowchart
# for Program DYNAPT

**1. SUBPROGRAM SDATPT FOR SUBPROGRAM VIBPT**
 a. Problem Identification

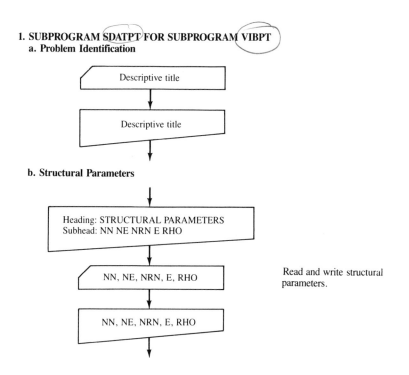

 b. Structural Parameters

Read and write structural parameters.

519

#### c. Nodal Coordinates

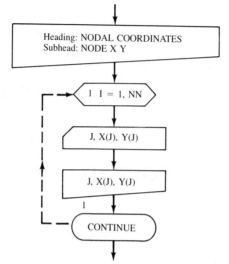

Read and write nodal coordinates.

#### d. Element Information

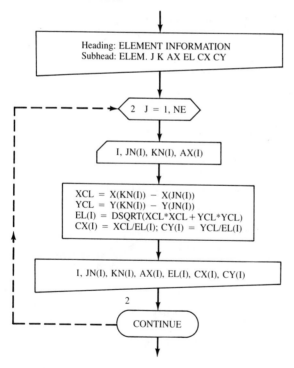

Read information for Element I.

Compute $x$ and $y$ components of element length XCL and YCL, the length EL, and the direction cosines CX and CY.

Write information for Element I.

## App. C  Flowchart for Program DYNAPT

**e. Nodal Restraints**

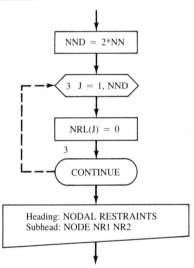

Calculate number of nodal displacements possible.

Clear nodal restraint list.

# Flowchart for Program DYNAPT   App. C

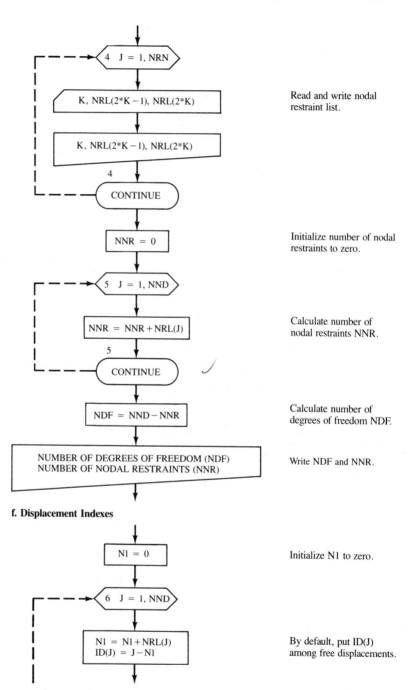

Read and write nodal restraint list.

Initialize number of nodal restraints to zero.

Calculate number of nodal restraints NNR.

Calculate number of degrees of freedom NDF.

Write NDF and NNR.

**f. Displacement Indexes**

Initialize N1 to zero.

By default, put ID(J) among free displacements.

App. C    Flowchart for Program DYNAPT                                523

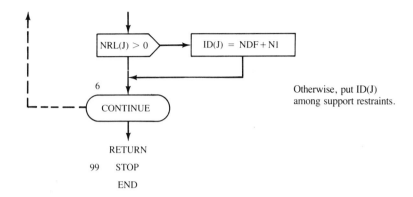

Otherwise, put ID(J) among support restraints.

## 2. SUBPROGRAM STIFPT FOR SUBPROGRAM VIBPT
### a. Clear Structure Stiffness Matrix

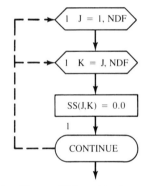

### b. Calculate Element Stiffness Matrix

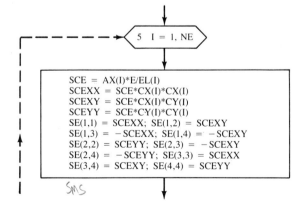

Compute stiffness constants and fill upper triangular part of SE, as shown by Eq. (3.5-26).

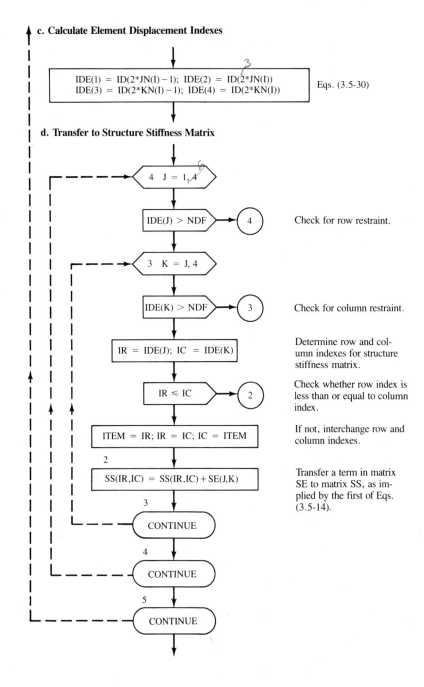

App. C   Flowchart for Program DYNAPT                                   **525**

**e. Fill Lower Triangle of Structure Stiffness Matrix**

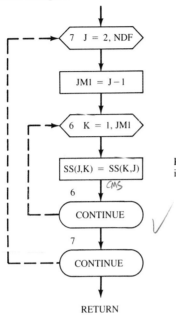

Place an upper triangular term into a lower triangular position.

**3. SUBPROGRAM CMASPT FOR SUBPROGRAM VIBPT**
This subprogram is similar to STIFPT, but the consistent mass matrix is generated instead of the stiffness matrix.

**4. SUBPROGRAM STASYM FOR SUBPROGRAM VIB**
  **a. Decompose Stiffness Matrix and Copy Mass Matrix to CMU**

Standard form
$(M_U - \lambda_i I)\Phi_{U_i} = 0$

$CMU = M_U = U^{-T} M U^{-1}$

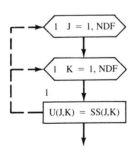

Transfer stiffness matrix to U.

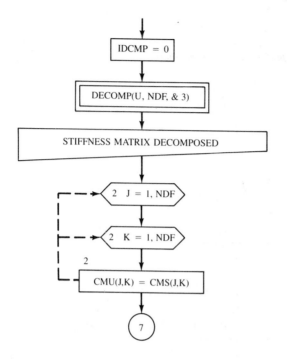

$S = U^T U$
Decompose stiffness matrix into the factored form $U^T U$, as given by Eq. (3.6-9).

Copy mass matrix to CMU.

**b. Decompose Mass Matrix and Copy Stiffness Matrix to CMU**

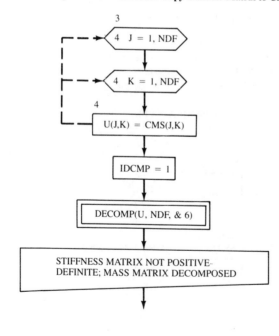

Transfer mass matrix to U.

$M = U^T U$
Decompose mass matrix into the factored form $U^T U$.

# App. C  Flowchart for Program DYNAPT

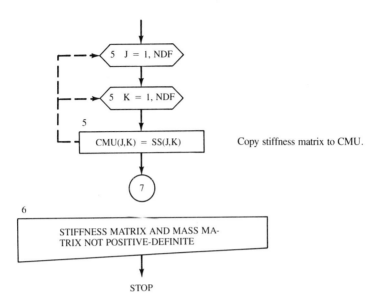

Copy stiffness matrix to CMU.

STIFFNESS MATRIX AND MASS MATRIX NOT POSITIVE-DEFINITE

STOP

### c. Calculate $U^{-T}$

Call Subprogram INVERU to obtain the inverse transpose of U.

### d. Transform to Standard, Symmetric Form
(1) Premultiply by $U^{-T}$

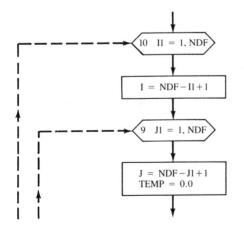

Set up decreasing index I.

Set up decreasing index J, and initialize TEMP to zero.

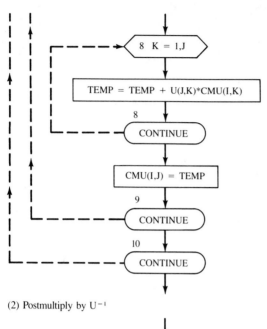

Premultiply CMU by $U^{-T}$, as indicated in the first of Eqs. (3.6-11).

Put result back into CMU.

(2) Postmultiply by $U^{-1}$

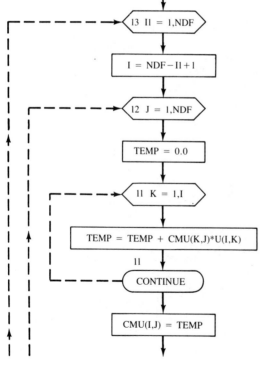

Set up decreasing index I.

Initialize TEMP to zero.

Postmultiply CMU by $U^{-1}$, as in the first of Eqs. (3.6-11).

Put result back into CMU.

## App. C  Flowchart for Program DYNAPT

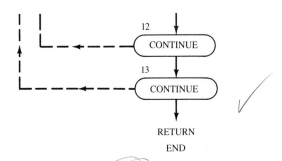

RETURN
END

### 5. SUBPROGRAM EIGEN2 FOR SUBPROGRAM VIB

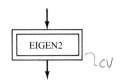

Calculate eigenvalues and eigenvectors to double precision.

### 6. SUBPROGRAM TRAVEC FOR SUBPROGRAM VIB
#### a. Calculate Angular Frequencies

$$(M_v - \lambda_i I)\Phi_{v_i} = 0$$

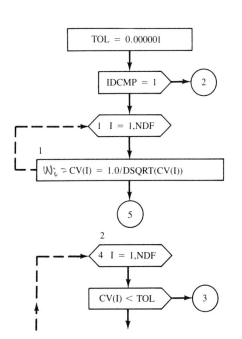

Set tolerance for zero eigenvalues.

If mass matrix was decomposed, go to 2.

Calculate angular frequencies from the first of Eqs. (3.6–12).

Check size of eigenvalue.

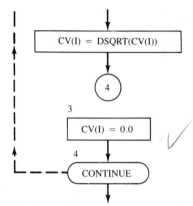

Calculate angular frequencies from the first of Eqs. (3.6–19).

Set small eigenvalue equal to zero.

**b. Back-Transform Eigenvectors**

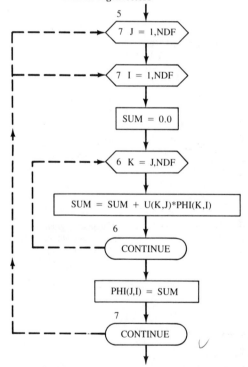

Initialize SUM to zero.

Back-transform eigenvectors, as in the second of Eqs. (3.6–12).

Put SUM into PHI.

## c. Normalize Eigenvectors with Respect to Largest Values

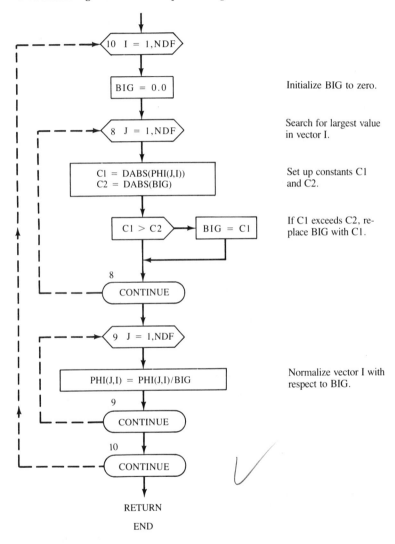

## 7. SUBPROGRAM RESIPT FOR SUBPROGRAM VIBPT
### a. Reorder Angular Frequencies and Eigenvectors

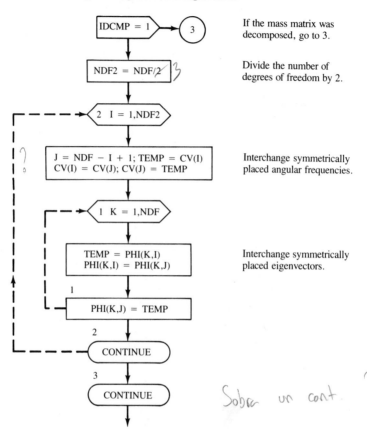

### b. Write Angular Frequencies and Expanded Eigenvectors

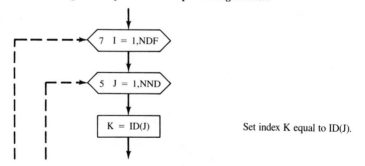

App. C  Flowchart for Program DYNAPT  533

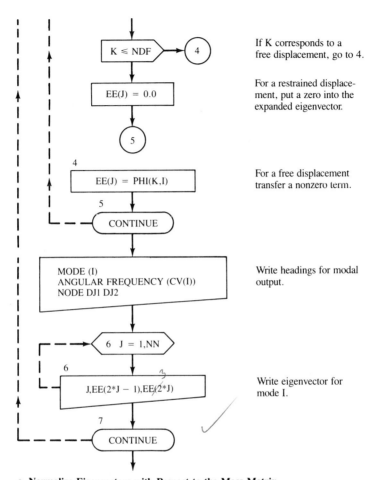

If K corresponds to a free displacement, go to 4.

For a restrained displacement, put a zero into the expanded eigenvector.

For a free displacement transfer a nonzero term.

Write headings for modal output.

Write eigenvector for mode I.

**c. Normalize Eigenvectors with Respect to the Mass Matrix**

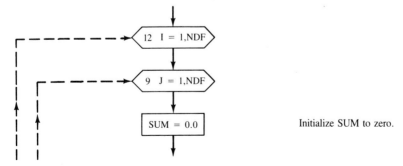

Initialize SUM to zero.

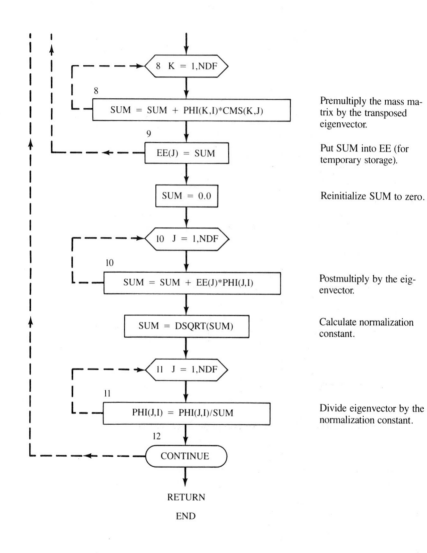

## 8. SUBPROGRAM DYLOPT FOR PROGRAM DYNAPT
### a. Dynamic Parameters

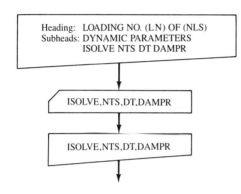

Read and write dynamic parameters.

### b. Initial Conditions

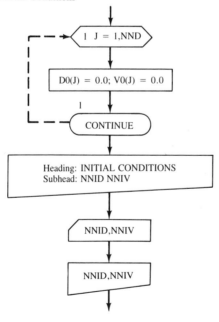

Clear initial displacement and velocity vectors.

Read and write initial condition parameters.

**Flowchart for Program DYNAPT    App. C**

(1) Initial Displacements

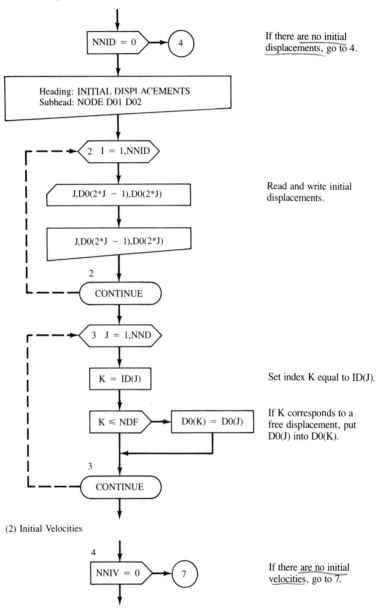

(2) Initial Velocities

## App. C Flowchart for Program DYNAPT

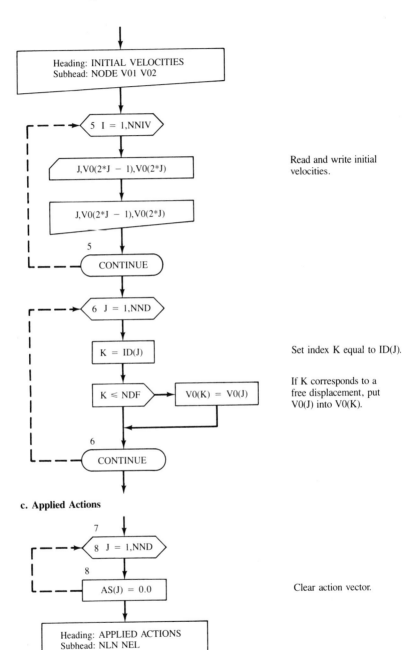

c. Applied Actions

**538**                      Flowchart for Program DYNAPT    App. C

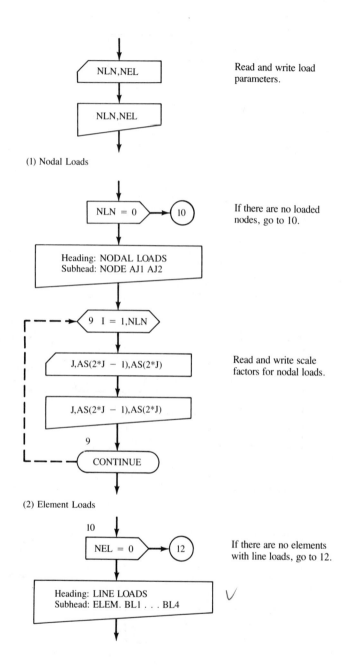

App. C   Flowchart for Program DYNAPT   **539**

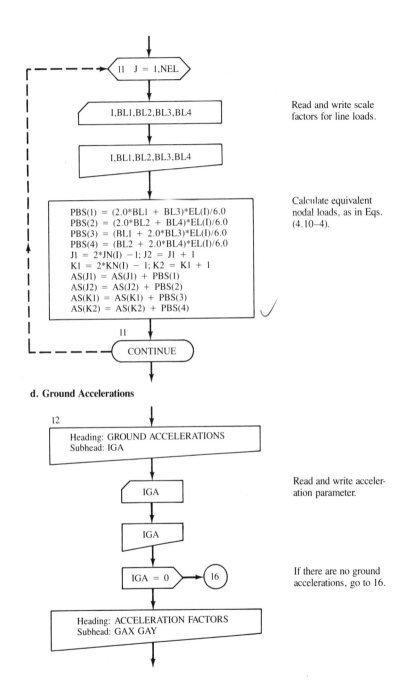

Read and write scale factors for line loads.

Calculate equivalent nodal loads, as in Eqs. (4.10-4).

**d. Ground Accelerations**

Read and write acceleration parameter.

If there are no ground accelerations, go to 16.

**540**          Flowchart for Program DYNAPT     App. C

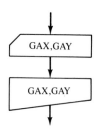

Read and write acceleration factors.

(1) Fill Vector GA with Ground Accelerations

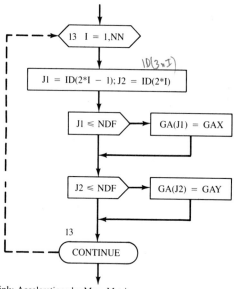

Calculate displacement indexes J1 and J2.

If J1 is free, put GAX into GA(J1).

If J2 is free, put GAY into GA(J2).

(2) Multiply Accelerations by Mass Matrix

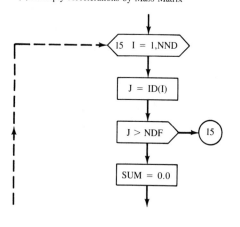

Set the displacement index J equal to ID(I).

If J is restrained, skip the multiplication.

Initialize SUM to zero.

## App. C  Flowchart for Program DYNAPT

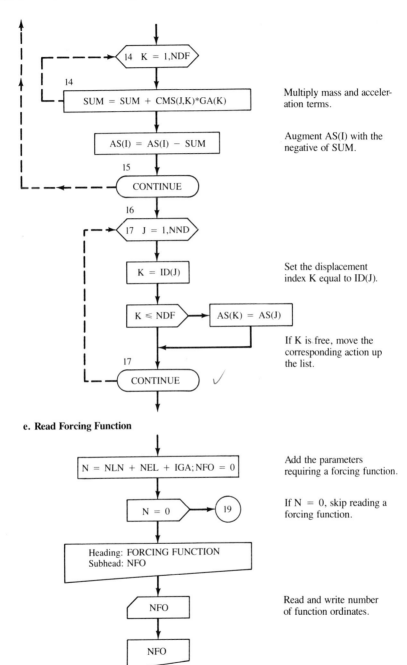

e. **Read Forcing Function**

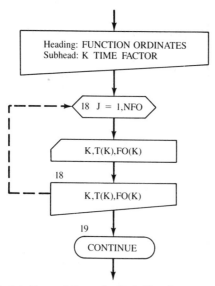

Read and write subscript, time, and function ordinate.

**f. Calculate Step and Ramp for Each Time Increment**

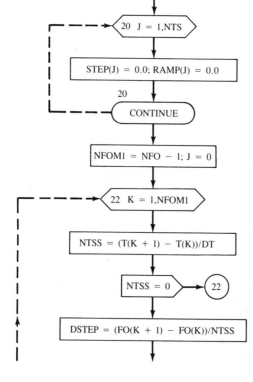

Clear step and ramp vectors.

Initialize NFOM1 and J.

Calculate number of time steps for a piecewise-linear segment.

If NTSS = 0, skip the step and ramp calculations.

Calculate the rate of change of ordinates with respect to time.

## App. C  Flowchart for Program DYNAPT    543

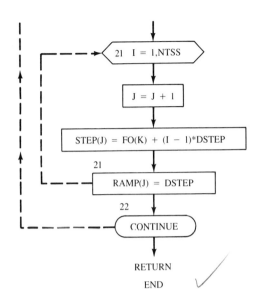

Increase J by 1.

Determine step and ramp for time increment within a piecewise-linear segment.

## 9. SUBPROGRAM TRANOR FOR SUBPROGAM NORMOD
### a. Read and Write Number of Modes

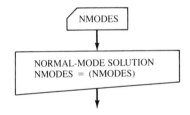

### b. Calculate Transformation Operator $\Phi_N^{-1}$

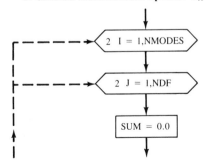

Initialize SUM to zero.

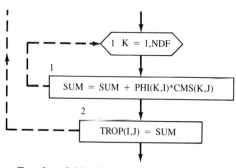

Calculate $\Phi_N^{-1} = \Phi_N^T \mathbf{M}$.

Put SUM into the transformation operator TROP.

**c. Transform Initial Conditions**

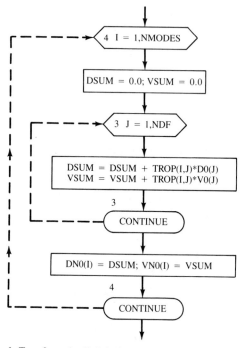

Initialize DSUM and VSUM to zero.

Premultiply D0 and V0 with TROP, as in Eqs. (4.3–2).

Put DSUM and VSUM into DN0 an VN0.

**d. Transform Applied Actions**

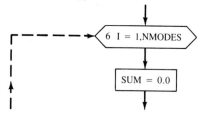

Initialize SUM to zero.

## App. C  Flowchart for Program DYNAPT

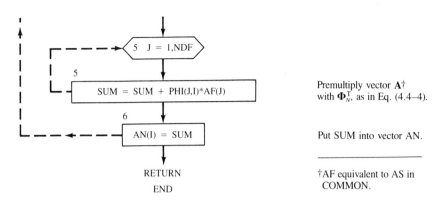

Premultiply vector $\mathbf{A}$†
with $\mathbf{\Phi}_N^T$, as in Eq. (4.4-4).

Put SUM into vector AN.

†AF equivalent to AS in COMMON.

### 10. SUBPROGRAM TIHIST FOR SUBPROGRAM NORMOD
**a. Calculate Response for Each Mode**

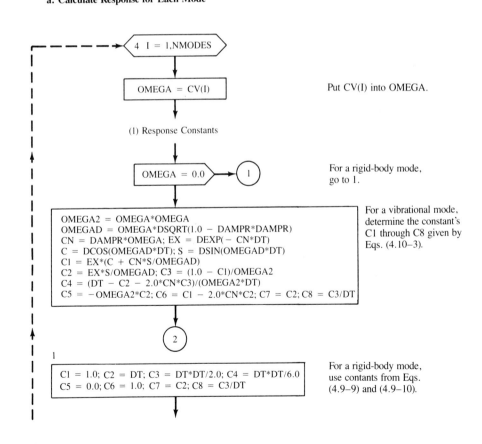

Put CV(I) into OMEGA.

(1) Response Constants

For a rigid-body mode, go to 1.

For a vibrational mode, determine the constant's C1 through C8 given by Eqs. (4.10-3).

For a rigid-body mode, use contants from Eqs. (4.9-9) and (4.9-10).

(2) Step-by-Step Response

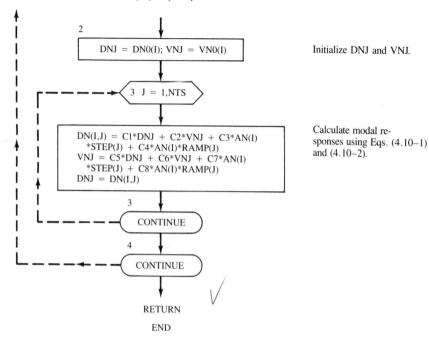

Initialize DNJ and VNJ.

Calculate modal responses using Eqs. (4.10-1) and (4.10-2).

## 11. SUBPROGRAM TRABAC FOR SUBPROGRAM NORMOD

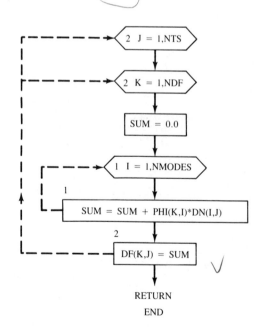

Initialize SUM to zero.

Back-transform displacements with Eq. (4.3-5).

Place SUM into the free-displacement matrix DF.

## 12. SUBPROGRAM NUMINT FOR PROGRAM DYNA
### a. Read and Write Integration Parameters

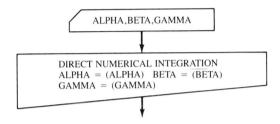

### b. Calculate Structure Damping Matrix

(1) Determine Factor SA

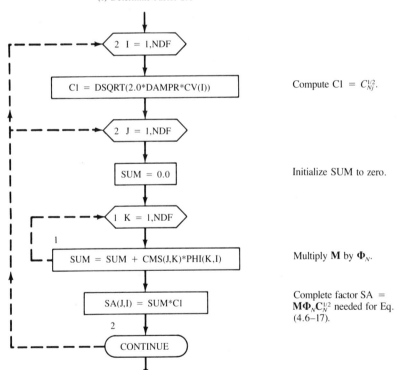

Compute $C1 = C_{Nj}^{1/2}$.

Initialize SUM to zero.

Multiply **M** by $\mathbf{\Phi}_N$.

Complete factor $SA = \mathbf{M\Phi}_N\mathbf{C}_N^{1/2}$ needed for Eq. (4.6–17).

(2) Multiply SA and Its Transpose

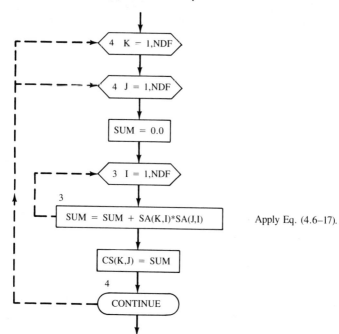

Apply Eq. (4.6–17).

c. Calculate Initial Accelerations

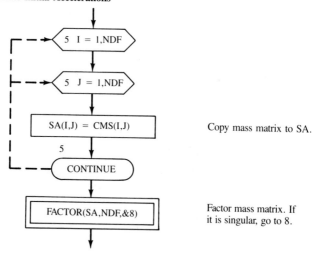

Copy mass matrix to SA.

Factor mass matrix. If it is singular, go to 8.

## App. C  Flowchart for Program DYNAPT

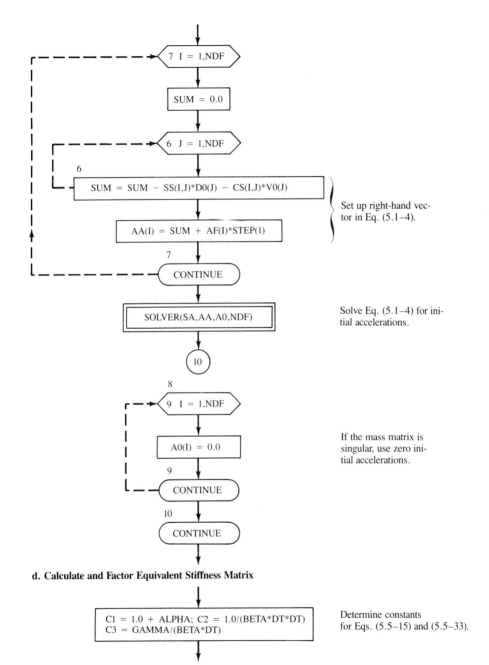

**d. Calculate and Factor Equivalent Stiffness Matrix**

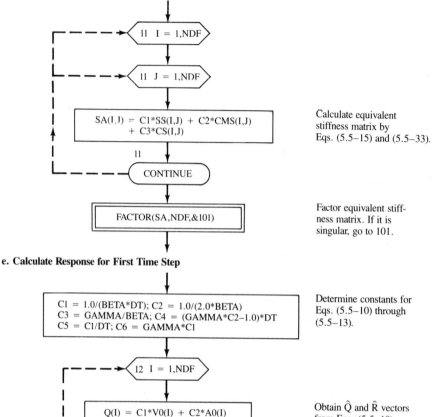

Calculate equivalent stiffness matrix by Eqs. (5.5-15) and (5.5-33).

Factor equivalent stiffness matrix. If it is singular, go to 101.

**e. Calculate Response for First Time Step**

Determine constants for Eqs. (5.5-10) through (5.5-13).

Obtain $\hat{Q}$ and $\hat{R}$ vectors from Eqs. (5.5-10) and (5.5-11).

Add mass and damping terms in Eq. (5.5-16).

App. C  Flowchart for Program DYNAPT    551

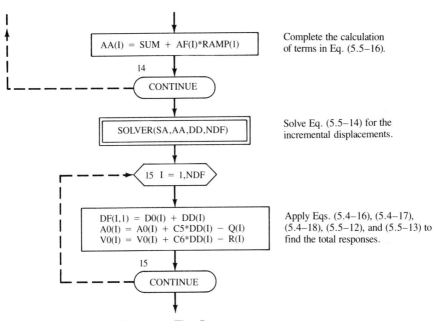

Complete the calculation of terms in Eq. (5.5–16).

Solve Eq. (5.5–14) for the incremental displacements.

Apply Eqs. (5.4–16), (5.4–17), (5.4–18), (5.5–12), and (5.5–13) to find the total responses.

**f. Calculate Responses for Subsequent Time Steps**

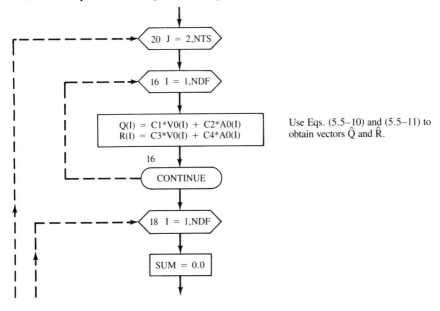

Use Eqs. (5.5–10) and (5.5–11) to obtain vectors $\hat{Q}$ and $\hat{R}$.

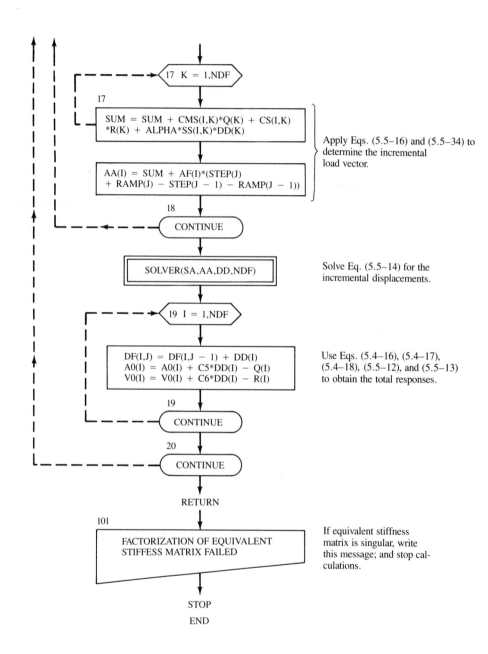

# App. C  Flowchart for Program DYNAPT

## 13. SUBPROGRAM RES2PT* FOR PROGRAM DYNAPT
### a. Read and Write Output Selections

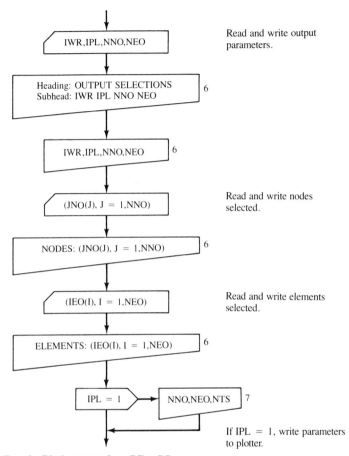

### b. Transfer Displacements from DF to DS

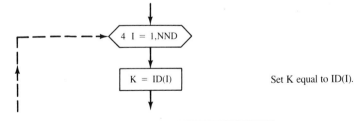

---

*For convenience, implied iterative control statements are used in this subprogram. Logical output units are 6 for printer and 7 for plotter.

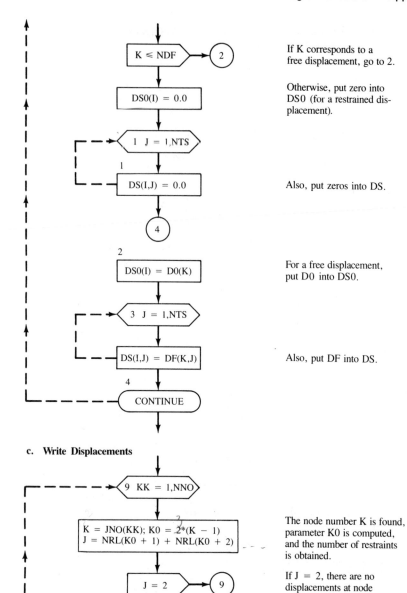

## App. C    Flowchart for Program DYNAPT

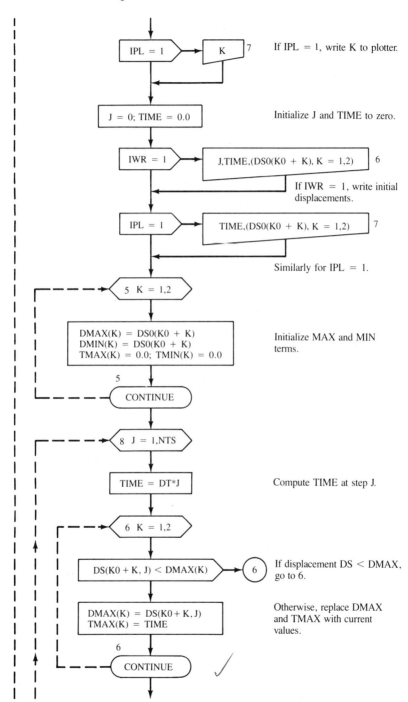

# Flowchart for Program DYNAPT    App. C

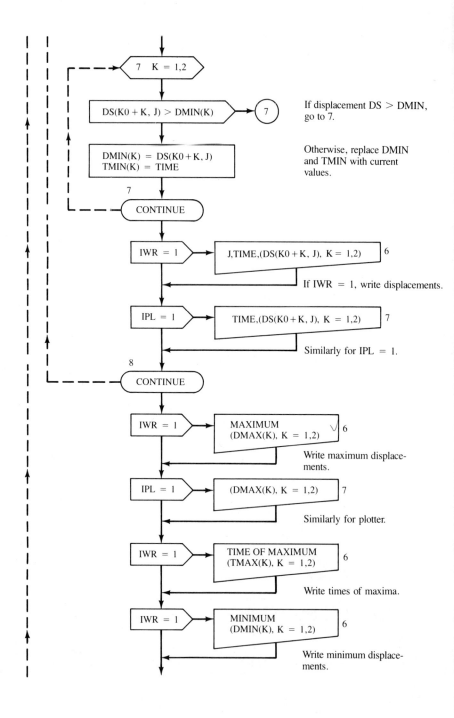

# App. C  Flowchart for Program DYNAPT

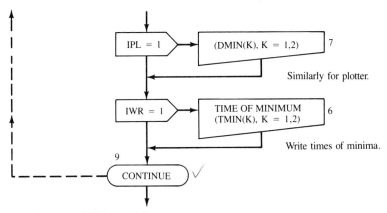

**d. Calculate and Write Axial Forces**

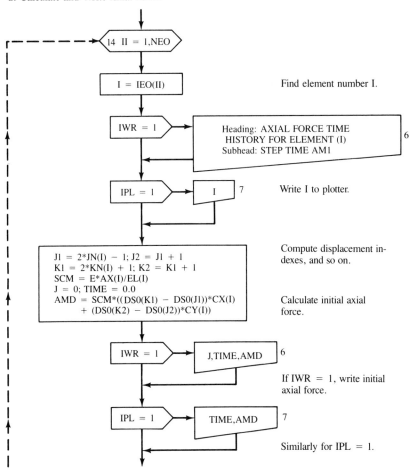

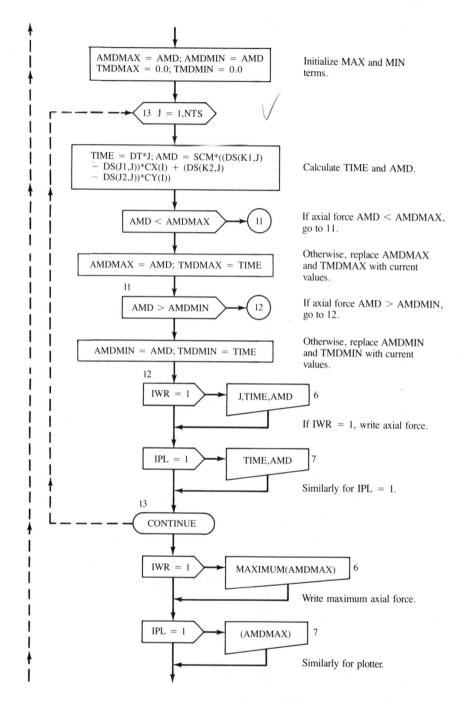

App. C   Flowchart for Program DYNAPT                                       559

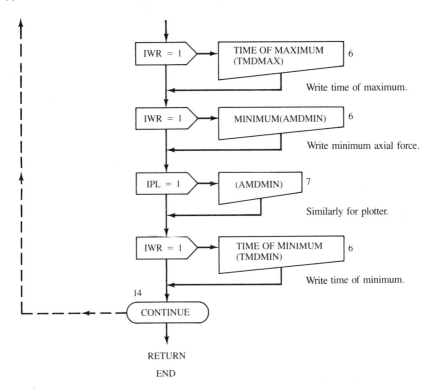

### 14. SUBPROGRAM DECOMP* FOR SUBPROGRAM STASYM

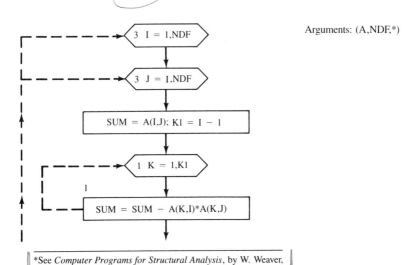

Arguments: (A,NDF,*)

*See *Computer Programs for Structural Analysis*, by W. Weaver, Jr., D. Van Nostrand, Princeton, N.J., 1967.

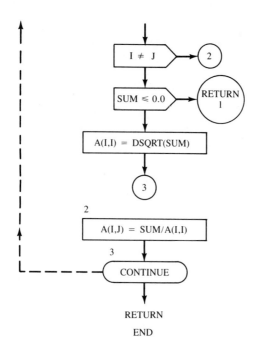

## 15. SUBPROGRAM INVERU* FOR SUBPROGRAM STASYM

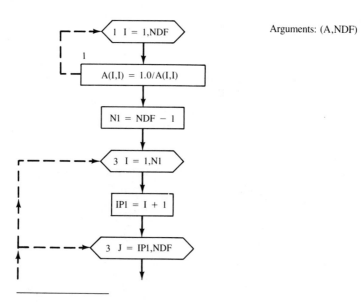

Arguments: (A,NDF)

---

*Ibid.

## App. C Flowchart for Program DYNAPT

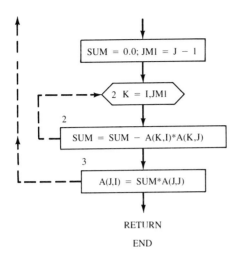

### 16. SUBPROGRAM FACTOR* FOR SUBPROGRAM NUMINT

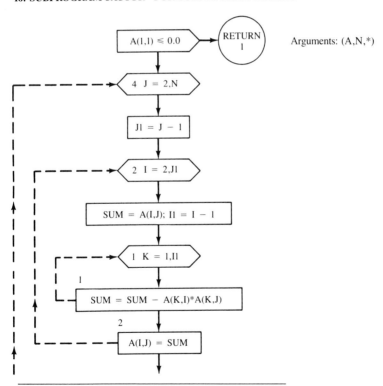

Arguments: (A,N,*)

---

*See *Matrix Analysis of Framed Structures*, 2nd ed., by W. Weaver, Jr. and J.M. Gere, Van Nostrand Reinhold, New York, 1980.

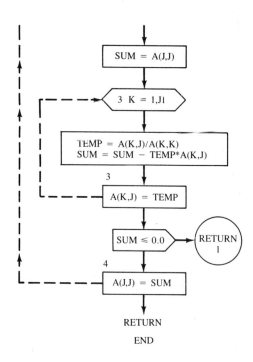

## 17. SUBPROGRAM SOLVER* FOR SUBPROGRAM NUMINT

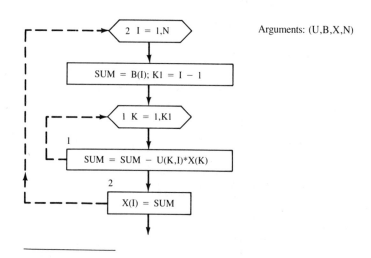

Arguments: (U,B,X,N)

---

*Ibid.

App. C  Flowchart for Program DYNAPT

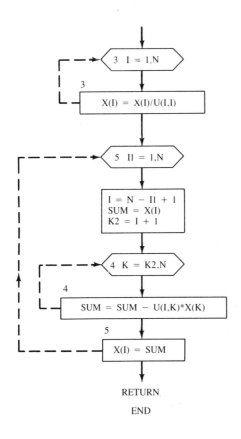

# Answers to Problems

**CHAPTER 2**

**2.2-1.** $\omega = \dfrac{1}{L}\sqrt{\dfrac{3EI}{mL}}$; $f = \dfrac{1}{2\pi L}\sqrt{\dfrac{3EI}{mL}}$; $T = 2\pi L\sqrt{\dfrac{mL}{3EI}}$

**2.2-2.** $\omega = \dfrac{4}{L}\sqrt{\dfrac{3EI}{mL}}$; $f = \dfrac{2}{\pi L}\sqrt{\dfrac{3EI}{mL}}$; $T = \dfrac{\pi L}{2}\sqrt{\dfrac{mL}{3EI}}$

**2.2-3.** $\omega = \dfrac{1}{L}\sqrt{\dfrac{3EI}{2mL}}$; $f = \dfrac{1}{2\pi L}\sqrt{\dfrac{3EI}{2mL}}$; $T = 2\pi L\sqrt{\dfrac{2mL}{3EI}}$

**2.2-4.** $\omega = \sqrt{\dfrac{AE_e}{mL}}$; $f = \dfrac{1}{2\pi}\sqrt{\dfrac{AE_e}{mL}}$; $T = 2\pi\sqrt{\dfrac{mL}{AE_e}}$

**2.2-5.** $\omega = \dfrac{2r^2}{L^2}\sqrt{\dfrac{3\pi G}{2\rho A}}$; $f = \dfrac{r^2}{\pi L^2}\sqrt{\dfrac{3\pi G}{2\rho A}}$; $T = \dfrac{\pi L^2}{r^2}\sqrt{\dfrac{2\rho A}{3\pi G}}$

**2.2-6.** $\omega = \dfrac{8}{L}\sqrt{\dfrac{3EI}{mL}}$; $f = \dfrac{4}{\pi L}\sqrt{\dfrac{3EI}{mL}}$; $T = \dfrac{\pi L}{4}\sqrt{\dfrac{mL}{3EI}}$

**2.3-1.** $u = \left[\dfrac{1}{1-(\Omega/\omega)^2}\right]d\cos\Omega t$  **2.3-2.** $u^* = -\left[\dfrac{1}{1-(\Omega/\omega)^2}\right]\dfrac{ma}{k}\cos\Omega t$

**2.3-3.** $D = \dfrac{PL^3}{144EI}\sin\Omega t$  **2.3-4.** $\delta\theta = -\dfrac{2ML}{3\pi r^4 G}\cos\Omega t$

**2.3-5.** $D = \dfrac{9}{5}d\sin\Omega t$  **2.3-6.** $D^* = \dfrac{maL^3}{45EI}\cos\Omega t$

564

# Answers to Problems

**2.3-7.** $D = \dfrac{16}{7} \delta\theta_g \sin \Omega t$  **2.3-8.** $D^* = \dfrac{LI_r \ddot{\delta\theta}_g}{12\pi r^4 G} \cos \Omega t$

**2.3-9.** $D = \dfrac{32}{15} d \sin \Omega t$  **2.3-10.** $D = -\dfrac{2}{35} d \cos \Omega t$

**2.4-1.** $u = \beta \dfrac{P}{k} \sin(\Omega t - \theta); \quad \theta = \tan^{-1}\left(\dfrac{2n\Omega}{\omega^2 - \Omega^2}\right)$

**2.4-2.** $u^* = -\beta \dfrac{ma}{k} \sin(\Omega t - \theta); \quad \theta = \tan^{-1}\left(\dfrac{2n\Omega}{\omega^2 - \Omega^2}\right)$

**2.4-3.** $u_{tr} = e^{-nt}\left(N \cos \omega_d t + \dfrac{M\Omega + Nn}{\omega_d} \sin \omega_d t\right)$

**2.4-4.** $D = 0.01445 \dfrac{PL^3}{EI} \sin(\Omega t - \theta)$  **2.4-5.** $\delta\theta = 9.086 \dfrac{ML}{\pi r^4 G} \cos(\Omega t - \theta)$

**2.4-6.** $D = -3.185 \dfrac{maL^3}{EI} \cos(\Omega t - \theta)$  **2.4-7.** $\beta_{res} = 50, 25, \ldots, 2.5$

**2.5-1.** $P(t) = \dfrac{4}{\pi} P_1\left(\sin \Omega t + \dfrac{1}{3} \sin 3\Omega t + \ldots\right)$

**2.5-2.** $P(t) = \dfrac{4}{\pi} P_1\left(\cos \Omega t - \dfrac{1}{3} \cos 3\Omega t + \ldots\right)$

**2.5-3.** $P(t) = \dfrac{2}{\pi} P_1\left(\sin \Omega t - \dfrac{1}{2} \sin 2\Omega t + \dfrac{1}{3} \sin 3\Omega t - \ldots\right)$

**2.5-4.** $P(t) = \dfrac{P_1}{2} - \dfrac{P_1}{\pi}\left(\sin \Omega t + \dfrac{1}{2} \sin 2\Omega t + \dfrac{1}{3} \sin 3\Omega t + \ldots\right)$

**2.5-5.** $u = \dfrac{a_0}{k} + \sum_{i=1}^{n} \dfrac{a_i \cos(i\Omega t - \theta_i) + b_i \sin(i\Omega t - \theta_i)}{k\sqrt{(1 - i^2\Omega^2/\omega^2)^2 + (2\gamma i\Omega/\omega)^2}}$

**2.5-6.** $u = 6.70 \dfrac{P_1}{k} \sin \Omega t$  **2.5-7.** $u = 13.06 \dfrac{P_1}{k} \cos \Omega t$

**2.5-8.** $u = -6.21 \dfrac{P_1}{k} \sin \Omega t$

**2.6-1.** $u = \dfrac{P_1}{k}\left(\sin \Omega t - \dfrac{\Omega}{\omega} \sin \omega t\right)\left[\dfrac{1}{1 - (\Omega/\omega)^2}\right]$

**2.6-2.** $u = \dfrac{P_1}{k}\left[1 - e^{-nt}\left(\cos \omega_d t + \dfrac{n}{\omega_d} \sin \omega_d t\right)\right]$

**2.6-3.** $u = \dfrac{P_1}{t_1 k \omega^2}\left[\omega^2 t - 2n + e^{-nt}\left(2n \cos \omega_d t - \dfrac{\omega_d^2 - n^2}{\omega_d} \sin \omega_d t\right)\right]$

**2.6-4.** $u = \dfrac{P_1}{k}(1 - \cos \omega t)$  $(0 \le t \le t_1)$

$u = \dfrac{P_1}{k}\left[\cos \omega(t - t_1) - \cos \omega t\right]$

$$-\frac{P_2}{k}\left[1 - \cos \omega(t - t_1)\right] \qquad (t_1 \leq t \leq t_2)$$

$$u = \frac{P_1}{k}\left[\cos \omega(t - t_1) - \cos \omega t\right]$$

$$-\frac{P_2}{k}\left[\cos \omega(t - t_2) - \cos \omega(t - t_1)\right] \qquad (t_2 \leq t)$$

**2.6-5.** $u = \dfrac{P_1}{k}\left(1 - \cos \omega t - \dfrac{t}{t_1} + \dfrac{\sin \omega t}{\omega t_1}\right) \qquad (0 \leq t \leq t_1)$

$$u = \frac{P_1}{k}\left[-\cos \omega t + \frac{\sin \omega t - \sin \omega(t - t_1)}{\omega t_1}\right] \qquad (t_1 \leq t)$$

**2.6-6.** $u = \dfrac{P_1}{k}(1 - \cos \omega t) \qquad (0 \leq t \leq t_1)$

$$u = \frac{P_1}{k}\left[1 - \cos \omega t - \frac{t - t_1}{t_2 - t_1} + \frac{\sin \omega(t - t_1)}{\omega(t_2 - t_1)}\right] \qquad (t_1 \leq t \leq t_2)$$

$$u = \frac{P_1}{k}\left[-\cos \omega t + \frac{\sin \omega(t - t_1) - \sin \omega(t - t_2)}{\omega(t_2 - t_1)}\right] \qquad (t_2 \leq t)$$

**2.6-7.** $u^* = -\dfrac{a_1}{\omega^2}\left(\dfrac{t}{t_1} - \dfrac{\sin \omega t}{\omega t_1}\right) \qquad (0 \leq t \leq t_1)$

$$u^* = -\frac{a_1}{\omega^2}\left[1 + \frac{\sin \omega(t - t_1) - \sin \omega t}{\omega t_1}\right] \qquad (t_1 \leq t)$$

**2.6-8.** $u^* = -\dfrac{a_0}{\omega^2}(1 - \cos \omega t) - \dfrac{a_1 - a_0}{\omega^2}\left(\dfrac{t}{t_1} - \dfrac{\sin \omega t}{\omega t_1}\right) \qquad (0 \leq t \leq t_1)$

$$u^* = -\frac{a_0}{\omega^2}\left[\cos \omega(t - t_1) - \cos \omega t\right]$$

$$-\frac{a_1 - a_0}{\omega^2}\left[\frac{\sin \omega(t - t_1) - \sin \omega t}{\omega t_1} + \cos \omega(t - t_1)\right] \qquad (t_1 \leq t)$$

**2.6-9.** $u^* = -\dfrac{a_1}{\omega^2}\left[\dfrac{t^2}{t_1^2} - \dfrac{2}{\omega^2 t_1^2}(1 - \cos \omega t)\right] \qquad (0 \leq t \leq t_1)$

$$u^* = -\frac{a_1}{\omega^2}\left\{\frac{2}{\omega^2 t_1^2}\left[\cos \omega t - \cos \omega(t - t_1) + \omega t_1 \sin \omega(t - t_1)\right]\right.$$

$$\left. + \cos \omega(t - t_1)\right\} \qquad (t_1 \leq t)$$

**2.7-1.** $u_{10} = 0$  **2.7-2.** $u_{10} = \dfrac{P_1}{k}$  **2.7-3.** Given  **2.7-4.** Given

**2.7-5.** $u_{20} = -2.546\dfrac{P_1}{k}$  **2.7-6.** $u_{20} = -4\dfrac{P_1}{k}$  **2.7-7.** $u_{10} = 0.9135\dfrac{P_1}{k}$

**2.7-8.** $u_{10} = 0.3186\dfrac{P_1}{k}$  **2.7-9.** $u_{10} = 1.379\dfrac{P_1}{k}$  **2.7-10.** $u_{10} = 1.102\dfrac{P_1}{k}$

# Answers to Problems

## CHAPTER 3

**3.4-1.** $\mathbf{p}_b(t) = \{2b_1 + b_2, b_1 + 2b_2\}\dfrac{L}{6}$   **3.4-2.** $\mathbf{p}_b(t) = \{1, 3\}\dfrac{b_2 L}{12}$

**3.4-3.** $\mathbf{K} = \dfrac{EA}{3L}\begin{bmatrix} 7 & -8 & 1 \\ -8 & 16 & -8 \\ 1 & -8 & 7 \end{bmatrix}$   **3.4-4.** $\mathbf{M} = \dfrac{\rho AL}{15}\begin{bmatrix} 2 & 1 & -\frac{1}{2} \\ 1 & 8 & 1 \\ -\frac{1}{2} & 1 & 2 \end{bmatrix}$

**3.4-5.** $\mathbf{p}_b(t) = \{1, 4, 1\}\dfrac{b_x L}{6}$   **3.4-6.** $\mathbf{p}_b(t) = \{5, 3\}\dfrac{m_{x1} L}{12}$

**3.4-7.** $\mathbf{p}_b(t) = \{L - x, x\}\dfrac{M_x}{L}$   **3.4-8.** $\mathbf{p}_b(t) = \{9, 2L, 21, -3L\}\dfrac{b_2 L}{60}$

**3.4-9.** $\mathbf{p}_b(t)_1 = \{2x^3 - 3x^2 L + L^3, x^3 L - 2x^2 L^2 + xL^3, -2x^3 + 3x^2 L,$

$$x^3 L - x^2 L^2\}\dfrac{P_y}{L^3}$$

$\mathbf{p}_b(t)_2 = \{6x^2 - 6xL, 3x^2 L - 4xL^2 + L^3, -6x^2 + 6xL, 3x^2 L - 2xL^2\}\dfrac{M_z}{L^3}$

**3.4-10.** $\mathbf{p}_b(t) = \{21b_1 + 9b_2, (3b_1 + 2b_2)L, 9b_1 + 21b_2, -(2b_1 + 3b_2)L\}\dfrac{L}{60}$

**3.4-11.** Given    **3.4-12.** Given

**3.5-1.** $S_{s11} = 0.64\dfrac{EA}{L}$; $S_{s12} = -0.48\dfrac{EA}{L}$; $S_{s13} = 0$; etc.

$M_{s11} = 2.72\dfrac{\rho AL}{6}$; $M_{s12} = 0$; $M_{s13} = 0.06\,\rho AL$; etc.

**3.5-2.** $S_{s11} = 1.64\dfrac{EA}{L}$; $S_{s12} = 0.48\dfrac{EA}{L}$; $S_{s13} = -\dfrac{EA}{L}$; etc.

$M_{s11} = 3.28\dfrac{\rho AL}{6}$; $M_{s12} = 0$; $M_{s13} = 0.64\dfrac{\rho AL}{6}$; etc.

**3.5-3.** $S_{s11} = \dfrac{3EA}{2L}$; $S_{s12} = -\dfrac{EA}{2L}$; $S_{s13} = -\dfrac{EA}{L}$; etc.

$M_{s11} = \dfrac{4\rho AL}{3}$; $M_{s12} = 0$; $M_{s13} = \dfrac{\rho AL}{6}$; etc.

**3.5-4.** $S_{s11} = \dfrac{EA}{L}$; $S_{s12} = 0$; $S_{s13} = -\dfrac{EA}{L}$; etc.

$M_{s11} = \dfrac{\rho AL}{3}$; $M_{s12} = 0$; $M_{s13} = 0.64\dfrac{\rho AL}{6}$; etc.

**3.5-5.** $S_{s11} = 1.64\dfrac{EA}{L}$; $S_{s12} = 0.48\dfrac{EA}{L}$; $S_{s13} = 0$; etc.

$M_{s11} = \dfrac{2\rho AL}{3}$; $M_{s12} = 0$; $M_{s13} = 0.06\,\rho AL$; etc.

**3.5-6.** $S_{s11} = \dfrac{3EA}{2L}$; $S_{s12} = -\dfrac{EA}{2L}$; $S_{s13} = 0$; etc.

$M_{s11} = \dfrac{4\rho AL}{3}$; $M_{s12} = 0$; $M_{s13} = \dfrac{\rho AL}{6}$; etc.

**3.5-7.** $S_{s11} = \dfrac{4EI}{L}$; $S_{s12} = \dfrac{2EI}{L}$; $S_{s13} = 0$; etc.

$M_{s11} = \dfrac{\rho AL^3}{105}$; $M_{s12} = -\dfrac{\rho AL^3}{140}$; $M_{s13} = 0$; etc.

**3.5-8.** $S_{s11} = \dfrac{8EI}{L}$; $S_{s12} = \dfrac{6EI}{L^2}$; $S_{s13} = \dfrac{2EI}{L}$; etc.

$M_{s11} = \dfrac{2\rho AL^3}{105}$; $M_{s12} = -\dfrac{13\rho AL^2}{420}$; $M_{s13} = -\dfrac{\rho AL^3}{140}$; etc.

**3.5-9.** $S_{s11} = \dfrac{12EI}{L^3}$; $S_{s12} = \dfrac{6EI}{L^2}$; $S_{s13} = \dfrac{6EI}{L^2}$; etc.

$M_{s11} = \dfrac{13\rho AL}{35}$; $M_{s12} = \dfrac{11\rho AL^2}{210}$; $M_{s13} = -\dfrac{13\rho AL^2}{420}$; etc.

**3.5-10.** $S_{s11} = \dfrac{8EI}{L}$; $S_{s12} = -\dfrac{6EI}{L^2}$; $S_{s13} = \dfrac{2EI}{L}$; etc.

$M_{s11} = \dfrac{2\rho AL^3}{105}$; $M_{s12} = \dfrac{13\rho AL^2}{420}$; $M_{s13} = -\dfrac{\rho AL^3}{140}$; etc.

**3.6-1.** Answer not provided.

**3.6-2.** $\omega_{1,2} = \dfrac{0.5412,\ 1.307}{L}\sqrt{\dfrac{E}{\rho}}$; $\Phi_1 = \begin{bmatrix} 2.414 \\ 1 \end{bmatrix}$; $\Phi_2 = \begin{bmatrix} -0.4142 \\ 1 \end{bmatrix}$

**3.6-3.** $\omega_{1,2} = \dfrac{0.9393,\ 3.529}{L}\sqrt{\dfrac{E}{\rho}}$; $\Phi_1 = \begin{bmatrix} -2 \\ 1 \end{bmatrix}$; $\Phi_2 = \begin{bmatrix} 1 \\ 2 \end{bmatrix}$

**3.6-4.** $\omega_{1,2} = \dfrac{0.6049,\ 1.815}{L}\sqrt{\dfrac{E}{\rho}}$; $\Phi_1 = \begin{bmatrix} 1 \\ 3 \end{bmatrix}$; $\Phi_2 = \begin{bmatrix} -3 \\ 1 \end{bmatrix}$

**3.6-5.** $\omega_{1,2} = \dfrac{5.546,\ 21.04}{L^2}\sqrt{\dfrac{EI}{\rho A}}$; $\Phi_1 = \begin{bmatrix} L \\ -0.3947 \end{bmatrix}$; $\Phi_2 = \begin{bmatrix} L \\ -26.61 \end{bmatrix}$

**3.6-6.** $\omega_{1,2} = \dfrac{15.14,\ 28.98}{L^2}\sqrt{\dfrac{EI}{\rho A}}$; $\Phi_1 = \begin{bmatrix} 1 \\ -1 \end{bmatrix}$; $\Phi_2 = \begin{bmatrix} 1 \\ 1 \end{bmatrix}$

**3.6-7.** $\omega_{1,2} = \dfrac{13.32,\ 34.79}{L^2}\sqrt{\dfrac{EI}{\rho A}}$; $\Phi_1 = \begin{bmatrix} 1 \\ -0.7071 \end{bmatrix}$; $\Phi_2 = \begin{bmatrix} 1 \\ 0.7071 \end{bmatrix}$

**3.7-1.** (a) $\omega_1 = \dfrac{22.74}{L^2}\sqrt{\dfrac{EI}{\rho A}}$; (b) $\omega_2 = \dfrac{81.98}{L^2}\sqrt{\dfrac{EI}{\rho A}}$

**3.7-2.** (a) $\omega_2 = \dfrac{81.98}{L^2}\sqrt{\dfrac{EI}{\rho A}}$; mode shape is unity.

(b) $\omega_{1,3} = \dfrac{53.28,\ 139.2}{L^2}\sqrt{\dfrac{EI}{\rho A}}$; $\Phi_1 = \begin{bmatrix} 1 \\ -0.7071 \end{bmatrix}$; $\Phi_3 = \begin{bmatrix} 1 \\ 0.7071 \end{bmatrix}$

Answers to Problems

**3.7-3.** (a) $\omega_{1,3} = \dfrac{16.64, 95.21}{L^2}\sqrt{\dfrac{EI}{\rho A}}$; $\Phi_1 = \begin{bmatrix} \ell \\ -0.5868 \end{bmatrix}$; $\Phi_3 = \begin{bmatrix} \ell \\ 16.86 \end{bmatrix}$

(b) $\omega_{2,4} = \dfrac{53.28, 139.2}{L^2}\sqrt{\dfrac{EI}{\rho A}}$; $\Phi_2 = \begin{bmatrix} 1 \\ -0.7071 \end{bmatrix}$; $\Phi_4 = \begin{bmatrix} 1 \\ 0.7071 \end{bmatrix}$

**3.7-4.** (a) $\omega_1 = \dfrac{0.9258}{L}\sqrt{\dfrac{E}{\rho}}$; mode shape is unity.

(b) $\omega_2 = \dfrac{1.134}{L}\sqrt{\dfrac{E}{\rho}}$; mode shape is unity.

**3.7-5.** (a) $\omega_{2,3} = \dfrac{1.485, 1.732}{L}\sqrt{\dfrac{E}{\rho}}$; $\Phi_2 = \begin{bmatrix} 1 \\ 0 \end{bmatrix}$; $\Phi_3 = \begin{bmatrix} 0 \\ 1 \end{bmatrix}$

(b) $\omega_{1,4} = \dfrac{0.6650, 2.909}{L}\sqrt{\dfrac{E}{\rho}}$; $\Phi_1 = \begin{bmatrix} 1 \\ 0.8801 \end{bmatrix}$; $\Phi_4 = \begin{bmatrix} 1 \\ -0.6091 \end{bmatrix}$

**3.7-6.** (a) $\omega_{1,4} = \dfrac{0.6581, 1.974}{L}\sqrt{\dfrac{E}{\rho}}$; $\Phi_1 = \begin{bmatrix} 1 \\ 1.344 \end{bmatrix}$; $\Phi_4 = \begin{bmatrix} 1 \\ -1.092 \end{bmatrix}$

(b) $\omega_{2,3} = \dfrac{1.271, 1.927}{L}\sqrt{\dfrac{E}{\rho}}$; $\Phi_2 = \begin{bmatrix} 1 \\ 0.08687 \end{bmatrix}$; $\Phi_3 = \begin{bmatrix} 1 \\ -7.365 \end{bmatrix}$

**3.7-7.** (a) $\omega_{3,4} = \dfrac{1.352, 2.218}{L}\sqrt{\dfrac{E}{\rho}}$; $\Phi_3 = \begin{bmatrix} 1 \\ -0.3242 \end{bmatrix}$; $\Phi_4 = \begin{bmatrix} 1 \\ 2.055 \end{bmatrix}$

(b) $\omega_{1,2} = \dfrac{0, 1.225}{L}\sqrt{\dfrac{E}{\rho}}$; $\Phi_1 = \begin{bmatrix} 1 \\ 1 \end{bmatrix}$; $\Phi_2 = \begin{bmatrix} 1 \\ -0.5 \end{bmatrix}$

## CHAPTER 4

**4.3-1.** $\mathbf{D} = \begin{bmatrix} (9\sin\omega_1 t)/\omega_1 + (\sin\omega_2 t)/\omega_2 \\ (3\sin\omega_1 t)/\omega_1 - (3\sin\omega_2 t)/\omega_2 \end{bmatrix} \dfrac{\dot{D}_{01}}{10}$

**4.3-2.** $\mathbf{D} = \begin{bmatrix} 1.961L\cos\omega_1 t + 0.03923L\cos\omega_2 t \\ 2.701\cos\omega_1 t + 0.2990\cos\omega_2 t \end{bmatrix} \dfrac{P_0 L^2}{6EI}$

**4.3-3.** $\mathbf{D} = \begin{bmatrix} 1.207\cos\omega_1 t - 0.2072\cos\omega_2 t \\ 0.5000\cos\omega_1 t + 0.5000\cos\omega_2 t \end{bmatrix} d$

**4.3-4.** $\mathbf{D} = \begin{bmatrix} (-2\sin\omega_1 t)/\omega_1 + (2\sin\omega_2 t)/\omega_2 \\ (\sin\omega_1 t)/\omega_1 + (4\sin\omega_2 t)/\omega_2 \end{bmatrix} \dfrac{\dot{D}_{01}}{5}$

**4.3-5.** $\mathbf{D} = -\begin{bmatrix} 27\cos\omega_1 t - 3\cos\omega_2 t \\ 81\cos\omega_1 t + \cos\omega_2 t \end{bmatrix} \dfrac{P_0 L}{18EA}$

**4.3-6.** $\mathbf{D} = \begin{bmatrix} (1.015L\sin\omega_1 t)/\omega_1 - (0.01506L\sin\omega_2 t)/\omega_2 \\ (-0.4006\sin\omega_1 t)/\omega_1 + (0.4006\sin\omega_2 t)/\omega_2 \end{bmatrix} \dfrac{\dot{D}_{01}}{L}$

**4.3-7.** $\mathbf{D} = \begin{bmatrix} 1 \\ -1 \end{bmatrix} \theta_{z0}\cos\omega_1 t$

**4.3-8.** $\mathbf{D} = \begin{bmatrix} 2.707\cos\omega_1 t + 1.293\cos\omega_2 t \\ -1.914\cos\omega_1 t + 0.9142\cos\omega_2 t \end{bmatrix} \dfrac{M_{z0}L}{14EI}$

**4.3-9.** $\mathbf{D} = \begin{bmatrix} t + (2\sin\omega_2 t)/\omega_2 \\ t - (\sin\omega_2 t)/\omega_2 \end{bmatrix} \dfrac{\dot{D}_{01}}{3}$

**4.4-1. (a)** $\mathbf{D} = \begin{bmatrix} 9(1-\cos\omega_1 t)/\omega_1^2 + (1-\cos\omega_2 t)/\omega_2^2 \\ 3(1-\cos\omega_1 t)/\omega_1^2 - 3(1-\cos\omega_2 t)/\omega_2^2 \end{bmatrix} \dfrac{P_1}{10m}$

**(b)** $\mathbf{D} = \begin{bmatrix} 3(1-\cos\omega_1 t)/\omega_1^2 - 3(1-\cos\omega_2 t)/\omega_2^2 \\ (1-\cos\omega_1 t)/\omega_1^2 + 9(1-\cos\omega_2 t)/\omega_2^2 \end{bmatrix} \dfrac{P_1}{10m}$

where $m = \dfrac{3.28\,\rho AL}{6}$; $(D_2)_{A_1} = (D_1)_{A_2}$

**4.4-2. (a)** $\mathbf{D} = \begin{bmatrix} (0.01943\beta_1/\omega_1^2 + 0.03775\beta_2/\omega_2^2)L \\ 0.02675\beta_1/\omega_1^2 + 0.2878\beta_2/\omega_2^2 \end{bmatrix} \dfrac{P_1\sin\Omega t}{mL}$

**(b)** $\mathbf{D} = \begin{bmatrix} (0.02675\beta_1/\omega_1^2 + 0.2878\beta_2/\omega_2^2)L \\ 0.03683\beta_1/\omega_1^2 + 2.193\beta_2/\omega_2^2 \end{bmatrix} \dfrac{M_2\sin\Omega t}{mL^2}$

where $m = \dfrac{\rho AL}{210}$; $(D_2)_{A_1} = (D_1)_{A_2}$

**4.4-3.** $\mathbf{D} = - \begin{bmatrix} 0.3536\left(t - \dfrac{1}{\omega_1}\sin\omega_1 t\right)/\omega_1^2 - 0.3536\left(t - \dfrac{1}{\omega_2}\sin\omega_2 t\right)/\omega_2^2 \\ 0.1465\left(t - \dfrac{1}{\omega_1}\sin\omega_1 t\right)/\omega_1^2 + 0.8536\left(t - \dfrac{1}{\omega_2}\sin\omega_2 t\right)/\omega_2^2 \end{bmatrix} \dfrac{P_1}{t_1 m}$

where $m = \rho AL$

**4.4-4.** $\mathbf{D} = \begin{bmatrix} -2(1-\cos\omega_1 t)/\omega_1^2 + 2(1-\cos\omega_2 t)/\omega_2^2 \\ (1-\cos\omega_1 t)/\omega_1^2 + 4(1-\cos\omega_2 t)/\omega_2^2 \end{bmatrix} \dfrac{P_1}{10m}$

where $m = \dfrac{2.72\,\rho AL}{6}$

**4.4-5.** $\mathbf{D} = \begin{bmatrix} \beta_1/\omega_1^2 + 9\beta_2/\omega_2^2 \\ 3\beta_1/\omega_1^2 - 3\beta_2/\omega_2^2 \end{bmatrix} \dfrac{P_1\cos\Omega t}{10m}$ where $m = \dfrac{3.28\,\rho AL}{6}$

**4.4-6.** $\mathbf{D} = \begin{bmatrix} -0.001745L(1-\cos\omega_1 t)/\omega_1^2 - 0.006828L(1-\cos\omega_2 t)/\omega_2^2 \\ 0.0006885(1-\cos\omega_1 t)/\omega_1^2 + 0.1816(1-\cos\omega_2 t)/\omega_2^2 \end{bmatrix} \dfrac{M_2}{m}$

where $m = \dfrac{\rho AL^3}{210}$

**4.4-7.** $\mathbf{D} = \begin{bmatrix} 1 \\ -1 \end{bmatrix} \dfrac{\beta_1 P_1 L \sin\Omega t}{88\omega_1^2 m}$ where $m = \dfrac{\rho AL^3}{420}$

**4.4-8.** $\mathbf{D} = \begin{bmatrix} -0.05776\left(t - \dfrac{1}{\omega_1}\sin\omega_1 t\right)/\omega_1^2 + 0.1882\left(t - \dfrac{1}{\omega_2}\sin\omega_2 t\right)/\omega_2^2 \\ 0.04084\left(t - \dfrac{1}{\omega_1}\sin\omega_1 t\right)/\omega_1^2 + 0.1331\left(t - \dfrac{1}{\omega_2}\sin\omega_2 t\right)/\omega_2^2 \end{bmatrix} \dfrac{M_2}{t_1 m}$

where $m = \dfrac{\rho AL^3}{420}$

Answers to Problems

**4.4-9.** $\mathbf{D} = -\begin{bmatrix} t^3 + 24\left(t - \dfrac{1}{\omega_2}\sin\omega_2 t\right)/\omega_2^2 \\ t^3 - 12\left(t - \dfrac{1}{\omega_2}\sin\omega_2 t\right)/\omega_2^2 \end{bmatrix} \dfrac{P_1}{108 t_1 m}$

where $m = \dfrac{\rho A L}{6}$

**4.5-1.** $\mathbf{D}^* = \begin{bmatrix} -9s_1 - s_2 \\ -3s_1 + 3s_2 \end{bmatrix} \dfrac{a_1}{10 t_1}$    $\begin{aligned} s_1 &= \left(t - \dfrac{1}{\omega_1}\sin\omega_1 t\right)/\omega_1^2 \\ s_2 &= \left(t - \dfrac{1}{\omega_2}\sin\omega_2 t\right)/\omega_2^2 \end{aligned}$

**4.5-2.** $\mathbf{D} = \begin{bmatrix} (1.000 - 1.221 c_1 + 0.2205 c_2) L \\ -1.681 c_1 + 1.681 c_2 \end{bmatrix} \dfrac{d}{L}$    $\begin{aligned} c_1 &= \cos\omega_1 t \\ c_2 &= \cos\omega_2 t \end{aligned}$

**4.5-3.** $\mathbf{D} = \begin{bmatrix} -1.000 + 1.207 c_1 + 0.2071 c_2 \\ -1.000 + 0.5000 c_1 + 0.5000 c_2 \end{bmatrix} L\theta_z$    $\begin{aligned} c_1 &= \cos\omega_1 t \\ c_2 &= \cos\omega_2 t \end{aligned}$

**4.5-4.** $\mathbf{D}^* = \begin{bmatrix} 2b_1 - 2b_2 \\ -b_1 - 4b_2 \end{bmatrix} \dfrac{a}{5}\cos\Omega t$    $\begin{aligned} b_1 &= \beta_1/\omega_1^2 \\ b_2 &= \beta_2/\omega_2^2 \end{aligned}$

**4.5-5.** $\mathbf{D} = \begin{bmatrix} (0.6s + 3m_1\Omega^2)b_1 - 3(1.8s + m_1\Omega^2)b_2 \\ 3(0.6s + 3m_1\Omega^2)b_1 + (1.8s + m_1\Omega^2)b_2 \end{bmatrix} \dfrac{d}{10m}\sin\Omega t$

where $s = EA/L$;  $m_1 = \rho A L/6$;  $m = 3.28 m_1$;  $b_1 = \beta_1/\omega_1^2$;  $b_2 = \beta_2/\omega_2^2$

**4.5-6.** $\mathbf{D} = \begin{bmatrix} (-1.000 + 1.053 c_1 - 0.0532 c_2)L \\ -1.000 - 0.4156 c_1 + 1.415 c_2 \end{bmatrix}\theta_z$    $\begin{aligned} c_1 &= \cos\omega_1 t \\ c_2 &= \cos\omega_2 t \end{aligned}$

**4.5-7.** $\mathbf{D}^* = -\begin{bmatrix} 1 \\ 1 \end{bmatrix} b_2 \theta_z \sin\Omega t$   where $b_2 = \beta_2/\omega_2^2$

**4.5-8.** $\mathbf{D} = \begin{bmatrix} 0.05776 b_1 - 0.1882 b_2 \\ -0.04084 b_1 - 0.1331 b_2 \end{bmatrix} \dfrac{\bar{s}\theta_z}{m}\sin\Omega t$    $\begin{aligned} \bar{s} &= s + 3m\Omega^2 \\ s &= 2EI/L \\ m &= \rho A L^3/420 \end{aligned}$

$b_1 = \beta_1/\omega_1^2$;  $b_2 = \beta_2/\omega_2^2$

## CHAPTER 5

**5.2-1.** and **5.2-11.** Given

**5.2-2.** $u_{10} = 0.9819 \dfrac{P_1}{k}$    **5.2-3.** $u_{20} = -3.063 \dfrac{P_1}{k}$

**5.2-4.** $u_{16} = 0.3622 \dfrac{P_1}{k}$    **5.2-5.** $u_{20} = -2.595 \dfrac{P_1}{k}$

**5.2-6.** $u_{20} = -4.004 \dfrac{P_1}{k}$    **5.2-7.** $u_{10} = 0.9182 \dfrac{P_1}{k}$

**5.2-8.** $u_{10} = 0.3146 \dfrac{P_1}{k}$    **5.2-9.** $u_{10} = 1.387 \dfrac{P_1}{k}$

**5.2-10.** $u_{10} = 1.107 \dfrac{P_1}{k}$

**5.2-12.** $(D_1)_{20} = -19.94\frac{P_1}{s}$; $(D_2)_{20} = -19.85\frac{P_1}{s}$

**5.2-13.** $(D_1)_{max} = -0.9992\frac{P_1}{s}$; $(D_2)_{max} = 0.7950\frac{P_1}{s}$

where $s = \frac{EA}{L}$

For problem sets 5.3 and 5.4, parts (a) and (b) denote average- and linear-acceleration methods.

**5.3-1.** and **5.3-11.** Given

**5.3-2.** (a) $u_{10} = 1.031\frac{P_1}{k}$ (b) $u_{10} = 1.016\frac{P_1}{k}$

**5.3-3.** (a) $u_{20} = -3.023\frac{P_1}{k}$ (b) $u_{20} = -3.038\frac{P_1}{k}$

**5.3-4.** (a) $u_{16} = 0.3676\frac{P_1}{k}$ (b) $u_{16} = 0.3659\frac{P_1}{k}$

**5.3-5.** (a) $u_{20} = -2.513\frac{P_1}{k}$ (b) $u_{20} = -2.541\frac{P_1}{k}$

**5.3-6.** (a) $u_{20} = -3.942\frac{P_1}{k}$ (b) $u_{20} = -3.963\frac{P_1}{k}$

**5.3-7.** (a) $u_{10} = 0.9087\frac{P_1}{k}$ (b) $u_{10} = 0.9119\frac{P_1}{k}$

**5.3-8.** (a) $u_{10} = 0.3191\frac{P_1}{k}$ (b) $u_{10} = 0.3176\frac{P_1}{k}$

**5.3-9.** (a) $u_{10} = 1.372\frac{P_1}{k}$ (b) $u_{10} = 1.377\frac{P_1}{k}$

**5.3-10.** (a) $u_{10} = 1.094\frac{P_1}{k}$ (b) $u_{10} = 1.098\frac{P_1}{k}$

**5.3-12.** (a) $(D_1)_{20} = -20.45\frac{P_1}{s}$ (b) $(D_1)_{20} = -20.32\frac{P_1}{s}$

$(D_2)_{20} = -19.47\frac{P_1}{s}$ $(D_2)_{20} = -19.51\frac{P_1}{s}$

**5.3-13.** (a) $(D_1)_{max} = -0.9976\frac{P_1}{s}$ (b) $(D_1)_{max} = -0.9994\frac{P_1}{s}$

$(D_2)_{max} = 0.7917\frac{P_1}{s}$ $(D_2)_{max} = 0.7860\frac{P_1}{s}$

**5.4-1** and **5.4-11** Given

**5.4-2.** (a) $u_{10} = 1.031\frac{P_1}{k}$ (b) $u_{10} = 1.016\frac{P_1}{k}$

**5.4-3.** (a) $u_{20} = -3.023\frac{P_1}{k}$ (b) $u_{20} = -3.038\frac{P_1}{k}$

Answers to Problems

**5.4-4.** (a) $u_{16} = 0.3676 \dfrac{P_1}{k}$  (b) $u_{16} = 0.3659 \dfrac{P_1}{k}$

**5.4-5.** (a) $u_{20} = -2.513 \dfrac{P_1}{k}$  (b) $u_{20} = -2.541 \dfrac{P_1}{k}$

**5.4-6.** (a) $u_{20} = -3.942 \dfrac{P_1}{k}$  (b) $u_{20} = -3.963 \dfrac{P_1}{k}$

**5.4-7.** (a) $u_{10} = 0.9087 \dfrac{P_1}{k}$  (b) $u_{10} = 0.9119 \dfrac{P_1}{k}$

**5.4-8.** (a) $u_{10} = 0.3191 \dfrac{P_1}{k}$  (b) $u_{10} = 0.3176 \dfrac{P_1}{k}$

**5.4-9.** (a) $u_{10} = 1.372 \dfrac{P_1}{k}$  (b) $u_{10} = 1.377 \dfrac{P_1}{k}$

**5.4-10.** (a) $u_{10} = 1.094 \dfrac{P_1}{k}$  (b) $u_{10} = 1.098 \dfrac{P_1}{k}$

**5.4-12.** (a) $(D_1)_{20} = -21.22 \dfrac{P_1}{s}$  (b) $(D_1)_{20} = -21.40 \dfrac{P_1}{s}$

$(D_2)_{20} = -19.28 \dfrac{P_1}{s}$   $(D_2)_{20} = -19.23 \dfrac{P_1}{s}$

**5.4-13.** (a) $(D_1)_{\max} = -0.9974 \dfrac{P_1}{s}$  (b) $(D_1)_{\max} = -0.9993 \dfrac{P_1}{s}$

$(D_2)_{\max} = 0.7917 \dfrac{P_1}{s}$   $(D_2)_{\max} = 0.7861 \dfrac{P_1}{s}$

**5.4-14.** $\overline{\mathbf{S}} = \mathbf{S} + \dfrac{4}{(\Delta t_j)^2}\mathbf{M} + \dfrac{2}{\Delta t_j}\mathbf{C}; \quad \overline{\mathbf{A}}_{j+1} = \mathbf{A}_{j+1} + \mathbf{M}\,\overline{\mathbf{Q}}_j + \mathbf{C}\,\overline{\mathbf{R}}_j$

where $\overline{\mathbf{Q}}_j = \dfrac{4}{(\Delta t_j)^2}\mathbf{D}_j + \dfrac{4}{\Delta t_j}\dot{\mathbf{D}}_j + \ddot{\mathbf{D}}_j; \quad \overline{\mathbf{R}}_j = \dfrac{2}{\Delta t_j}\mathbf{D}_j + \dot{\mathbf{D}}_j$

**5.4-15.** $\overline{\mathbf{C}} = \mathbf{C} + \dfrac{2}{\Delta t_j}\mathbf{M} + \dfrac{\Delta t_j}{2}\mathbf{S}; \quad \overline{\mathbf{A}}_{j+1} = \mathbf{A}_{j+1} + \mathbf{M}\,\overline{\mathbf{Q}}_j + \mathbf{S}\,\overline{\mathbf{R}}_j$

where $\overline{\mathbf{Q}}_j = \dfrac{2}{\Delta t_j}\dot{\mathbf{D}}_j + \ddot{\mathbf{D}}_j; \quad \overline{\mathbf{R}}_j = -\mathbf{D}_j - \dfrac{\Delta t_j}{2}\dot{\mathbf{D}}_j$

**5.4-16.** $\overline{\mathbf{M}} = \mathbf{M} + \dfrac{\Delta t_j}{2}\mathbf{C} + \dfrac{(\Delta t_j)^2}{4}\mathbf{S}; \quad \overline{\mathbf{A}}_{j+1} = \mathbf{A}_{j+1} + \mathbf{C}\,\overline{\mathbf{Q}}_j + \mathbf{S}\,\overline{\mathbf{R}}_j$

where $\overline{\mathbf{Q}}_j = -\dot{\mathbf{D}}_j - \dfrac{\Delta t_j}{2}\ddot{\mathbf{D}}_j; \quad \overline{\mathbf{R}}_j = -\mathbf{D}_j - \Delta t_j\,\dot{\mathbf{D}}_j - \dfrac{(\Delta t_j)^2}{4}\ddot{\mathbf{D}}_j$

**5.4-17.** $\overline{\mathbf{C}} = \mathbf{C} + \dfrac{2}{\Delta t_j}\mathbf{M} + \dfrac{\Delta t_j}{2}\mathbf{S}; \quad \Delta\overline{\mathbf{A}}_j = \Delta\mathbf{A}_j + \mathbf{M}\,\overline{\mathbf{Q}}_j + \mathbf{S}\,\overline{\mathbf{R}}_j$

where $\overline{\mathbf{Q}}_j = 2\ddot{\mathbf{D}}_j; \quad \overline{\mathbf{R}}_j = -\Delta t_j\,\dot{\mathbf{D}}_j$

**5.4-18.** $\overline{\mathbf{M}} = \mathbf{M} + \dfrac{\Delta t_j}{2}\mathbf{C} + \dfrac{(\Delta t_j)^2}{4}\mathbf{S}; \quad \Delta\overline{\mathbf{A}}_j = \Delta\mathbf{A}_j + \mathbf{C}\,\overline{\mathbf{Q}}_j + \mathbf{S}\,\overline{\mathbf{R}}_j$

where $\overline{\mathbf{Q}}_j = -\Delta t_j\,\ddot{\mathbf{D}}_j; \quad \overline{\mathbf{R}}_j = -\Delta t_j\,\dot{\mathbf{D}}_j - \dfrac{(\Delta t_j)^2}{2}\ddot{\mathbf{D}}_j$

**5.4-19.** $S^* = S + \dfrac{6}{(\Delta t_j)^2} M + \dfrac{3}{\Delta t_j} C; \quad A^*_{j+1} = A_{j+1} + M Q^*_j + C R^*_j$

where $Q^*_j = \dfrac{6}{(\Delta t_j)^2} D_j + \dfrac{6}{\Delta t_j} \dot{D}_j + 2\ddot{D}_j; \quad R^*_j = \dfrac{3}{\Delta t_j} D_j + 2\dot{D}_j + \dfrac{\Delta t_j}{2} \ddot{D}_j$

**5.4-20.** $C^* = C + \dfrac{2}{\Delta t_j} M + \dfrac{\Delta t_j}{3} S; \quad A^*_{j+1} = A_{j+1} + M Q^*_j + S R^*_j$

where $Q^*_j = \dfrac{2}{\Delta t_j} \dot{D}_j + \ddot{D}_j; \quad R^*_j = -D_j - \dfrac{2\Delta t_j}{3} \dot{D}_j - \dfrac{(\Delta t_j)^2}{6} \ddot{D}_j$

**5.4-21.** $M^* = M + \dfrac{\Delta t_j}{2} C + \dfrac{(\Delta t_j)^2}{6} S; \quad A^*_{j+1} = A_{j+1} + C Q^*_j + S R^*_j$

where $Q^*_j = -\dot{D}_j - \dfrac{\Delta t_j}{2} \ddot{D}_j; \quad R^*_j = -D_j - \Delta t_j \dot{D}_j - \dfrac{(\Delta t_j)^2}{3} \ddot{D}_j$

**5.4-22.** $C^* = C + \dfrac{2}{\Delta t_j} M + \dfrac{\Delta t_j}{3} S; \quad \Delta A^*_j = \Delta A_j + M Q^*_j + S R^*_j$

where $Q^*_j = 2\ddot{D}_j; \quad R^*_j = -\Delta t_j \dot{D}_j - \dfrac{(\Delta t_j)^2}{6} \ddot{D}_j$

**5.4-23.** $M^* = M + \dfrac{\Delta t_j}{2} C + \dfrac{(\Delta t_j)^2}{6} S; \quad \Delta A^*_j = \Delta A_j + C Q^*_j + S R^*_j$

where $Q^*_j = -\Delta t_j \ddot{D}_j; \quad R^*_j = -\Delta t_j \dot{D}_j - \dfrac{(\Delta t_j)^2}{2} \ddot{D}_j$

**5.5-1.** $\hat{S} = S + \dfrac{1}{\beta(\Delta t_j)^2} M + \dfrac{\gamma}{\beta \Delta t_j} C; \quad \hat{A}_{j+1} = A_{j+1} + M \hat{Q}_j + C \hat{R}_j$

where $\hat{Q}_j = \dfrac{1}{\beta(\Delta t_j)^2} D_j + \dfrac{1}{\beta \Delta t_j} \dot{D}_j + \dfrac{1-2\beta}{2\beta} \ddot{D}_j$

and $\hat{R}_j = \dfrac{\gamma}{\beta \Delta t_j} D_j - \left(1 - \dfrac{\gamma}{\beta}\right)\dot{D}_j - \left(1 - \dfrac{\gamma}{2\beta}\right)\Delta t_j \ddot{D}_j$

**5.5-2.** $\hat{C} = C + \dfrac{1}{\gamma \Delta t_j} M + \dfrac{\beta \Delta t_j}{\gamma} S; \quad \hat{A}_{j+1} = A_{j+1} + M \hat{Q}_j + S \hat{R}_j$

where $\hat{Q}_j = \dfrac{1}{\gamma \Delta t_j} \dot{D}_j + \dfrac{1-\gamma}{\gamma} \ddot{D}_j$

and $\hat{R}_j = -D_j - \left(1 - \dfrac{\beta}{\gamma}\right)\Delta t_j \dot{D}_j - \left(\dfrac{1}{2} - \dfrac{\beta}{\gamma}\right)(\Delta t_j)^2 \ddot{D}_j$

**5.5-3.** $\hat{M} = M + \gamma \Delta t_j C + \beta(\Delta t_j)^2 S; \quad \hat{A}_{j+1} = A_{j+1} + C \hat{Q}_j + S \hat{R}_j$

where $\hat{Q}_j = -\dot{D}_j - (1 - \gamma) \Delta t_j \ddot{D}_j$

and $\hat{R}_j = -D_j - \Delta t_j \dot{D}_j - \left(\dfrac{1}{2} - \beta\right)(\Delta t_j)^2 \ddot{D}_j$

**5.5-4.** $\hat{C} = C + \dfrac{1}{\gamma \Delta t_j} M + \dfrac{\beta \Delta t_j}{\gamma} S; \quad \Delta \hat{A}_j = \Delta A_j + M \hat{Q}_j + S \hat{R}_j$

where $\hat{Q}_j = \dfrac{1}{\gamma} \ddot{D}_j; \quad \hat{R}_j = -\Delta t_j \dot{D}_j - \left(\dfrac{1}{2} - \dfrac{\beta}{\gamma}\right)(\Delta t_j)^2 \ddot{D}_j$

Answers to Problems

**5.5-5.** $\hat{\mathbf{M}} = \mathbf{M} + \gamma \Delta t_j \mathbf{C} + \beta(\Delta t_j)^2 \mathbf{S}$; $\Delta \hat{\mathbf{A}}_j = \Delta \mathbf{A}_j + \mathbf{C} \hat{\mathbf{Q}}_j + \mathbf{S} \hat{\mathbf{R}}_j$
where $\hat{\mathbf{Q}}_j = -\Delta t_j \dot{\mathbf{D}}_j$; $\hat{\mathbf{R}}_j = -\Delta t_j \dot{\mathbf{D}}_j - \dfrac{(\Delta t_j)^2}{2} \ddot{\mathbf{D}}_j$

# CHAPTER 6

**6.2-1.** $\mathbf{p}'_{bj}(t) = \begin{bmatrix} 0 \\ 78 \\ 11L \end{bmatrix} \dfrac{Lb_{y'}(t)}{192}$; $\mathbf{p}'_{bk}(t) = \begin{bmatrix} 0 \\ 18 \\ -5L \end{bmatrix} \dfrac{Lb_{y'}(t)}{192}$

$\mathbf{p}_{bj}(t) = \begin{bmatrix} 234 \\ 312 \\ 55L \end{bmatrix} \dfrac{Lb_{y'}(t)}{960}$; $\mathbf{p}_{bk}(t) = \begin{bmatrix} 54 \\ 72 \\ -25L \end{bmatrix} \dfrac{Lb_{y'}(t)}{960}$

**6.2-2.** $\mathbf{p}'_{Pj}(t) = \begin{bmatrix} 0 \\ 10 \\ 3L \end{bmatrix} \dfrac{P_{y'}(t)}{64}$; $\mathbf{p}'_{Pk}(t) = \begin{bmatrix} 0 \\ 54 \\ -9L \end{bmatrix} \dfrac{P_{y'}(t)}{64}$

$\mathbf{p}_{Pj}(t) = \begin{bmatrix} -10 \\ 10 \\ 3\sqrt{2}L \end{bmatrix} \dfrac{P_{y'}(t)}{64\sqrt{2}}$; $\mathbf{p}_{Pk}(t) = \begin{bmatrix} -54 \\ 54 \\ -9\sqrt{2}L \end{bmatrix} \dfrac{P_{y'}(t)}{64\sqrt{2}}$

**6.2-3.** $\mathbf{p}'_M(t) = \{0, -6, -L, 0, 6, -L\} \dfrac{M_z(t)}{4L}$

$\mathbf{p}_M(t) = \{-12, -6, -\sqrt{5}L, 12, 6, -\sqrt{5}L\} \dfrac{M_z(t)}{4\sqrt{5}L}$

**6.2-4.** $\mathbf{p}'_b(t) = \{0, 9, 2L, 0, 21, -3L\} \dfrac{Lb_2(t)}{60}$

$\mathbf{p}_b(t) = \{-36, 27, 10L, -84, 63, -15L\} \dfrac{Lb_2(t)}{300}$

**6.2-5.** $\mathbf{p}'_P(t) = \{0, 9, 2L, 0, 9, -2L\} \dfrac{P_{y'}(t)}{9}$

$\mathbf{p}_P(t) = \{-18, -9, 2\sqrt{5}L, -18, -9, -2\sqrt{5}L\} \dfrac{P_{y'}(t)}{9\sqrt{5}}$

**6.2-6.** $\mathbf{p}'_M(t) = \{0, -18, -L, 0, 18, -L\} \dfrac{M_z(t)}{8L}$

$\mathbf{p}_M(t) = \{18, -18, -\sqrt{2}L, -18, 18, -\sqrt{2}L\} \dfrac{M_z(t)}{8\sqrt{2}L}$

**6.2-7.** $\mathbf{p}'_b(t) = \{1, 0, 0, 3, 0, 0\} \dfrac{Lb_2(t)}{12}$

$\mathbf{p}_b(t) = \{2, -1, 0, 6, -3, 0\} \dfrac{Lb_2(t)}{12\sqrt{5}}$

**6.3-1.** $\mathbf{p}_b'(t) = \{0, -5L, 18, 0, 11L, 78\}\dfrac{Lb_z(t)}{192}$

$\mathbf{p}_b(t) = \{20L, -15L, 90, -44L, 33L, 390\}\dfrac{Lb_z(t)}{960}$

**6.3-2.** $\mathbf{p}_P'(t) = \{0, -4L, 20, 0, 2L, 7\}\dfrac{P_z(t)}{27}$

$\mathbf{p}_P(t) = \{-4L, -12L, 20\sqrt{10}, 2L, 6L, 7\sqrt{10}\}\dfrac{P_z(t)}{27\sqrt{10}}$

**6.3-3.** $\mathbf{p}_M'(t) = \{0, 3L, 18, 0, -5L, -18\}\dfrac{M_{y'}(t)}{16L}$

$\mathbf{p}_M(t) = \{-3L, 3L, 18\sqrt{2}, 5L, -5L, -18\sqrt{2}\}\dfrac{M_{y'}(t)}{16\sqrt{2}L}$

**6.3-4.** $\mathbf{p}_M'(t) = \{12L, 3L, 18, 4L, -5L, -18\}\dfrac{M_y(t)}{16\sqrt{2}L}$

$\mathbf{p}_M(t) = \{9L, 15L, 18\sqrt{2}, 9L, -L, -18\sqrt{2}\}\dfrac{M_y(t)}{32L}$

**6.3-5.** $\mathbf{p}_M'(t) = \{12L, -3L, -18, 4L, 5L, 18\}\dfrac{M_x(t)}{16\sqrt{2}L}$

$\mathbf{p}_M(t) = \{15L, 9L, -18\sqrt{2}, -L, 9L, 18\sqrt{2}\}\dfrac{M_x(t)}{32L}$

**6.3-6.** $\mathbf{p}_b'(t) = \{0, -3L, 21, 0, 2L, 9\}\dfrac{Lb_1(t)}{60}$

$\mathbf{p}_b(t) = \{-9L, -12L, 105, -6L, -8L, 45\}\dfrac{Lb_1(t)}{300}$

**6.3-7.** $\mathbf{p}_P'(t) = \{0, -3L, 16, 0, 3L, 16\}\dfrac{P_z(t)}{16}$

$\mathbf{p}_P(t) = \{9L, 12L, 80, -9L, -12L, 80\}\dfrac{P_z(t)}{80}$

**6.3-8.** $\mathbf{p}_M'(t) = \{0, 1, 0, 0, -1, 0\}\dfrac{M_{y'}(t)}{3}$

$\mathbf{p}_M(t) = \{2, 1, 0, -2, -1, 0\}\dfrac{M_{y'}(t)}{3\sqrt{5}}$

**6.4-1.** $\mathbf{p}_b'(t) = \{0, 1, 0, 0, 1, 0\}\dfrac{Lb_{y'}(t)}{4}$; $L = \sqrt{14}$

$\mathbf{p}_{bj}(t) = \mathbf{p}_{bk}(t) = \{-0.3043, 0.7161, -0.5192\}b_{y'}(t)$

**6.4-2.** $\mathbf{p}_b'(t) = \{0, 0, 1, 0, 0, 1\}\dfrac{Lb_{z'}(t)}{6}$; $L = \sqrt{14}$

$\mathbf{p}_{bj}(t) = \mathbf{p}_{bk}(t) = \{-0.3126, 0.2233, 0.4912\}b_{z'}(t)$

**6.4-3.** $\mathbf{p}_P'(t) = \{0, 0, 3, 0, 0, 2\}\dfrac{P_{z'}(t)}{5}$

$\mathbf{p}_{Pj}(t) = \tfrac{3}{2}\mathbf{p}_{Pk}(t) = \{-0.4899, 0.2449, 0.2449\}P_{z'}(t)$

# Answers to Problems

**6.4-4.** $\mathbf{p}_P'(t) = \{0, 2, 0, 0, 3, 0\}\dfrac{P_{y'}(t)}{5}$

$\mathbf{p}_{Pj}(t) = \tfrac{2}{3}\mathbf{p}_{Pk}(t) = \{0, 0.2828, -0.2828\}P_{y'}(t)$

**6.4-5.** $\mathbf{p}_M'(t) = \{0, -1, 0, 0, 1, 0\}\dfrac{M_{z'}(t)}{L}; \; L = \sqrt{29}$

$\mathbf{p}_{Mj}(t) = -\mathbf{p}_{Mk}(t) = \{-0.07773, -0.1184, -0.1201\}M_{z'}(t)$

**6.4-6.** $\mathbf{p}_M'(t) = \{0, 0, 1, 0, 0, -1\}\dfrac{M_{y'}(t)}{L}; \; L = \sqrt{29}$

$\mathbf{p}_{Mj}(t) = -\mathbf{p}_{Mk}(t) = \{-0.1332, -0.03805, 0.1327\}M_{y'}(t)$

**6.4-7.** $\mathbf{p}_{bj}'(t) = \mathbf{p}_{bk}'(t) = \{-1.000, 0.4242, 0\}b(t)$

$\mathbf{p}_{bj}(t) = \mathbf{p}_{bk}(t) = \{-0.5992, -0.1141, -0.8988\}b(t)$

**6.4-8.** $\mathbf{p}_{bj}'(t) = \tfrac{1}{2}\mathbf{p}_{bk}'(t) = \{-0.3334, 0.6009, 0\}b_2(t)$

$\mathbf{p}_{bj}(t) = \tfrac{1}{2}\mathbf{p}_{bk}(t) = \{-0.3234, 0.3637, -0.4851\}b_2(t)$

**6.5-1.** $\mathbf{p}_{bj}(t) = \{-1.057, 2.439, -0.4269, -1.769, -0.4913, 1.572\}b_{y'}(t)$

$\mathbf{p}_{bk}(t) = \{-1.057, 2.439, -0.4269, 1.769, 0.4913, -1.572\}b_{y'}(t)$

**6.5-2.** $\mathbf{p}_{bj}(t) = \{-1.601, -0.4447, 1.423, 0.6523, -1.505, 0.2634\}b_{y'}(t)$

$\mathbf{p}_{bk}(t) = \{-0.3695, -0.1026, 0.3284, -0.2965, 0.6842, -0.1197\}b_{y'}(t)$

**6.5-3.** $\mathbf{p}_{Pj}(t) = \{-0.3675, 0.4492, 0.6125, 0.3811, -0.1429, 0.3335\}P_{y'}(t)$

$\mathbf{p}_{Pk}(t) = \{-0.06805, 0.08318, 0.1134, -0.1270, 0.04764, -0.1112\}P_{y'}(t)$

**6.5-4.** $\mathbf{p}_{Pj}(t) = \{0.1132, -0.04244, 0.09902, 0.07639, -0.09336, -0.1273\}P_{z'}(t)$

$\mathbf{p}_{Pk}(t) = \{0.6111, -0.2292, 0.5347, -0.2292, 0.2801, 0.3819\}P_{z'}(t)$

**6.5-5.** $\mathbf{p}_{Mj}(t) = \{-4, 0, 0, 0, 0, 0\}\dfrac{M_{y'}(t)}{15}$

$\mathbf{p}_{Mk}(t) = \{4, 0, 0, 0, -4, 3\}\dfrac{M_{y'}(t)}{15}$

**6.5-6.** $\mathbf{p}_{Mj}(t) = \{0, -16, 12, 5, 0, 0\}\dfrac{M_{z'}(t)}{75}$

$\mathbf{p}_{Mk}(t) = \{0, 16, -12, 0, 0, 0\}\dfrac{M_{z'}(t)}{75}$

**6.5-7.** $\mathbf{p}_{bj}(t) = \{1.5, 0, 0, 0, 0.5, 0.5\}b_x(t)$

$\mathbf{p}_{bk}(t) = \{1.5, 0, 0, 0, -0.5, -0.5\}b_x(t)$

**6.5-8.** $\mathbf{p}_{bj}(t) = \{-0.01111, 0.02222, 1.028, -0.3000, -0.1500, 0\}b_1(t)$

$\mathbf{p}_{bk}(t) = \{0.01111, -0.02222, 0.4722, 0.2000, 0.1000, 0\}b_1(t)$

**6.7-1.** $\omega_1 = \dfrac{15.69}{L^2}\sqrt{\dfrac{EI_z}{\rho A}}$

**6.7-2.** $\omega_{1,2} = 5.603, 31.19 \dfrac{1}{L^2}\sqrt{\dfrac{EI_z}{\rho A}}; \; \Phi = \begin{bmatrix} 1.000 & 1.000 \\ 0.5435 & -0.9364 \end{bmatrix}$

**6.7-3.** $\omega_1 = \dfrac{9.941}{L^2}\sqrt{\dfrac{EI_z}{\rho A}}$

**6.7-4.** $\omega_{1,2,3} = 0, 0, 22.47\dfrac{1}{L^2}\sqrt{\dfrac{EI_z}{\rho A}}$; $\Phi = \begin{bmatrix} 1 & 1 & 1 \\ 1 & 0 & -0.6 \\ 1 & -1 & 1 \end{bmatrix}$

**6.7-5.** $\omega_{1,2} = 1.648, 5.540\dfrac{1}{L^2}\sqrt{\dfrac{EI_z}{\rho A}}$; $\Phi = \begin{bmatrix} 0.8844 & -0.4109 \\ 1.000 & 1.000 \end{bmatrix}$

**6.7-6.** $\omega_{1,2} = 0.9044, 3.840\dfrac{1}{L^2}\sqrt{\dfrac{EI_z}{\rho A}}$; $\Phi = \begin{bmatrix} 1 & 0 \\ 0 & 1 \end{bmatrix}$

**6.7-7.** $\omega_1 = \dfrac{6.149}{L^2}\sqrt{\dfrac{GJ}{\rho A}}$  **6.7-8.** $\omega_1 = \dfrac{5.943}{L^2}\sqrt{\dfrac{GJ}{\rho A}}$

# Index

**Absolute damping,** 166
Absolute system of units, 495
Accelerations:
   ground, 21: arbitrary, 42; periodic, 38; rigid-body, 159
   incremental, 212
   relative, 23
Accelerometer, 22
Accuracy, 208
   numerical, 223
Action equations of motion, 140
Action vector:
   applied, 152
   reduced, 282
Actual nodal loads, 98
Adjoint matrix, 106
Algorithmic damping, 218
Amplification matrix, 223
Amplitude:
   forced vibration, 21
   free vibration, 16
   suppression, 225, 234
Angular frequency:
   damped free vibrations, 28
   undamped free vibrations, 14, 105

Anisotropic materials, 311
Antisymmetric modes, 112
Applied actions, 4
   normal-mode response, 152
Applied body forces, 80
Arbitrary forcing functions, 38, 172
Arbitrary ground accelerations, 42, 159
Assembly:
   of elements, 95
   of substructures: in component-mode method, 470; in tridiagonal method, 448
Average-acceleration method, 219
   direct linear extrapolation, 212
   iteration with, 203
Axes:
   body, principal, 420
   rotation of, 95
   stress, principal, 315
   translation of, 157, 414
Axial element, 82
Axial rigidity, 94
Axial strains, 83
   constraints against, 290

Axisymmetric loads, 357
Axisymmetric shells:
　element AXSH3, 394
　nonaxisymmetric loads, 401
　Program DYAXSH, 406
Axisymmetric solids, 357
　element AXQ4, 359
　element AXQ8, 359
　nonaxisymmetric loads, 361
　Program DYAXSO, 365

**Backward substitution**, 461
Beams:
　continuous, 243,
　frequency coefficients (*table*), 288
Blast load, 50
Block mass submatrices, 437
Block stiffness submatrices, 437
Body axes, principal, 420
Body forces, 78
Body-oriented method, 434
Boundary nodes, 446
Bracing in tier buildings, 468
Buildings, multistory, 425

**Center of mass**, 418, 427, 434
Central-difference predictor, 197
Characteristic equation, 105, 498
Characteristic matrix, 106
Characteristic-value matrix, 141
Cholesky method, modified, 498
Cholesky square-root method, 106
Coefficients:
　damping, 165
　flexibility, 145
　for Gaussian Quadrature (*table*), 328
　frequency, 287, (*table*), 288
　mass, 82
　stiffness, 82
Component-mode method, 468
　assembly of substructures, 470
　dynamic coupling in, 470
　for plane trusses, 471: Program COMOPT, 477
　for space trusses, 477: Program COMOST, 477
　substructure equations of motion, 468
Computer programs, 9
Condensation, matrix, 282
Conditional stability, 208
Consistent-mass matrix:
　for axial element, 84
　for axisymmetric shell element with nonaxisymmetric loads, 405
　for axisymmetric solid element with nonaxisymmetric loads, 364
　for element, 82
　for element AXQ4, 359
　for element AXSH3, 401
　for element H8, 347
　for element PBQ8, 378
　for element Q4, 336
　for element SHQ8, 389
　for flexural element, 91
　for grid member, 251
　for plane frame member, 246
　for plane truss member, 103: in component-mode method, 474
　for space frame member, 261
　for space truss member, 255
　for structure, 98
　for torsional element, 87
Consistent systems of units (*table*), 496
Constant-acceleration method, 219
Constraint, modal, 500
Constraint matrix, 291, 372, 382, 394
Constraints:
　against axial strains, 290
　rigid-body, 411, 427
Continua:
　stresses and strains in, 310

two- and three-dimensional, 310–69
Continuous beams, 243
  loads, 265
  Program DYNACB, 265
Convergence, rate of, 206
Conversion factors, 496
Conversion of US units to SI units (*table*), 496
Coordinates:
  natural, 318
  normal, 139
  relative, 22, 159
Corrector, 203
Critical damping, 30
Critical time step, 198, 224
Curvature, 90

**D'Alembert's principle,** 3, 13
Damped forced vibrations, 30
Damped free vibrations, 27
Damped response of MDOF systems:
  to arbitrary forcing functions, 172
  to periodic forcing functions, 168
Damping:
  absolute, 166
  critical, 30
  effects of, 25
  in MDOF systems, 164
  modal, 167
  numerical, 218
  proportional, 165
  relative, 166
Damping coefficients, 165
Damping constant, 3, 27
  modal, 165
Damping matrix, 164
Damping ratio, 31
  modal, 165
Decomposition:
  Fourier, 361
  spectral, 223
Deflation, 500

Degrees of freedom, number of, 105
Density, mass, 80, 497
Dependent displacements, 282, 291
Determinant of Jacobian matrix, 330, 332
Diagonalization, 140
Differential equations of motion, 75
Dimensionless coordinates, 319
Direct iteration, 498
Direct linear extrapolation, 211
  average-acceleration method, 212
  linear-acceleration method, 213
Direct numerical integration methods, 195–240
Direct stiffness method, 98
Direction cosines, 96, 101
Discretization of structures, 6
Displacement equations of motion, 145
Displacements:
  dependent, 282, 291
  generic, 75, 78
  ground, 21: rigid-body, 157
  incremental, 38, 212,
  independent, 282, 291
  initial, 14, 147
  kinematically equivalent, 414
  nodal, 75, 79: free, 99; restrained, 99
  relative, 23, 159
Displacement shape functions, 74, 79
  for axial element, 82
  for element AXSH3, 398
  for element H8, 345
  for element H20, 349
  for element PBQ8, 375
  for element Q4, 333
  for element Q8, 338
  for element SHQ8, 386
  for flexural element, 89
  for torsional element, 85
Dissipative force, 3
Duhamel's integral, 39, 152, 173
Dynamic equilibrium, 1

Dynamic force, 1
Dynamic influences, 4
Dynamic load data:
 for multistory plane frames (*table*), 465
 for plane frames (*table*), 269
 for plane trusses (*table*), 180
 for Program DYAXSH (*table*), 407
 for Program DYNAPB (*table*), 380
 for Program DYNAPS (*table*), 341
 for Program DYNASH (*table*), 391
 for Program DYNASO (*table*), 352
 for space frames (*table*), 279
Dynamic loads, 1
 blast, 50
 reciprocal theorem, 154
Dynamic reduction, 282

**Eigenvalue matrix,** 141
Eigenvalue problem, 105, 314
Eigenvalues, 106, 498
 repeated, 107
Eigenvector matrix, 140, 315
Eigenvectors, 106, 498
 normalization of, 142
Elasticity, modulus, 497
Elasticity force, 3
Element AXQ4, 359
Element AXQ8, 359
Element AXSH3, 394
Element AXSR3, 394
Element H8, 345
Element H20, 349
Element PBQ8, 371
Element PQR8, 371
Element Q4, 333
Element Q8, 338
Element R4, 333
Element R8, 338
Element RS8, 345
Element RS20, 349
Element SHQ8, 382
Elimination:
 forward, 461

Gauss-Jordan, 291
Elimination method, parallel, 455
Equations of motion, 3, 13, 75, 78
 action, 140
 displacement, 145
 for finite elements, 78
 incremental, 212
 structural, 98
Equivalent nodal loads:
 for element, 82
 nonproportional, 267
 for structure, 98
 for surface pressure, 366, 408
Equivalent viscous damping, 26
Euler's extrapolation formula, 205
Explicit formulas for extrapolation, 197
Extrapolation, direct linear, 211

**Factorization:**
 Cholesky, 106, 498
 Givens, 514
Factors, conversion, (*table*), 496
Finite-difference formulations, 195
Flexibilities, 10
Flexibility coefficients, 145
Flexible-body motions, 138
Flexural deformations, 90
 in plate-bending, 378
Flexural element, 88
Flexural rigidity, 94
Flexural strains, 89
Forced vibrations, 19, 30
Forces, body, 78
Force systems, types of, 241
Forcing functions:
 arbitrary, 38, 172
 harmonic, 19
 periodic, 35, 168
 piecewise-linear, 46, 175, 182
Form factor, 377
FORTRAN, 9
Forward elimination, 461
Forward iteration, 498
Fourier coefficients, 36

Fourier series, 35, 169, 361, 401
Framed structures, 6, 241–309
Frames:
  plane, 243, 244
  space, 244, 259
Free nodal displacements, 99
Free vibrations, 12, 27
Frequency, 14, 28
Frequency coefficients, 287, (table), 288

**Gaussian quadrature,** 328
  coefficients for (table), 328
Gauss-Jordan elimination, 291
Generalized acceleration method, 217
Generalized rotation matrix, 505
Generalized stresses and strains, 92
General shells:
  element SHQ8, 382
  Program DYNASH, 390
General solids:
  element H8, 345
  element H20, 349
  Program DYNASO, 351
Generic displacements, 75, 78
Geometric center, 318
Geometric interpolation functions, 320, 324
  for element AXSH3, 396
  for element H8, 345
  for element H20, 350
  for element PBQ8, 374
  for element Q4, 333
  for element Q8, 338
  for element SHQ8, 383
Geometric transformations, 431
Givens factorization, 514
Givens method, 509
Global axes, 95
Gram-Schmidt orthogonalization, 107, 500
Gravitational system of units, 495
Grids, 243, 249

Program DYNAGR, 270
Ground accelerations, 21
  arbitrary, 42
  periodic, 38
  rigid-body, 159
Ground displacements, 21
  rigid-body, 157
Guyan reduction, 282, 445

**Harmonic forcing functions,** 19
Hexahedral coordinates, 323
  infinitesimal volume in, 330
Hilbert-$\alpha$ method, 222
Householder method, 511

**Immobility,** static, 145
Implicit formulas for iteration, 203
Incremental accelerations, 212
Incremental actions, 212
Incremental displacements, 38, 212
Incremental equations of motion, 212
Incremental impulse, 38
Incremental velocities, 38, 212
Independent displacements, 282, 291
Independent motions of support restraints, 162
Inertia, moment of, 91
Inertial body forces, 80
Inertial force, 3
Inertial moment per unit length, 87
Inertias, translational and rotational (rotary), 91, 378, 390, 401
Infinitesimal area in quadrilateral coordinates, 329
Infinitesimal volume in hexahedral coordinates, 330
Initial conditions, 4, 14
  normal-mode response, 147
Integration, numerical, 326
Integration points:
  for hexahedron, 332
  for quadrilateral, 330
Interior nodes, 446

Interpolation functions:
  geometric, 320, 325 (*see also* Geometric interpolation functions)
  piecewise-linear, 46, 175, 182
Inverse iteration, 498
Inverse of Jacobian matrix, 321, 325
Inverse of normalized modal matrix, 143
Isoparametric elements, 310–69
  element AXQ4, 359
  element AXQ8, 359
  hexahedron H8, 345
  hexahedron H20, 349
  quadrilateral Q4, 333
  quadrilateral Q8, 338
Isotropic materials, 77, 311
Iteration (for eigenvalues), 498
Iteration with implicit formulas, 203

**Jacobian matrix,** 321, 324
  determinant of, 321, 325
  inverse of, 321, 325
Jacobi method, 505
Joints, 241

**Kilogram** (kg), 495
Kilopound (kip), 497
Kinematically equivalent displacements, 414

**Lag time,** 17
Linear-acceleration method, 219
  direct linear extrapolation, 213
  iteration with, 206
Linear differential operator, 79
Linear extrapolation, direct, 211
Load operator, 223
Loads:
  axisymmetric, 357
  on continuous beams, 265
  dynamic, 1
    reciprocal theorem, 154

  equivalent nodal:
    for element, 82
    for structure, 98
  nonaxisymmetric, 361
  normal-mode, 152
  on grids, 271
  on hexahedra, 351
  on plane frames, 267
  on plane trusses, 179
  on quadrilaterals, 340
  on space frames, 278
  on space trusses, 274
Local axes, 95
Lumped mass matrix, 99

**Magnification factor,** 20, 32, 169
Main program:
  for DYNA (*flow chart*), 226
  for NOMO (*flow chart*), 178
  for VIB (*flow chart*), 119
Mass
  center of, 418, 427, 434
  of a body, 419
Mass coefficients, 82
Mass density, 80, 497
Mass matrix:
  for element, 82
  for structure, 98
  lumped, 99
  principal, 140
  reduced, 283
Mass moments of inertia of a rigid body, 419
  principal, 420
Mass products of inertia of a rigid body, 419
Master displacements, 283, 10.2-19
Material properties (*table*), 497
Materials:
  anisotropic, 311
  isotropic, 77, 311
  orthotropic, 311
  properties of (*table*), 497
Matrix condensation, 282

MDOF systems, 74
Mechanical analogue, 12
Member-oriented approach, 416
Members, 241
Meter (m), 495
Modal constraint, 500
Modal damping, 167
Modal damping constant, 165
Modal damping ratio, 165
Modal matrix, 140, 505
  normalized, 142: inverse of, 143
Modal truncation, 139, 143
Modes:
  antisymmetric, 112
  flexible-body, 138
  rigid-body, 138, 148
  symmetric, 112
  vibrational, 138
Mode shape, 105
Modified Cholesky method, 498
Modified tridiagonal method, 451
  for multistory buildings, 457
Modulus of elasticity, 497
Motion, equations of, 3, 13, 75, 78
Moving load:
  on a beam, 266
  on a plate, 381
Multiplicity (of repeated eigenvalues), 107
Multistory buildings:
  modified tridiagonal method, 457
  with rigid laminae, 425
Multistory plane frames:
  dynamic load data (*table*), 465
  Program DYMSPF, 463
  structural data (*table*), 464

**Natural coordinates,** 318
Natural frequency, 14, 28
Natural period, 4, 14, 28
Newmark-$\beta$ method, 218
Newton, 495
Nodal actions, 79
Nodal circles, 357

Nodal coordinates:
  for element H8 (*table*), 347
  for element H20 (*table*), 351
  for element Q4 (*table*), 335
  for element Q8 (*table*), 340
Nodal displacements, 75, 79
  free, 99
  restrained, 99
Nodal loads:
  actual, 98
  equivalent, 153
    for element, 82
    for structure, 98
Nodal vectors:
  for element AXSH3, 396, 406
  for element SHQ8, 385, 391
Nodes:
  boundary, 446
  interior, 446
  of finite elements, 6, 73
Nodewise solution, 198
Nonaxisymmetric loads:
  on axisymmetric shells, 401
  on axisymmetric solids, 361
Nonproportional equivalent nodal loads, 267
Normal coordinates, 139
Normalization of eigenvectors, 142
Normal-mode load, 152
Normal-mode method, 138–94
Normal-mode response:
  Program NOMO for, 177
  to applied actions, 152
  to initial conditions, 147
  to support motions, 157
Normal stresses and strains, 75
Notation, 483
  for matrices and vectors, 483
  for programs, 487
  Greek letters, 486
  simple variables, 485
  subscripts for matrices and vectors, 484
Numerical accuracy, 223
Numerical damping, 218

Numerical integration, 326
Numerical stability, 223

**One-dimensional elements,** 82
Orthogonality relationships, 140
Orthogonalization, Gram-Schmidt, 107, 500
Orthotropic materials, 311
Outline of Program DYMSPF, 463
Outline of Programs DYPFAC and DYSFAC, 299
Output selection data (*table*), 182
Overdamping, 29
Overlay technique, 453, 461

**Parallel elimination method,** 455
Parent rectangle:
   element AXSR3, 394
   element PQR8, 371
   element R4, 333, 336
   element R8, 338
Parent rectangular solid:
   element RS8, 345, 348
   element RS20, 349
Pascal, 496
Period, 4, 14, 28
Period elongation, 225, 234
Periodic forcing functions, 35
   damped response of MDOF systems, 168
Periodic ground acceleration, 38
Phase angle, 16, 32, 105, 169
Phase plane, 14
Piecewise-linear forcing functions, 46, 175, 182
Plane frames, 243, 244
   axial constraints, 290: program DYPFAC, 299
   dynamic load data (*table*), 269
   Program DYNAPF, 267
   rigid bodies: Program DYRBPF, 424
   structural data (*table*), 268

Plane stress and plane strain, 313
   Program DYNAPS, 340
Plane trusses, 243
   component-mode method, 471: Program COMOPT, 477
   dynamic load data (*table*), 180
   Program DYNAPT, 227
   structural data for (*table*), 120
Plate-bending:
   element PBQ8, 371
   flexural and shearing deformations, 378
Plates, 370–410
Poisson's ratio, 497
Polar moment of inertia, 87
Pound (lb), 495
Predictor, central-difference, 197
Predictor-corrector method, 203
Pressure, surface, 366, 408
Principal body axes, 420
Principal coordinates, 139
Principal damping matrix, 165
Principal flexibility matrix, 145
Principal mass matrix, 140
Principal mass moments of inertia, 420
Principal normal strains, 316
Principal normal stresses, 314
Principal planes of bending:
   for space frame member, 259
   for space truss member, 255, 477
Principal stiffness matrix, 140
Principal stress axes, 315
Program COMOPT for plane trusses by component-mode method, 477
Program COMOST for space trusses by component-mode method, 477
Program DYAXSH for axisymmetric shells, 406
   dynamic load data (*table*), 407
   structural data (*table*), 406
Program DYAXSO for axisymmetrical solids, 365

Index

Program DYMSPF for multistory plane frames, 463
Program DYMSTB for multistory tier buildings, 463
Program DYNA for dynamic response, 225
  main program (*flow chart*), 226
Program DYNACB for continuous beams, 265
Program DYNAGR for grids, 270
Program DYNAPB for plates in bending, 379
  dynamic load data (*table*), 380
  structural data (*table*), 380
Program DYNAPF for plane frames, 267
  dynamic load data (*table*), 269
  structural data (*table*), 268
Program DYNAPS for plane stress and plane strain, 340
  dynamic load data (*table*), 341
  structural data (*table*), 341
Program DYNAPT for plane trusses, 227
  flow chart, 519–63
Program DYNASF for space frames, 277
  dynamic load data (*table*), 279
  structural data (*table*), 276
Program DYNASH for general shells, 390
  dynamic load data (*table*), 391
  structural data (*table*), 391
Program DYNASO for general solids, 351
  dynamic load data (table), 352
  structural data (*table*), 352
Program DYNAST for space trusses, 273
Program DYPFAC for plane frames with axial constraints, 299
Program DYRBPB for rigid bodies in plate-bending continua, 438
  rigid-body data (*table*), 440
  rigid-body dynamic-load data (*table*), 440
Program DYRBPF for rigid bodies in plane frames, 424
Program DYSFAC for space frames with axial constraints, 299
Program NOMO for normal-mode response, 177
  main program (*flow chart*), 178
Program NOMOPT, 179
  dynamic load data (*table*), 180
Program notation, 487
Programs for framed structures, 264
Program VIB for vibrational analysis, 118
  main program (*flow chart*), 119
Program VIBPT, 118
  structural data (*table*), 120
Proportional damping, 165
Properties of materials (*table*), 497
Proportional damping, 165
Proportional loads, 182
Pseudostatic problems, 212

**Quadrature,** Gaussian, 328
Quadrilateral coordinates, 318
  infinitesimal area in, 329
QR algorithm, 514

**Radius of gyration:**
  for cross section of grid member, 251
  for rigid body, 433
Ramp function, 43
Ramp-step function, 55
Rate of convergence, 206
Rayleigh quotient, 500
Reactions at supports, 99
Reciprocal theorem for dynamic loads, 154
Rectangular impulse, 41
Rectangular parent:
  element AXSR3, 394
  element PQR8, 371

## 588 Index

Rectangular parent (*cont.*)
  element R4, 333, 336
  element R8, 338
Rectangular solid parent:
  element RS8, 345, 348
  element RS20, 349
Reduced action vector, 282
Reduced equations of motion, 448
Reduced mass matrix, 283
Reduced stiffness matrix, 282
Reduction:
  dynamic, 282
  Guyan, 282, 445
  static, 282, 420
Redundant constraints, 290
Reference point on a rigid body, 413, 416
Relative acceleration, 23
Relative coordinates, 22, 159
Relative damping, 166
Relative displacement, 23
Repeated eigenvalues, 107
Resonance, 21
Response, 4
  steady-state, 20, 31
  transient, 25, 34
Response calculations, step-by-step, 45, 175
Response to harmonic forcing function:
  forced part, 20
  free part, 20
Response spectra, 51
Restrained nodal displacements, 99
Retained modes, 469, 472
Reverse iteration, 498
Rigid-bodies, 8, 411–43
  equations of motion, 418, 437
  in finite-element networks, 434
  in framed structures, 413
  in plane frames, Program DYRBPF, 424
  in plate-bending continua, Program DYRBPB, 438
  mass and mass-moment-of-inertia

Rigid-body center of mass, 418
Rigid-body constraints, 411, 427
Rigid-body data for Program DYRBPB (*table*), 440
Rigid-body dynamic-load data for Program DYRBPB (*table*), 440
Rigid-body ground accelerations, 159
Rigid-body ground displacements, 157
Rigid-body modes, 148
Rigid-body motions, 138, 153
Rigid-body reference point, 413, 427
Rigidity:
  axial, 94
  flexural, 94
  torsional, 93
Rigid laminae in multistory buildings, 427
Ring element, 357
Rotary inertias (*see* Rotational inertias)
Rotating vectors, 14
Rotation of axes:
  for grid member, 252
  for plane frame member, 247
  for plane truss member, 102: in component-mode method, 475
  for space frame member, 263
  for space truss member, 256
  for stresses and strains, 314
Rotation-of-axes transformations, 95
Rotation matrix, 97
  generalized, 505
Rotational (or rotary) inertias, 91, 378, 390, 401

**SDOF systems,** 10
Second moment of area, 91
Second (s), 495
Selection data, output (*table*), 182
Series:
  Fourier, 35
  substructures in, 445
Setbacks in tier buildings, 468

Index

Shape functions, displacement, 74, 79 (*see also* Displacement shape functions)
Shape functions and derivatives for element Q8 (*table*), 340
Shear cores in tier buildings, 468
Shearing deformations:
  in plate-bending, 378
Shearing stresses and strains, 75
Shell element SHQ8, 382
Shells, 370–410
  axisymmetric: element AXSH3, 394; Program DYAXSH, 406
  general: element SHQ8, 382; Program DYNASH, 390
Shifting, spectral, 503
SI units, 495
Slave displacements, 283, 457
Slug, 495
Solid of revolution (*see* Axisymmetric solids)
Solids:
  axisymmetric: isoparametric elements, 357; Program DYAXSO, 365
  general: isoparametric elements, 345; Program DYNASO, 351
Space frames, 244, 259
  axial constraints, Program DYSFAC, 299
  dynamic load data (*table*), 279
  Program DYNASF, 277
  structural data (*table*), 276
Space trusses, 243, 253
  component-mode method, 477: Program COMOST, 477
  Program DYNAST, 273
Spectra, response, 51
Spectral decomposition, 223
Spectral matrix, 141, 315, 470, 505
Spectral radius, 223
Spectral shifting, 503
Square-root method, Cholesky, 106
Stability:
  conditional, 208
  numerical, 223
  unconditional, 208
Stability criterion, 224
Standard, symmetric form (of eigenvalue problem), 106, 505
Static determinacy, 145
Static equilibrium, 1
Static immobility, 145
Static reduction, 282, 420
Statically equivalent actions, 414
Steady-state forced vibration, 20
Step function, 40
Step-by-step response calculations, 45, 175
Stiffness coefficients, 82
Stiffness matrix:
  for axial element, 84
  for axisymmetric shell element with nonaxisymmetric loads, 405
  for axisymmetric solid element with nonaxisymmetric loads, 364
  for element, 82
  for element AXQ4, 359
  for element AXSH3, 400
  for element H8, 347
  for element PBQ8, 377
  for element Q4, 335
  for element SHQ8, 389
  for flexural element, 91
  for grid member, 249
  for plane frame member, 244
  for plane truss member, 101: in component-mode method, 474
  for space frame member, 259
  for space truss member, 255
  for structure, 98
  for substructure in multistory building, 460
  for torsional element, 87
  for x-beam, 428
  for y-beam, 428
  for z-column, 430
  principal, 140

Stiffness matrix (*cont.*)
  reduced, 282
  reduced and assembled, 448
  tridiagonal, 446
Stiffnesses, 11
Stiffness method, direct, 98
Strain-displacement relationships, 79, 310
Strain energy, virtual, 80
Strain energy density, virtual, 316
Strain-stress relationships, 77
Strain transformation matrix, 316
Stress-strain relationships, 77, 80, 311
Stress transformation matrix, 316
Stresses and strains, 75, 310
  generalized, 92
Structural damping, 26
Structural data:
  for multistory plane frames (*table*), 464
  for plane frames (*table*), 268
  for plane trusses (*table*), 120
  for Program DYAXSH (*table*), 406
  for Program DYNAPB (*table*), 380
  for Program DYNAPS (*table*), 341
  for Program DYNASH (*table*), 391
  for Program DYNASO (*table*), 352
  for space frames (*table*), 276
Structural equations of motion, 98
Structural mass matrix, 98
Structural stiffness matrix, 98
Subparametric elements, 333
Subprogram CMAS, 118
Subprogram DECOMP (*flow chart*), 559
Subprogram DYLO, 177, 225
Subprogram DYLOPT, (*flow chart*), 535
Subprogram EIGEN2, 118

Subprogram FACTOR (*flow chart*), 561
Subprogram INVERU (*flow chart*), 560
Subprogram NORMOD, 227
Subprogram NUMINT, 227, (*flow chart*), 547
Subprogram RES1, 118
Subprogram RES1PT (*flow chart*), 532
Subprogram RES2, 179, 227
Subprogram RES2PT (*flow chart*), 553
Subprogram SDAT, 118
Subprogram SDATPT (*flow chart*), 519
Subprogram SOLVER (*flow chart*), 562
Subprogram STASYM, 118, (*flow chart*), 525
Subprogram STIF, 118
Subprogram STIFPT (*flow chart*), 523
Subprogram TIHIST, 177, (*flow chart*), 545
Subprogram TRABAC, 179, (*flow chart*), 546
Subprogram TRANOR, 177, (*flow chart*), 543
Subprogram TRAVEC, 118, (*flow chart*), 529
Subprogram VIB, 177, 225
Substitution, backward, 461
Substructure equations of motion, 446
  in component-mode method, 468
Substructures, 8, 444–82
Substructures in parallel, 455
Substructures in series, 445
  modified tridiagonal method, 451
  tridiagonal method, 445
Superparametric elements, 333
Support motions, 4
  independent, 162

# Index

normal-mode response, 157
Support reactions, 99
Support restraints, independent motions of, 162
Surface pressure, equivalent nodal loads for, 366, 408
Symmetric, standard form (of eigenvalue problem), 106, 505
Symmetric modes, 112
Systems of units:
  absolute and gravitational, 495
  consistent (*table*), 496

**Three-dimensional continua,** 310–69
Tier buildings, 425, 461
  bracing, setbacks, and shear cores, 468
  Program DYMSTB, 463
Time step, critical, 198, 224
Torsional element, 84
Torsional rigidity, 93
Torsional strains, 85
Transformation matrices for framed structures (*table*), 415
Transformation methods for eigenvalue problems, 505
Transformations, geometric, 431
Transient response:
  with damping, 34
  without damping, 25
Translational inertias, 91, 378, 390, 401
Translation of axes, 157, 414
Trapezoidal rule, 203
Triangular impulse, 45
Tridiagonal matrix, 509
Tridiagonal method, 445
  modified, 451
Trigonometric series, 35

Truncation, modal, 139, 143
Trusses:
  component-mode method, 471
  plane, 243, 471
  space, 243, 253, 477
Twist, 85
Two-dimensional continua, 310–69
Types of force systems, 241
Types of framed structures, 241

**Unconditional stability,** 208
Undamped forced vibrations, 19
Undamped free vibrations, 12
Underconstrained frames, 290
Unit-load method, 146
Units, systems of (SI and US), 495
Unit vectors, 96

**Velocities:**
  incremental, 38, 212
  initial, 14, 147
Vibrational analysis, 105
  Program VIB, 118
Vibrational motion, 152
Vibrations:
  forced, damped, 30
  forced, undamped, 19
  free, damped, 27
  free, undamped, 12
Virtual strain energy, 80
  density, 316
Virtual work, 80
Viscous damping, 26, 164

**Wilson-$\theta$ method,** 220
Work, virtual, 80
Working points, 416